The Perfect Slime

The Perfect Slime

Microbial Extracellular Polymeric Substances (EPS)

Edited by
Hans-Curt Flemming, Thomas R. Neu and
Jost Wingender

IWA PUBLISHING

Published by **IWA Publishing**
Alliance House
12 Caxton Street
London SW1H 0QS, UK
Telephone: +44 (0)20 7654 5500
Fax: +44 (0)20 7654 5555
Email: publications@iwap.co.uk
Web: www.iwapublishing.com

First published 2017

British Library Cataloguing in Publication Data
A CIP catalogue record for this book is available from the British Library

ISBN: 9781780407418 (Paperback)
ISBN: 9781780407425 (eBook)

DOI: 10.2166/9781780407418

Printed by Bell & Bain Ltd, Glasgow.

Cover legend
Laser microscopy maximum intensity projection of a river biofilm developed on a pebble stone. The biofilm with a thickness of 186 µm was examined for its autofluorescence and stained for nucleic acids as well as for glycoconjugates. Take notice of one class of biofilm matrix glycoconjugates shown in red. In addition, a symbiotic interaction between algal filaments (blue) and filamentous bacteria (green) can be seen.

Colour allocation: Nucleic acid stain (SybrGreen) = green; lectin stain (AAL-Alexa568) = red; autofluorescence of algae (chlorophyll A) = blue; autofluorescence of cyanobacteria = purple/white. Image dimensions: 246 × 246 µm.

Reference: Kamjunke N, Herzsprung P, Neu TR (2015) Quality of dissolved organic matter affects planktonic but not biofilm bacterial production in streams. Science of the Total Environment 506–507: 353–360.

Contents

Chapter 3

Preface

In 1999, the editors published the first book on microbial extracellular polymeric substances (EPS) which by then was a bit ahead of its time (Wingender *et al.* 1999). Meanwhile, the EPS matrix has acquired more and more interest as the material which keeps biofilm cells together and fills and forms the space between them. Although a considerable proportion of the scientific community still thinks of polysaccharides while speaking of the matrix, is now increasingly acknowledged that the biofilm matrix is a complex structure, formed by a multitude of biopolymers, including significant amounts, e.g., of proteins and nucleic acids, and self-organized as the "house of biofilm cells". In an excellent review, Hobley *et al.* (2015) state EPS that *"the biofilm matrix is not simply a 'slime' that surrounds the cells. It is in fact a highly ordered structure, with extensive amounts of protein localization, and interactions between components, that result in a robust and protective 'biofilm coat'"*. That is quite an advancement in recognition for something which is regarded as simple "slime", with a disgusted connotation by the public.

In this book, we attempt to reveal the surprising properties, dynamics and functions of the matrix, and in recognition of the role of the matrix as the fundamental component of biofilms, the oldest and most successful form of life on Earth, we gave it the title "The Perfect Slime".

But: who reads a foreword? Just go ahead and enjoy the book!

April 2016

Hans-Curt Flemming
Thomas R. Neu
Jost Wingender

REFERENCES

Wingender J., Neu T. and Flemming H.-C. (eds). (1999). Microbial Extracellular Polymeric Substances. Springer, Berlin, Heidelberg, New York.

Hobley L., Harkins C. MacPhee C. E. and Stanley-Wall N. R. (2015). Giving structure to the biofilm matrix: an overview of individual strategies and emerging common themes. *FEMS Microbiol. Rev.,* **39**, 649–669.

Chapter 1

The perfect slime – and the "dark matter" of biofilms

Hans-Curt Flemming[1,2,3]

[1]*Biofilm Centre, Aquatic Microbiology, University of Duisburg-Essen, Universitätsstrasse 5, 45141 Essen*

[2]*IWW Water Centre, Mülheim, Moritzstrasse 26, 45456 Mülheim*

[3]*Singapore Center for Life Science Engineering (SCELSE), Nanyang Technological University, 60 Nanyang Drive, SBS-01N-27, Singapore 637551*

ABSTRACT

In biofilms, the extracellular polymeric substances (EPS) provide the matrix in which the biofilm organisms are retained and develop their interactions. The properties of the EPS molecules are crucial for the living conditions of these organisms. This matrix allows for the emergence of new properties of biofilms as a whole which are not predictable from the study of planktonic cells, e.g., the formation of stable microconsortia, new dimensions of resource capture, retention of extracellular enzymes which turns the matrix into an external digestion system, and enhanced tolerance to antimicrobial effects. Nevertheless, there are still fundamental questions open, such as how to manage extent and type of EPS in biotechnology and antifouling applications; mechanisms of sorption of hydrophobic substances or retention of water; and which EPS components contribute to matrix stability and how we can predict it. Therefore, the matrix still can be termed the "black matter of biofilms" although it is utterly important it is for understanding the biofilm mode of life.

Keywords: EPS, matrix, resource capture, emergent properties, stress tolerance

1.1 WHAT IS 'PERFECT'?

The biofilm matrix, consisting of extracellular polymeric substances (EPS), a.k.a. "slime", is a completely underrated and generally perceived as a disgusting phenomenon in every-day life. Usually, it tends to be associated with hygienic problems because it can harbour smelly and even pathogenic microorganisms.

Nevertheless, it is the material which supports the first, most widely distributed and most successful form of life on Earth and is called "The perfect slime" in the title of this book. Of course, it is debatable what "perfect" means. My friend and colleague Jean-Claude Block (Univ. of Lorraine, Nancy, France) suggested to add a question mark to the title because he interprets "perfect" as "finished to perfection", so that it cannot be improved any further. That, of course, does not apply to the EPS matrix – it is constantly changing.

However, there are wider definitions, which somehow justify the use of the word "perfect". The Oxford Dictionary, for example, lists in point 1.3 among the meanings of "perfect": *"Highly suitable for someone or something; exactly right"*. This comes closer to the acknowledgment of the multifunctional versatility and adaptability of the biofilm matrix: in fact, it is perfectly suited for the biofilm mode of life under extremely different conditions. If we consider that microorganisms have developed the oldest and most successful form of life, the matrix is the underlying principle. And it is highly dynamic and suited for adaptation to an extreme range of conditions of life. A huge variety of EPS components is available (Sutherland, 2002; Fong & Yildiz, 2015) from an also huge variety of microorganisms which produce further variations of their EPS repertoire during life cycle and/or under changing conditions. To make things even more complex, even recycled cytoplasmic proteins released from live cells, e.g., *S. aureus,* during the development of the biofilm and remained cell associated at pH 5 but were reversibly released when the pH-value was increased (Foulston *et al.* 2014).

Although as the matrix of some biofilms may resemble a stochastic mess of contributions from the inhabiting organisms (see, e.g., the lectin staining images of lotic "river snow" as presented by the group of Thomas Neu). However, there are examples of strikingly complex functional structures, as in microbial mats (see Decho, this volume), the complex cellulose network in *E. coli* biofilms (Serra *et al.* 2013, 2014) or dental plaque biofilms (Welch *et al.* 2016); Schaudinn *et al.* (2007) even found regular honeycomb patterns in the EPS matrix of *Pseudomonas aeruginosa*. Hobley *et al.* (2015) acknowledge in an excellent review on EPS that *"the biofilm matrix is not simply a 'slime' that surrounds the cells. It is in fact a highly ordered structure, with extensive amounts of protein localization, and interactions between components, that result in a robust and protective 'biofilm coat'"*.

This chapter is intended as an introduction to the various aspects of the EPS matrix covered in this book.

1.2 THE MATRIX: BASIS FOR THE EMERGENT PROPERTIES OF BIOFILMS

In a recent publication, the success of the biofilm mode of life has been associated with the emergent properties which arise from aggregated cells, kept together by the biofilm matrix (Flemming *et al.* 2016). Emergent properties are novel structures, activities, patterns and properties, which arise during the process, and as a consequence, of self-organization in complex systems (Corning, 2002). For microbiological communities, emergent properties of a community are characteristics not identifiable by analyzing the component organisms in isolation (Konopka, 2009).

In biofilms, the cells are encased by a self-produced EPS matrix, mainly composed of polysaccharides, proteins, lipids and DNA (Flemming & Wingender, 2010). The formation of the matrix is a continuous, dynamic process, depending, e.g., on nutrients available, shear stress, competition and grazing. Not surprisingly, the production of this matrix comes with a considerable energetic cost (Saville *et al.* 2011; van Gestel *et al.* 2014).

It is this matrix which provides the stage for the emergence of all further properties of biofilms. A consequence is the biogenic formation of the biofilm habitat, characterized by strong gradients, high biodiversity, and complex, dynamic and synergistic interactions, including enhanced horizontal gene transfer and true features of multicellularity. Biofilm development shows clear characteristics of self-organization. The primary emergent property of biofilms is based on the fact that all of these characteristics are present concomitantly with the matrix as a scaffold. The result is a differentiated, highly heterogeneous, resilient and flexible biological system whose particular properties cannot be predicted from those of single cell (Figure 1.1).

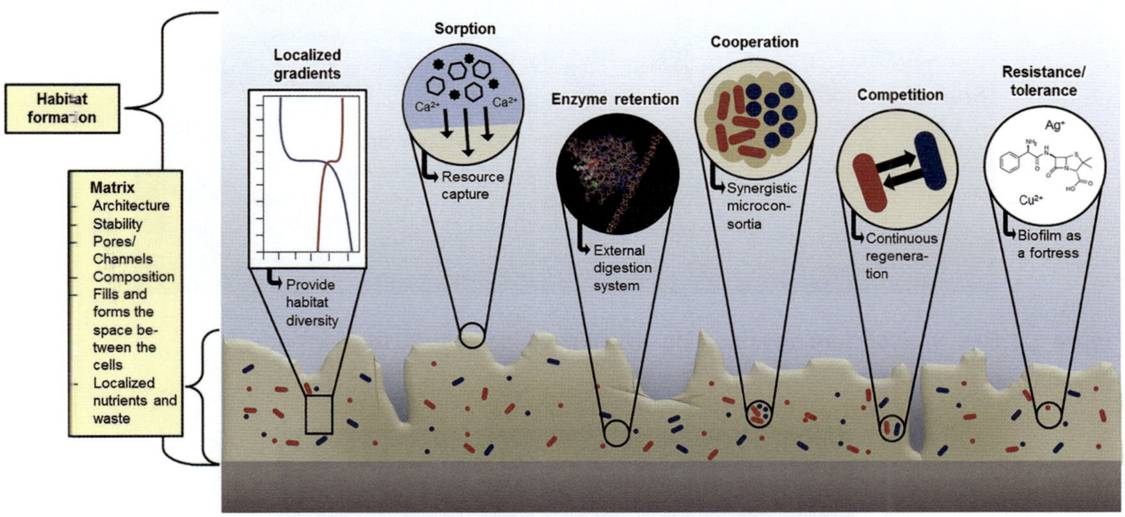

Figure 1.1 Emergent properties of biofilms (after Flemming *et al.* 2016, modified).

The biofilm matrix acts as the "house of biofilm cells" (Flemming *et al.* 2007) and represents an activated reaction space, ideally suited for resource capture. This is given by the sorptive properties of biofilms as well as by the retention and accumulation of extracellular enzymes, resulting in an acquired nutrient reservoir with a digestion system external to the cells due to retention of enzymes in the matrix. In addition to degrading sorbed substances, the extracellular enzymes can also attack the substratum, leading both to biodegradation of particulate matter and contribute to the global self-purification processes on Earth. Such processes, however, can occur in the wrong place and at the wrong time from a human perspective and are known as biodeterioration (Flemming, 2010). The matrix represents an ultimate recycling yard which contributes to optimal resource capture and exploitation. This complex and very effective system is symbolized in Figure 1.2.

An example for the enzymatic activation of the matrix is the interaction between lipase and alginate in *P. aeruginosa* as demonstrated by molecular modelling in Figure 1.3 (Tielen *et al.* 2013). Positively-charged amino acids are located opposite to the active centre and bind to the negatively-charged groups of the alginate moiety.

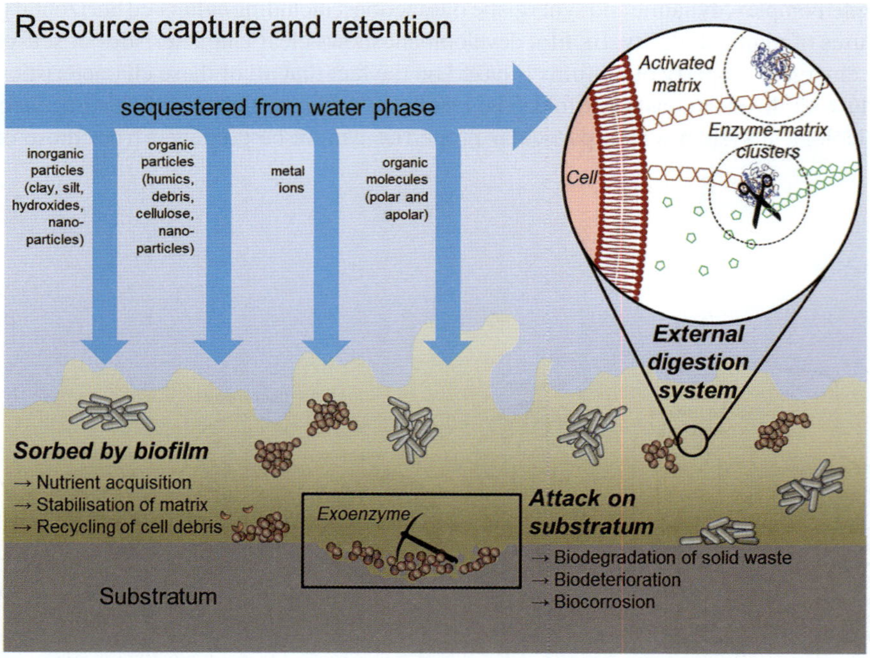

Figure 1.2 Resource capture and retention (after Flemming *et al.* 2016, modified).

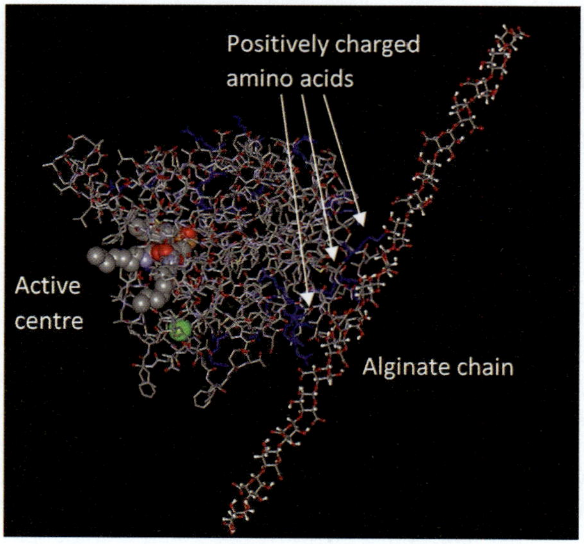

Figure 1.3 Model of interaction between lipase A and alginate from *P. aeruginosa*. Left: Lipase protein with active centre of the enzyme enlarged. Furthermore, the co-factor molecule Ca^{2+} is indicated in green. Site chains of positively-charged amino acids are shown in blue. Right: Section of an alginate molecule composed of negatively-charged uronic acids in ball and stick representation (Tielen *et al.* 2013).

The multiple functions of the matrix can be summarized as in Table 1.1:

Table 1.1 Functions attributed to EPS (after Flemming *et al.* 2007, modified).

Function of EPS Component	Relevance for Biofilm Organism	EPS Components Involved
Adhesion	Initial steps in colonization of abiotic and biotic surfaces by planktonic cells, long-term attachment of whole biofilms to surfaces	Polysaccharides, proteins (e.g., fimbriae), eDNA
Aggregation of bacterial cells	Bridging between cells, (temporary) immobilization of bacterial populations, basis for development of high cell densities, cell-cell recognition	Polysaccharides, proteins, DNA
Cohesion of biofilms	Structural elements forming a hydrated polymer network (biofilm matrix), mediation of mechanical stability of biofilms (frequently in conjunction with multivalent cations or hydrophobic interactions), determination of EPS structure (capsule, slime, sheath) and biofilm architecture, generation of matrix	Neutral and charged polysaccharides, proteins (e.g., amyloids, lectins), DNA
Retention of water	Maintenance of highly hydrated microenvironment around biofilm organisms, dessication tolerance in water-deficient environments	Hydrophilic polysaccharides and Proteins; skin-forming hydrophobic proteins (BslA, Hobley *et al.* 2013)
Protective barrier against antimicrobials	Resistance to nonspecific and specific host defenses during infection, tolerance to various antimicrobial agents (e.g., disinfectants, antibiotics), protection of cyanobacterial nitrogenase from harmful effects of oxygen; protection against some (but not all!) grazers	Polysaccharides, proteins
Sorption of polar organic compounds	Accumulation of nutrients from the environment, sorption of xenobiotics (detoxification)	Charged or hydrophobic polysaccharides and proteins
Sorption of inorganic ions	Promotion of polysaccharide gel formation, ion exchange, mineral formation, accumulation of toxic metal ions (detoxification)	Charged polysaccharides and proteins, including inorganic substituents such as phosphate and sulphate
Sorption of apolar organic substances	Resource capture	Proteins, not yet defined hydrophobic pockets in matrix

(Continued)

Table 1.1 Functions attributed to EPS (after Flemming *et al.* 2007, modified) (*Continued*).

Function of EPS Component	Relevance for Biofilm Organism	EPS Components Involved
Sorption of particles	Resource capture	Sticky matrix components
Enhanced access to resources captured in the matrix	Providing additional enzymatic competence and capacity in the matrix; killer-vesicles as a weapon in competition (Schooling *et al.* 2006)	Membrane vesicles (contain nucleic acids, enzymes, proteins, LPS etc.)
Enzymatic activity	Digestion of exogenous macromolecules for nutrient acquisition, degradation of structural EPS allowing release of cells from biofilms, utilization of substratum as substrate	Proteins
Nutrient source	Source of C, N and P compounds for utilization by biofilm community	Potentially all EPS components
Genetic information	Horizontal gene transfer between biofilm cells	DNA
Intercellular information	Regulation of biofilm dynamics and responses, regulating c-di-GMP concentration	Polysaccharides
Electron donor or acceptor	Redox activity in biofilm matrix, electron transport mediation to surfaces	Proteins (e.g., pili, nanowires), humic substances
Resource capture by export of enzymes into matrix	Providing additional enzymatic competence and capacity in the matrix	Outer membrane vesicles (contain nucleic acids, enzyme proteins, lipopolysaccharides, phospholipids)
Sink for excess energy	Sink for excess carbon under unbalanced C:N:P metabolic conditions	Polysaccharides
Binding of enzymes	Accumulation, retention and stabilization of enzymes through their interaction with polysaccharides	Polysaccharides, enzyme proteins

After excretion by the cells, the EPS can be modified along the biofilm life cycle – the biofilm cells continuously remodel their immediate environment. Some of the possible modifications are listed in Table 1.2 (Flemming, 2011).

These examples illustrate the variability and versatility of the EPS components. Of course, the microorganisms present in a biofilm contribute to the matrix. In environmental biofilms, the integration of new cells into a given biofilm appears quite stochastic, leading to a high biodiversity of the communities. This can be understood as "life insurance" for the biofilm as a whole, because biofilms can be exposed to a wide range of environmental conditions, and there are always organisms for whom they may be favourable. In fact, diversification seems so advantageous that it occurs even in pure cultures – but only in biofilms, not in planktonic cultures (Boles *et al.* 2004).

Table 1.2 Overview of modification of EPS and matrix structure after excretion (after Flemming *et al.* 2007, modified).

Microbial Influence	Effect
Degradation of EPS components by hydrolases, esterases, lipases, proteases and other lytic enzymes	Shortening of chain length, degradation of EPS, change of matrix structure and stability, formation of pores and channels Destabilization of matrix, dispersion, release of biofilm organisms
Variation of EPS composition in mixed biofilms during development	EPS of different properties, resistance to EPS-lysing enzymes
Post-excretioinal addition of substituents to polysaccharides	Influence on shape, charge, hydrophobicity of polymer, surface activity
Molecular structure suitable for protein-polysaccharide interaction	Retention, possible protection and activation of extracellular enzymes
Excretion of rhamnolipids	Increase of porosity, favouring of cell motility, influencing mass transport
Movement of "stealth swimmers" (Houry *et al.* 2014)	Formation of channels, improvement of convective mass transport
Environmental Influence	**Effect**
Shear forces	Washout of loosely bound EPS, accumulation of strongly bound EPS, increase of stability of remaining matrix, sloughing off, erosion
Grazing by higher organisms (protozoa, larvae, snails etc.)	Selective removal of EPS and EPS producing organisms, formation of channels, destabilization of matrix

In general, we look at a multitude of highly diverse EPS molecules with most of them providing a hydrated environment for biofilm organisms. This system has additional advantages which result in the emergent properties of the biofilm as a whole. The role of components that are less water soluble, such as amyloids or cellulose should not be underestimated, because they provide further advantages such as mechanical stability (van Gerven *et al.* 2015; Dueholm & Nielsen, 2016, this volume) and tolerance to antimicrobial agents and desiccation (Flemming *et al.* 2016). Figure 1.4 may illustrate the enormous role of EPS in diffusion-reaction resistance against antibiotics (Oubekka *et al.* 2012; Harrison *et al.* 2007; Olsen *et al.* 2015), as well as in desiccation tolerance (which is much further explained by Decho (2016), this volume).

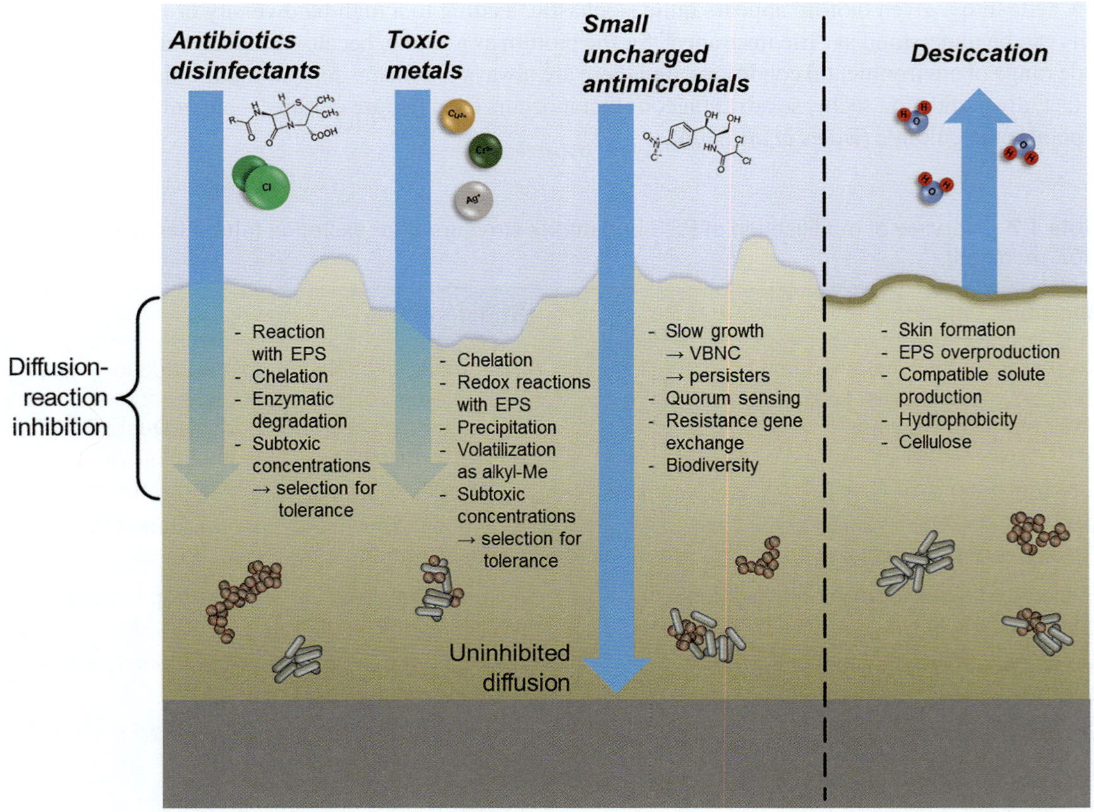

Figure 1.4 Role of EPS in tolerance to antimicrobials and desiccation (after Flemming *et al.* 2016, with permission).

Nevertheless, there is still very much information missing about the EPS matrix. In their chapter, Neu and Lawrence (2016, this volume) termed them positively as "challenges", and there are quite a few of them which currently look almost impossible to meet.

1.3 THE "DARK MATTER OF BIOFILMS"

In spite of the enormous relevance of the matrix for the biofilm life style and a lot of intense research, still so little is known about it that Thomas Neu termed it "the dark matter of biofilms" in a keynote lecture. Decho (2016, this volume) called it "the least understood component of biofilms". Interestingly, little is known about the matrix of archaeal biofilms; it seems that it has escaped the attention of the archaeal scientific community, called "the great unexplored" (Orell *et al.* 2013), and come to attention only recently (Fröls, 2013; Perras *et al.* 2015; Benninghoff *et al.* 2016, this volume); in all of these studies, mainly the polysaccharide moiety of the EPS matrix is addressed (e.g., Sutherland 2016, this volume) while proteins, glycoproteins, lipids, glycolipids, and eDNA remain much less explored. Some of the open questions have concrete practical relevance and are presented here; the reader is recommended to refer also to the "challenges" mentioned in the chapter of Neu and Lawrence (2016, this volume).

1.3.1 What influences EPS production and how can it be managed?

It remains unknown which EPS component and how much of it is produced in environmental biofilms. Related to this is a practical question: how can the extent of EPS in an environmental biofilm be managed? This is of great importance, e.g., for quenching filter or membrane fouling in water treatment. It seems as if an imbalance in the C:N:P ratio might play a role in over- or underproduction of EPS, however, any systematic research on this topic is missing in spite of its obvious importance. The hydraulic permeability of porous media is also controlled by EPS production and fundamentally influences water and the hydraulic conductivity in soils (Rossi *et al.* 2012). Also, the fate and transport of contaminants in soils depends upon the EPS, as shown for colloid mobility (Leon-Morales *et al.* 2007). The matrix clearly responds to changes in pH-value, ionic species and strength, shear forces, grazing, temperature, pressure, radiation, disinfectants etc., not only short-term but also by acclimatization over time. For the time being, research mainly consists of manipulating external parameters such as the nature and amount of carbon, nitrogen and phosphorous sources, but rather in a "black-box"- manner. It is striking that this kind of research is still in such an early stage of infancy.

1.3.2 Where and by which mechanisms are hydrophobic substances sorbed in the matrix?

Hydrophobic pollutants are a concern because they may be accumulated in sewage sludge, impeding its land use. Considering the sorption properties, capacities and mechanisms, the accumulation of hydrophobic substances in the matrix (e.g., Späth *et al.* 1998) is not understood at all. Decho (2000) suggested "hydrophobic domains" in microbial mats, however, without explaining which EPS molecules may contribute hydrophobicity. Aldeek *et al.* (2013) reported hydrophobic domains in the matrix, but again, without any further characterization. Accumulation of hydrophobic substances in biofilms plays an important role in contamination of sewage sludge by apolar toxic substances and the impairment of the use of such sludge as a fertilizer. On the other hand, biofilms with a high capacity for accumulation of hydrophobic pollutants could be used for cleaning up waste water streams.

1.3.3 What are the mechanisms behind water retention?

Another important property of biofilms is the retention of water. This is addressed by Decho (2016, this volume). However, a precise definition of "bound" water refers to the water shell of biopolymers. This water is less mobile than the rest of the biofilm water phase, but it constitutes only a fraction of much less than one percent of the entire water content. Maybe it can be understood as an entropic effect, similar to a sponge in which the water is not specifically bound either, while the mixture between EPS and water yields a higher entropy.

1.3.4 Can the permeability of biofouling layer be engineered in membrane technology?

In membrane systems, water is pressed through the biofilm matrix. Interestingly, the spatial organization of the biopolymers in the matrix responds to pressure and, thus, is compressible upon vertical flux as occurring in biofouling layers on separation membranes. This leads to an increase of pressure resistance ("hair-in-sink-effect", Vrouwenvelder *et al.* (2016, this volume). Is this an inevitable effect or can it be mitigated, possibly increasing biofilm permeability by chemical means in a longer term, as suggested by McDonogh *et al.* already in 1994 in their work about understanding the effect of cleaners? Improvement of biofouling layer permeability could mitigate fouling effects in cases when cleaning fails.

1.3.5 Which are the interactions of the various EPS components?

Furthermore, despite the many functions that were associated with the biofilm matrix (see Table 1.1), little is known about the interactions between the constituent macromolecules. The study of matrix components and their interactions at the molecular level forms the basis to understand the dynamics of the biofilm matrix, its remodeling in response to environmental or community changes, and finally the biofilm disassembly mechanisms that allow the dispersion of resident cells and the colonization of new habitats (Hobley *et al.* 2015). In that sense, the work of Zhang *et al.* (2015) is particularly interesting, revealing a dynamic picture of matrix-associated proteins during biofilm development as an integral and well-regulated system that contributes to stress resistance, nutrient acquisition and the stability of the biofilm. Polysaccharides are equally important, providing structure to biofilms, allowing stratification of the bacterial community and establishing gradients of nutrients and waste products (Limoli *et al.* 2015). Even the formation of hydrophobic coatings, caused by a self-assembling bacterial hydrophobin, has been observed, e.g., in *B. subtilis* biofilms (Hobley *et al.* 2013).

1.3.6 What is the function of complex EPS components?

For example, the function of amyloids, proteoglycans, glycoproteins and other complex EPS components still leaves a lot of open questions. Looking at amyloids, it is quite probable that they do more than stabilize the matrix. We know little about diversity and biological roles of these fascinating structures. The abundance of amyloids in nature raises some fundamental questions regarding structure and function. In order to address these questions, new functional microbial amyloids should be isolated and their function determined (Dueholm & Nielsen, 2016, this volume).

1.3.7 Which EPS components contribute to matrix stability and how can we predict it?

The interaction between biofilm structure and chemical composition with their viscoelastic properties remains one of the fundamental questions concerning biofilm research. This question has stimulated modellers to generate mathematical models with the aim of expanding our basic science understanding of the physics of biofilms and leading to new insight between the structure/function interactions (De Leenheer & Cogan, 2009). In addition, mechanical modes will help the design of more effective strategies to remove biofilms as well as to better control and predict their attachment and growth in natural, industrial and medical environments (Fabbri & Stoodley, 2016, this volume).

1.4 CHAOS AND FUNCTION – SELF-ORGANIZATION

So many fundamental questions still remain open – and "perfect"? And it is easy to predict that further research will reveal many new features and functions of the matrix. Therefore, one thing is for sure: an understanding of the matrix is far from being "finished to perfection".

However, it is the material upon which the huge evolutional success of the biofilm mode of life is based. It allows life to extend to the most extreme conditions (Horikoshi, 2016) and it seems that life itself is not at all endangered on this planet, except the existence of certain eukaryotic species.

The key for all this is the nature of the interactions between the EPS molecules: they have no covalent bonds but are kept together and to substrates by weak physico-chemical interactions and entanglement (Mayer *et al.* 1999). Due to the interaction of fluctuating binding points (Flemming *et al.* 2000), they provide a network strong enough to serve as a scaffold for self-organization of organisms and molecules and at the same time, loose enough to allow for cellular movement in the biofilm as well as for escaping, because the matrix is no prison although it may have its restrictions for some of its inhabitants, as illustrated in Figure 1.5 (© James C. Bryers, Univ. of Washington).

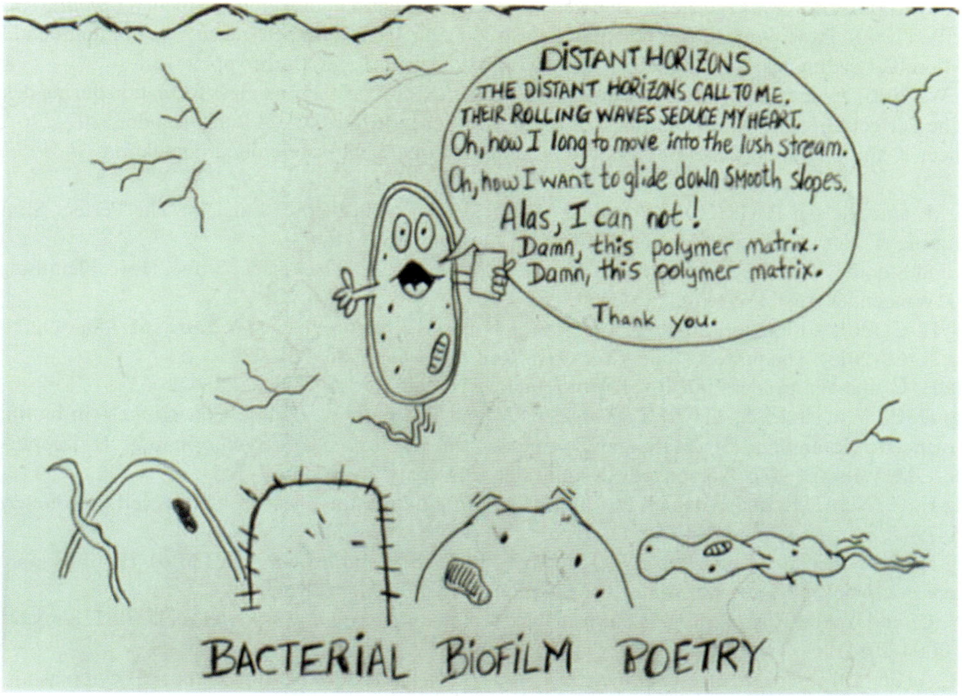

Figure 1.5 The dammed polymer matrix (cartoon: J.C. Bryers, Univ. Washington).

This flexibility allows for easy remodelling and flexibility of the matrix and, thus, adapting to the living conditions of the biofilm organisms. And the fact that so many different types of molecules – from polysaccharides to proteins to eDNA – can fulfil the same purpose, namely to form the "house of biofilm cells", offers a biological system perfect to cope with the situations biofilm organisms may be exposed to on Earth.

ACKNOWLEDGEMENTS

This work is based on a long period of cooperation with physico-chemists: Wiebren Veeman, Christian Mayer, Werner Borchard, Hubert Kuhn and it would not have been possible without the inspiring, sustainable, and loyal discussions and support of my colleague Jost Wingender.

REFERENCES

Aldeek F., *et al.* (2013). Patterned hydrophobic domains in the exopolymer matrix of *Shewanella oneidensis* MR-1 biofilms. *Appl. Environ. Microbiol.*, **79**, 1400–1402.

Benninghof J., Wingender J. and Siebers B. (2016). Biofilms x-treme – EPS composition and exopolysaccharide synthesis in the thermoacidophilic archaeon *Sulfolobus acidocaldarius*. In: The Perfect Slime, H.-C. Flemming, T. Neu and J. Wingender (eds), IWA Publ., London, 301–317.

Boles B. R., Thoendel M. and Singh P. K. (2004). Self-generated diversity produces "insurance effects" in biofilm communities. *Proc. Natl. Acad. Sci.*, **101**, 16630–16635.

Corning P. A. (2002). The re-emergence of "emergence": a venerable concept in search of a theory. *Complexity*, **7**, 18–30.

Decho A. W. (2000). Exopolymer microdomains as a structuring agent for heterogeneity within microbial biofilms. In: Microbial Sediments, R. E. Riding, S. M. Awramik (eds), Springer, Berlin, pp. 9–15.

Decho A. W. (2016). Unique and baffling aspects of the matrix: EPS syneresis and glass formation during desiccation. In: The Perfect Slime. H.-C. Flemming, T. Neu and J. Wingender (eds), IWA Publ., London, 207–226.

De Leenheer P. and Cogan N. G. (2008). Failure of antibiotic treatment in microbial populations. *J Math Biol*, **59**, 563–579.

Dueholm M. and Nielsen P. H. (2016). Amyloids – a neglected child of the slime. In: The Perfect Slime, H.-C. Flemming, T. Neu, J. Wingender (eds), IWA Publ., London, 113–134.

Fabbri S. and Stoodley P. (2016). Mechanical properties of biofilms. In: The Perfect Slime, H.-C. Flemming, T. Neu and J. Wingender (eds), IWA Publ., London, 153–178.

Flemming H.-C. (2010). Biodeterioration of synthetic polymers – a brief review. *Mat. Corr.*, **61**, 986–992.

Flemming H.-C. (2011). The perfect slime. *Coll. Surf. B: Biointerfaces*, **86**, 251–259.

Flemming H.-C. and Wingender J. (2010). The biofilm matrix. *Nat. Rev. Microbiol.*, **8**, 623–633.

Flemming H.-C., Wingender J., Mayer C., Körstgens V. and Borchard W. (2000). Cohesiveness in biofilm matrix polymers. In: Community Structure and Co-operation in Biofilms. SGM Symposium 59, H. Lappin-Scott, P. Gilbert, M. Wilson and D. Allison (eds), Cambridge University Press, pp. 87–105.

Flemming H. C., Neu T. R. and Wozniak D. (2007). The EPS matrix: the house of biofilm cells. *J. Bacteriol.*, **189**, 7945–7947.

Flemming H.-C., Wingender J., Kjelleberg S., Steinberg P., Rice S. and Szewzyk U. (2016). Biofilms: an emergent form of microbial life. *Nat. Rev. Microbiol.*, **14**, doi: 10.1038/nrmicro.2016.94.

Fong J. N. C. and Yildiz F. H. (2015). Biofilm matrix proteins. *Microbiol. Spectr.*, **3**, doi: 10.1128/microbiolspec.MB-0004-2014.

Foulston L., *et al.* (2014). The extracellular matrix of *Staphylococcus aureus* biofilms comprises cytoplasmic proteins that associate with the cell surface in response to decreasing pH. *mBio*, **5**, e01667–14.

Fröls S. (2013). Archeal biofilms: widespread and complex. *Mol. Biol. Archaea*, **41**, 393–398.

Harrison J. and Turner R. J. (2007). Multimetal resistance and tolerance in microbial biofilms. *Nat. Rev. Microbiol.*, **5**, 928–939.

Hobley L., *et al.* (2013). BsIA is a self-assembling bacterial hydrophobin that *coats* the *Bacillus subtilis* biofilm.

Hobley L., Harkins C., MacPhee C. E. and Stanley-Wall N. R. (2015). Giving structure to the biofilm matrix: an overview of individual strategies and emerging common themes. *FEMS J. Microbiol. Rev.*, **39**, 649–669.

Horikoshi K. (2016). Extremophiles: How It All Began. Springer, Heidelberg, New York, 172pp.

Houry A., *et al.* (2012). Bacterial swimmers that infiltrate and take over the biofilm matrix. *Proc. Natl. Acad. Sci. USA*, **109**, 13088–13093.

Konopka A. (2009). What is microbial community ecology? *ISME J.*, **3**, 1223–1230.

Leon-Morales C. F., Strathmann M., Flemming H.-C. (2007). Influence of biofilms on the movement of colloids in porous media. Implications for colloid facilitated transport in subsurface environments. *Wat. Res.*, **41**, 2069–2078.

Limoli D. H., Jones C. J. and Wozniak D. J. (2015). Bacterial extracellular polysaccharides in biofilm formation and function. *Microbiol. Spectr.*, **3**, doi: 10.1128/microbiolspec.MB-0011-2014.

Mayer C., Moritz R., Kirschner C., Borchard W., Maibaum R., Wingender J. and Flemming H.-C. (1999). The role of intermolecular interactions: studies on model systems for bacterial biofilms. *Int. J. Biol. Macromol.*, **26**, 3–16.

McDonogh R., Schaule G. and Flemming H.-C. (1994). The permeability of biofouling layers on membranes. *J. Membr. Sci.*, **87**, 199–217.

Neu T. R. and Lawrence J. (2016). The extracellular matrix – an inextractable part of biofilm systems. In: The Perfect Slime, H.-C. Flemming, T. Neu and J. Wingender, (eds), IWA Publ., London, 25–60.

Olsen I. (2015). Biofilm-specific antibiotic tolerance and resistance. *Eur. J. Clin. Microbiol. Infect. Dis.*, **34**, 877–886.

Orell A., Fröls S. and Albers S.-V. (2013). Archeal biofilms: the great unexplored. *Annu. Rev. Microbiol.*, **67**, 337–354.

Oubekka S. D., Briandet R., Fontaine-Aupart M.-P. and Steenkeste K. (2012). Correlative timeresolved fluorescence microscopy to assess antibiotic diffusion-reaction in biofilms. *Antimicrob. Ag. Chemother.*, **56**, 3349–3358.

Perras A. K., *et al.* (2015). Grappling archaea: ultrastructural analyses of an uncultivated, cold-loving archaeon and its biofilm. *Front. Microbiol.*, **5**, article 397.

Rossi F., Potrafka R. M., Pichel G. F. and de Philippis R. (2012). The role of the exopolysaccharides in enhancing hydraulic conductivity in biological soil crusts. *Soil Biol. Biochem.*, **46**, 33–40.

Saville R. M., *et al.* (2011). Energy-dependent stability of *Shewanella oneidensis* MR-1 biofilms. *J Bact*, **193**, 3257–3264.

Schaudinn C., *et al.* (2007). Bacterial biofilms, other structures seen as mainstream concepts. *Microbe*, **5**, 231–237.

Schooling S. R. and Beveridge T. J. (2006). Membrane vesicles: an overlooked component of matrices of biofilms. *J. Bact.*, **188**, 5945–5957.

Serra D. O. and Hengge R. (2014). Stress responses go three dimensional – the spatial order of physiological differentiation in bacterial macrocolony biofilms. *Environ Microbiol*, **16**, 1455–1471.

Serra D. O., Richter A. M., Klauck G., Mika F. and Hengge R. (2014). Cellulose as an architectural element in spatially structured *Escherichia coli* biofilms. *J. Bact.*, **195**, 5540–5554.

Späth R., Flemming H.-C. and Wuertz S. (1998). Sorption properties of biofilms. *Wat. Sci. Technol.*, **37**, 207–210.

Sutherland I. W. (2007). Bacterial exopolysaccharides. In: Comprehensive Glycoscience Vol. 2, J. P. Kamerling (ed.), Elsevier, Doordrecht, pp. 521–558.

Sutherland I. W. (2016). EPS – a complex mixture. In: The Perfect Slime – Microbial Extracellular Polymeric Substances (EPS), H.-C. Flemming, T. Neu, J. Wingender (eds), IWA Publ., London, in press.

Tielen P., Rosenau F., Wilhelm S., Jaeger K.-C., Flemming H.-C. and Wingender J. (2013). Extracellular enzymes affect biofilm formation in mucoid *Pseudomonas aeruginosa*. *BMC Microbiol.*, **159**, 221–228.

Van Gerven N., Klein R. D., Hultgren S. J. and Remaut H. (2015). Bacterial amyloid formation: Structural insights into curli biogenesis. *Trends Microbiol.*, **23**, 693–706.

Van Gestel J., Vlamakis H. and Kolter R. (2015). From cell differentiation to cell collectives: *Bacillus subtilis* uses division of labour to migrate. *PLoS Biol.*, **13**, e1002141.

Vrouwenvelder *et al.* (2016). Why and how biofilms cause biofouling on membranes – the hair-in-sink effect. In: The Perfect Slime, H.-C. Flemming, T. Neu and J. Wingender (eds), IWA Publ., London, 193–206.

Welch J. M., Rossetti B. J., Rieken C. W., Dewhirst F. E. and Borisy G. G. (2016). Biogeography of a human oral microbiome at the micron scale. *Proc. Nat. Acad. Sci.*, doi: pnas.org/cgi/doi/10.1073/pnas.1522149113.

Zhang W., *et al.* (2015). Extracellular matrix-associated proteins form an integral and dynamic system during *Pseudomonas aeruginosa* biofilm development. *Front. Microbiol.*, **5**, art. 40.

Chapter 2

EPS – a complex mixture

Ian W. Sutherland

School of Biological Sciences, University of Edinburgh, Edinburgh, Scotland

ABSTRACT

Our extensive knowledge of the composition, structure and physical properties of microbial polysaccharides can usefully be applied to our determination of their roles in biofilm composition and structure. We now know that the polysaccharides are essential components of biofilms and that they have major roles in determining their structure and function. However, biofilms present such a huge range of differing composition and structure that care must be applied in extrapolation of that knowledge. While new methods have been developed for the determination of biofilm structures in the laboratory, the natural environment presents both greater complexity and greater difficulties. We are still a long way from understanding how stable microbial biofilms, presumably mainly composed of Archaea, are able to withstand the extremes of temperature, pH, shear and other physical stress which are found in thermal pools and vents. Even the ephemeral types observed in small pools or on stone and other structures present technical difficulties and cannot be easily reconstructed in the laboratory. Currently much less is known about Cyanobacterial and algal polysaccharides than those associated with bacteria. It is probably safe to predict that novel polysaccharide components and structures may be found but it is then more difficult to predict their physical properties. At the same time, methods from other scientific disciplines may be of value and assist the understanding of biofilm structures and the role of their component macromolecules.

Keywords: EPS, exopolysaccharides, physical properties, polysaccharide interactions

2.1 INTRODUCTION

Biofilms comprise the normal environment for many microbial cells in both natural and artificial habitats and are found in diverse range of ecosystems (Stoodley *et al.* 2002). They can be regarded as assemblies of microorganisms and their extracellular products attached to surfaces (Davey & O'Toole, 2000). EPS – extracellular polymeric substances, form the bulk of the solid matrix of biofilms and provide the 'cement' holding together the cells and associated material. The matrices are indeed basic to biofilm genesis and development and provide the possibility for a wide range of architectural types. They are complex and very diverse mixtures of products (Branda *et al.* 2005). They are excreted from the microbial cells within the biofilm, comprising secretion and degradation products from these cells, and material, both organic and inorganic, incorporated from the environment. The properties conferred on biofilms by their matrix composition and structure thus far exceed those attributable to the single microbial cells. Although there is a wide variety of matrix composition, the polymers which comprise it have the same general function. As a result of the matrix composition and its properties, it plays a key role in biofilm development and stability; in attachment of the biofilm to natural or synthetic surfaces; in protection of the cells within the biofilm; and in the other interactions between biofilms and the surrounding milieu. The exact structure and composition of any biofilm is probably unique to the environment in which it develops and is a reflection of chemical and physical conditions prevailing. Stoodley *et al.* (1999a) showed in laboratory biofilms how their nature was greatly affected by the nutritional and physical conditions in which the biofilm was generated. Natural biofilms are equally the unique products of any specific environmental conditions. The biofilm is thus normally a complex structure formed by the interaction of various macromolecules within which cells, water, ions and soluble low- and high- molecular mass components are trapped. As yet, relatively few studies have been made on the polysaccharides which, in most biofilms form the major macromolecular component of the extracellular matrix (Sutherland, 2001a). Only in pure cultures grown in the laboratory, in relatively simple biofilms found under some industrial conditions or in the biofilms formed when catheters or other materials are implanted into the human body, is one likely to find simpler and more consistent composition. These last two types may be composed almost entirely of one microbial species with a single associated polysaccharide but the potential ability of many bacterial species to synthesise and excrete several polysaccharides should not be forgotten. This capability may not necessarily be expressed simultaneously; it is possible that one polysaccharide is formed during growth and another in the homeostatic state found later in biofilms. Some bacterial species require specific substrates from which to produce extracellular polysaccharides, as is the case for the homopolysaccharide mutan formed by *Streptococcus mutans* when grown on sucrose and commonly found as a constituent of dental biofilms. More commonly the polysaccharide composition and structure is independent of the carbohydrate substrate. The biofilm matrix should not be considered as a homogeneous entity but rather as a complex series of microenvironments which change as the biofilm develops and ages despite presenting the appearance of uniformity. Within this wide range of highly hydrated medium, a series of nutritional gradients will be influenced by the action of intracellular and extracellular enzymes as well as by physical factors such as shear, temperature, etc. (Stoodley *et al.* 1999b).

There is no apparent difference in composition between the polysaccharides found in pure culture and the material recovered from biofilms. Such observations should perhaps be regarded with caution as alterations in mass, acylation or side-chain composition might not necessarily be recognised. Again the limited number of studies must be recognised and further experimentation may yet reveal variations either over time or with changes in physiological conditions. When multiple species or natural biofilms are under consideration, the situation is much more complex as polysaccharides interact in various ways with each other and with other macromolecules. Another possibility is that the polymers formed may be of identical composition to those formed by the free-living bacteria but, due to minor structural differences such as the degree of acylation or the molecular mass, differ in their physical properties. These differences may result in altered viscosity or gel-forming capacity under biofilm conditions. Thus Tielen *et al.* (2005) demonstrated that in *Pseudomonas aeruginosa*, alginate acetylation influenced the initial colonisationof surfaces. The impetus of signals generated within the complex biofilm and adaptation to some of the unique physiological conditions prevailing, might stimulate the bacteria to synthesise polymers of novel composition and physical properties totally dissimilar from those found in the free-living state, or more probably derepress synthesis of a polymer found only in highly specialised environmental conditions.

The polysaccharides in biofilms play a number of very significant roles in addition to their contribution to the structure of the biofilm. They bind ions, proteins and of course allow the complex to retain water, which is the major component amounting to perhaps 97% of the matrix (Zhang *et al.* 1998). The insoluble or poorly soluble polysaccharides such as 'mutan' from *Streptococcus mutans* and cellulose from other bacteria enable the biofilm to attach firmly to an inert matrix. Some of these biofilms are very strongly attached even under very harsh environments as is seen in the hot springs found in geothermal locations. Exopolysaccharides (EPS) may also be involved in coaggregation, a process which is integral to the development of multispecies biofilms. Although this was originally observed in dental biofilms by Kolenbrander (1991), it clearly has much wider application.

While biofilms with their associated EPS and other material are effectively immobilised, the matrix of cells, polymers etc. appears to reach homeostasis to provide a system optimally organised to utilise all available nutrients. Such a matrix shows not only microheterogeneity but should be regarded as a dynamic environment in which the EPS may well be constantly subjected to change (Sutherland, 2001b). Thus its components may be dissolved, enzymatically degraded and lost, while new material and cells are inserted into the complex. Other interactions occur with bacteriophage, protozoa and of course with higher organisms.

2.2 MICROBIAL POLYSACCHARIDE COMPOSITION

Microbial polysaccharides from Eubacteria range from homopolysaccharides synthesised from a single monosaccharide species, to complex structures formed from 2–5 (or very occasionally more) sugars. The products from Cyanobacteria may be more complex, although suggestions that some may be composed of 5–12 monosaccharides should be treated with caution as the polysaccharides which have been analysed from these sources are almost certainly impure and few complete structures have been reported. Further complexity is provided by organic or inorganic substituents. Most of the common sugars such as the hexoses – D-glucose, D-galactose and D-mannose are frequently found in bacterial polysaccharides. Much less common are pentoses including D-arabinose and D-xylose although these appear to be found more often in the material from Cyanobacteria. (De Philippis & Vincenzini, 1998). Rare monosaccharides from these sources include a range of *O*-methylated sugars and also a few L-isomers of the list below. Interestingly, (Kaczynski *et al.* 2006) identified a linear D-rhamnan from a *Burholderia sp.* in which α- and β-linked residues of D-rhamnose formed the structure seen in Figure 2.1.

→3)-β-D-Glc*p*-(1→3)-β-D-Glc*p*-(1→3)-β-D-Glc*p*-(1→3)-β-D-Glc*p*-(1→

Curdlan from *Agrobacterium* spp.

→3)-α-D-Glc*p*-(1→3)-α-D-Glc*p*-(1→3)-α-D-Glc*p*-(1→3)-α-D-Glc*p*-(1→

Mutan from *Streptococcus mutans*

→4)-β-D-Glc*p*-(1→4)-β-D-Glc*p*-(1→4)-β-D-Glc*p*-(1→4)-β-D-Glc*p*-(1→

Bacterial cellulose (Produced by various species)

→4)- α-D-Rha*p*-(1→ 3)- α-D-Rha*p*-(1→ 3)- α-D-Rha*p*-(1→ 3)-β-D-Rha*p*-(1 →

The structure of an unusual linear homopolysaccharide composed of D-rhamnose as opposed to the more commonly detected L-rhamnose

Figure 2.1 The structures of some bacterial linear homopolysaccharides.

Table 2.1 Some monosaccharide and non-carbohydrate substituents of exopolysaccharides.

Substituent	Linkage	Charge Conferrred	Occurrence
Monosaccharides			
D-Glucose	–	None	Very common
D-Galactose	–	None	Very common
D-Mannose	–	None	Very common
L-Fucose	–	None	Very common
L-Rhamnose	–	None	Very common
D-Rhamnose	–	None	Rare
N-acetyl-D-Glucosamine	–	None	Very common
N-acetyl-D-Galacosamine	–	None	Very common
N-acetyl-D-Mannosamine	–	None	Less common
D-Glucosamine	–	Positive	Rare
D-Glucuronic acid	–	Negative	Very common
D-Galacturonic acid	–	Negative	Very common
D-Mannuronic acid	–	Negative	In bacterial alginates, otherwise not very common
Organic Acids			
Acetate	Ester	None	Very common - e.g., *Klebsiella* spp.; colanic acid
Glycerate	Ester	Negative	*Sphingomonas elodea*
Hydroxybutanoate	Ester	None	*Rhizobium trifolii*; *R. leguminosarum*, etc.
Propionate	Ester	None	Rare - some *Escherichia coli*
Pyruvate	Ketal	Negative	Very common - e.g., *Klebsiella* spp.; colanic acid
Succinate	Half ester	Negative	*Rhizobium* spp.; *Agrobacterium* spp.
Inorganic acids			
Phosphate		Negative	Common in some genera especially Gram positive spp.
Sulphate		Negative	Cyanobacteria; Archae; *Haloferax mediterranea*

The polysaccharides may be neutral if these sugars are the only constituents but most also contain acid sugars – commonly D-glucuronic or D-galacturonic acid and more rarely D-mannuronic acid and other uronic acids (Table 2.1). These provide a negative charge rendering the polymers polyanionic. Other monosaccharides frequently found in these polysaccharides are acetylamino sugars. Of these, *N*-acetyl-D-glucosamine has recently been found in adhesive polymers from a range of bacterial species (e.g., Itoh *et al.* 2008) and is probably the most commonly found sugar of this type. Very occasionally, non-acetylated amino sugars have been shown to be structural components of adhesive polysaccharides, thus conferring a positive charge to the macromolecule. This is exemplified by D-glucosamine found in the adhesion of some *Staphylococcus epidermidis* and related species (Mack *et al.* 1996). Additionally, this polyaminosugar is a product of unrelated Gram positive and Gram negative bacterial species including, *E. coli*, *Actinobacillus*, *Yersinia* and *Bordetella* (Itoh *et al.* 2005; Mack *et al.* 2009). Other non-carbohydrate components of microbial polysaccharides are either organic or inorganic (Table 2.2). The commonest are ester-linked acetyl residues which confer no charge on the polysaccharides, and ketal-linked pyruvate which does provide a negative charge. Ester-linked *O*-acetyl groups are very common indeed, but do not contribute to the overall charge on the macromolecule, although as will be discussed below, they may induce some complex and interesting effects on polysaccharide conformation and interactions. The acetyl groups are normally present in stoichimetric amounts, linked to a single sugar, but in products such as bacterial alginates, monosaccharide residues may be multiply acetylated and the total ester content may be relatively high. In some exopolysaccharides, several monosaccharide residues in the repeat unit may be acetylated. Also common in bacterial polysaccharides are ketal-linked pyruvate groups. These too, are normally present in stoichimetric amounts. Examples of polysaccharides in which more than one sugar residue is pyruvylated are also known.

In a few bacterial polysaccharides, pyruvate alone may render the polymer anionic in the absence of any uronic acids, but in most pyruvate-containing types, uronic acids are also present. As well as other ester-linked substituents, the amino acids L-glutamic acid and L-serine have been found. All these substituents may affect the physical properties of the polymers, conferring ion-binding capacity, lipophilicity and other attributes as indicated below. They may also render polysaccharides water-soluble which in their absence would be insoluble.

Most of the biofilm polysaccharides appear to be specific to one species or even one strain. Two major exceptions are bacterial cellulose which is a potential product of numerous bacteria and poly-*N*-acetyl-D-glucosamine (PNAG) which is found in diverse species including *Burholderia* (Yakandahala *et al.* 2011). Each of these polymers is found in diverse range of bacteria. A further feature which they have in common, is the synthesis of several polysaccharides by single species of these bacteria.

2.3 POLYSACCHARIDE STRUCTURE

The carbohydrate structures of microbial polysaccharides are usually unique to a particular bacterial genus or species, however a few examples are know of similar or even identical products from unrelated species. The polysaccharides excreted extracellularly by biofilm bacteria may be linear molecules as exemplified by a number of polymers involved in biofilm formation. These include mutan (Figure 2.1), bacterial cellulose and the poly-*N* -acetylglucosamine adhesin of *Escherichia coli* K12, *Burkholderia* spp. and other bacteria (Figure 2.2) (Itoh *et al.* 2008; Yakandahala *et al.* 2011). Also in this group is bacterial alginate from *Pseudomonas aeruginosa* and related species, which is unique in being a linear polysaccharide composed of random sequences of D-mannuronic acid and L-guluronic acid, together with O-acetyl groups (Figure 2.3). As seen in the polymer from *Haloferax mediterranea* (Figure 2.4), inorganic substituents such as sulphate groups do occur, albeit rarely in the polysaccharides of Eubacteria which have been examined to date. Sulphate substituents appear more commonly in the products of Archae and Cyanobacteria. Alternatively the main chain may be carry regular side-chains usually consisting of one to three or more sugars. A further possibility is the formation of a highly branched homopolysaccharide macromolecule as is found in levans and dextrans, polysaccharides formed by a number of bacterial species when grown specifically on sucrose as a carbon substrate. The adhesive polymers from *Staphylococcus epidermidis* (Figure 2.2) are unusual in that two appear to be formed simultaneously. One is composed entirely of *N*-acetyl-D-glucosamine while the other is partially composed of non-acetylated aminosugar residues. It thus confers a positive charge on the macromolecule (Sutherland, 2007).

Table 2.2 Effect of polysaccharide composition on physical properties.

Polysaccharide Component	Effect	Properties Affected	Example
Neutral sugars	Uncharged polymer	Insoluble	Bacterial Cellulose, Mutan, Biofilm EPS
Uronic acids	Polyanionic	Solubility, Ion binding	Xanthan, Alginates
Pyruvate	Polyanionic	Ion binding, Transition	Xanthan, Galactoglucans
Methylpentoses	Lipophilicity	Solubility	Biofilm EPS
Acetylation	Solubility	Gelation, Reduced ion binding	Alginates, Gellan, etc
Side-Chains	Various	Solubility	Scleroglucan, xanthan
1,3 or 1,4 Linkages	Rigidity	(In)Solubility	Mutan, Curdlan, Cellulose
1,2 Linkages	Flexibility	Solubility, Stability	Dextrans

EPS I

→ [-6-β-D-Glc*p*NAc-(1→6)-β-D-Glc*p*NAc-(1→6)-β-D-Glc*p*N-(1→ 6)-β-D-Glc*p*NAc(1→6)-β-D-Glc*p*NAc-(1→]

EPS II

→ [6-β-D-Glc*p*NAc-(1→ 6)-β-D-Glc*p*NAc-(1→6)-β-D-Glc*p*N-(1→6)-β-D-Glc*p*NAc(1→6)-β-D-Glc*p*NAc-(1→

Figure 2.2 *Staphylococcus epidermidis* biofilm exopolysaccharides.

→ [-4-β-D-Man*p*A-(1→ 4)-β-D-Man*p*A-(1→ 4)-β-D-Man*p*A-(1→ 4)-β-D-Man*p*A-(1→ 4)-β-D-Man*p*A-(1→ 4)-β-D-Man*p*A-(1→ 4)-α-L-Gul*p*A (1] →

↑ ↑ ↑

$CH_3CO.O$ $CH_3CO.O$ $CH_3CO.O$

Bacterial alginates have no regular structure and only D-mannuronosyl residues carry O-acetyl groups; some of these monosaccharides are multiply acetylated. In *Pseudomonas* polymers there are normally only single L-guluronosyl residues.

Figure 2.3 *Pseudomonas auruginosa* alginate.

These may be linear molecules as exemplified by *Streptococcus oralis* exopolysaccharides.
Alternatively they may carry side-chains, varying in length from single monosaccharides to trisaccharides or pentasaccharides. Additionally acyl substituents may be present.

-[α-L-Rha*p*-(1→2)-α-L-Rha*p*-(1→3)-α-D-Gal*p*-(1→3)-β-D-Gal*p*(1→4)-β-D-Glc*p*-(1→3)-α/β-D-Gal*p* -]-

Streptococcus oralis H1

Each hexasaccharide repeat unit carries a phosphate substituent

-[α-D-Gal*p*NAc-(1→3)-β-L-Rha*p* (1→4)-β-D-Glc*p*-(1→6)-β-D-Gal*f*(1→6)-β-D-Gal*p*NAc-(1→3)-α-D-Gal*p* -]$_n$-→PO_4'

Streptococcus oralis 34

Results of Kolenbrander (1991)

Figure 2.4 The structures of several more complex bacterial heteropolysaccharides from bacteria associated with biofilms.

2.4 PHYSICAL PROPERTIES

As exopolysaccharides synthesized by microbial cells vary so greatly in their composition, their chemical and physical properties also demonstrate a very wide of attributes. They may be very soluble or only poorly soluble or totally insoluble. Curdlan (Figure 1) is an example of a homopolysaccharide produced by a number of bacterial strains, including Agrobacterium and Rhizobium species, each of which also produce several distinct exopolysaccharides. Curdlan, is a neutral gel-forming 1,3-β-D-glucan of relatively low molecular mass (~74,000) which is insoluble in cold water. It is also produced exceptionally by one bacterial species as an extracellular storage polymer by the bacteria which form it, probably the only bacterial exopolysaccharide known to fit this role. Depending on their structure, the exopolysaccharides may be able to absorb and bind relatively large amounts of water. Some structures, however, tend to exclude water due to the tight binding of the polymer strands.

Depending on their flexibility they may be highly viscous under all physical conditions or they may be susceptible to changes in the physical environment. A very significant property affecting the viability and indeed the preservation of the bacterial cells within the biofilm is the state of hydration of the matrix. We do not know whether there are major changes to the polysaccharides during the life cycle of the biofilm, although this is certainly probable. Variations in composition are possible although probably unlikely. A more plausible scenario is alteration of the substituents or the length of side-chains, perhaps due to the excretion of polysaccharases, polysaccharide lyases, esterases or other enzymes from intact and degraded microbial cells within the biofilm. The effect of such enzyme action, could lead to changes in the physical attributes of the polysaccharides of the biofilm matrix. In some cases, the polymer could be degraded and the biofilm dissipated. Alternatively local solubilisation of the polymer might have a more restricted action. From earlier work in our laboratory we know that specific extracellular polysaccharases from bacteriophage can readily lead to complete destruction of biofilm structures in both single species and multispecies preparations (Hughes *et al.* 1998). While the polysaccharides within the biofilm matrix are not in themselves substrates for the bacteria from which they are produced, any degradation to oligosaccharides or single sugars could yield substrates for heterologous microbial cells associated with the biofilm. The original nutrients which supported the development of the biofilm could thus be efficiently recycled if the degradation products are retained within the matrix or are at least accessible to enzymes in its proximity.

The physical properties of biofilm polysaccharides are determined by their primary composition but are also the result of ordered secondary configuration. In some polymers, backbone sequences of 1,4-β- or 1,3-β-linkages may confer considerable rigidity. Polysaccharides formed solely from these structures, without any side-chains, are normally insoluble and thus capable of contributing to the stability of biofilms. This is exemplified in insoluble bacterial cellulose and mutan, but is also seen in other examples. Those polysaccharides with the same backbone structure but having additional side-chains tend to yield soluble macromolecules. The molecular mass also plays an important role. Thus bacterial alginates, some of which can be essentially homopolymers, are usually highly soluble despite their very high mass. Chemical deacetylation does however produce a highly insoluble product. When 1,2-α- or 1,6-α- linkages, found for example in highly branched bacterial dextrans, are present, the structures are likely to be much more flexible. Relatively small alterations in structure may cause significant changes in polysaccharide properties (Sutherland, 1997).

2.5 POLYSACCHARIDE INTERACTIONS

As pointed out by Burmølle *et al.* (2014) synergistic interactions affect the bacterial distribution and the biomass produced within biofilms. While this applies to all the components within a biofilm, it can also be focussed on the individual macromolecular species. The whole structure and integrity of biofilms depends not only on the composition and physical characteristics of the component macromolecules but also on the way in which they interact with one another. Polysaccharides are essentially very long, thin molecular chains with molecular mass ranging from $0.5–2.0 \times 10^6$ Da or greater. They can interact with other polysaccharides but also with proteins and if they are present in significant amounts, with nucleic acids. In some of these interactions, the presence of multivalent cations is necessary to achieve a stable complex but this may not always be necessary. The side-chains and polyanionic segments of exopolysaccharides are likely to play important roles as are both organic and inorganic substituents. Because of the complexity of these interactions and the range of molecules involved, it may well be difficult to attribute the biofilm properties to specific macromolecules.

Polysaccharide:polysaccharide interactions may be of several types. Entanglement of polysaccharide chains either during synthesis or post-synthesis may form gels or other changes in physical chatacteristics (e.g., Wloka *et al.* 2004). Gels may also result from ionic bridging between the macromolecules. If hydrogen bonding can develop between linear molecules, water may be excluded and insoluble complexes result. This may account for some of the highly stable biofilms found in high temperature, high ionic environments which are found in thermal pools. In the case of polysaccharides such as xanthan, the polymer chains align to form random coil or ordered helical structures. Transition to ordered helical aggregates is often greatly influenced by the presence or absence of acyl substituents such as O-acetyl or O-succinyl esters or by pyruvate ketals (Sutherland, 1997). The presence of multivalent cations also influences the structure which is adopted. In the great majority of natural environments at relatively low temperatures, and in the presence of salts, ordered conformations are adopted. Mayer *et al.* (1999) have suggested that electrostatic and hydrogen bonds are the dominant forces involved. More subtle formation of chain-chain complexes, in which one chain 'fits' into another, may yield flocs or networks which are poorly soluble in aqueous environments.

Interactions between polysaccharides and proteins may be especially significant. There may be mutual repulsion if the polymers are segregative and incompatible. Alternatively, helical aggregates can form complexes with properties not seen in the individual components. If coacervation occurs, in which a polysaccharide effectively bridges separate protein molecules, flocculation can result and produce complexes which are insoluble and of varying stability. It is highly likely that interactions of these types promote the bacterial coaggregation identified by Rickard *et al.* (2003) as an integral process in multispecies biofilm development.

In the natural environment, biofilms are exposed to biological products as well as to physical and chemical stress. Among these are bacteriophage, *Bdellovibrio* and amoebae. Exposure to specific polysaccharases may cause destruction of biofilms as exemplified by the depolymerisation of the poly-β-*N*-acetyl –D-glucosamine adhesin common to *Escherchia coli*, *Staphylococcus epidermidis*, *Yersinia pestis* and *Pseudomonas fluorescens* (Itoh *et al.* 2005). Many phages acting on exopolysaccharide-forming bacteria are also able to synthesise polysaccharide-degrading enzymes (hydrolases or polysaccharide lyases) which are found both as integral components of the viral particles and as soluble proteins released on lysis of the host bacteria. Thus, disruption of a biofilm was caused by a combination of polysaccharide degradation by the soluble enzymes and infection and subsequent cell lysis by the viral particles (Hughes *et al.* 1998). A rapid loss of cells from the biofilms in this particular study, was observed even before lysis of the susceptible infected bacteria occurred, supporting the major role of the soluble polysaccharide depolymerises in biofilm disruption. Phage-free enzyme alone, caused considerable disruption of the biofilms. However, in some laboratory systems using enteric bacteria, the enzymes from certain bacteriophage specific for *Enterobacter* strains were shown to degrade biofilms (Tait *et al.* 2002) but not remove them completely. Some bacteria persisted within a dual species biofilm subjected to phage attack. This might be due to the inaccessibitity of some bacterial cells within the structural heterogeneity of the matrix. This does demonstrate the difficulty in extrapolating results from one system to another. Similar results were obtained in an attempt to control *Salmonella enterica* (Sklar & Joerger, 2001). It indicates that bacteriophage control of biofilms may well be difficult to achieve. However interaction between algal viruses and their hosts may represent natural systems in which virus with or probably without associated enzymes, play a significant role in biofilms as well as in their destruction of algal blooms (Suttle, 2007). Co-existence between phage and bacterial hosts within the biofilm is another possible outcome or the viral particles may become trapped within the biofilm matrix without significantly affecting its structure (Sutherland *et al.* 2004). The bacteriophage may even be concentrated to yield a reservoir of viral particles (Flood & Ashbolt, 2000). An alternative use for genetically modified T7 phage in biofilm dispersion was developed by Pei and Lamas-Samanamud (2014). These authors demonstrated that disruption of signals within the biofilm rather than polysaccharide degradation could be effective.

Several studies indicate the complexity of interactions between the biofilm matrix and amoebae. A role was proposed for alginate in the resistance to grazing protozoa (Matz *et al.* 2004) but other factors were also involved. Alginate-forming *Pseudomonas aeruginosa* showed resistance to some but not all protozoa (Weitere *et al.* 2005). Bacterial polysaccharide capsules were demonstrated to present no barrier to attack by *Bdellovibrio* (Koval & Bayer, 1997). However it has been shown more recently that attack by *Bdellovibrio* can lead to significant destruction of a biofilm, but not its complete destruction (Kadouri & O'Toole, 2005). In these interactions, the fate of the complex depends on the structure and composition of the biofilm matrix and on the armoury of the invading microorganisms.

It is dangerous to generalise when so many variables are possible. As pointed out by Stoodley *et al.* (1999b) biofilm structure is also greatly influenced by both hydrodynamics and by its nutritional status. A further role of the matrix may be the exclusion of antimicrobial agents, although whether the exopolysaccharides function in this role is unclear (Mah & O'Toole, 2001). Thus, there are still many questions to be posed and answered in relation to the biofilm matrix and the many roles which it plays.

REFERENCES

Branda S. S., Vik S., Friedman L. and Kolter R. (2005). Biofilms; The matrix revisited. *Trends in Microbiol.*, **13**, 20–26.

Burmølle M., Ren D., Bjarnsholt B. and Sørensen S. (2014). Interactions in multispecies biofilms: do they actually matter?. *Trends in Microbiol.*, **22**, 84–91.

Deavey M. E. and O'Toole G. A. (2000). Micobial biofilms: from ecology to molecular genetics. *Microbiol. and Molecular Biol. Rev.*, **64**, 847–867.

DePhillippis R. and Vincenzini M. (1998). Exocellular polysaccharides from cyanobacteria and their possible applications. *FEMS Microbiol. Rev.*, **22**, 151–175.

Flood J. A. and Ashbolt N. J. (2000). Virus-sized particles can be entrapped and concentrated one hundred fold within wetland biofilms. *Adv. Environ. Res.*, **3**, 403–411.

Hughes K. A., Sutherland I. W. and Jones M. V. (1998). Biofilm susceptibility to bacteriophage attack: the role of phage-borne polysaccharide depolymerase. *Microbiol.*, **144**, 3039–3047.

Itoh Y., Rice J. D., Goller C., Pannuri A., Taylor J., Meisner J., Beveridge T. J. Preston J. F. and Romeo T. (2005). Depolymerization of β-1,6-*N*-Acetyl-D-glucosamine disrupts the integrity of diverse bacterial biofilms. *J. Bacteriol.*, **190**, 3670–3680.

Itoh Y., Wang X., Hinnebusch B. J., Preston J. F. and Romeo T. (2008). Roles of *pgaABCD* genes in synthesis, modification and export of the *Escherchia coli* biofilm adhesion polyβ-1,6-*N*-Acetyl-D-glucosamine. *J. Bacteriol.*, **187**, 382–387.

Kaczynski Z., Karapetyan G., Evidente A., Iacobellis N. S. and Holst O. (2006). The structure of a putative exopolysaccharides of *Burholderia gladioli* pv. *agaricicola*. *Carbohydr. Res.*, **341**, 285–288.

Kadoori D., and O'Toole G. A. (2005). Susceptibility of biofilms to *Bdellovibrio bacteriovorus* attack. *Appl. Env. Microbiol.*, **71**, 4044–4051.

Kolenbrander P. E. 1991. Coaggregation: adherance in the human oral microbial system. In: Microbial Cell-Cell Interactions, M. Dworkin (ed.), ASM, Washington, pp. 303–329, Chapt. 10.

Koval S. and Bayer M. E. (1997). Bacterial capsules: no barrier against *Bdellovibrio*. *Microbiol.*, **143**, 749–753.

Mack D., Fischer W., Krokotsch A., Leopold K., Hartmann R., Egge H. and Laufs R. (1996). The intercellular adhesion involved in biofilm accumulation in *Staphylococcus epidermidis* is a linear β-1,6-linked glucosaminoglycan: purification and structural analysis. *J. Bacteriol.*, **178**, 175–183.

Mack D., Davies A. P., Harris L. G., Knobloch J. K. and Rohde H. (2009). *Staphylococcus epidermidis* biofilms: functional molecules, relation to virulence and vaccine potential. *Topics Curr. Chem.*, **288**, 157–182.

Mah T.-F. and O'Toole G. A. (2001). Mechanisms of biofilm resistance to antimicrobial agents. *Trends in Microbiol.*, **9**, 34–39.

Matz C., Bergfeld T., Rice S. A. and Kjelleberg S. (2004). Microcolonies, quorum sensing and cytotoxicity determine the survival of *Pseudomonas aeruginosa* biofilms. *Env. Microbiol.*, **6**, 218–226.

Mayer C., Moritz R., Kirschner C., Borchard W., Maibaum, R., Wingender, J. and Flemming H.-C. (1999). The role of intermolecular interactions: studies on model systems for bacterial biofilms. *Int. J. Biol. Macromol.*, **26**, 3–16.

Pei, R. and Lamas-Samanamud G. R. (2014). Inhibition of biofilm formation by T7 bacteriophages producing quorum-quenching enzymes. *Appl. Environ. Microbiol.*, **80**, 5340–5348.

Rickard A. H., Leach, S. A., Buswell, C. M., High N. J. and Handley P. M. (2000). Coaggregation between aquatic bacteria is mediated by specific growth phase dependent lectin saccharide interactions. *Appl. Env. Microbiol.*, **66**, 431–434.

Sklar I. B. and Joerger R. D. (2001). Attempts to utilize bacteriophage to combat *Salmonella enterica* serovar *enteritidis* infection in chickens. *J. Food Saf.*, **21**, 15–29.

Stoodley P., Boyle J. D., DeBeer D. and Lappin-Scott H. M. (1999a). Evolving perspectives of biofilm structure. *Biofouling*, **14**, 75–90.

Stoodley P., Dodds I., Boyle J. D. and Lappin-Scott H. M. (1999b). Influence of hydrodynamic s and nutrients on biofilm structure. *J. Appl. Microbiol.*, **85**, S19–S28.

Stoodley P., Sauer K., Davies D. G. and Costerton J. W. (2002). Biofilms as complex differentiated communities. *Ann. Rev. Microbiol.*, **56**, 187–209.

Sutherland I. W. (1997). Microbial Exopolysaccharides - Structural subtleties and their consequences. *Pure & Appl. Chem.*, **69**, 1911–1917.

Sutherland I. W. (2001a). Biofilm exoplysaccharides: a strong and sticky framework. *Microbiol.*, **147**, 3–9.

Sutherland I. W. (2001b). The biofilm matrix – an immobilized but dynamic microbial environment. *Trends in Microbiol.*, **9**, 222–227.

Sutherland I. W., Hughes K. A., Skillman L. C. and Tait K. (2004). The interaction of phage and biofilms. *FEMS Microbiol. Letters*, **232**, 1–6.

Suttle C. A. (2007). Marine viruses – major players in the global ecosystem. *Nature Rev. Microbiol.*, **5**, 801–812.

Tait K., Skillman L. and Sutherland (2002). The Efficacy of Bacteriophage as a Method of Biofilm Eradication. *Biofouling*, **18**, 305–311.

Tielen P., Strathmann M., Jaeger K.-E., Flemming H.-C. and Wingender J. (2005). *Microbiol. Res.*, **160**, 165–176.

Weitere M., Bergfeld T., Rice S. A., Matz C. and Kjelleberg S. (2005). Grazing resistance of *Pseudomonas aeruginosa* biofilms depends on type of protective mechanism, developmental stage and protozoa feeding mode. *Env. Microbiol.*, **7**, 1593–1601.

Wloka M., Rehage H., Flemming H.-C. and Wingender J. (2004). Structure and rheological phenomena of mucoid *Pseudomonas aeruginosa* biofilms. *Coll. Polymer Sci.*, **282**, 1067–1076.

Yakandahala N., Gawande P. V., LoVetri K., Cardona T., Romeo T., Nitz M. and Madhyastha S. (2011). Characterization of the poly-β-*N*-acetylglucosamine component of *Burkholderia* biofilms. *Appl. Env. Microbiol.*, **77**, 8303–8309.

Zhang, X. Q., Bishop, P. L., Kupferle, M. J. (1998). Measurement of polysaccharides and proteins in biofilm extracellular polymers. *Water Sci. Technol.*, **37**, 345–348.

Chapter 3

The extracellular matrix – an intractable part of biofilm systems

Neu T. R.[1] *and Lawrence J. R.*[2]

[1]*Helmholtz Centre for Environmental Research, Magdeburg, Germany*

[2]*Environment and Climate Change Canada, Saskatoon, Saskatchewan*

ABSTRACT

In an attempt to highlight major aspects and issues in biofilm matrix research from a new perspective, a linkage between matrix-analysis, -constituents, and -function in relation to well-studied biofilm species is discussed. This chapter represents a comprehensive update and extension of a previous review on the extracellular polymeric substances in microbial biofilms (Neu & Lawrence, 2009). The topics covered in the following sections comprise: general matrix challenges and open questions, main approaches to biofilm matrix analysis, established biofilm matrix constituents, extracellular microbiology of important biofilm species, the evolving concept of biofilm matrix functionality and emerging views of the matrix.

Keywords: biofilm matrix, matrix analysis, matrix constituents, extracellular microbiology, matrix functionality

3.1 INTRODUCTION

The biofilm matrix in environmental microbial communities represents a continuous scientific challenge which arises from the matrix's intricate makeup produced by numerous diverse microorganisms. As a result, the matrix still represents the "dark matter" of microbial biofilms due to its physicochemical characteristics related to its polymeric and complex identity. The matrix has been investigated classically by extraction and chemical analysis as well as using modern approaches such as molecular techniques and in situ imaging. Nevertheless, with current methods and techniques the biofilm matrix continues to be intractable. Despite this situation new and innovative approaches have been suggested in order to shed light into the "dark matter" of microbial biofilms.

The polymer matrix of biofilms was for many years linked to the acronym EPS (extracellular polymeric substances). However, EPS as an acronym in microbiology was translated by different authors in several ways (Allison *et al.* 2003; Neu & Lawrence, 2009). Recently the term extracellular matrix with the acronym ECM has been used more frequently. This seems to be justified as more and more matrix constituents are identified which are not necessarily purely polymeric.

This manuscript is built on an earlier one in which different aspects of the biofilm matrix were discussed (Neu & Lawrence, 2009). The previous topics covered included a critical assessment of the review literature published, the use of the acronym EPS, general issues regarding the biofilm matrix and a first draft of a novel concept on EPS functionality. In the present chapter we would like to update these aspects, discussing the new literature published since 2009 and adding an extended view on biofilm matrix functionality.

In order to provide the reader with access to the broader state of the art on this topic a few review articles published more recently are worth mentioning. Flemming and Wingender (2010) summarized the various topics of the biofilm matrix and pointed to open questions. In a second manuscript the matrix was discussed as being the perfect slime for microorganisms (Flemming, 2011). This article covers less known aspects of the matrix including modification after excretion, the wet and sticky matrix, interaction of the matrix with light and matrix at the water-air interface. At this point an excellent overview on the mechanical properties of the biofilm matrix has to be mentioned (Guelon *et al.* 2011). Topics such as elasticity, viscoelasticity and plasticity are elaborated on with respect to biofilm stability. In addition, the various methods used so far for measuring mechanical biofilm characteristics are described. In a textbook on cyanobacteria one chapter covers the matrix of this specific group of phototrophic microorganisms (Helm & Potts, 2012). The authors provide a critical appraisal of the different matrix compounds in terms of cyanobacteria ecology and in relation to the sometimes extreme habitats cyanobacteria live in. Another article put the focus on the role of the extracellular matrix in geomicrobiology (Tourney & Ngwenya, 2014). The authors discussed four basic processes in relation to the extracellular matrix: cell surface adsorption, metabolic uptake, biomineralisation as well as microbial oxidation and reduction.

3.2 CHALLENGES

In the following section some of the issues, challenges and questions regarding the biofilm matrix are briefly touched upon.

1) Is there always a polymer matrix? The common view of matrix embedded biofilm bacteria may be questioned by the observation of "naked" bacterial filaments without any obvious matrix in aquatic biofilms (Figure 3.1). For example in river biofilms two types of filamentous bacteria are often present. One group of microorganisms are attached by one cell pole and grow into long filaments sticking out of a base biofilm. A second group of very thin filamentous bacteria form mesh-like associations as a top layer of certain biofilms. In both cases the cells certainly possess glycoconjugates at their cell surface, but they do not produce a matrix covering or integrating the entire cellular structures.
2) What is the nature of the matrix in relation to the genotype, phenotype and environmental cues? How is the dynamic of matrix expression controlled?

(a)

(b)

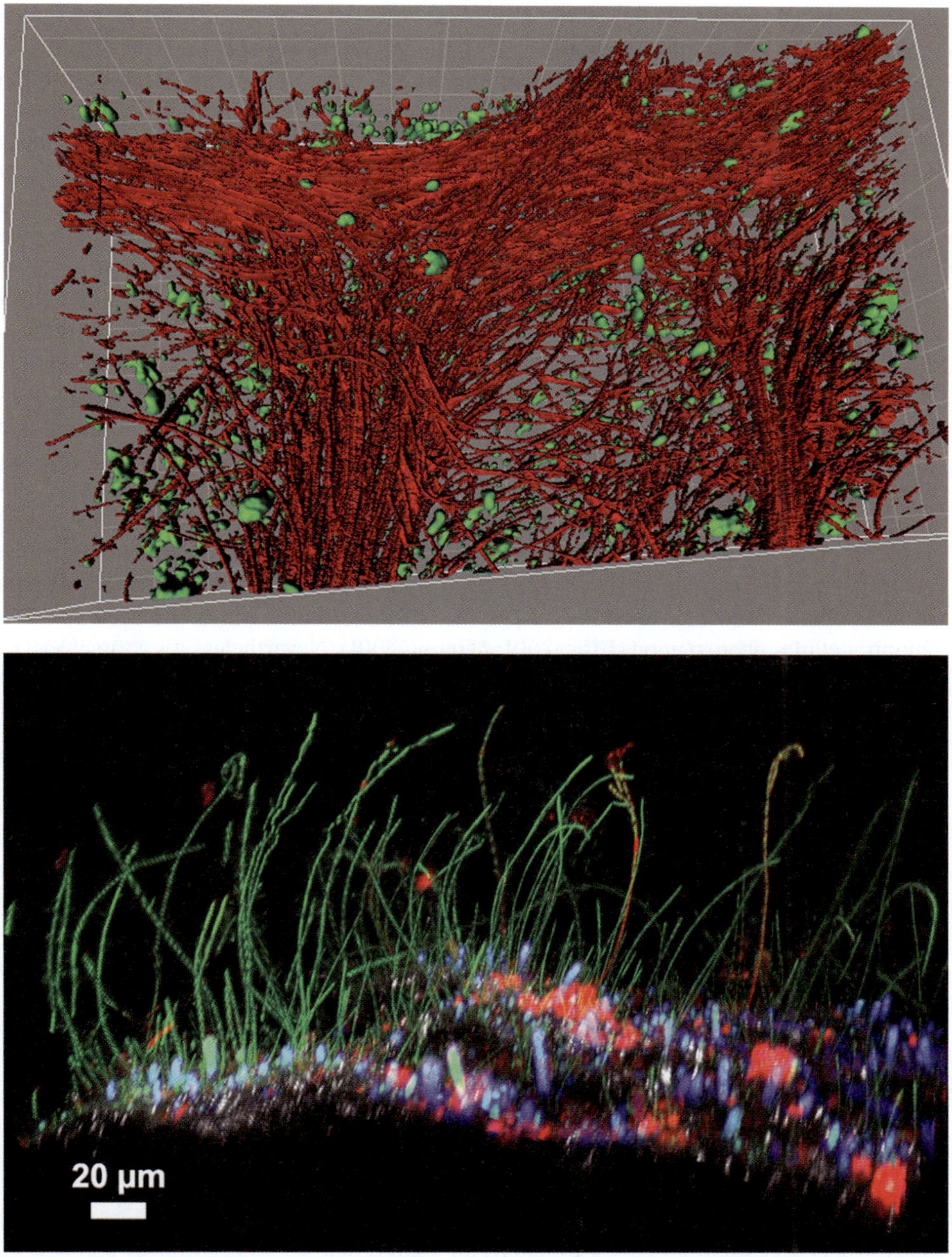

Figure 3.1 Biofilms from aquatic habitats showing bacterial filaments without an extensive matrix. a) Activated sludge reactor biofilm showing filamentous bacteria (nucleic acid stain) covering a basal biofilm layer (glycoconjugate stain). Colour allocation: green – lectin stain, red – nucleic acid stain. Grid size = 10 μm. b) Hairy biofilm from a river with a basal biofilm layer and long filamentous bacteria protruding into the water phase. Colour allocation: green – nucleic acid stain, red – lectin stain, blue – autofluorescence of Chl A. Sample courtesy: Marian Haesner and Norbert Kamjunke.

3) What is the distance over which the microbial cell is able to control its external microhabitat? This includes extracellular enzymes and the products of enzyme activity, the control of matrix production and degradation of the matrix as well as the assembly and release of extracellular vesicles.
4) What is the fate of released polymers in terms of lifetime, turnover and recycling? This aspect has been studied e.g. in microbial stromatolites using ^{14}C tracer experiments (Decho *et al.* 2005). The results showed that natural biofilm matrix represents a "more refractory remnant" of the original polymers produced. In addition a layer specific difference in EPS turnover could be established. This aspect was further defined for hypersaline microbial mats by studying the turnover of polymers and suggesting a conceptual model (Braissant *et al.* 2009). Similarly, the reuse of extracellular organic carbon by enzymatic activities of cyanobacteria in microbial mats indicates a dynamic pool of C resources (Stuart *et al.* 2016).
5) Are there general matrix regulators such as c-di-GMP or others (Ha *et al.* 2015)?
6) What is the distribution of reactive groups? There might be cationic, anionic and hydrophobic groups present. How much are they affected by ions, chelators and surface active compounds?
7) What are the physicochemical properties of the matrix compounds? Are there directed strands or randomly entangled polymers. One issue related to this aspect is on the one hand a limited penetration of e.g. antibiotics through the matrix (Tseng *et al.* 2013), while on the other there are so-called matrix swimmers (Houry *et al.* 2012; Piard *et al.* 2016; this volume). One consideration at this point is molecular crowding of polymeric biomolecules which has been discussed only as an intracellular phenomenon (Ellis, 2001; Minton, 2001). Nevertheless it's a topic very relevant for biofilm matrix characteristics.
8) What is the fractal dimension of the overall matrix and its individual compounds? This issue comprises the matrix boundary as well as biofilm internal boundaries e.g. surrounding bacterial cells, bacterial microcolonies and larger cell clusters of different species including eukaryotic microorganisms. In addition the individual matrix compounds may show a fractal dimension in terms of their spatial organisation. So far only the fractal dimensions of cellular biofilm constituents have been estimated. Nevertheless, to the best of our knowledge, fractal aspects of the matrix have been discussed for sludge polymers and wastewater aggregates in only two studies (Yuan *et al.* 2014; Ding *et al.* 2015).
9) Are there common biofilm polymers? Some bacterial groups produce cellulose, others polysaccharide intercellular adhesin (PIA), alginate or alginate-like polysaccharides. On the one hand there might be a huge variety of polymer types (Staudt *et al.* 2003), but perhaps the matrix in certain communities and habitats is dominated by a specific class of polymers (Seviour *et al.* 2012).
10) What is the interaction of the ECM with dissolved organic carbon (DOM) and self-assembled microgels (SAGs)? This aspect has been specifically addressed for the marine habitat (Decho, 1990; Verdugo, 2012). The presence of an ECM may dramatically change the availability of DOM (Decho & Lopez, 1993).
11) To what degree are mineralisation processes important for the assembly of a biofilm matrix? This aspect was highlighted very recently in a study with *Bacillus subtilis* and *Mycobacterium smegmatis*. The authors demonstrated the production of extracellular calcite forming a mineral scaffold with an effect on biofilm structural organisation (Oppenheimer-Shaanan *et al.* 2016).

In order to set the stage for biofilm matrix discussion, a drawing is provided showing some of the many options regarding how matrix constituents could be associated and interlinked at the molecular level (Figure 3.2).

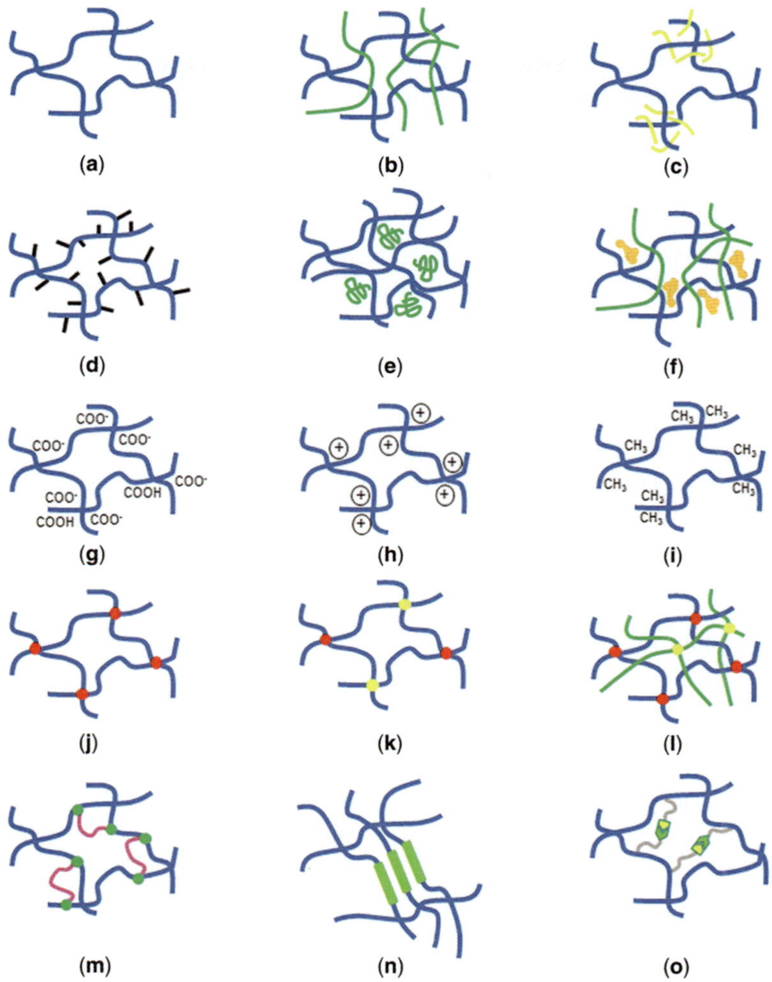

Figure 3.2 Theoretical considerations of polymer entanglements and cross-linking between different polymer types in the biofilm matrix. The various polymers and compounds are indicated by different colour and shapes. Linkages between polymers are indicated by solid circles and lock-key symbols. a) Interaction between identical polymers, e.g. polysaccharides. b) Interaction between different types of polymers e.g. charged and neutral polysaccharide. c) Interaction between large and small polymers, e.g. polysaccharide and lipopolysaccharide. d) Interaction between polymers carrying substituents e.g. fatty acids or amino acids. e) Polymer network with non-interacting secondary polymer. f) Interaction between different types of polymers plus an embedded compound, e.g. charged and neutral polysaccharide plus protein. g) Interactions between proteinaceous compounds. h) Interaction of negatively charged polymers via counter ions. i) Interaction between "hydrophobic" polymers substituted with methyl groups. j) Linkage of polymers by another compound, e.g. polysaccharides linked by a protein. k) Linkage of polymers by two other compounds, e.g. polysaccharides linked by two different proteins. l) Interaction between different types of polymers each with a different protein link, e.g. charged and neutral polysaccharide plus two different proteins. m) Interaction between different types of polymers linked via a secondary compound, e.g. polysaccharide plus fibre-like protein. n) Interactions of polymers by parallel alignment of homogeneous structures. o) Interaction between polymers by specific binding, e.g. polysaccharide and lectins.

3.3 BIOFILM MATRIX ANALYSIS

The biofilm matrix may be analysed by three different approaches. Firstly, classically via extraction and biochemical analysis; secondly, by using various imaging approaches; and thirdly, by means of molecular techniques.

The classical way of analysis describing the physical and chemical methods as well as the pitfalls was published some time ago (Nielsen & Jahn, 1999). In the meantime the various extraction methods have been used in many studies but mainly on activated sludge. Several of the reports made a comparison between the different techniques, see Table 3.2 in (Neu & Lawrence, 2014). A major problem with all of these methods is the question of contamination from the cell interior. Another critical aspect is the huge variety of biochemically different polymers. The major fractions extracted may comprise proteins, extracellular DNA (eDNA) and polysaccharides. The separation of these three fractions already represents an analytical challenge. Furthermore, even if the polysaccharide fraction can be isolated, it will contain a broad mixture of different polysaccharide types (Staudt *et al.* 2003). The polysaccharides may contain in their repeating unit neutral monosaccharides only, they may carry positive or negative charges or they are can be highly charged. The latter would be typically an alginate-like polysaccharide. This type of gel-forming polysaccharides was isolated from aerobic, granular sludge as a key component for a strong but elastic granule structure (Lin *et al.* 2010; Lin *et al.* 2013). In any case, it is still a matter of debate to what extent specific polysaccharides exist in biofilms and bioaggregates.

Imaging may be applied for biofilm matrix analysis using a range of techniques including synchrotron imaging e.g. by scanning transmission X-ray microscopy (STXM), chemical imaging e.g. by Raman microscopy or fluorescence techniques in combination with confocal laser scanning microscopy (CLSM). By means of CLSM the network of matrix glycoconjugates can be probed by using immunofluorescence techniques in pure cultures e.g. in (Chalmers *et al.* 2007) or fluorescence lectin-binding analysis (FLBA) in environmental samples (Neu *et al.* 2001; Neu & Lawrence, 2014). Raman microscopy may reveal certain chemical groups within the matrix which then can be attributed to certain polymers (Ivleva *et al.* 2009; Ivleva *et al.* 2010). Raman may also be combined with confocal laser microscopy in order to take advantage of both techniques (Wagner *et al.* 2009; Janissen *et al.* 2015). Using STXM the simultaneous analysis of elemental and biochemical constituents within the biofilm matrix becomes possible. The proof of principle has been demonstrated for the matrix of river biofilms (Lawrence *et al.* 2003). Nevertheless with some of the techniques there are limitations in terms of sample mounting and as a result CLSM remains the most flexible approach which is also more readily available (Neu *et al.* 2010). Currently new laser based imaging approaches are explored for biofilm examination at the meso-, micro- and nano-scale (Neu & Lawrence, 2015).

Molecular techniques for biofilm matrix analysis were mostly applied to defined bacterial species. In the age of "omics" it is most obvious to examine matrix complexity by genomics, transcriptomics and proteomics. In nearly all cases the matrix was subjected to proteomic analysis in order to identify proteinaceous matrix constituents. This was specifically done for *Haemophilus influenzae* (Gallaher *et al.* 2006), *Myxococcus xanthus* (Curtis *et al.* 2007), *Listeria monocytogenes* (Dumas *et al.* 2008), *Shewanella* sp. (Cao *et al.* 2011) and a mixed community from dental plaque (Paes Leme *et al.* 2008). In only one manuscript, both the metagenomic and proteomic approach were combined for the analysis of an activated sludge matrix (Albertsen *et al.* 2013). With respect to functionality, apart from proteomics the second major aspect that was so far not dealt with would be – glycomics. The missing link is the combination of all four approaches: genomics, transcriptomics, proteomics and glycomics. In order to go even one step further the surfome (complete set of secretion systems involved in surface colonisation and biofilm formation) and secretome (complete set of surface exposed proteinaceous moieties) have been suggested (Dang & Lovell, 2016). However, in this respect a manuscript on "badomics" may have to be considered

(Eisen, 2012). The overall progress in studying the extracellular matrix will be discussed in a later section on "bacterial extracellular biology" with examples of common biofilm species investigated in detail.

3.4 BIOFILM MATRIX CONSTITUENTS

3.4.1 Polysaccharides

The traditional view looked at the extracellular space in biofilms and bioaggregates as a matrix composed of polysaccharides. A major contribution towards extracellular polysaccharides originates from Ian Sutherland who kicked off the research field with his monograph in 1977 (Sutherland, 1977). He published a long list of original articles with focus on microbial extracellular polysaccharides. In one of the latest reviews a comprehensive overview was presented. Emphasis is put on chemical composition and physical properties versus functionality and physiological aspects (Sutherland, 2007). In parallel he discussed this topic with respect to biofilms in a manuscript critically assessing the biofilm matrix components and their dynamics. Sutherland suggested the matrix was comparable to an immobilized enzyme system controlled by intrinsic and extrinsic factors. As a result the matrix represents a highly dynamic polymeric structure at the molecular level creating numerous bio/chemically different microhabitats (Sutherland, 2001).

In an attempt to present current knowledge on biofilm polysaccharide formation and function, polysaccharides were categorized into aggregative, protective and architectural (Limoli *et al.* 2015). The polysaccharides are described as being aggregative with respect to initial attachment, micro/macrocolony formation and detachment or disassembly. Examples include the polysaccharide intercellular adhesion (PIA) in *Staphylococcus aureus* as well as the Pel (pellicle polysaccharide) and Psl (polysaccharide synthesis locus) polysaccharides in *Pseudomonas aeruginosa*. Polysaccharides are discussed as being protective against environmental and host factors. This includes phagocytosis and in relation to alginate, protection from desiccation and antibiotics. In addition, polysaccharides support viability, gene expression and lead to a stable microbial population. On the other hand, levan seems to serve mainly as a nutrient storage. Finally architectural polysaccharides are suggested to play a key role in the structure of many biofilm systems. For example colanic acid was described as important for building a mature 3-dimensional biofilm structure. Vibrio polysaccharide promotes environmental biofilm development as well as the host infectious cycle. *Bacillus subtilis* polysaccharide has a role in pellicle formation and colony surface structures. Cellulose is also involved in pellicle formation and matrix-fimbriae interactions and also functions as a UV protective compound. At the end of their review the authors provide future perspectives in which they identify the need for new analytical approaches and techniques, improved animal models for in vivo studies and advanced understanding of matrix molecular interactions (Limoli *et al.* 2015).

Apart from general publications reviewing the role of extracellular polysaccharides, progress has been made in detailed studies using selected microbial species. Most basic research on polysaccharide synthesis, export and presence on the cell surface was probably done with *Escherichia coli* (Whitfield, 2006). However, polysaccharide research related to biofilms had and still has a prominent proxy namely *Pseudomonas aeruginosa*. More details of the *Escherichia coli* and *Pseudomonas aeruginosa* matrix are discussed below in the section on "bacterial extracellular biology".

3.4.2 Proteins

Extracellular proteins of microorganisms fall into several different classes. These comprise cell surface adhesins, protein subunits of pili and fimbriae, secreted extracellular proteins, proteins of outer membrane vesicles and extracellular enzymes (Fong & Yildiz, 2015). The cell surface associated proteins are involved in initial adhesion and migration on surfaces and as such in colonisation of interfaces.

A protein directly related to the biofilm matrix is represented by the biofilm associated protein (Bap) which was first identified in Staphylococci biofilms. Bap proteins are important for adhesion and biofilm development but they also serve as virulence factors. Orthologs of this protein were also found in other Gram-positive and Gram-negative bacteria. Thus it remains a question of how widespread this protein family in biofilm formation by different groups of bacteria is (Fong & Yildiz, 2015).

In relation to the biofilm matrix and its functionality other proteins which show a binding affinity to matrix compounds such as polysaccharides or eDNA have to be considered. Through this affinity these proteins can form a crosslink in the polymer scaffold of the matrix. They may form protein-eDNA or protein-polysaccharide complexes contributing substantially to the stability of the biofilm matrix (Fong & Yildiz, 2015). An example of such a protein has been described for *Vibrio fischeri* having a role in biofilm maturation (Ray *et al.* 2015). More examples for this kind of protein are mentioned below in the section on "bacterial extracellular biology". Interestingly the protein-polysaccharide interaction was hypothesized as a platform for protein delivery in applied systems with abiotic surfaces as a target (Absalon *et al.* 2012).

An established functional component, are enzymes which are important for degrading polymeric compounds in order to make them amenable for metabolism. This implies that enzymes must have a recognition and binding functionality for the substrate. For more information on extracellular enzymes in biofilms the reader is referred to e.g. (Hoffman & Decho, 1999; Sutherland, 1999). A major aspect of extracellular enzymes relates to their deployment, in other words, are they cell-surface-bound or freely localised in the near microhabitat. As a result the microbial extracellular enzyme strategy will create different substrate fields with implications on the availability and efficiency of nutrient utilisation (Stocker, 2012; Traving *et al.* 2015).

3.4.3 Amyloids

One of the first reports on curli-like structure and their role in biofilm formation was described for *Escherichia coli* (Hammar *et al.* 1996) and *Salmonella enteritidis* (Austin *et al.* 1998). Later the biogenesis of amyloids in the form of curli produced by *Escherichia coli* was investigated and suggested as a general model system for amyloid formation (Chapman *et al.* 2002; Barnhart & Chapman, 2006). Amyloids as a functional coating for microorganisms including bacteria and fungi were discussed in an early review (Gebbink *et al.* 2005). The first reports on the universal presence of amyloids as extracellular matrix components in biofilms came from the group of Per Halkjaer Nielsen (Larsen *et al.* 2007). Subsequent studies describing various aspects of amyloids were published for bacterial communities from different habitats and defined species (Table 3.1). For amyloids different terms were used including curli, chaplins, microcins, pili, modulins, tafi, hydrophobins or just the letter code of the protein e.g. Fap, TasA. etc.. In fact curli, Fap and TasA are the best studied amyloid systems in relation to biofilms. For more details on *Pseudomonas aeruginosa* and *Bacillus subtilis* multiple matrix compounds see the section on "bacterial extracellular biology". A review on functional amyloids detailed the assembly of curli as well as their functions including biofilm and floc formation, role as adhesins, as cell surface modifiers and as toxins (Blanco *et al.* 2012). In addition, functional amyloids were described as crucial determinants for adhesion, biofilm and bioaggregate formation. Amyloids can also serve as cell surface modifiers, especially when microorganisms grow from the hydrophilic water phase into the hydrophobic aerial phase. This was reported for chaplins in Streptomyces. Functionally similar compounds are produced by fungi where they are often named hydrophobins. Sometimes amyloids act as toxins e.g. as microcins in Klebsiella or as harpins in plant pathogenic bacteria. Another review article from the Chapman group goes into more details and suggests amyloids as structural material, toxins, surface active fibres, genetic material, adhesins, host mimetics and speculates on more functions (DePas & Chapman, 2012). In an opinion paper the characteristics and functions of known amyloids were discussed and compiled in a table (Schwartz *et al.* 2012). The authors conclude a role for different types of amyloids as structural material, adhesin, toxin and protection against host response. An excellent update on amyloids in bacteria gives an overview on confirmed functional bacterial amyloids (FuBA) (Dueholm *et al.* 2013a). They discussed the functions of amyloids as a structural component in the biofilm matrix, as a virulence factor of human and plant pathogens, as oligomeric toxins and as coatings of spores. See also Dueholm and Nielsen (2016, this volume).

Table 3.1 Reports on amyloids in microbial communities and specific strains with respect to their functionality and interaction with interfaces.

Species	Amyloid Function	References
Natural aquatic biofilms	Presence and abundance in several phyla	(Larsen *et al.* 2007)
Activated sludge	Adhesion, sheath strengthening, backbone of flocs, hydrophobicity	(Larsen *et al.* 2008)
Escherichia coli	Requirement of curli for development of 3d biofilm architecture, most studied amyloid	(May & Okabe, 2008)
Gram-positive bacteria	Widespread abundance in various species	(Jordal *et al.* 2009)
Pseudomonas sp.	Functional amyloid involved in biofilm formation	(Dueholm *et al.* 2010)
Bacillus subtilis	Protein component of the matrix provides structural integrity and interacts with exopolysaccharide	(Romero *et al.* 2010; Ostrowski *et al.* 2011; Romero *et al.* 2011)
Saccharomyces cerevisiae, Candida albicans	Needed for cell aggregation and biofilm formation	(Garcia *et al.* 2011)
Staphylococcus aureus	Formation and disassembly of biofilms	(Schwartz *et al.* 2012)
Candida albicans	Formation of adhesion nanodomains	(Lipke *et al.* 2012)
Streptococcus mutans	Contribution to oral biofilm formation	(Oli *et al.* 2012)
Pseudomonas	Fap amyloids are important for aggregation and biofilm formation	(Dueholm *et al.* 2013b)
Pseudomonas aeruginosa	Effect of Fap on *Pseudomonas aeruginosa* proteome and physiology	(Herbst *et al.* 2015)
Pseudomonas aeruginosa	Interaction of amyloids with quorum sensing molecules	(Seviour *et al.* 2015)
Pseudomonas	Fap makes bacteria more hydrophobic and increases biofilm stiffness	(Zeng *et al.* 2015)

3.4.4 Extracellular nucleic acids

Extracellular DNA (eDNA) was identified as an important component in the ECM and mentioned in very early reports as well as several studies at the beginning of the first decade of the 21st century, see (Neu & Lawrence, 2009) for earlier literature. Since then eDNA has been studied in different bacterial species often with a focus on functionality. The recent literature was compiled in a table outlining the functions suggested for various bacterial species (Table 3.2). The overall functions of eDNA in adhesion, maintenance of biofilms, gene transfer and protection were compiled in a comprehensive overview (Okshevsky & Meyer, 2013). The authors also address the origin of eDNA and the question of lysis or active secretion. In another review, eDNA is discussed with respect to release mechanisms, binding with other matrix compounds and its influence on cell surface properties and biofilm development. It is suggested that eDNA serves as a scaffold providing structural integrity for biofilms in medical and technical habitats (Das *et al.* 2013b). From its wide distribution it became obvious to look at eDNA as a target for biofilm control. This particular topic was summarized in a very recent review discussing the potential strategies (Okshevsky *et al.* 2015). Apart from eDNA the presence of extracellular RNA (eRNA) in *Escherichia coli* was reported to be associated with outer membrane vesicles but also as free molecules (Ghosal *et al.* 2015).

Table 3.2 Selected literature published since 2008 on the function and roles of extracellular DNA in microbial communities.

Species	eDNA Function	References
Staphylococcus aureus/ epidermis	Differential roles for polysaccharides and eDNA	(Izano *et al.* 2008)
Strepococcus pneumoniae	Important matrix compound	(Hall-Stoodley *et al.* 2008)
Shewanella sp.	Utilization of eDNA as sole source of P and C	(Pinchuk *et al.* 2008)
Pseudomonas aeruginosa	Needed for mushroom cap formation in flow cells	(Barken *et al.* 2008)
Pseudomonas aeruginosa	eDNA chelates cations	(Mulcahy *et al.* 2009)
Staphylococcus aureus	Biofilm formation and maintenance	(Mann *et al.* 2009)
Bacillus cereus	Required as an adhesive and part of the matrix	(Vilain *et al.* 2009)
Enterococcus faecalis	Needed for biofilm development	(Guiton *et al.* 2009)
Rhodovulum sulfidophilum	Role in floc stabilisation	(Suzuki *et al.* 2009)
Staphylococcus aureus	Protein-crosslinking in the presence of eDNA	(Huseby *et al.* 2010)
Neisseria meningitis	Used in one of two biofilm development strategies	(Lappmann *et al.* 2010)
Caulobacter crescentus	Inhibition of progeny settling in biofilms	(Berne *et al.* 2010)
Listeria monocytogenes	Central component in initial attachment and early biofilm formation	(Harmsen *et al.* 2010)
Streptococcus sp.	Role in adhesion and surface aggregation	(Das *et al.* 2010)
Azotobacter vinelandii	Role in transformation	(Lu *et al.* 2010)
Escherichia coli	Proteins responsible for the integrity of an eDNA containing matrix	(Goodman *et al.* 2011)
Mixed activated sludge biofilms	Identification of eDNA producers, sludge stability	(Dominiak *et al.* 2011)
Vibrio cholerae	Matrix compound involved in development, architecture, nutrient acquisition, detachment, colonisation fitness	(Seper *et al.* 2011)

(*Continued*)

Table 3.2 Selected literature published since 2008 on the function and roles of extracellular DNA in microbial communities (*Continued*).

Species	eDNA Function	References
Myxococcus xanthus	eDNA interacting with polysaccharides strengthen the matrix	(Hu *et al.* 2012a)
Shewanella oneidensis	Role of extracellular nucleases for nutrient supply and detachment	(Goedeke *et al.* 2011a; Heun *et al.* 2012)
Pseudomonas aeruginosa	Pyocyanin as trigger for eDNA release and cell-cell interactions	(Das & Manefield, 2012; Das *et al.* 2013a)
Pseudomonas aeruginosa	Trail formation and cell movement	(Gloag *et al.* 2013)
Defined mixed biofilms	Role in viscoelastic relaxation	(Peterson *et al.* 2013)
Enterococcus feacium	Biofilm stability assessed by autolysins	(Paganelli *et al.* 2013)
Various model strains	Control of eDNA by proteins related to degradation but also stability	(Jakubovics *et al.* 2013)
Streptococcus mutans	eDNA contributing to structural integrity interacting with polysaccharides	(Liao *et al.* 2014)
Shewanella oneidensis	Iron as a trigger for phage and eDNA release	(Binnenkade *et al.* 2014)
Clostridium phytofermentans	Biofilm structure and stability	(Zuroff *et al.* 2014)

3.4.5 Amphiphilic compounds

Amphiphilic compounds may range from biosurfactants to polymeric amphiphiles. Their biochemistry may be represented by carbohydrates or proteins. Examples are glycolipids and peptidolipids as described for Pseudomonas and Bacillus or emulsan described for Acinetobacter. In fact rhamnolipid surfactants of *Pseudomonas aeruginosa* were found to affect the structure of biofilms, control the swarming behaviour of bacteria and have a role in detachment from biofilms (Davey *et al.* 2003; Boles *et al.* 2005; Caiazza *et al.* 2005; Morris *et al.* 2011; Kim *et al.* 2015). Similarly the peptidolipids such as viscosin (*Pseudomonas fluorescens*) and surfactin (*Bacillus subtilis*) are discussed with respect to interfacial processes and biofilm formation (Neu *et al.* 1990; Neu & Poralla, 1990; Neu, 1996; Raaijmakers *et al.* 2010). In addition, an amphiphilic protein (BslA) is produced by *Bacillus subtilis* which is responsible for surface repellency of biofilms and colonies (Kobayashi & Iwano, 2012). Emulsan is a polyanionic heteropolysaccharide bioemulsifier produced by *Acinetobacter calcoaceticus*. Due to its properties it was patented as a measure to clean tankers from shipping oil (Rosenberg & Ron, 1997). Hydrophobic compounds were also described for other bacterial species. For example in *Staphylococcus aureus* a surfactant-peptide was identified as a biofilm structuring factor which is also involved in detachment processes (Periasamy *et al.* 2012). Thereby the phenol-soluble modulin (PSM) indirectly controls *Staphylococcus aureus* biofilm structure. Apart from isolated and identified compounds, hydrophobic domains were detected in the matrix of biofilms. In a study with *Shewanella oneidensis*, amphiphilic quantum dots were used as a probe for the biofilm matrix. By this means a pattern of hydrophobic domains were found within the matrix which may restrict diffusion of hydrophobic compounds through the matrix (Aldeek *et al.* 2013).

3.4.6 Membrane vesicles

Membrane vesicles are released by bacteria, archaea and eukaryotic microorganisms (Deatherage & Cookson, 2012). They were named: membrane vesicles, outer membrane vesicles (OMV), extracellular vesicles (EV), exosomes, shedding microvesicles, etc.. They have been known for some time and were brought into the biofilm focus by the group of Terry Beveridge. His contribution was honoured in a short overview on outer membrane vesicles (Mashburn-Warren *et al.* 2008). For the earlier literature on membrane vesicles see (Neu & Lawrence, 2009), the newer literature was compiled in a table listing their significance and function (Table 3.3). In a comprehensive summary with a focus on Gram-negative bacteria, the biogenesis and control together with the biological functions of outer membrane vesicles were discussed (Kulp & Kuehn, 2010). Another overview discussed membrane vesicles having offensive and defensive capabilities in inter-microbial and microbe-host interactions (MacDonald & Kuehn, 2012). Offensive functions comprised: activation of toxins, killing of competing bacteria, delivery of virulence factors, disruption of tissue and inflammatory response. Defensive functions included: sharing of antibiotic resistance, biofilm formation, suppression of immune responses, decoy function for antimicrobial peptides and phages as well as relief of membrane stress. A mini-review with a focus on *Pseudomonas aeruginosa* highlights the role of membrane vesicles in bacteria-bacteria cooperation and competition, as a pathogen in microbe-host interactions and as a key element for survival in various environments (Tashiro *et al.* 2012). The authors discuss the constituents of membrane vesicles such as proteins, LPS, phospholipids, DNA and quinolones. Further they explain the current understanding of membrane vesicle biogenesis including the mechanisms for stimulation of membrane vesicle formation. Finally a relationship between membrane vesicle function and membrane vesicle components is established. The latest reports on extracellular vesicles were summarized in two reviews, first focusing on OMV biogenesis and function in Gram-negative bacteria (Schwechheimer & Kuehn, 2015) and second by looking at EV's in thick-walled microorganisms such as Gram-positive bacteria, mycobacteria and fungi (Brown *et al.* 2015).

Table 3.3 Selected recent publications on outer membrane vesicles, their significance and function.

Species	OMV Significance/Function	References
Pseudomonas aeruginosa	Delivery of virulence factors: – β-lactamase – Alkaline phosphatase – Haemolytic phosphatase – Cif toxin	(Bomberger *et al.* 2009)
Pseudomonas aeruginosa	Interaction with eDNA	(Schooling *et al.* 2009)
Pseudomonas aeruginosa	Growth phase dependent physicochemical properties	(Tashiro *et al.* 2010)
Antarctic bacteria	Enrichment of outer membrane and periplasmic proteins related to nutrient processing and transport	(Frias *et al.* 2010)
Escherichia coli	Bacterial defense by adsorption: – Antimicrobial peptides – Bacteriophages	(Manning & Kuehn, 2011)
Pseudomonas putida	Stress response due to: – Long chain alcohols – Osmotic stress – EDTA – Heat shock	(Baumgarten *et al.* 2012)
Shewanella vesiculosa	DNA transfer in double bi-layer vesicles	(Pérez-Cruz *et al.* 2013)

(*Continued*)

Table 3.3 Selected recent publications on outer membrane vesicles, their significance and function (*Continued*).

Species	OMV Significance/Function	References
Xylella fastidiosa	Modulation of exploration and attachment to plant cells	(Ionescu *et al.* 2014)
Pseudomonas aeruginosa	Control of protein content by O polysaccharide	(Murphy *et al.* 2014)
Bacteroides	Identification of specific packed proteins: – Glycosidases – Proteases	(Elhenawy *et al.* 2014)
Vibrio cholerae	– Identification of specific proteins – DegP protease controlling Protein content and matrix formation	(Altindis *et al.* 2014)
Prochlorococcus	– Contain: proteins, DNA, RNA – Support growth of heterotrophs	(Biller *et al.* 2014)
Escherichia coli	RNA in microbe-microbe and microbe-host intercellular communication	(Ghosal *et al.* 2015)

3.4.7 Refractory compounds

Very little is known about microbially derived refractory compounds. This class of substances is represented by remains of bacterial cells which are not readily degraded after cell death. As high molecular weight compounds they will be present within microbial communities contributing to the polymeric matrix. In an early report the bacterial derived dissolved organic matter (DOM) was investigated in the marine environment (Ogawa *et al.* 2001). The significance of this refractory fraction was also studied in soil as a source of organic carbon originating from microorganisms (Miltner *et al.* 2012; Schurig *et al.* 2013). To the best of our knowledge, the fraction of refractory matrix compounds is so far fully neglected if the biofilm matrix literature is screened for this particular topic.

3.5 BACTERIAL EXTRACELLULAR BIOLOGY

The heading of this section was taken from a special issue in FEMS Microbiology Reviews in order to differentiate the intracellular metabolism from extracellular matrix processes (Chater, 2010). In this section the microbial biofilm species which were studied in depth in an effort to advance understanding of the biofilm matrix are briefly discussed.

The standard biofilm species in numerous investigations has been *Pseudomonas aeruginosa*. Examination of the available literature reveals a more detailed understanding of polysaccharide production and control as well as an evolving concept of matrix molecular organisation. The characteristics of the *Pseudomonas aeruginosa* biofilm matrix with several strong contributions from the group around Dan Wozniak, are summarized as a table (see Table 3.4). Apart from polysaccharides only few extracellular proteins are known including amyloids, two lectins (LecA and LecB) as well as a Psl-binding protein (Borlee *et al.* 2010; Fong & Yildiz, 2015). For the functional amyloid of Pseudomonas (Fap) multiple functions have been demonstrated (Table 3.1). Current knowledge of biofilm development by various Pseudomonas species was summarized recently by Tolker-Nielsen. Overall the biofilm structural features found seem to be an adaptive response of individual bacteria to micro-environmental cues (Tolker-Nielsen, 2015).

Table 3.4 Recent publications on the extracellular polysaccharides produced by *Pseudomonas aeruginosa* in relation to biofilm development.

Focus of Study	References
Current state of the art on the role of alginate, Pel and Psl in the biofilm matrix	(Ryder *et al.* 2007)
Real time matrix formation and visualisation of Psl showing its role in biofilm development	(Ma *et al.* 2009)
Biosynthesis of alginate, Pel and Psl suggesting a predictive model	(Franklin *et al.* 2011)
Review on biofilm matrix discussing polysaccharides, eDNA and proteins	(Mann & Wozniak, 2012)
Role of Pel and Psl in matrix indicating their redundancy	(Colvin *et al.* 2012)
Regulation and modulation of polysaccharide synthesis	(Ma *et al.* 2012)
Matrix penetration of antibiotics with different properties	(Tseng *et al.* 2013)
Pili-Psl interaction causes a spider web feature in biofilms	(Wang *et al.* 2013)
Role of Pel (reduced cross-linking) and Psl (increased elasticity and cross-linking) in microrheology and dual species biofilms	(Chew *et al.* 2014)
Coordination of biosurfactant synthesis and matrix production	(Wang *et al.* 2014)
Psl-eDNA interaction results in a biofilm skeleton structure	(Wang *et al.* 2015)
Crosslinking of eDNA by the cationic polysaccharide Pel	(Jennings *et al.* 2015)
Pel production has a differential role in pellicle and agar grown cultures	(Madsen *et al.* 2015)

A second model organism in biofilm research is *Bacillus subtilis*. It shows a rather complex biofilm assembly mechanism which includes several of the above mentioned matrix constituents. The matrix of *Bacillus subtilis* may contain various matrix proteins, enzymes, different polysaccharides, amyloids, hydrophobins and biosurfactants. Although a basic understanding is currently available it remains rather obscure how these matrix components interact with each other and with the interface (liquid-solid/liquid-gas) at which the microbes develop their biofilm. The major original results and reviews many of which originate from the group of Roberto Kolter are compiled in a table (Table 3.5).

Table 3.5 Recent publications on the biofilm matrix constituents of *Bacillus subtilis*.

Focus of Study	References
Polysaccharide and surfactin are needed for fruiting body formation	(Branda *et al.* 2001)
Protein TasA and a polysaccharide as major matrix components	(Branda *et al.* 2006)
Surface spreading by the *B. subtilis* biosurfactant surfactin	(Angelini *et al.* 2009)
Review on multiple cell types and their differentiation	(Lopez *et al.* 2009)
TasA is an amyloid and an important matrix component	(Romero *et al.* 2010)
Extracellular signals in cell differentiation and matrix production	(López & Kolter, 2010)
Surface motility by extracellular polysaccharide	(Seminara *et al.* 2012)
Water repellence by hydrophobic protein, protection from environmental cues	(Kobayashi & Iwano, 2012; Hobley *et al.* 2013)
General review on biofilm assembly and disassembly	(Vlamakis *et al.* 2013)
Review on regulation and assembly with focus on matrix composition and function	(Cairns *et al.* 2014)
Review on molecular mechanisms of biofilm formation, differentiation and disassembly	(Mielich-Süss & Lopez, 2015)

Pathogenic strains of *Vibrio cholerae* are the focus of many studies. The formation of *Vibrio cholerae* biofilms was described as a sequence starting with scanning a surface by orbiting, followed by attachment and microcolony formation (Berk *et al.* 2012; Teschler *et al.* 2015). During this process the cells produce a series of extracellular compounds including vibrio polysaccharide (VPS), the biofilm structure modulator (RbmA), the biofilm matrix protein (Bap1), the biofilm matrix protein (RmbC) and the chitin-binding protein (GbpA). The matrix associated protein Bap1 is concentrated on surfaces forming a conditioning film, whereas RbmA strengthens intercellular attachments of cells. From these results a new matrix paradigm was suggested as the polysaccharide scaffold specifically interacts differentially with the individual matrix proteins (Absalon *et al.* 2011). Another study defined the role of secreted proteases in more detail. It was demonstrated that RbmA undergoes proteolysis which results through a cascade of events in a reinforcement of the biofilm (Smith *et al.* 2015). In addition, eDNA and outer membrane vesicles carrying proteases are released from biofilm bacteria and become associated with the mature biofilm. Both eventually will contribute to *Vibrio cholerae* biofilm reinforcement. Recent advances expanded the knowledge on *Vibrio cholerae* biofilms and the major matrix compounds and regulators were identified, however several questions remain open. These comprise among others: dispersal mechanisms, growth in mixed biofilms, predation by protists, phage infection and in vivo biofilm characteristics.

Shewanella oneidensis is another biofilm model species which was shown to produce highly structured biofilms. It was demonstrated that prophage-mediated lysis contributed to the release of biofilm promoted compounds identified as eDNA (Goedeke *et al.* 2011b). Later it was found that extracellular endonucleases have a role in biofilm formation and to a certain degree in eDNA degradation for improved nutrient supply (Goedeke *et al.* 2011a). A subsequent study identified a further extracellular nuclease which had a major role in eDNA degradation (Heun *et al.* 2012). The investigation of iron as a trigger demonstrated the fine-tuned link between bacterial iron requirement, iron-induced DNA damage, prophage-mediated lysis and eDNA release (Binnenkade *et al.* 2014).

Streptococcus as a pathogenic biofilm species is involved in a wide range of disease and consequently the interaction with tissues as well as inert surfaces is of high interest. In a comprehensive review the adherence and colonisation of streptococci was summarized (Nobbs *et al.* 2009). The cell surfaces of streptococci contain a wide collection of constituents involved in environmental sensing, virulence, adherence and biofilm formation. They also produce so-called anchorless proteins which are mainly enzymes. Polysaccharides were mostly studied in oral streptococci where they have a crucial role in adherence and dental plaque formation (Koo *et al.* 2010; Klein *et al.* 2015). In fact they produce a water-soluble and a water-insoluble glucan. Both serve different functions, the water-insoluble glucan is responsible for adherence and biofilm formation, whereas the water soluble one is important for nutrient scavenging. The review concluded that streptococci actively and continuously monitor their environment and as a result fine-tune their expression of adhesins, production of effector molecules, communication system and metabolic pathways (Nobbs *et al.* 2009). Other streptococcal strains and their matrix components are of high interest as a cause of pneumonia (Domenech *et al.* 2013).

Staphylococci play a role in wound infection as well as in a number of diseases. State of the art knowledge of Staphylococcus biofilms was compiled by Otto (Otto, 2013). The most significant extracellular compound produced during biofilm development is the polysaccharide intercellular adhesin (PIA) also known as poly-N-acetyl-glucosamin (PNAG). This polymer is de-acetylated which is important for anchoring at the cell surface. Instead of PIA other strains produce so-called biofilm-adhesive proteins. Additional matrix compounds include teichoic acids, eDNA and a cascade of proteins which seem to be important for biofilm development (Foster *et al.* 2014; Hobley *et al.* 2015). New findings suggested an electrostatic net model for eDNA interacting with a variety of matrix proteins required for connecting biofilm cells (Dengler *et al.*

2015). Furthermore phenol-soluble modulins (PSMs) and enzymes are present and contribute to biofilm maturation and detachment (Periasamy *et al.* 2012).

Candida albicans represents a fungal pathogen which frequently forms biofilms. The genetic control and regulation of biofilm formation has been the subject of several studies e.g. (Finkel & Mitchell, 2011; Desai & Mitchell, 2015; Fox *et al.* 2015). Furthermore, biofilm architecture and matrix assembly was investigated (Chandra *et al.* 2001). The matrix produced contains polysaccharides (α-mannan, β-1,6 glucan, β-1,3 glucan) but also specific matrix proteins and eDNA. A recent study by Mitchell *et al.* demonstrated that matrix biogenesis is a coordinated extracellular action suggesting cooperation within a mixed community of different mutants (Mitchell *et al.* 2015).

In the special issue of FEMS Microbiology Reviews (2010, volume 34, issue 2) the topic "bacterial extracellular biology" was discussed for selected microorganisms. The examples presented included: *Myxococcus xanthus*, Xanthomonas, *Bacillus subtilis*, Rhizobia, Streptomyces and lactobacilli. For *Myxococcus xanthus* polysaccharides and proteins were described as major extracellular components. It could be shown that pilin is able to directly interact with extracellular polysaccharides under native conditions (Hu *et al.* 2012b). Recently extracellular vesicles were identified as a crucial compound. In addition, *Myxococcus xanthus* does take advantage of two different motility systems for gliding (Konovalova *et al.* 2010). Xanthomonas as a plant pathogen uses a range of compounds such as adhesins, polysaccharides, lipopolysaccharides (LPS) and enzymes for successful infection (Büttner & Bonas, 2010). Rhizobia also rely on a number of matrix compounds such as polysaccharides and secreted proteins. The major compounds are a glucomannan and a specific protein called rhicadhesin. Furthermore rhizobia produce Nod factors which are oligosaccharides with a number of substituents. Interestingly these include deoxy-sugars as well as acyl groups and thus impart a hydrophobic property or even biosurfactant characteristics (Downie, 2010). Streptomyces as a soil organism produces an exceptionally large number of secreted proteins. These include non-catalytic proteins as well as hydrolytic enzymes. Some compounds are amphipathic such as chaplins and rodlins and contribute to the complex cascade of extracellular compounds produced (Chater *et al.* 2010). Lactobacilli as important food fermenters are known to produce many different proteins and polysaccharides. These compounds are of interest due to their health-promoting effects (Kleerebezem *et al.* 2010).

Last but not least *Escherichia coli* has served as a model in basic studies of polymer synthesis, transport and extracellular location. Current knowledge of *Escherichia coli* biofilms including adhesion, development and control was compiled in a review article by the group of Jean-Marc Ghigo (Beloin *et al.* 2008). The organisation and complexity of uropathogenic *Escherichia coli* biofilms was described with curli, cellulose and pili as major matrix constituents (Hung *et al.* 2013). In general overviews on the *Escherichia coli* matrix the main ECM compounds, curli, cellulose, and flagella were discussed including their presence in discrete zones of the biofilm and control by the regulator CsgD (Hobley *et al.* 2015; Hufnagel *et al.* 2015).

3.6 BIOFILM MATRIX FUNCTIONALITY

In this section the evolving concept of matrix functionality is briefly introduced. A first draft of this concept was published several years ago (Neu & Lawrence, 2009). A summary of the functionality is also provided in the form of a table in order to show the multi-functionality of matrix compounds (Table 3.6).

3.6.1 Architecture

Constructive matrix compounds are those that produce the scaffold into which the microorganisms are embedded. These compounds also form what was called the "house" of biofilm microorganisms (Flemming *et al.* 2007). These polysaccharides are often represented by neutral polysaccharides which are produced during microcolony formation and biofilm development. In addition proteins represent a second major compound which can make up a large proportion of the matrix. In more recent publications eDNA was also suggested as a constructive matrix compound. An example of an embedding matrix from a reactor biofilm shows a lectin staining which covered most of the matrix glycoconjugates as there are hardly any bacteria left outside the overall matrix signal (Figure 3.3).

Table 3.6 Extended view of biofilm matrix multi-functionality including the allocation of biochemical matrix constituents.

Constructive – Polysaccharides – Proteins – eDNA	**Protective** – Polysaccharides – Proteins – eDNA	**Sorptive** – Charged polysaccharides – Hydrophobic polymers	**Precipitative** – Charged polysaccharides – Proteins
Adhesive – Polysaccharides – Proteins	**Repulsive** – Amphiphiles – Polysaccharides	**Cohesive** – Proteins (amyloids)	**Connective** – Linkage proteins e.g. lectins – Other proteins
Active – Proteins (enzymes) – Vesicles	**Surface-active** – Amphiphiles – Vesicles – Proteins	**Informative** – Nucleic acids – Vesicles containing nucleic acids	**Competitive** – Polysaccharides
Nutritive – Polysaccharides – Proteins – Nucleic acids	**Transportive** – Vesicles carrying various constituents	**Communicative** – Peptidoglycan fragments - RNA	**Conductive** – Proteins – Nucleic acids
Redox-active – Bacterial refractory compounds	**Cryo-protective** – Anti-freeze proteins – Polysaccharides	**Dispersive** – Proteins (enzymes) – Amphiphiles	**Translocative** – Polysaccharides – Amphiphiles

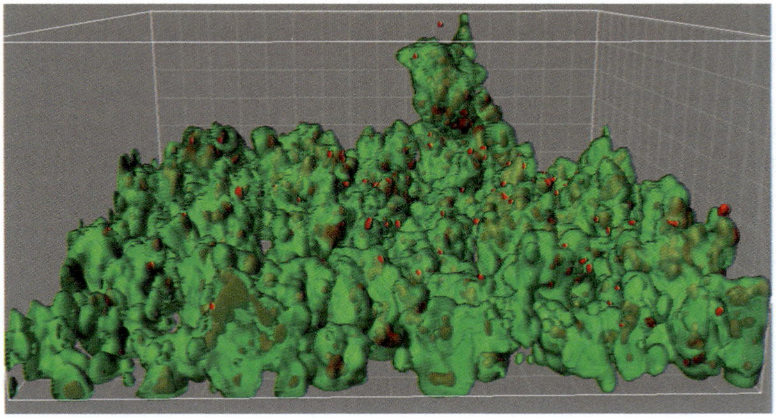

Figure 3.3 Rotating annular reactor biofilm developed from activated sludge showing bacteria more or less completely embedded into a glycoconjugate matrix stained with a fluor-labelled lectin. Colour allocation: green – lectin (semi-transparent), red – nucleic acid stain. Grid size = 10 μm. Sample courtesy Christian Staudt.

3.6.2 Protection

Protective matrix compounds are important in neutralising the exposure of biofilms against chemicals from the environment. This aspect comprises a wide variety of substances including e.g. ions, heavy metals, hydrocarbons, biocides, antibiotics, etc. In addition the matrix has a key role regarding the mechanical properties of biofilms (Guelon *et al.* 2011). As a consequence the viscoelasticity is considered to have a crucial role in protecting the biofilm from mechanical and chemical challenges (Peterson *et al.* 2015).

3.6.3 Cryo- and Osmo-protection

By studying the microbial ecology of the cryosphere it is apparent that many bacteria are psychrophilic with a cold adapted metabolism (Boetius *et al.* 2015). Further it became clear that specific matrix compounds such as anti-freeze proteins and carbohydrate based extracellular polymeric substances are produced as a measure for cryo- and osmo-protection (Marx *et al.* 2009). As a result these specific extracellular polymers alter the physical properties of sea ice with consequences for ice colonisation and survival (Krembs *et al.* 2011).

3.6.4 Sorption

Sorptive matrix compounds are present as charged (positive or negative) or hydrophobic polymers. Examples of charged polymers are polysaccharides which are composed of repeating units often carrying one or more monosaccharides with a positive or negative charge. The repeating unit of polysaccharides may also contain deoxy-monosaccharides having a hydrophobic CH_3-group or they may carry a substitute such as a methyl- or acetyl- group which adds a further hydrophobic moiety. Hydrophobic characteristics may also be encountered in proteins due to their amino acid composition. Thereby the sorptive matrix compounds can be involved in exchange processes of dissolved anionic or cationic and organic compounds (Kurniawan *et al.* 2012).

3.6.5 Precipitation

Precipitative matrix compounds are involved in biogeochemical precipitation of e.g. calcium carbonate. This process represents the major precipitation process in freshwater, marine and hypersaline habitats (Dupraz *et al.* 2009). There are a number of coupled biotic-abiotic processes that precipitate or dissolve calcium carbonate (Visscher & Stolz, 2005). In this mineralisation process biopolymers have been identified as key players (Decho, 2010). In another study by means of STXM, calcium carbonate nucleation was found to occur within cyanobacterial EPS layers (Obst *et al.* 2009). An applied aspect where this process comes into play is the microbial approach to self-healing of concrete, a process in which microbial extracellular polymeric substances have a crucial role (Seifan *et al.* 2016).

3.6.6 Adhesion

Adhesive matrix compounds are the real sticky polymers forming the glue in between bacterial cells and the interface. It is well known that these compounds are produced sometimes at one cell pole only. Furthermore this material can be identified as microbial "footprints" after detachment of bacteria from solid interfaces (Neu & Marshall, 1991; Neu, 1992). An example of microbial footprints is presented for diatoms (Figure 3.4). The innovative aspect of these compounds is their function as an under-water glue that is in some ways similar to the adhesive produced by mussels (Maier *et al.* 2015; Wilker, 2015).

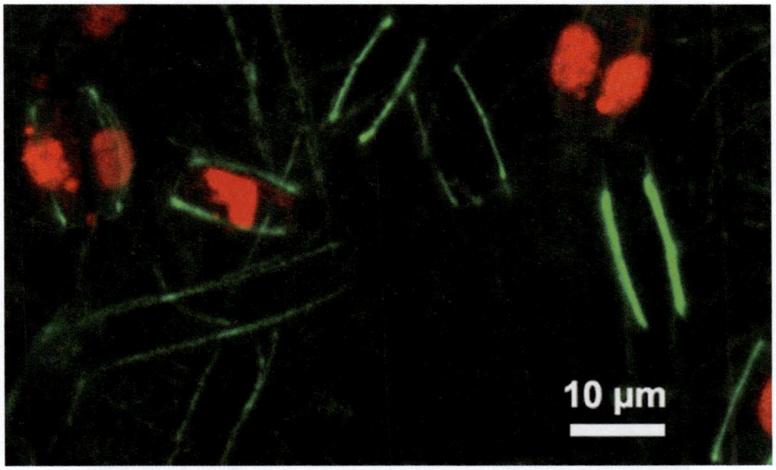

Figure 3.4 Glycoconjugate footprints of diatoms left on a surface. The two parallel footprints are identical to the raphe of the diatom *Amphora coffeaeformis*. Colour allocation: green – lectin, red – autofluorescence of Chl A. Sample courtesy: Matthias Buhmann.

3.6.7 Repellent

Repulsive matrix compounds are those produced by one bacterial species preventing the colonisation of surfaces by other species. This aspect of anti-biofilm polysaccharides was summarized in a review describing state of the art and potential applications (Rendueles *et al.* 2013).

3.6.8 Cohesion

Cohesive matrix compounds are represented by the wide variety of amyloids which are frequently found in biofilms from very different habitats. As already explained they are known under various names but all of them share common properties (see above section on amyloids).

3.6.9 Connectivity

Connective matrix compounds are the ones which specifically link one polymer type or biochemically different polymers. This function has not been demonstrated directly but it is very apparent that proteins with a binding affinity for polysaccharides can link matrix polymers thereby forming a true polymer-network. Potential candidates include extracellular enzymes, lectins (Sharon & Lis, 2003) and carbohydrate-binding modules (Boraston *et al.* 2004; Shoseyov *et al.* 2006). However it is also possible that other specific matrix constituents (proteins) are produced for the purpose of crosslinking.

3.6.10 Activity

Active matrix compounds are represented by extracellular enzymes. They may have a dual role. Firstly they are released for biodegradation of polymeric substrates to deliver monomeric compounds to the cell. Secondly they may have a function in controlling the matrix polymers produced. Thereby they can control and limit biofilm development and they might be involved in the release of embedded bacteria from the biofilm. The example presented takes advantage of the ELF97 enzyme kit. This kit is ideally suited for laser microscopy as the product of enzyme activity is precipitated at the location of activity (Figure 3.5).

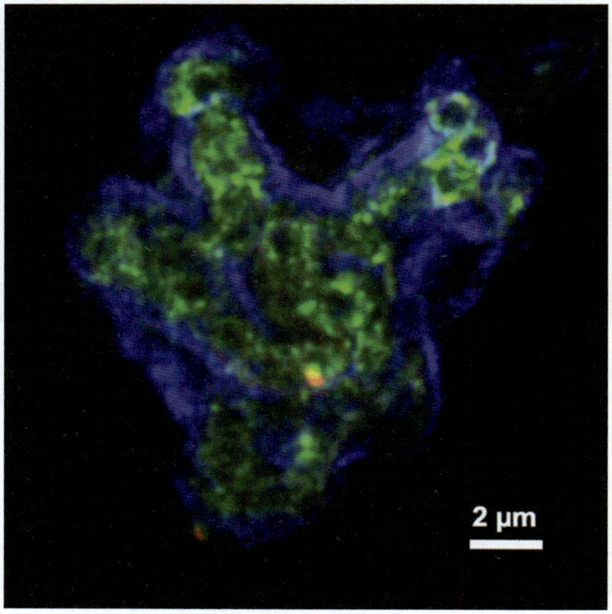

Figure 3.5 Microcolony of a river biofilm developed in a rotating annular reactor fed with river water. The dual channel data set shows alkaline phosphatase activity in combination with glycoconjugate staining. Dataset was subjected to blind deconvolution indicating cell surface localisation of enzyme activity. Colour allocation: green – ELF97, blue – lectin stain.

3.6.11 Surface-activity

Surface-active compounds of bacteria have been known for some time and their significance in interaction with of bacteria with interfaces was discussed (Neu, 1996). They are represented by different constituents such as amphiphilic polymers (Ding *et al.* 2008), biosurfactants of various types (Pamp & Tolker-Nielsen, 2007) as well as hydrophobins and phenol-soluble modulins. Although not all of them are polymeric, these compounds appear to have a function in shaping the biofilm architecture of various microorganisms. Possible roles in surface coating and detachment are outlined (Figure 3.6).

3.6.12 Information

Informative matrix compounds are nucleic acids released by cell lysis or actively excreted eDNA. Despite its acceptance as an important matrix compound, there is still discussion on the individual functions of eDNA in biofilm processes (see Table 3.2). One example shown is the release of eDNA by an aggregate-associated river isolate (Böckelmann *et al.* 2006). This species appears to produce eDNA as an aggregation mechanism or as a guide for cells moving in a directed way through a very dynamic habitat (Figure 3.7).

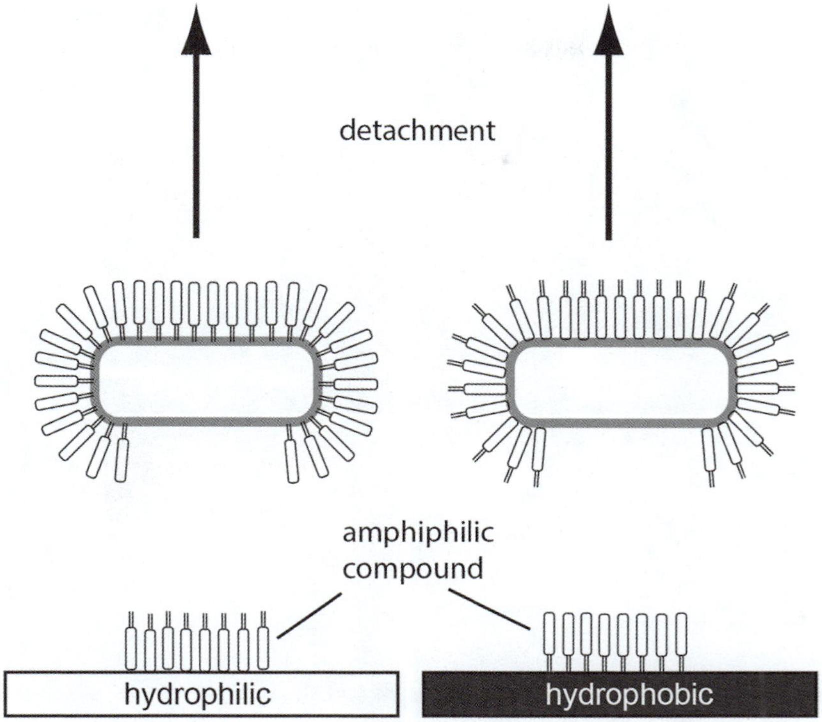

Figure 3.6 Role of surface-active matrix compounds in conditioning of surfaces and detachment processes. The sketch indicates two different bacterial cells with hydrophilic and hydrophobic cell surface layers. These layers of amphiphilic compounds, depending upon their orientation, will allow adhesion to hydrophilic or hydrophobic surfaces and may be left during detachment in the form of a microbial footprint. Printed and modified with permission by Elsevier via Copyright Clearance Center.

3.6.13 Competition

Competitive matrix compounds are those polymers identified in dual species biofilms. It has been shown that a polysaccharide producing strain can out-compete a non-producing strain (Xavier & Foster, 2007). Thereby the polymers produced by one species can provide a competitive advantage and contribute to the displacement of the second species.

3.6.14 Nutrition

Nutritive matrix compounds are all the ones released which can be made available to the cell by means of extracellular enzymes. These polymers comprise polysaccharides (carbon), proteins (nitrogen) and nucleic acids (phosphorus). This aspect with focus on extracellular polysaccharides was described in stromatolites and hypersaline microbial mats (Decho *et al.* 2005; Braissant *et al.* 2009). Another prominent example is the presence of eDNA at high concentration in marine sediments. It was shown that eDNA is an important source of phosphorous in oligotrophic deep sea sediments (Dell'Anno & Danovaro, 2005).

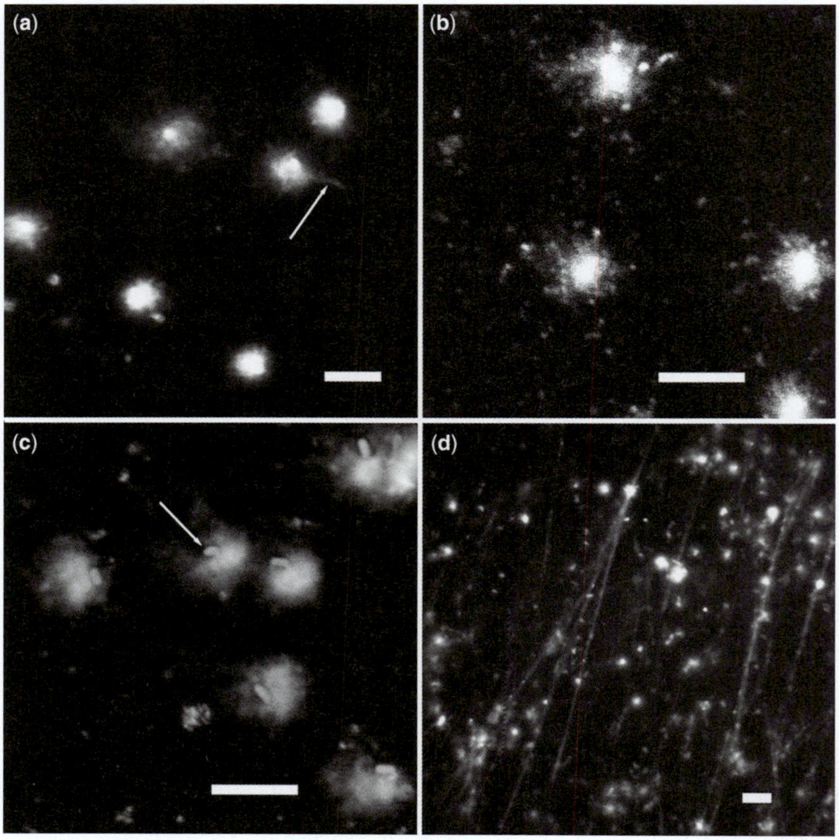

Figure 3.7 Extracellular DNA produced by a bacterial species isolated from lotic aggregates (river snow). The sequence of eDNA release starts with local enrichment around bacterial cells and ends with extended long filamentous fibres carrying bacterial cells (Böckelmann *et al.* 2006). The eDNA was stained with Syto 9. Scale bar = 5 μm.

3.6.15 Motility

Translocative matrix compounds are the ones involved in motility. In fact procaryotes move in several diverse ways (Jarrell & McBride, 2008). Movement without flagella was attributed to some extracellular constituents such as polysaccharides and surface-active compounds. Although gliding based on these compounds was suggested for different species many years ago (Burchard, 1981), currently the mechano-physical models are favoured as for example in Myxobacteria (Nan & Zusman, 2011). Some bacteria can also move on a surface by rolling thereby leaving a polymer track behind (Peltola *et al.* 2008) (Figure 3.8).

3.6.16 Transportation

Transportive matrix compounds are represented by the membrane vesicles shed from the cell surface. These vesicles are not only produced by Gram-negative and Gram-positive bacteria but also by mycobacteria and fungi. The vesicles carry determinants and antigens outside the cell surface into remote locations and thus extend the cell surface. Inside the vesicles they transport intracellular constituents such as proteins, nucleic acids or other compounds into the nearby microhabitat. See also above section on MV's.

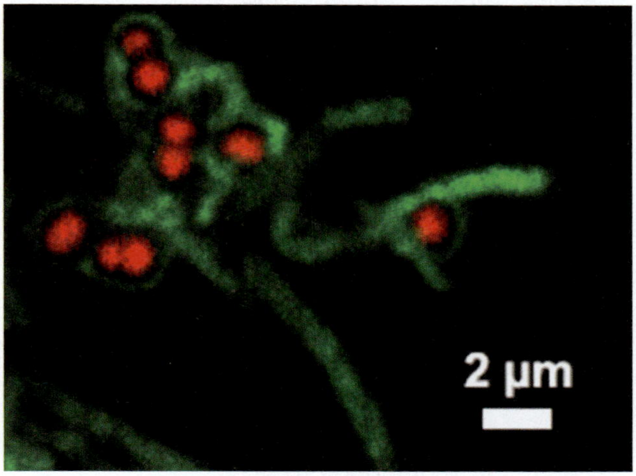

Figure 3.8 *Deinococcus geothermalis* produces a series of glycoconjugates showing multiple binding patterns based on a lectin screening. The image demonstrates one type showing trails produced by *Deinococcus geothermalis* during rolling on a metal surface (Peltola *et al.* 2008). Colour allocation: green – lectin stain, red – nucleic acid stain. Sample courtesy: Minna Peltola.

3.6.17 Communication

Communicative matrix compounds are of different origins. For example one class of signalling compounds are derived from peptidoglycan. This aspect of the matrix has been recognised only recently. Peptidoglycan fragments have been shown to serve as a signal for non-growing cells and as a signal between bacteria and eukaryotic organisms in pathogenic and symbiotic interactions (Dworkin, 2014). Another class of extracellular polymers which has been suggested to mediate intercellular communication are represented by RNA in free form or associated with extracellular vesicles (Ghosal *et al.* 2015).

3.6.18 Conduction

Conductive matrix compounds are the polymers which are involved in electron transport. To date several components have been suggested as being involved in electron transport including pili (Reguera *et al.* 2005; Gorby *et al.* 2006), redox proteins (Magnuson, 2011), extracellular enzymes (Deutzmann *et al.* 2015) and in principal eDNA (Giese, 2002).

3.6.19 Redox-activity

Redox-active matrix compounds are speculative although there is a strong indication of their presence based on a study of anaerobic respiration (Lovley & Coates, 2000; Shimizu *et al.* 2013). Consequently it may well be that similar to plant derived humics, bacterial derived refractory compounds function in a similar way. The possible reaction is outlined below (Figure 3.9).

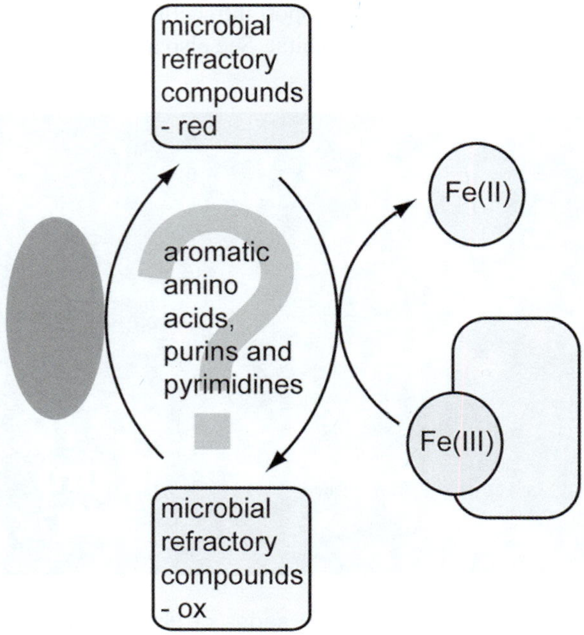

Figure 3.9 Potential function of redox-active matrix compounds. Printed and modified with permission by Elsevier via Copyright Clearance Center.

3.6.20 Dispersion

Dispersive matrix compounds break the bonds within the matrix. They include enzymes degrading the matrix polymers, proteins altering the adhesive appendages as well as amphiphilic molecules reducing the surface tension. An overview of different dispersal aspects have been provided recently (McDougald *et al.* 2012; Barraud *et al.* 2015).

3.7 EMERGING VIEWS OF THE BIOFILM MATRIX

By looking back to the first book publication on biofilm EPS (Wingender *et al.* 1999) we can ask where are we standing currently and what progress has been made? The prevailing view on the biofilm matrix shows a mosaic of various new aspects. These comprise firstly, the acceptance of additional matrix constituents such as: eDNA, proteins other than enzymes, amyloids, amphiphiles, membrane vesicles and bacterial refractory compounds. Secondly, it has become clear that the matrix contains defined clusters of constituents having particular characteristics in terms of their chemical makeup as well as their physical properties. Thirdly, specific interactions and linkages between different matrix compounds have been identified. Thereby the scaffold of the matrix is stabilised and controlled. Fourthly, the matrix is accepted as an active player having a role in controlling bacterial behaviour at different stages of biofilm development. This includes ECM signalling during 1) exploration of the interface, 2) three-dimensional structural development and 3) late stages of the biofilm cycle. Fifthly, mineral scaffolds were suggested to influence biofilm structure and appearance. Several of these aspects were summarized and discussed in two excellent reviews (Hobley *et al.* 2015; Steinberg & Kolodkin-Gal,

2015) with examples of the known biofilm species already mentioned in the section on "bacterial extracellular biology". Finally, the biofilm and its matrix maybe regarded as a complex fluid with gradients of polymers and surface-active compounds affecting the structure, mechanics and dynamics of biofilms (Wilking *et al.* 2011). It may also be worthwhile to consider the spatiotemporal organisation of the biofilm matrix in relation to its involvement in the chemical reactivity of the extracellular space (de Lorenzo *et al.* 2015). Consequently the ECM represents a key signalling element and active player at different stages of biofilm and bioaggregate development thereby controlling both extracellular as well as cellular functionality in microbial films and aggregates.

REFERENCES

Absalon C., Van Dellen K. and Watnick P. I. (2011). A communal bacterial adhesin anchors biofilm and bystander cells to surfaces. *PLoS Pathogens*, **7**, e1002210.

Absalon C., Ymele-Leki P. and Watnick P. I. (2012). The bacterial biofilm matrix as a platform for protein delivery. *mBio*, **3**(4).

Albertsen M., Stensballe A., Nielsen K. L. and Nielsen P.-H. (2013). Digging into the extracellular matrix of a complex microbial community using a combined metagenomic and metaproteomic approach. *Water Science and Technology*, **67**, 1650–6.

Aldeek F., Schneider R., Fontaine-Aupart M. P., Mustin C., Lécart S., Merlin C. and Block J. C. (2013). Patterned hydrophobic domains in the exopolymer matrix of *Shewanella oneidensis* MR-1 biofilms. *Applied and Environmental Microbiology*, **79**(4), 1400–2.

Allison D. G., Sutherland I. W. and Neu T. R. (2003). EPS: what's in an acronym? In: Biofilm Communities: Order from Chaos? A. Mc Bain, D. G. Allison, M. Brading, A. Rickard, J. Verran and J. Walker (eds), BioLine, Cardiff, pp. 381–7.

Altindis E., Fu Y. and Mekalanos J. J. (2014). Proteomic analysis of *Vibrio cholerae* outer membrane vesicles. *Proceedings of the National Academy of Sciences*, **111**(15), E1548–E56.

Angelini T. E., Roper M., Kolter R., Weitz D. A. and Brenner M. P. (2009). Bacillus subtilis spreads by surfing on waves of surfactant. *Proceedings of the National Academy of Sciences*, **106**(43), 18109–13.

Austin J. W., Sanders G., Kay W. W. and Collinson S. K. (1998). Thin aggregative fimbriae enhance *Salmonella enteritidis* biofilm formation. *FEMS Microbiology Letters*, **162**, 295–301.

Barken K. B., Pamp S. J., Yang L., Gjermansen M., Bertrand J. J., Klausen M., Givskov M., Whitchurch C. B., Engel J. N. and Tolker-Nielsen T. (2008). Roles of type IV pili, flagellum-mediated motility and extracellular DNA in the formation of mature multicellular strutures in *Pseudomonas aeruginosa* biofilms. *Environmental Microbiology*, **10**, 2331–43.

Barnhart M. M. and Chapman M. R. (2006). Curli biogenesis and function. *Annual Review of Microbiology*, **60**(1), 131–47.

Barraud N., Kjelleberg S. and Rice S. A. (2015). Dispersal from biofilms. In: Microbial Biofilms, M. Ghannoum, M. Parsek, M. Whiteley and P. K. Mukherjee (eds), American Society for Microbiology, Washington, pp. 343–62.

Baumgarten T., Sperling S., Seifert J., von Bergen M., Steiniger F., Wick L. Y. and Heipieper H. J. (2012). Membrane vesicle formation as a multiple-stress response mechanism enhances *Pseudomonas putida* DOT-T1E cell surface hydrophobicity and biofilm formation. *Applied and Environmental Microbiology*, **78**(17), 6217–24.

Beloin C., Roux A. and Ghigo J. M. (2008). *Escherichia coli* biofilms. In: Bacterial Biofilms, T. Romeo (ed.), Springer, Berlin, Heidelberg, pp. 249–89.

Berk V., Fong J. C. N., Dempsey G. T., Develioglu O. N., Zhuang X., Liphardt J., Yildiz F. H. and Chu S. (2012). Molecular architecture and assembly principles of *Vibrio cholerae* biofilms. *Science*, **337**(6091), 236–9.

Berne C., Kysela D. T. and Brun Y. V. (2010). A bacterial extracellular DNA inhibits settling of motile progeny cells within a biofilm. *Molecular Microbiology*, **77**(4), 815–29.

Biller S. J., Schubotz F., Roggensack S. E., Thompson A. W., Summons R. E. and Chisholm S. W. (2014). Bacterial vesicles in marine ecosystems. *Science*, **343**(6167), 183–6.

Binnenkade L., Teichmann L. and Thormann K. M. (2014). Iron triggers λSo prophage induction and release of extracellular DNA in *Shewanella oneidensis* MR-1 biofilms. *Applied and Environmental Microbiology*, **80**(17), 5304–16.

Blanco L. P., Evans M. L., Smith D. R., Badtke M. P. and Chapman M. R. (2012). Diversity, biogenesis and function of microbial amyloids. *Trends in Microbiology*, **20**(2), 66–73.

Böckelmann U., Janke A., Kuhn R., Neu T. R., Wecke J., Lawrence J. R. and Szewzyk U. (2006). Bacterial extracellular DNA forming a defined network like structure. *FEMS Microbiology Letters*, **262**, 31–8.

Boetius A., Anesio A. M., Deming J. W., Mikucki J. A. and Rapp J. Z. (2015). Microbial ecology of the cryosphere: sea ice and glacial habitats. *Nature Reviews Microbiology*, **13**(11), 677–90.

Boles B. R., Thoendel M. and Singh P. K. (2005). Rhamnolipids mediate detachment of *Pseudomonas aeruginosa* from biofilms. *Molecular Microbiology*, **57**, 1210–23.

Bomberger J. M., MacEachran D. P., Coutermarsh B. A., Ye S., O'Toole G. A. and Stanton B. A. (2009). Long-distance delivery of bacterial virulence factors by *Pseudomonas aeruginosa* outer membrane vesicles. *PLoS Pathogens*, **5**(4).

Boraston A. B., Bolam D. N., Gilbert H. J. and Davies G. J. (2004). Carbohydrate-binding modules: fine-tuning polysaccharide recognition. *Biochemical Journal*, **382**(3), 769–81.

Borlee B. R., Goldman A. D., Murakami K., Samudrala R., Wozniak D. J. and Parsek M. R. (2010). Pseudomonas aeruginosa uses a cyclic-di-GMP-regulated adhesin to reinforce the biofilm extracellular matrix. *Molecular Microbiology*, **75**(4), 827–42.

Braissant O., Decho A. W., Przekop K. M., Gallagher K. L., Glunk C., Dupraz C. and Visscher P. T. (2009). Characteristics and turnover of exopolymeric substances in a hypersaline microbial mat. *FEMS Microbiology Ecology*, **67**, 293–307.

Branda S. S., Chu F., Kearns D. B., Losick R. and Kolter R. (2006). A major protein component of the *Bacillus subtilis* biofilm matrix. *Molecular Microbiology*, **59**(4), 1229–38.

Branda S. S., González-Pastor J. E., Ben-Yehuda S., Losick R. and Kolter R. (2001). Fruiting body formation by *Bacillus subtilis*. *Proceedings of the National Academy of Sciences*, **98**(20), 11621–6.

Brown L., Wolf J. M., Prados-Rosales R. and Casadevall A. (2015). Through the wall: extracellular vesicles in Gram-positive bacteria, mycobacteria and fungi. *Nature Reviews Microbiology*, **13**(10), 620–30.

Burchard R. P. (1981). Gliding motility of procaryotes: ultrastructure, physiology, and genetics. *Annual Review of Microbiology*, **35**, 497–529.

Büttner D. and Bonas U. (2010). Regulation and secretion of Xanthomonas virulence factors. *FEMS Microbiology Reviews*, **34**(2), 107–33.

Caiazza N. C., Shanks R. M. Q. and O'Toole G. A. (2005). Rhamnolipids modulate swarming motility patterns of *Pseudomonas aeruginosa*. *Journal of Bacteriology*, **187**, 7351–61.

Cairns L. S., Hobley L. and Stanley-Wall N. R. (2014). Biofilm formation by *Bacillus subtilis*: new insights into regulatory strategies and assembly mechanisms. *Molecular Microbiology*, **93**(4), 587–98.

Cao B., Shi L., Brown R. N., Xiong Y., Fredrickson J. K., Romine M. F., Marshall M. J., Lipton M. S. and Beyenal H. (2011). Extracellular polymeric substances from *Shewanella sp.* HRCR-1 biofilms: characterization by infrared spectroscopy and proteomics. *Environmental Microbiology*, **13**(4), 1018–31.

Chalmers N. I., Palmer R. J., Jr., Du-Thumm L., Sullivan R., Shi W. and Kolenbrander P. E. (2007). Use of quantum dot luminescent probes to achieve single-cell resolution of human oral bacteria in biofilms. *Applied and Environmental Microbiology*, **73**, 630–6.

Chandra J., Kuhn D. M., Mukherjee P. K., Hoyer L. L., McCormick T. and Ghannoum M. A. (2001). Biofilm formation by the fungal pathogen *Candida albicans*: development, architecture and drug resistance. *Journal of Bacteriology*, **183**, 5385–94.

Chapman M. R., Robinson L. S., Pinkner J. S., Roth R., Heuser J., Hammar M., Normark S. and Hultgren S. J. (2002). Role of *Escherichia coli* curli operons in directing amyloid fiber formation. *Science*, **295**, 851–5.

Chater K. F. (2010). Bacterial extracellular biology. *FEMS Microbiology Reviews*, **34**(2), 87–8.

Chater K. F., Biró S., Lee K. J., Palmer T. and Schrempf H. (2010). The complex extracellular biology of Streptomyces. *FEMS Microbiology Reviews*, **34**(2), 171–98.

Chew S. C., Kundukad B., Seviour T., van der Maarel J. R. C., Yang L., Rice S. A., Doyle P. and Kjelleberg S. (2014). Dynamic remodeling of microbial biofilms by functionally distinct exopolysaccharides. *mBio*, **5**(4).

Colvin K. M., Irie Y., Tart C. S., Urbano R., Whitney J. C., Ryder C., Howell P. L., Wozniak D. J. and Parsek M. R. (2012). The Pel and Psl polysaccharides provide *Pseudomonas aeruginosa* structural redundancy within the biofilm matrix. *Environmental Microbiology*, **14**(8), 1913–28.

Curtis P. D., Atwood III J., Orlando R. and Shimkets L. J. (2007). Proteins associated with the *Myxococcus xanthus* extracellular matrix. *Journal of Bacteriology*, **189**, 7634–42.

Dang H. and Lovell C. R. (2016). Microbial surface colonization and biofilm development in marine environments. *Microbiology and Molecular Biology Reviews*, **80**(1), 91–138.

Das T., Kutty S. K., Kumar N. and Manefield M. (2013a). Pyocyanin facilitates extracellular DNA binding to *Pseudomonas aeruginosa* influencing cell surface properties and aggregation. *PLoS ONE*, **8**, e58299.

Das T. and Manefield M. (2012). Pyocyanin promotes extracellular DNA release in *Pseudomonas aeruginosa*. *PLoS ONE*, **7**, e46718.

Das T., Sehar S. and Manefield M. (2013b). The roles of extracellular DNA in the structural integrity of extracellular polymeric substance and bacterial biofilm development. *Environmental Microbiology Reports*, **5**(6), 778–86.

Das T., Sharma P. K., Busscher H. J., van der Mei H. C. and Krom B. P. (2010). Role of extracellular DNA in initial bacterial adhesion and surface aggregation. *Applied and Environmental Microbiology*, **76**, 3405–8.

Davey M. E., Caiazza N. C. and O'Toole G. A. (2003). Rhamnolipid surfactant production affects biofilm architecture in *Pseudomonas aeruginosa* PAO1. *Journal of Bacteriology*, **185**, 1027–36.

de Lorenzo V., Sekowska A. and Danchin A. (2015). Chemical reactivity drives spatiotemporal organisation of bacterial metabolism. *FEMS Microbiology Reviews*, **39**(1), 96–119.

Deatherage B. L. and Cookson B. T. (2012). Membrane vesicle release in bacteria, eukaryotes, and archaea: a conserved yet underappreciated aspect of microbial life. *Infection and Immunity*, **80**(6), 1948–57.

Decho A. (1990). Microbial exopolymer secretions in ocean environments: their role(s) in food webs and marine processes. *Oceanography and Marine Biology an Annual Review*, **28**, 73–153.

Decho A. W. (2010). Overview of biopolymer-induced mineralization: what goes on in biofilms? *Ecological Engineering*, **36**, 137–44.

Decho A. W. and Lopez G. R. (1993). Exopolymer microenvironments of microbial flora: multiple and interactive effects on trophic relationships. *Limnology and Oceanography*, **38**, 1633–45.

Decho A. W., Visscher P. T. and Reid R. P. (2005). Production and cycling of natural microbial exopolymers (EPS) within a marine stromatolite. *Palaeogeography, Palaeoclimatology, Palaeoecology*, **219**(1–2), 71–86.

Dell'Anno A. and Danovaro R. (2005). Extracellular DNA plays a key role in deep-sea ecosystem functioning. *Science*, **309**, 2179.

Dengler V., Foulston L., DeFrancesco A. S. and Losick R. (2015). An electrostatic net model for the role of extracellular DNA in biofilm formation by *Staphylococcus aureus*. *Journal of Bacteriology*, **197**(24), 3779–87.

DePas W. H. and Chapman M. R. (2012). Microbial manipulation of the amyloid fold. *Research in Microbiology*, **163**(9–10), 592–606.

Desai J. V. and Mitchell A. P. (2015). *Candida albicans* biofilm development and its genetic control. *Microbiology Spectrum*, **3**(3).

Deutzmann J. S., Sahin M. and Spormann A. M. (2015). Extracellular enzymes facilitate electron uptake in biocorrosion and bioelectrosynthesis. *mBio*, **6**(2).

Ding Y.-X., Chin W.-C., Rodriguez A., Hung C.-C., Santschi P. H. and Verdugo P. (2008). Amphiphilic exopolymers from *Sagittula stellata* induce DOM self-assembly and formation of marine microgels. *Marine Chemistry*, **112**(1–2), 11–9.

Ding Z., Bourven I., Guibaud G., van Hullebusch E., Panico A., Pirozzi F. and Esposito G. (2015). Role of extracellular polymeric substances (EPS) production in bioaggregation: application to wastewater treatment. *Applied Microbiology and Biotechnology*, **99**(23), 9883–905.

Domenech M., García E., Prieto A. and Moscoso M. (2013). Insight into the composition of the intercellular matrix of *Streptococcus pneumoniae* biofilms. *Environmental Microbiology*, **15**(2), 502–16.

Dominiak D. M., Nielsen J. L. and Nielsen P. H. r. (2011). Extracellular DNA is abundant and important for microcolony strength in mixed microbial biofilms. *Environmental Microbiology*, **13**(3), 710–21.

Downie J. A. (2010). The roles of extracellular proteins, polysaccharides and signals in the interactions of rhizobia with legume roots. *FEMS Microbiology Reviews*, **34**(2), 150–70.

Dueholm M. S., Nielsen P. H., Chapman M. and Otzen D. (2013a). Functional amyloids in bacteria. In: Amyloid Fibrils and Prefibrillar Aggregates, Wiley-VCH Verlag GmbH & Co. KGaA, pp. 411–38.

Dueholm M. S., Petersen S. V., Soenderkaer M., Larsen P., Christiansen G., Hein K. L., Enghild J. J., Nielsen J. L., Nielsen K. L., Nielsen P. H. and Otzen D. E. (2010). Functional amyloid in Pseudomonas. *Molecular Microbiology*, **77**(4), 1009–20.

Dueholm M. S., Søndergaard M. T., Nilsson M., Christiansen G., Stensballe A., Overgaard M. T., Givskov M., Tolker-Nielsen T., Otzen D. E. and Nielsen P. H. (2013b). Expression of Fap amyloids in *Pseudomonas aeruginosa, P. fluorescens, and P. putida* results in aggregation and increased biofilm formation. *MicrobiologyOpen*, **2**(3), 365–82.

Dueholm, M. S. and Nielsen, P. H. (2016). Amyloids – a neglected child of the slime. In: The Perfect Slime, H.-C. Flemming, T. R. Neu, J. Wingender (eds), IWA Publishing, London, ch. 6, pp. 113–133.

Dumas E., Meunier B., Berdague J.-L., Chambon C., Desvaux M. and Hebraud M. (2008). Comparative analysis of extracellular and intracellular proteomes of *Listeria monocytogenes* strains reveals a correlation between protein expression and serovar. *Applied and Environmental Microbiology*, **74**, 7399–409.

Dupraz C., Reid R. P., Braissant O. and Decho A. W. (2009). Processes of carbonate precipitation in modern microbial mats. *Earth-Science Reviews*, **96**, 141–62.

Dworkin J. (2014). The medium is the message: interspecies and interkingdom signaling by peptidoglycan and related bacterial glycans. *Annual Review of Microbiology*, **68**(1), 137–54.

Eisen J. A. (2012). Badomics words and the power and peril of the ome-meme. *GigaScience*, **1**(1), 1–4.

Elhenawy W., Debelyy M. O. and Feldman M. F. (2014). Preferential packing of acidic glycosidases and proteases into bacteroides outer membrane vesicles. *mBio*, **5**(2).

Ellis R. J. (2001). Macromolecular crowding: obvois but underappreciated. *Trends in Biochemical Sciences*, **26**, 597–604.

Finkel J. S. and Mitchell A. P. (2011). Genetic control of *Candida albicans* biofilm development. *Nature Reviews Microbiology*, **9**(2), 109–18.

Flemming H.-C. (2011). The perfect slime. *Colloids and Surfaces B: Biointerfaces*, **86**(2), 251–9.

Flemming H.-C., Neu T. R. and Wozniak D. (2007). The EPS matrix: the "house of biofilm cells". *Journal of Bacteriology*, **189**, 7945–7.

Flemming H.-C. and Wingender J. (2010). The biofilm matrix. *Nat Rev Micro*, **8**(9), 623–33.

Fong J. N. C. and Yildiz F. H. (2015). Biofilm matrix proteins. *Microbiology Spectrum*, **3**(2).

Foster T. J., Geoghegan J. A., Ganesh V. K. and Hook M. (2014). Adhesion, invasion and evasion: the many functions of the surface proteins of *Staphylococcus aureus*. *Nature Reviews Microbiology*, **12**(1), 49–62.

Fox E. P., Bui C. K., Nett J. E., Hartooni N., Mui M. C., Andes D. R., Nobile C. J. and Johnson A. D. (2015). An expanded regulatory network temporally controls *Candida albicans* biofilm formation. *Molecular Microbiology*, **96**(6), 1226–39.

Franklin M. J., Nivens D. E., Weadge J. T. and Howell P. L. (2011). Biosynthesis of the *Pseudomonas aeruginosa* extracellular polysaccharides, alginate, Pel, and Psl. *Frontiers in Microbiology*, **2**.

Frias A., Manresa A., de Oliveira E., Lopez-Iglesias C. and Mercade E. (2010). Membrane vesicles: a common feature in the extracellular matter of cold-adapted Antarctic bacteria. *Microbial Ecology*, **59**, 476–86.

Gallaher T. K., Wu S., Webster P. and Aguilera R. (2006). Identification of biofilm proteins in non-typeable *Haemophilus influenzae*. *BMC Microbiology*, **6**, 1–9.

Garcia M. C., Lee J. T., Ramsook C. B., Alsteens D., Dufrene Y. F. and Lipke P. N. (2011). A role for amyloid in cell aggregation and biofilm formation. *PLoS ONE*, **6**, e17632.

Gebbink M. F. B. G., Claessen D., Bouma B., Dijkhuizen L. and Wösten H. A. B. (2005). Amyloids - A functional coat for microorganisms. *Nature Reviews Microbiology*, **3**, 333–41.

Ghosal A., Upadhyaya B. B., Fritz J. V., Heintz-Buschart A., Desai M. S., Yusuf D., Huang D., Baumuratov A., Wang K., Galas D. and Wilmes P. (2015). The extracellular RNA complement of Escherichia coli. *MicrobiologyOpen*, **4**(2), 252–66.

Giese B. (2002). Long-distance electron transfer through DNA. *Annual Review of Biochemistry*, **71**, 51–70.

Gloag E. S., Turnbull L., Huang A., Vallotton P., Wang H., Nolan L. M., Mililli L., Hunt C., Lu J., Osvath S. R., Monahan L. G., Cavaliere R., Charles I. G., Wand M. P., Gee M. L., Prabhakar R. and Whitchurch C. B. (2013). Self-organization of bacterial biofilms is facilitated by extracellular DNA. *Proceedings of the National Academy of Sciences*, **110**(28), 11541–6.

Goedeke J., Heun M., Bubendorfer S., Paul K. and Thormann K. M. (2011a). Roles of two *Shewanella oneidensis* MR-1 extracellular endonucleases. *Applied and Environmental Microbiology*, **77**(15), 5342–51.

Goedeke J., Paul K., Lassak J. and Thormann K. M. (2011b). Phage-induced lysis enhances biofilm formation in *Shewanella oneidensis* MR-1. *ISME J*, **5**(4), 613–26.

Goodman S. D., Obergfell K. P., Jurcisek J. A., Novotny L. A., Downey J. S., Ayala E. A., Tjokro N., Li B., Justice S. S. and Bakaletz L. O. (2011). Biofilms can be dispersed by focusing the immune system on a common family of bacterial nucleoid-associated proteins. *Mucosal Immunol*, **4**(6), 625–37.

Gorby Y. A., Yanina S., McLean J. S., Rosso K. M., Moyles D., Dohnalkova A., Beveridge T. J., Chang I. S., Kim B. H., Culley D. E., Reed S. B., Romine M. F., Saffarini D. A., Hill E. A., Shi L., Elias D. A., Kennedy D. W., Pinchuk G., Watanabe K., Ishii S., Logan B., Nealson K. H. and Frederickson J. K. (2006). Electrically conductive bacterial nanowires produced by *Shewanella oneidensis* strain MR-1 and other microorganisms. *Proceedings of the National Academy of Sciences USA*, **103**, 11358–63.

Guelon T., Mathias J.-D. and Stoodley P. (2011). Advances in biofilm mechanics. In: Biofilm Highlights, H.-C. Flemming, J. Wingender and U. Szewzyk (eds), Springer, Berlin, pp. 111–39.

Guiton P. S., Hung C. S., Kline K. A., Roth R., Kau A. L., Hayes E., Heuser J., Dodson K. W., Caparon M. G. and Hultgren S. J. (2009). Contribuation of autolysin and sortase A during *Enterococcus faecalis* DNA-dependent biofilm development. *Infection and Immunity*, **77**, 3626–38.

Ha D.-G. and O'Toole G. A. (2015). c-di-GMP and its effects on biofilm formation and dispersion: a *Pseudomonas aeruginosa* review. *Microbiology Spectrum*, **3**(2).

Hall-Stoodley L., Nistico L., Sambanthamoorthy K., Dice B., Nguyen D., Mershon W. J., Johnson C., Hu F. Z., Stoodley P., Ehrlich G. and Post J. C. (2008). Characterisation of biofilm matrix, degradation by DNase treatment and evidence of capsule downregulation in *Streptococcus pneumoniae* clinical isolates. *BMC Microbiology*, **8**(173).

Hammar M., Bian Z. and Normark S. (1996). Nucleator-dependent intercellular assembly of adhesive curli organelles in Escherichia coli. *Proceedings of the National Academy of Sciences*, **93**(13), 6562–6.

Harmsen M., Lappann M., Knochel S. and Molin S. (2010). Role of extracellular DNA during biofilm formation by *Listeria monocytogenes*. *Applied and Environmental Microbiology*, **76**(7), 2271–9.

Helm R. F. and Potts M. (2012). Extracellular matrix (ECM). In: Ecology of Cyanobacteria II, B. A. Whitton (ed.), Springer, Berlin, pp. 461–79.

Herbst F.-A., Søndergaard M. T., Kjeldal H., Stensballe A., Nielsen P. H. and Dueholm M. S. (2015). Major proteomic changes associated with amyloid-induced biofilm formation in *Pseudomonas aeruginosa* PAO1. *Journal of Proteome Research*, **14**(1), 72–81.

Heun M., Binnenkade L., Kreienbaum M. and Thormann K. M. (2012). Functional specificity of extracellular nucleases of *Shewanella oneidensis* MR-1. *Applied and Environmental Microbiology*, **78**(12), 4400–11.

Hobley L., Harkins C., MacPhee C. E. and Stanley-Wall N. R. (2015). Giving structure to the biofilm matrix: an overview of individual strategies and emerging common themes. *FEMS Microbiology Reviews*, **39**(5), 649–69.

Hobley L., Ostrowski A., Rao F. V., Bromley K. M., Porter M., Prescott A. R., MacPhee C. E., van Aalten D. M. F. and Stanley-Wall N. R. (2013). BslA is a self-assembling bacterial hydrophobin that coats the *Bacillus subtilis* biofilm. *Proceedings of the National Academy of Sciences*, **110**(33), 13600–5.

Hoffman M. and Decho A. W. (1999). Extracellular enzymes within microbial biofilms and the role of the extracellular polymer matrix. In: Microbial Extracellular Polymeric Substances, J. Wingender, T. R. Neu and H.-C. Flemming (eds), Springer, New York, pp. 217–30.

Houry A., Gohar M., Deschamps J., Tischenko E., Aymerich S. p., Gruss A. and Briandet R. (2012). Bacterial swimmers that infiltrate and take over the biofilm matrix. *Proceedings of the National Academy of Sciences*, **109**, 13088–93.

Hu W., Li L., Sharma S., Wang J., McHardy I., Lux R., Yang Z., He X., Gimzewski J. K., Li Y. and Shi W. (2012a). DNA builds and strengthens the extracellular matrix in *Myxococcus xanthus* biofilms by interacting with exopolysaccharides. *PLoS ONE*, **7**(12), e51905.

Hu W., Yang Z., Lux R., Zhao M., Wang J., He X. and Shi W. (2012b). Direct visualization of the interaction between pilin and exopolysaccharides of *Myxococcus xanthus* with eGFP-fused PilA protein. *FEMS Microbiology Letters*, **326**(1), 23–30.

Hufnagel D. A., Depas W. H. and Chapman M. R. (2015). The biology of the *Escherichia coli* extracellular matrix. *Microbiology Spectrum*, **3**(3).

Hung C., Zhou Y., Pinkner J. S., Dodson K. W., Crowley J. R., Heuser J., Chapman M. R., Hadjifrangiskou M., Henderson J. P. and Hultgren S. J. (2013). *Escherichia coli* biofilms have an organized and complex extracellular matrix structure. *mBio*, **4**(5).

Huseby M. J., Kruse A. C., Digre J., Kohler P. L., Vocke J. A., Mann E. E., Bayles K. W., Bohach G. A., Schlievert P. M., Ohlendorf D. H. and Earhart C. A. (2010). Beta toxin catalyzes formation of nucleoprotein matrix in staphylococcal biofilms. *Proceedings of the National Academy of Sciences*, **107**(32), 14407–12.

Ionescu M., Zaini P. A., Baccari C., Tran S., da Silva A. M. and Lindow S. E. (2014). *Xylella fastidiosa* outer membrane vesicles modulate plant colonization by blocking attachment to surfaces. *Proceedings of the National Academy of Sciences*, **111**(37), E3910–E8.

Ivleva N., Wagner M., Horn H., Niessner R. and Haisch C. (2009). Towards a nondestructive chemical characterization of biofilm matrix by Raman microscopy. *Analytical and Bioanalytical Chemistry*, **393**(1), 197–206.

Ivleva N. P., Wagner M., Horn H., Niessner R. and Haisch C. (2010). Raman microscopy and surface-enhanced Raman scattering (SERS) for in situ analysis of biofilms. *Journal of Biophotonics*, **3**(8–9), 548–56.

Izano E. A., Amarante M. A., Kher W. B. and Kaplan J. B. (2008). Differential roles of poly-N-acetylglucosamine surface polysaccharide and extracellular DNA in *Staphylococcus aureus* and *Staphylococcus epidermidis* biofilms. *Applied and Environmental Microbiology*, **74**(2), 470–6.

Jakubovics N. S., Shields R. C., Rajarajan N. and Burgess J. G. (2013). Life after death: the critical role of extracellular DNA in microbial biofilms. *Letters in Applied Microbiology*, **57**(6), 467–75.

Janissen R., Murillo D. M., Niza B., Sahoo P. K., Nobrega M. M., Cesar C. L., Temperini M. L. A., Carvalho H. F., de Souza A. A. and Cotta M. A. (2015). Spatiotemporal distribution of different extracellular polymeric substances and filamentation mediate Xylella fastidiosa adhesion and biofilm formation. *Scientific Reports*, **5**, 9856.

Jarrell K. F. and McBride M. J. (2008). The surprisingly diverse ways that procaryotes move. *Nature Reviews Microbiology*, **6**, 466–76.

Jennings L. K., Storek K. M., Ledvina H. E., Coulon C., Marmont L. S., Sadovskaya I., Secor P. R., Tseng B. S., Scian M., Filloux A., Wozniak D. J., Howell P. L. and Parsek M. R. (2015). Pel is a cationic exopolysaccharide that cross-links extracellular DNA in the *Pseudomonas aeruginosa* biofilm matrix. *Proceedings of the National Academy of Sciences*, **112**(36), 11353–8.

Jordal P. B., Dueholm M. S., Larsen P., Petersen S. V., Enghild J. J., Christiansen G., Hoejrup P., Nielsen P. H. and Otzen D. E. (2009). Widespread abundance of functional bacterial amyloid in mycolata and other gram-positive bacteria. *Applied and Environmental Microbiology*, **75**, 4101–10.

Kim L. H., Jung Y., Yu H. W., Chae K. J. and Kim I. S. (2015). Physicochemical interactions between rhamnolipids and *Pseudomonas aeruginosa* biofilm layers. *Environmental Science & Technology*, **49**(6), 3718–26.

Kleerebezem M., Hols P., Bernard E., Rolain T., Zhou M., Siezen R. J. and Bron P. A. (2010). The extracellular biology of the lactobacilli. *FEMS Microbiology Reviews*, **34**(2), 199–230.

Klein M. I., Hwang G., Santos P. H. S., Campanella O. H. and Koo H. (2015). Streptococcus mutans-derived extracellular matrix in cariogenic oral biofilms. *Frontiers in Cellular and Infection Microbiology*, **5**, 10.

Kobayashi K. and Iwano M. (2012). BslA(YuaB) forms a hydrophobic layer on the surface of *Bacillus subtilis* biofilms. *Molecular Microbiology*, **85**(1), 51–66.

Konovalova A., Petters T. and Søgaard-Andersen L. (2010). Extracellular biology of *Myxococcus xanthus*. *FEMS Microbiology Reviews*, **34**(2), 89–106.

Koo H., Xiao J., Klein M. I. and Jeon J. G. (2010). Exopolysaccharides produced by *Streptococcus mutans* glucosyltransferases modulate the establishment of microcolonies within multispecies biofilms. *The Journal of Bacteriology*, **192**(12), 3024–32.

Krembs C., Eicken H. and Deming J. W. (2011). Exopolymer alteration of physical properties of sea ice and implications for ice habitability and biogeochemistry in a warmer Arctic. *Proceedings of the National Academy of Sciences.*

Kulp A. and Kuehn M. J. (2010). Biological functions and biogenesis of secreted bacterial outer membrane vesicles. *Annual Review of Microbiology*, **64**(1), 163–84.

Kurniawan A., Yamamoto T., Tsuchiya Y. and Morisaki H. (2012). Analysis of the ion adsorption-desorption characteristics of biofilm matrices. *Microbes and Environments*, **27**, 399–406.

Lappmann M., Claus H., van Alen T., Harmsen M., Elias J., Molin S. and Vogel U. (2010). A dual role of extracellular DNA during biofil formation of *Neisseria meningitis. Molecular Microbiology*, **75**, 1355–71.

Larsen P., Nielsen J. L., Dueholm M. S., Wetzel R., Otzen D. and Nielsen P. H. (2007). Amyloid adhesins are abundant in natural biofilms. *Environmental Microbiology*, **9**, 3077–90.

Larsen P. L., Nielsen J. L., Otzen D. and Nielsen P. H. (2008). Amyloid-like adhesins produced by floc-forming and filamentous bacteria in activated sludge. *Applied and Environmental Microbiology*, **74**, 1517–26.

Lawrence J. R., Swerhone G. D. W., Leppard G. G., Araki T., Zhang X., West M. M. and Hitchcock A. P. (2003). Scanning transmission x-ray, laser scanning, and transmission electron microscopy mapping of the exopolymeric matrix of microbial biofilms. *Applied and Environmental Microbiology*, **69**, 5543–54.

Liao S., Klein M. I., Heim K. P., Fan Y., Bitoun J. P., Ahn S.-J., Burne R. A., Koo H., Brady L. J. and Wen Z. T. (2014). *Streptococcus mutans* extracellular DNA is upregulated during growth in biofilms, actively released via membrane vesicles, and influenced by components of the protein secretion machinery. *Journal of Bacteriology*, **196**(13), 2355–66.

Limoli D. H., Jones C. J. and Wozniak D. J. (2015). Bacterial extracellular polysaccharides in biofilm formation and function. *Microbiology Spectrum*, **3**(3).

Lin Y., de Kreuk M., van Loosdrecht M. C. M. and Adin A. (2010). Characterization of alginate-like exopolysaccharides isolated from aerobic granular sludge in pilot-plant. *Water Research*, **44**(11), 3355–64.

Lin Y. M., Sharma P. K. and van Loosdrecht M. C. M. (2013). The chemical and mechanical differences between alginate-like exopolysaccharides isolated from aerobic flocculent sludge and aerobic granular sludge. *Water Research*, **47**(1), 57–65.

Lipke P. N., Garcia M. C., Alsteens D., Ramsook C. B., Klotz S. A. and Dufrene Y. F. (2012). Strengthening relationships: Amyloids create adhesion nanodomains in yeasts. *Trends in Microbiology*, **20**(2), 59–65.

López D. and Kolter R. (2010). Extracellular signals that define distinct and coexisting cell fates in Bacillus subtilis. *FEMS Microbiology Reviews*, **34**(2), 134–49.

Lopez D., Vlamakis H. and Kolter R. (2009). Generation of multiple cell types in *Bacillus subtilis. FEMS Microbiology Reviews*, **33**(1), 152–63.

Lovley D. R. and Coates J. D. (2000). Novel forms of anaerobic respiration of environmental relevance. *Current Opinion in Microbiology*, **3**, 252–6.

Lu N., Zilles J. L. and Nguyen T. H. (2010). Adsorption of extracellular chromosomal DNA and Its effects on natural transformation of *Azotobacter vinelandii. Applied and Environmental Microbiology*, **76**(13), 4179–84.

Ma L., Conover M., Lu H., Parsek M. R., Bayles K. and Wozniak D. J. (2009). Assembly and development of the *Pseudomonas aeruginosa* biofilm matrix. *PLoS Pathogens*, **5**(3).

Ma L., Wang J., Wang S., Anderson E. M., Lam J. S., Parsek M. R. and Wozniak D. J. (2012). Synthesis of multiple *Pseudomonas aeruginosa* biofilm matrix exopolysaccharides is post-transcriptionally regulated. *Environmental Microbiology*, **14**(8), 1995–2005.

MacDonald I. A. and Kuehn M. J. (2012). Offense and defense: microbial membrane vesicles play both ways. *Research in Microbiology*, **163**(9–10), 607–18.

Madsen J. S., Lin Y.-C., Squyres G. R., Price-Whelan A., de Santiago Torio A., Song A., Cornell W. C., Sørensen S. J., Xavier J. B. and Dietrich L. E. P. (2015). Facultative control of matrix production optimizes competitive fitness in *Pseudomonas aeruginosa* PA14 biofilm models. *Applied and Environmental Microbiology*, **81**(24), 8414–26.

Magnuson T. S. (2011). How the xap locus put electrical "zap" in *Geobacter sulfurreducens* biofilms. *The Journal of Bacteriology*, **193**(5), 1021–2.

Maier G. P., Rapp M. V., Waite J. H., Israelachvili J. N. and Butler A. (2015). Adaptive synergy between catechol and lysine promotes wet adhesion by surface salt displacement. *Science*, **349**(6248), 628–32.

Mann E. E., Rice K. C., Boles B. R., Endres J. L., Ranjit D., Chandramohan L., Tsang L. H., Smeltzer M. S., Horswill A. R. and Bayles K. W. (2009). Modulation of eDNA release and degradation affects *Staphylococcus aureus* biofilm maturation. *PLoS ONE*, **4**(6), e5822.

Mann E. E. and Wozniak D. J. (2012). Pseudomonas biofilm matrix composition and niche biology. *FEMS Microbiology Reviews*, **36**(4), 893–916.

Manning A. J. and Kuehn M. J. (2011). Contribution of bacterial outer membrane vesicles to innate bacterial defense. *BMC Microbiology*, **11**(258).

Marx J. G., Carpenter S. D. and Deming J. W. (2009). Production of cryoprotectant extracellular polysaccharide substances (EPS) by the marine psychrophilic bacterium *Colwellia psychrerythraea* strain 34H under extreme conditions. *Canadian Journal of Microbiology*, **55**(1), 63–72.

Mashburn-Warren L., McLean R. J. C. and Whiteley M. (2008). Gram-negative outer membrane vesicles: beyond the cell surface. *Geobiology*, **6**, 214–9.

May T. and Okabe S. (2008). *Escherichia coli* harboring a natural IncF conjugative F plasmid develops complex matrue biofilms by stimulating synthesis of colanic acid and curli. *Journal of Bacteriology*, **190**, 7479–90.

McDougald D., Rice S. A., Barraud N., Steinberg P. D. and Kjelleberg S. (2012). Should we stay or should we go: mechanisms and ecological consequences for biofilm dispersal. *Nature Reviews Microbiology*, **10**(1), 39–50.

Mielich-Süss B. and Lopez D. (2015). Molecular mechanisms involved in *Bacillus subtilis* biofilm formation. *Environmental Microbiology*, **17**(3), 555–65.

Miltner A., Bombach P., Schmidt-Brücken B. and Kästner M. (2012). SOM genesis: microbial biomass as a significant source. *Biogeochemistry*, **111**(1–3), 41–55.

Minton A. P. (2001). The Influence of macromolecular crowding and macromolecular confinement on biochemical reactions in physiological media. *Journal of Biological Chemistry*, **276**(14), 10577–80.

Mitchell K. F., Zarnowski R., Sanchez H., Edward J. A., Reinicke E. L., Nett J. E., Mitchell A. P. and Andes D. R. (2015). Community participation in biofilm matrix assembly and function. *Proceedings of the National Academy of Sciences*, **112**(13), 4092–7.

Morris J. D., Hewitt J. L., Wolfe L. G., Kamatkar N. G., Chapman S. M., Diener J. M., Courtney A. J., Leevy W. M. and Shrout J. D. (2011). Imaging and analysis of *Pseudomonas aeruginosa* swarming and rhamnolipid production. *Applied and Environmental Microbiology*, **77**(23), 8310–7.

Mulcahy H., Charron-Mazenod L. and Lewenza S. (2009). Extracellular DNA chelates cations and induces antibiotic resistance in *Pseudomonas aeruginosa* biofilms. *PLoS Pathogens*, **4**(11).

Murphy K., Park A. J., Hao Y., Brewer D., Lam J. S. and Khursigara C. M. (2014). Influence of O polysaccharides on biofilm development and outer membrane vesicle biogenesis in *Pseudomonas aeruginosa* PAO1. *Journal of Bacteriology*, **196**(7), 1306–17.

Nan B. and Zusman D. R. (2011). Uncovering the mystery of gliding motility in the Myxobacteria. *Annual Review of Genetics*, **45**(1), 21–39.

Neu T. R. (1992). Microbial "footprints" and the general ability of microorganisms to label interfaces. *Canadian Journal of Microbiology*, **38**, 1005–8.

Neu T. R. (1996). Significance of bacterial surface-active compounds in interaction of bacteria with interfaces. *Microbiological Reviews*, **60**, 151–66.

Neu T. R., Härtner T. and Poralla K. (1990). Surface active properties of viscosin: a peptidolipid antibiotic. *Applied Microbiology and Biotechnology*, **32**, 518–20.

Neu T. R. and Lawrence J. R. (2009). Extracellular polymeric substances in microbial biofilms. In: Microbial Glycobiology: Structures, Relevance and Applications, A. Moran, P. Brenan, O. Holst and M. von Itzstein (eds), Elsevier, San Diego, pp. 735–58.

Neu T. R. and Lawrence J. R. (2014). Advanced techniques for in situ analysis of the biofilm matrix (structure, composition, dynamics) by means of laser scanning microscopy. In: Microbial Biofilms - Methods and Protocols, G. Donelli (ed.), Springer, New York, pp. 43–64.

Neu T. R. and Lawrence J. R. (2015). Innovative techniques, sensors, and approaches for imaging biofilms at different scales. *Trends in Microbiology*, **23**(4), 233–42.

Neu T. R., Manz B., Volke F., Dynes J. J., Hitchcock A. P. and Lawrence J. R. (2010). Advanced imaging techniques for assessment of structure, composition and function in biofilm systems. *FEMS Microbiology Ecology*, **72**, 1–21.

Neu T. R. and Marshall K. C. (1991). Microbial "footprints" - A new approach to adhesive polymers. *Biofoulin*, **3**, 101–12.

Neu T. R. and Poralla K. (1990). Emulsifying agents from bacteria isolated during screening for cells with hydrophobic surfaces. *Applied Microbiology and Biotechnology*, **32**, 521–5.

Neu T. R., Swerhone G. D. W. and Lawrence J. R. (2001). Assessment of lectin-binding analysis for in situ detection of glycoconjugates in biofilm systems. *Microbiology*, **147**, 299–313.

Nielsen P. H. and Jahn A. (1999). Extraction of EPS. In: Microbial Extracellular Polymeric Substances, J. Wingender, T. R. Neu and H.-C. Flemming (eds), Springer, Berlin, pp. 49–72.

Nobbs A. H., Lamont R. J. and Jenkinson H. F. (2009). Streptococcus adherence and colonization. *Microbiology and Molecular Biology Reviews*, **73**, 407–50.

Obst M., Dynes J. J., Lawrence J. R., Swerhone G. D. W., Benzerara K., Karunakaran C., Kaznatcheev K., Tyliszczak T. and Hitchcock A. P. (2009). Precipitation of amorphous $CaCO_3$ (aragonite-like) by cyanobacteria: a STXM study of the influence of EPS on the nucleation process. *Geochimica et Cosmochimica Acta*, **73**, 4180–98.

Ogawa H., Amagai Y., Koike I., Kaiser K. and Benner R. (2001). Production of refractory organic matter by bacteria. *Science*, **292**, 917–20.

Okshevsky M. and Meyer R. L. (2013). The role of extracellular DNA in the establishment, maintenance and perpetuation of bacterial biofilms. *Critical Reviews in Microbiology*, **41**(3), 341–52.

Okshevsky M., Regina V. R. and Meyer R. L. (2015). Extracellular DNA as a target for biofilm control. *Current Opinion in Biotechnology*, **33**, 73–80.

Oli M. W., Otoo H. N., Crowley P. J., Heim K. P., Nascimento M. M., Ramsook C. B., Lipke P. N. and Brady L. J. (2012). Functional amyloid formation by *Streptococcus mutans*. *Microbiology*, **158**, 2903–16.

Oppenheimer-Shaanan Y., Sibony-Nevo O., Bloom-Ackermann Z., Suissa R., Steinberg N., Kartvelishvily E., Brumfeld V. and Kolodkin-Gal I. (2016). Spatio-temporal assembly of functional mineral scaffolds within microbial biofilms. *Npj Biofilms And Microbiomes*, **2**, 15031.

Ostrowski A., Mehert A., Prescott A., Kiley T. B. and Stanley-Wall N. R. (2011). YuaB functions synergistically with the exopolysaccharide and TasA amyloid fibers to allow biofilm formation by *Bacillus subtilis*. *The Journal of Bacteriology*, **193**(18), 4821–31.

Otto M. (2013). Staphylococcal infections: mechanisms of biofilm maturation and detachment as critical determinants of pathogenicity. *Annual Review of Medicine*, **64**(1), 175–88.

Paes Leme A. F., Bellato C. M., Bedi G., Del Bel Cury A. A., Koo H. and Cury J. A. (2008). Effects of Sucrose on the extracellular matrix of plaque-like biofilm formed in vivo, studied by proteomic analysis. *Caries Research*, **42**, 435–43.

Paganelli F. L., Willems R. J. L., Jansen P., Hendrickx A., Zhang X., Bonten M. J. M. and Leavis H. L. (2013). *Enterococcus faecium* biofilm formation: identification of major autolysin $AtlA_{Efm}$, associated Acm surface localization, and $AtlA_{Efm}$-independent extracellular DNA release. *mBio*, **4**(2).

Pamp S. J. and Tolker-Nielsen T. (2007). Multiple roles of biosurfactants in structural biofilm development by *Pseudomonas aeruginosa*. *Journal of Bacteriology*, **189**, 2531–9.

Peltola M., Neu T. R., Kanto-Oqvist L., Raulio M., Kolari M. and Salkinoja-Salonen M. S. (2008). Architecture of *Deinococcus geothermalis* biofilms on glass and steel: a lectin study. *Environmental Microbiology*, **10**, 1752–9.

Pérez-Cruz C., Carrión O., Delgado L., Martinez G., López-Iglesias C. and Mercade E. (2013). New type of outer membrane vesicle produced by the Gram-negative bacterium *Shewanella vesiculosa* M7T: implications for DNA content. *Applied and Environmental Microbiology*, **79**(6), 1874–81.

Periasamy S., Joo H. S., Duong A. C., Bach T. H., Tan V. Y., Chatterjee S. S., Cheung G. Y. C. and Otto M. (2012). How *Staphylococcus aureus* biofilms develop their characteristic structure. *Proceedings of the National Academy of Sciences*, **109**(4), 1281–6.

Peterson B. W., He Y., Ren Y., Zerdoum A., Libera M. R., Sharma P. K., van Winkelhoff A.-J., Neut D., Stoodley P., van der Mei H. C. and Busscher H. J. (2015). Viscoelasticity of biofilms and their recalcitrance to mechanical and chemical challenges. *FEMS Microbiology Reviews*, **39**(2), 234–45.

Peterson B. W., van der Mei H. C., Sjollema J., Busscher H. J. and Sharma P. K. (2013). A distinguishable role of eDNA in the viscoelastic relaxation of biofilms. *mBio*, **4**(5).

Picard, J. C., Kim, S.-Y., Deschamps, J., Li, Y., Dorel, C., Gruss, A., Trubuil, A. and Briandet, R. (2016). Travelling through slime – bacterial movements in the EPS matrix. In: The Perfect Slime, H.-C. Flemming, T. R. Neu and J. Wingender (eds), IWA Publishing, London, ch. 9, pp. 179–191.

Pinchuk G. E., Ammons C., Culley D. E., Li S.-M. W., McLean J. S., Romine M. F., Nealson K. H., Fredrickson J. K. and Beliaev A. S. (2008). Utilization of DNA as a sole source of phosphorus, carbon, and energy by *Shewanella* spp: ecological and physiological implications for dissimilatory metal reduction. *Applied and Environmental Microbiology*, **74**, 1198–208.

Raaijmakers J. M., De Bruijn I., Nybroe O. and Ongena M. (2010). Natural functions of lipopeptides from Bacillus and Pseudomonas: more than surfactants and antibiotics. *FEMS Microbiology Reviews*, **34**(6), 1037–62.

Ray V. A., Driks A. and Visick K. L. (2015). Identification of a Novel Matrix Protein That Promotes Biofilm Maturation in Vibrio fischeri. *Journal of Bacteriology*, **197**(3), 518–28.

Reguera G., McCarthy K. D., Metha T., Nicoll J. S., Tuominen M. T. and Lovley D. R. (2005). Extracellular electron transfer via microbial nanowires. *Nature*, **435**, 1098–101.

Rendueles O., Kaplan J. B. and Ghigo J. M. (2013). Antibiofilm polysaccharides. *Environmental Microbiology*, **15**(2), 334–46.

Romero D., Aguilar C., Losick R. and Kolter R. (2010). Amyloid fibers provide structural integrity to *Bacillus subtilis* biofilms. *Proceedings of the National Academy of Sciences*, **107**(5), 2230–4.

Romero D., Vlamakis H., Losick R. and Kolter R. (2011). An accessory protein required for anchoring and assembly of amyloid fibres in *B. subtilis* biofilms. *Molecular Microbiology*, **80**(5), 1155–68.

Rosenberg E. and Ron E. Z. (1997). Bioemulsans: microbial polymeric emulsifiers. *Current Opinion in Biotechnology*, **8**, 313–6.

Ryder C., Byrd M. and Wozniak D. J. (2007). Role of polysaccharides in *Pseudomonas aeruginosa* biofilm development. *Current Opinion in Microbiology*, **10**, 644–8.

Schooling S. R., Hubley A. and Beveridge T. J. (2009). Interaction of DNA with biofilm-derived membrane vesicles. *Journal of Bacteriology*, **191**, 4097–102.

Schurig C., Smittenberg R., Berger J., Kraft F., Woche S., Goebel M.-O., Heipieper H., Miltner A. and Kaestner M. (2013). Microbial cell-envelope fragments and the formation of soil organic matter: a case study from a glacier forefield. *Biogeochemistry*, **113**(1–3), 595–612.

Schwartz K., Syed A. K., Stephenson R. E., Rickard A. H. and Boles B. R. (2012). Functional amyloids composed of phenol soluble modulins stabilize *Staphylococcus aureus* biofilms. *PLoS Pathogens*, **8**, e1002744.

Schwechheimer C. and Kuehn M. J. (2015). Outer-membrane vesicles from Gram-negative bacteria: biogenesis and functions. *Nat Rev Micro*, **13**(10), 605–19.

Seifan M., Samani A. K. and Berenjian A. (2016). Bioconcrete: next generation of self-healing concrete. *Applied Microbiology and Biotechnology*, **100**(6), 2591–602.

Seminara A., Angelini T. E., Wilking J. N., Vlamakis H., Ebrahim S., Kolter R., Weitz D. A. and Brenner M. P. (2012). Osmotic spreading of *Bacillus subtilis* biofilms driven by an extracellular matrix. *Proceedings of the National Academy of Sciences*, **109**(4), 1116–21.

Seper A., Fengler V. H. I., Roier S., Wolinski H., Kohlwein S. D., Bishop A. L., Camilli A., Reidl J. and Schild S. (2011). Extracellular nucleases and extracellular DNA play important roles in *Vibrio cholerae* biofilm formation. *Molecular Microbiology*, **82**(4), 1015–37.

Seviour T., Hansen S. H., Yang L., Yau Y. H., Wang V. B., Stenvang M. R., Christiansen G., Marsili E., Givskov M., Chen Y., Otzen D. E., Nielsen P. H., Geifman-Shochat S., Kjelleberg S. and Dueholm M. S. (2015). Functional amyloids keep quorum-sensing molecules in check. *Journal of Biological Chemistry*, **290**(10), 6457–69.

Seviour T., Yuan Z., Van Loosdrecht M. C. M. and Lin Y. (2012). Aerobic sludge granulation: a tale of two polysaccharides? *Water Research*, **46**(15), 4803–13.

Sharon N. and Lis H. (2003). Lectins. Kluwer Academic Publishers, Dordrecht.

Shimizu M., Zhou J., Schr+Âder C., Obst M., Kappler A. and Borch T. (2013). Dissimilatory Reduction and Transformation of Ferrihydrite-Humic Acid Coprecipitates. *Environmental Science & Technology*, **47**(23), 13375–84.

Shoseyov O., Shani Z. and Levy I. (2006). Carbohydrate binding modules: biochemical properties and novel applications. *Microbiology and Molecular Biology Reviews*, **70**(2), 283–95.

Smith D. R., Maestre-Reyna M., Lee G., Gerard H., Wang A. H.-J. and Watnick P. I. (2015). In situ proteolysis of the *Vibrio cholerae* matrix protein RbmA promotes biofilm recruitment. *Proceedings of the National Academy of Sciences*, **112**(33), 10491–6.

Staudt C., Horn H., Hempel D. C. and Neu T. R. (2003). Screening of lectins for staining lectin-specific glycoconjugates in the EPS of biofilms. In: Biofilms in Medicine, Industry and Environmental Technology, P. Lens, A. P. Moran, T. Mahony, P. Stoodley and V. O'Flaherty (eds), IWA Publishing, UK, pp. 308–27.

Steinberg N. and Kolodkin-Gal I. (2015). The matrix reloaded: how sensing the extracellular matrix synchronizes bacterial communities. *Journal of Bacteriology*, **197**(13), 2092–103.

Stocker R. (2012). Marine microbes see a sea of gradients. *Science*, **338**(6107), 628–33.

Stuart R K., Mayali X., Lee J. Z., Craig Everroad R., Hwang M., Bebout B. M., Weber P. K., Pett-Ridge J. and Thelen M. P. (2016). Cyanobacterial reuse of extracellular organic carbon in microbial mats. *ISME J*, **10**(5), 1240–51.

Sutherland I. (1999). Polysaccharases in biofilms – sources – action – consequences. In: Microbial Extracellular Substances, J. Wingender, T. R. Neu and H.-C. Flemming (eds), Springer, Berlin, pp. 201–16.

Sutherland I. (2007). Bacterial exopolysaccharides. In: Comprehensive Glycoscience, J. P. Kamerling (ed.), Elsevier, Doordrecht, pp. 521–58.

Sutherland I. W. (1977). Surface Carbohydrates of the Procaryotic Cell. Academic Press, New York.

Sutherland I. W. (2001). The biofilm matrix – an immobilized but dynamic microbial environment. *Trends in Microbiology*, **9**, 222–7.

Suzuki H , Daimon M., Awano T., Umekage S., Tanaka T. and Kikuchi Y. (2009). Characterization of extracellular DNA production and flocculation of the marine photosynthetic bacterium *Rhodovulum sulfidophilum. Applied Microbiology and Biotechnology*, **84**, 349–56.

Tashiro Y , Ichikawa S., Shimizu M., Toyofuku M., Takaya N., Nakajima-Kambe T., Uchiyama H. and Nomura N. (2010). Variation of physiochemical properties and cell association activity of membrane vesicles with growth phase in *Pseudomonas aeruginosa. Applied and Environmental Microbiology*, **76**(11), 3732–9.

Tashiro Y., Uchiyama H. and Nomura N. (2012). Multifunctional membrane vesicles in Pseudomonas aeruginosa. *Environmental Microbiology*, **14**(6), 1349–62.

Teschler J. K., Zamorano-Sanchez D., Utada A. S., Warner C. J. A., Wong G. C. L., Linington R. G. and Yildiz F. H. (2015). Living in the matrix: assembly and control of *Vibrio cholerae* biofilms. *Nature Reviews Microbiology*, **13**(5), 255–68.

Tolker-Nielsen T. (2015). Biofilm development. *Microbiology Spectrum*, **3**(2).

Tourney J. and Ngwenya B. T. (2014). The role of bacterial extracellular polymeric substances in geomicrobiology. *Chemical Geology*, **386**(0), 115–32.

Traving S. J., Thygesen U. H., Riemann L. and Stedmon C. A. (2015). A model of extracellular enzymes in free-living microbes: which strategy pays off? *Applied and Environmental Microbiology*, **81**(21), 7385–93.

Tseng B. S. Zhang W., Harrison J. J., Quach T. P., Song J. L., Penterman J., Singh P. K., Chopp D. L., Packman A. I. and Parsek M. R. (2013). The extracellular matrix protects Pseudomonas aeruginosa biofilms by limiting the penetration of tobramycin. *Environmental Microbiology*, **15**(10), 2865–78.

Verdugo P. (2012). Marine Microgels. *Annual Review of Marine Science*, **4**(1), 375–400.

Vilain S., Pretorius J. M., Theron J. and Brözel V. S. (2009). DNA as an adhesin: *Bacillus cereus* requires extracellular DNA to form biofilms. *Applied and Environmental Microbiology*, **75**, 2861–8.

Visscher P. T. and Stolz J. F. (2005). Microbial mats as bioreactors: populations, processes, and products. *Palaeogeography, Palaeoclimatology, Palaeoecology*, **219**(1–2), 87–100.

Vlamakis H., Chai Y., Beauregard P., Losick R. and Kolter R. (2013). Sticking together: building a biofilm the *Bacillus subtilis* way. *Nature Reviews Microbiology*, **11**(3), 157–68.

Wagner M., Ivleva N. P., Haisch C., Niessner R. and Horn H. (2009). Combined use of confocal laser scanning microscopy (CLSM) and raman microscopy (RM): investigations on EPS-matrix. *Water Research*, **43**, 63–76.

Wang S., Liu X., Liu H., Zhang L., Guo Y., Yu S., Wozniak D. J. and Ma L. Z. (2015). The exopolysaccharide Psl-eDNA interaction enables the formation of a biofilm skeleton in *Pseudomonas aeruginosa. Environmental Microbiology Reports*, **7**(2), 330–40.

Wang S., Parsek M. R., Wozniak D. J. and Ma L. Z. (2013). A spider web strategy of type IV pili-mediated migration to build a fibre-like Psl polysaccharide matrix in *Pseudomonas aeruginosa* biofilms. *Environmental Microbiology*, **15**(8), 2238–53.

Wang S., Yu S., Zhang Z., Wei Q., Yan L., Ai G., Liu H. and Ma L. Z. (2014). Coordination of swarming motility, biosurfactant synthesis, and biofilm matrix exopolysaccharide production in *Pseudomonas aeruginosa. Applied and Environmental Microbiology*, **80**(21), 6724–32.

Whitfield C. (2006). Biosynthesis and assembly of capsular polysaccharides in *Escherichia coli. Annual Review of Biochemistry*, **75**, 39–68.

Wilker J. J. (2015). Positive charges and underwater adhesion. *Science*, **349**(6248), 582–3.

Wilking J. N., Angelini T. E., Seminara A., Brenner M. P. and Weitz D. A. (2011). Biofilms as complex fluids. *MRS Bulletin*, **36**(05), 385–91.

Wingender J., Neu T. R. and Flemming H.-C. (1999). Microbial Extracellular Polymeric Substances. Springer, Heidelberg.

Xavier J. B. and Foster K. R. (2007). Cooperation and conflict in microbial biofilms. *Proceedings of the National Academy of Sciences*, **104**(3), 876–81.

Yuan D. Q., Wang Y. L. and Feng J. (2014). Contribution of stratified extracellular polymeric substances to the gel-like and fractal structures of activated sludge. *Water Research*, **56**(0), 56–65.

Zeng G., Vad B. S., Dueholm M. S., Christiansen G., Nilsson M., Tolker-Nielsen T., Nielsen P. H., Meyer R. L. and Otzen D. E. (2015). Functional bacterial amyloid increases Pseudomonas biofilm hydrophobicity and stiffness. *Frontiers in Microbiology*, **6**.

Zuroff T. R., Gu W., Fore R. L., Leschine S. B. and Curtis W. R. (2014). Insights into *Clostridium phytofermentans* biofilm formation: aggregation, microcolony development and the role of extracellular DNA. *Microbiology*, **160**, 1134–43.

Chapter 4

The transition from bacterial adhesion to the production of EPS and biofilm formation

Vera Carniello[1], Jiapeng Hou[1], Henny C. van der Mei[2,3] and Henk J. Busscher

University of Groningen and University Medical Center Groningen, Department of BioMedical Engineering, Antonius Deusinglaan 1, 9713 AV Groningen, Netherlands

[1]These authors contributed equally to this work

[2,3]Biomedical Engineering, University Medical Center Groningen, Antonius Deusinglaan 1, 9731 AV Groningen, The Netherlands

[3]Corresponding author: h.c.van.der.mei@umcg.nl

ABSTRACT

Planktonic life is dangerous for most bacterial strains and species and adhesion to surfaces is often considered a survival mechanism. Once adhering, a cascade of events is triggered that involves amongst others the production of an EPS-matrix leading to the formation of what is generally called a "biofilm". In this chapter, we hypothesize about the way bacteria become aware of their adhering state. We consider that bacteria deform under the influence of the adhesion forces exerted upon them when adhering to a substratum surface. Surface enhanced fluorescence has convincingly demonstrated the existence of minor cell wall deformation upon adhesion that can act as a trigger for an adhering bacterium to start EPS-matrix production, as a hallmark of the transition between planktonic and sessile phenotypes. For staphylococci, EPS-matrix production appears related with the magnitude of the adhesion forces felt. The importance of the EPS-matrix is ubiquitously present throughout the entire process of biofilm development, from facilitating initial bacterial adhesion to maintaining biofilm structural integrity during growth and offering "back up" resources in case of nutrient depletion during biofilm aging. During the entire biofilm life cycle, EPS protects the biofilm against chemical attacks such as from antimicrobial treatment and through its viscoelastic properties, against mechanical stresses. The viscoelastic response of biofilms to external mechanical stresses can be modeled using three Maxwell elements representing the flow of water, more viscous EPS and bacteria repositioning in a deformed biofilm. Bacterial repositioning is more prominent in open structured biofilms than in more condensed ones with an impact on antimicrobial penetration and detachment. The main advantages of stress relaxation to determine biofilm structure and composition over microscopic techniques are that it yields quantitative data covering an area of several square millimeters. Biofilms are also slippery due to their EPS-matrix, but whether the slipperiness of biofilms serves any physiological role or not is unknown. Summarizing, this chapter reviews the events occurring during the transition from bacterial adhesion to EPS-matrix production and biofilm formation from a physico-chemical perspective, offering new concepts like bacterial adhesion force-sensing and cell wall deformation as a trigger for the development of sessile bacterial phenotypes.

Keywords: Microbial adhesion, surface sensing, cell wall deformation, atomic force microscopy, surface thermodynamics, surface enhanced fluorescence, biofilm, EPS, lubrication

4.1 INTRODUCTION

Biofilm formation is initiated by bacterial adhesion to a substratum surface. Bacteria adhere to surfaces because the nutrient concentration at a surface is usually higher than in suspension (Van Loosdrecht *et al.* 1990; Stoodley *et al.* 2002; Zobell, 1943) and in addition, adhesion often serves as a survival strategy. Oral microorganisms for instance need to adhere to oral hard or soft tissues in order to avoid being swallowed with saliva and being killed in the gastrointestinal tract. Marine bacteria adhere to ship hulls to have easy access to nutrients and find shelter against predation (Dang & Lovell, 1999).

Upon adhesion, bacteria produce a matrix of extracellular polymeric substances (EPS), which consists of polysaccharides, proteins, nucleic acids and lipids in which bacteria embed themselves (Figure 4.1). EPS production is a survival strategy as well for the microorganisms, since it provides higher mechanical stability to the biofilm, as well as higher resistance to antimicrobials and protection against the host immune system (Flemming & Wingender, 2010). When biofilms form on biomaterials implanted or introduced in the human body, such as catheters or joint implants, antibiotic resistance increases up to 1000 times the resistance of planktonic bacteria (Olsen, 2015; Lee *et al.* 2014; Mah, 2012; Rodriguez-Martinez & Pascual, 2006). As a consequence, these biomaterial-associated infections can mostly only be treated by implant replacement, involving additional invasive surgery for the patient and high costs for the health care system (Busscher & Van der Mei, 2012). To prevent biofilm formation, reduction of initial bacterial adhesion has been the main aim of worldwide investigation (Beloin *et al.* 2014; Chauhan *et al.* 2014; Hook *et al.* 2012). Promising options for decreasing the number of bacteria to surfaces are hydrophilic polymer brush coatings, acting as steric barriers that prevent direct bacterial contact with a surface. However, the approach of reducing bacterial adhesion has limitations, since even adhesion of a small number of bacteria can lead to biofilm formation (Nejadnik *et al.* 2008). Moreover it can be doubted whether interference in bacterial adaptation to their adhering state would not be a better way to go, leaving bacteria in their susceptible planktonic state, also when adhering.

Hitherto, the mechanisms of bacterial adaptation to their adhering state are not completely unraveled. Bacteria have only limited possibilities to sense a surface. Since bacteria are known to respond differently to different biomaterials (Gomes *et al.* 2015; Ludecke *et al.* 2014), we have hypothesized that adhering bacteria sense a surface through cell wall deformation under the influence of the prevailing adhesion forces as a first step in their adaptation to an adhering state and the formation of a biofilm. In this chapter, we review the available literature on adhesion force sensing and cell wall deformation as well as the main techniques for studying cell wall deformation and adhesion force sensing. Also, we will highlight the role of EPS production as triggered through adhesion force sensing in biofilm formation. Further research aimed at modifying the bacterial response upon adhesion to a surface in order to block bacterial transition from a "planktonic" to a "biofilm" mode of growth may yield novel strategies to prevent biofilm formation.

Figure 4.1 Steps in the transition from initial bacterial adhesion to biofilm formation. When a surface is in contact with body fluids, proteins and other compounds are adsorbed first onto the surface, forming the so-called "conditioning film" to which bacteria mostly adhere rather than to a bare substratum surface (depending on environmental conditions). Upon adhesion to a surface, bacteria form a micro-colony by cell division and EPS production. Subsequently, colonies expand by adhesion of other bacteria, growth and the continued production of EPS (adapted from Costerton & Stewart, 2001).

4.2 THE TRANSITION FROM BACTERIAL ADHESION TO BIOFILM FORMATION

Adhesion of the first arriving bacteria on a substratum surface is of primary importance, as these initial colonizers provide the link between the substratum surface and the entire biofilm. It has been hypothesized that upon first contact with a surface, these "linking" bacteria sense the surface (Karatan & Watnick, 2009) through the magnitude of the adhesion force exerted by the substratum. Therewith the adhering bacteria become aware of their adhering state as a result of membrane deformation occurring during adhesion (Busscher & Van der Mei, 2012). Membrane deformation is minor in most naturally occurring strains and species, since it is counteracted by the rigid peptidoglycan layer surrounding a bacterium, but still sufficient to allow communication between a bacterium and its environment.

The mechanism by which a bacterium becomes aware of its adhering state has been named "surface sensing". This term has been applied to motile bacteria, whose flagella can sense the surface so that a bacterium can choose between adhering to the surface or remaining in the planktonic state (Belas, 2014; Karatan & Watnick, 2009). Other authors used the term "surface sensing" with regard to non-motile bacteria, to indicate the "sense of touch" that takes place before adhesion to the surface and provides a bacterium with information about the type of material it adheres to (Lower *et al.* 2010; Tuson & Weibel, 2013).

The first steps in biofilm formation always involve adhesion and surface sensing. The transition from a planktonic to a biofilm mode of growth is characterized by gene upregulation or downregulation and by the production of EPS to make bacteria more resilient to different stresses and to promote irreversible adhesion. The biofilm phenotype that develops after adhesion to a surface is often referred to as the "sessile phenotype" (Stoodley *et al.* 2002). In the first ten minutes after contact with a surface, *Escherichia coli* was found to decrease its cell respiration level compared to planktonic bacteria (Geng *et al.* 2014) and this phenomenon has been interpreted as a strategy to ensure survival of adhering bacteria in the eventuality of oxidative stress conditions during the biofilm formation and growth (Aoki *et al.* 2009). In motile bacteria, the sessile phenotype often involves the repression of flagellum synthesis, facilitating irreversible cell attachment on the surface (Abdallah *et al.* 2014; Karatan & Watnick, 2009). In *Caulobacter crescentus*, adhesin production is stimulated by adhesion on a substratum surface, in order to create irreversible attachment (Li *et al.* 2012). *Staphylococcus aureus* also shows an accumulation of adhesins in contact with a surface (Lower *et al.* 2010).

In the examples above, adhesion forces prevail in the range of the "interaction regime", as described in a regimented "three adhesion force" model (Busscher & Van der Mei, 2012). According to this model, the "interaction regime" describes changes in phenotypes that gradually occur with increasing adhesion forces. Weak adhesion forces are related to the planktonic regime in which surface sensing is very weak and bacteria behave like planktonic cells. Without the necessary changes in gene expression and subsequent production of EPS, these cells remain susceptible to antimicrobials. This regime occurs for hydrogels and polymer-brush coated surfaces (Jucker *et al.* 1996; Nejadnik *et al.* 2008). In the lethal regime, strong adhesion forces lead to membrane stresses that deactivate the adhering bacteria and cause cell death. In *Bacillus subtilis* it was shown how mechanical stresses on the cell membrane provoked cell death without disrupting the membrane integrity of the cells (Sahoo *et al.* 2006). For this effect, the term "stress-deactivation" has been proposed and a number of studies have focused on this topic (Liu *et al.* 2008).

4.3 BACTERIAL SURFACE SENSING AND CELL WALL DEFORMATION

Motile bacteria can sense surfaces through appendages like flagella or pili. The contact of these bacterial structures with a substratum surface causes a drag on the appendage and a subsequent local stress that provokes a signal inducing the transition to the biofilm phenotype (Karatan & Watnick, 2009). In *B. subtilis*, contact of a flagellum with a surface inhibits flagellar rotation and this surface signal activates the two-component regulatory system DegS–DegU that triggers the change in gene expression towards the sessile phenotype (Belas, 2013; Cairns *et al.* 2013).

In absence of surface appendages, non-motile bacteria sense the surface through cell wall deformation and the magnitude of adhesion forces. Following adhesion to a surface, cell wall deformation causes minor deformations of the lipid bilayer, provoking a change in intra-bilayer pressure profile in the lipid membrane (Perozo *et al.* 2002). This variation in pressure profile can be sensed by the bacterial cell through gating of mechanosensitive channels or through stress sensitive proteins (Harapanahalli *et al.* 2015b). Mechanosensitive channels represent an interesting response to surface adhesion and a mechanism for surface sensing. Cell wall deformation and variations in membrane curvature imply thinning of the membrane and wider spaces between lipids in the bilayer, which allow water molecules to insert within the lipids. This provokes energetically an unfavorable hydrophobic mismatch between lipids and water molecules and the energy necessary to compensate this is provided by the adhesion forces between a bacterium and a substratum surface (Harapanahalli *et al.* 2015b). Mechanosensitive channels gate in response to the variation in membrane pressure profile, and both membrane thinning and hydrophobic mismatch were found to facilitate channel gating (Iscla & Blount, 2012; Perozo *et al.* 2002; Perozo & Rees, 2003). Mechanosensitive channel gating results in the opening of a pore in the bacterial membrane, allowing a solute flux between the cytoplasm and the outer environment (Powl *et al.* 2008), which might be the surface signal for a bacterium to become aware of its adhering state.

Stress sensitive proteins are located on the bacterial membrane and trigger complex signaling pathways after adhesion on a substratum surface. In *E. coli* the outer membrane lipoprotein NlpE can sense surface adhesion and transmit the information to the stress sensitive protein CpxA (Vogt & Raivio, 2012). After detecting this signal, CpxA activates the Cpx pathway that regulates the expression of gene clusters related to the biofilm mode of growth (Keller *et al.* 2015; Otto & Silhavy, 2002).

4.4 METHODS TO STUDY BACTERIAL CELL WALL DEFORMATION

Considering the role of cell wall deformation in making an adhering bacterium aware of its adhering state, we will now briefly review techniques to study bacterial cell wall deformation.

4.4.1 Macroscopic bio-optical fluorescence imaging

Bio-optical fluorescent imaging is a macroscopic technique, suitable for *in vitro* as well as *in vivo* investigations. Its most relevant component is a highly sensitive CCD camera, which can measure the radiance from fluorescent probes. Images are obtained by trans-illumination or reflection at single or multiple excitation wavelengths (Ntziachristos *et al.* 2005).

Bio-optical fluorescence imaging can be used to analyze cell wall deformation, exploiting the effect of surface enhanced fluorescence (SEF). SEF involves an increase in fluorescence occurring when a fluorophore is close to a reflecting metal surface and it is due to two effects: surface plasmon resonance and a mirror effect (Lee *et al.* 2011). Surface plasmon resonance takes place when the absorption and emission spectra of the fluorophore overlap with the scattering spectrum of the metal surface and this resonance condition determines a greater radiance from the metal-fluorophore system. The mirror effect takes place when the fluorophore is illuminated by the Raman-scattered field reflected by the conducting metal. Under these conditions, the emitted intensity can be amplified up to four times (Le Moal *et al.* 2007; Lee *et al.* 2011). The effect of SEF decreases exponentially with the distance between the fluorescent dye and the metal surface and the fluorescence enhancement is not detectable for distances larger than about 30 nm. For this reason SEF cannot only be exploited to measure the distance between a bacterium and a substratum surface (Lee *et al.* 2011), but also to measure bacterial cell wall deformation. After adhesion of a fluorescent bacterium on a reflecting metal surface, cell wall deformation will gradually bring more fluorophores within a bacterium closer to a substratum, i.e. in the range of SEF (Figure 4.2) (Li *et al.* 2014) enhancing the fluorescence signal. The advantage of SEF over other techniques to measure cell wall deformation is that several millions of bacteria can be analyzed within one experiment, taking only several minutes.

4.4.2 Atomic force microscopy

AFM allows to obtain topographical images of bacteria, cells or other particles and surfaces by scanning a sharp tip over the sample and measuring the interaction forces between the tip and the sample. This tip is mounted on a cantilever, which can be moved by a piezoelectric scanner in the x, y and z directions with high accuracy. When the cantilever bends in response to the interaction forces between tip and sample, the cantilever deflection is detected by the position of a laser beam reflected by the cantilever on a photodiode (Dufrene, 2003; Dufrene, 2008; Heinisch *et al.* 2012; Pillet *et al.* 2014). AFM offers several modes for imaging living cells and microorganisms, as well as synthetic surfaces. In its contact mode, the tip is in direct contact with the sample and cantilever deflection is kept constant to maintain the same force between tip and sample at all times (Kailas *et al.* 2009; Zhang *et al.* 2013). In tapping mode, the cantilever oscillates close to its resonance frequency, allowing the tip to come in contact with the sample at the end of its downwards movement (Dufrene, 2003; Turner *et al.* 2013). Another mode of AFM application is PeakForce quantitative nanomechanical mapping (PeakForce-QNM) with which it is possible to simultaneously record the topographical image as well as mapping other parameters of the sample, such as adhesion force and stiffness (Torrent-Burgues & Sanz, 2014; Yeroslavsky *et al.* 2015).

AFM topographical imaging has been exploited to measure bacterial dimensions, i.e. height, width and length. Bolshakova and co-workers (2001) employed AFM in both tapping and contact mode to measure the size of *E. coli* in air, water and bacterial culture medium. The authors concluded that the bacterial height measured in tapping mode in liquid was lower than in contact mode and that both height and width are smaller if the bacteria are dried to perform AFM measurements in air. Moreover, they observed that, when *E. coli* was treated with lysozyme, an enzyme that lyses the outer cell wall, the size of the Gram-negative bacteria decreased, their shape became more compressed and perforations of the inner cell membrane developed. Using the AFM in tapping mode, other authors found a significant height reduction in *E. coli* after treatment with the antibacterial drugs streptomycin and EDTA in combination with lysozyme (Li *et al.* 2015). Similarly, measurement of the heights and widths of adhering bacteria can be done to quantify cell wall deformation after adhesion on a substratum surface (Chen *et al.* 2014; Touhami *et al.* 2004) using the AFM PeakForce-QNM imaging mode. By comparing the polar radius of individual, adhering bacteria with the radius of planktonic bacteria measured using dynamic light scattering the extent of cell wall deformation could be quantified and was found to be higher for Δ*pbp4* mutants lacking crosslinking of their peptidoglycan layer (Chen *et al.* 2014).

The major drawback of AFM imaging to investigate bacterial cell wall deformation is the need to measure individual, adhering bacteria, making it difficult to obtain statistically significant data. Furthermore, it can be argued that the loading force applied by the AFM cantilever can contribute to cell wall deformation. However, this drawback can be limited by using the PeakForce-QNM mode of AFM, which allows to image adhering bacteria by applying a low loading force (Chen *et al.* 2014).

4.4.3 Surface thermodynamic approach

The surface thermodynamic approach to bacterial adhesion is based on a surface free energy analysis of the bacterial and substratum surface along with the surface tension of the suspending liquid (Absolom *et al.* 1983; Hermansson, 1999; Van Oss, 1993). Different methods to calculate substratum surface free energies exist using contact angle data with different liquids as input (Kwok *et al.* 1996; Owens & Wendt, 1969; Van Oss, 1995). Common to all methods is the Gibbs free energy of adhesion ΔG_{adh} as an outcome parameter, defined as:

$$\Delta G_{adh} = \gamma_{BS} - \gamma_{BL} - \gamma_{SL} \tag{4.1}$$

where γ_{BS}, γ_{BL}, γ_{SL} are the interfacial free energies of the bacterium-surface, bacterium-liquid and surface-liquid interfaces, respectively (Absolom *et al.* 1983; Hermansson, 1999). Positive values of the Gibbs free energy of adhesion ($\Delta G_{adh} > 0$) correlate with unfavorable thermodynamic conditions for microorganisms to adhere on a surface, while energetically favorable conditions are characterized by $\Delta G_{adh} < 0$ (Absolom *et al.* 1983).

The Gibbs free energy of adhesion can be divided in Lifshitz -Van der Waals and acid-base components (Van Oss, 1995):

$$\Delta G_{adh} = \Delta G_{adh}^{LW} + \Delta G_{adh}^{AB} \tag{4.2}$$

Lifshitz-Van der Waals interactions are due to attractive forces which are present as soon as a bacterium approaches a surface and can extend up to hundreds of nanometers (Chen *et al.* 2011; Liu *et al.* 2008). The summation of all pair-wise Lifshitz-Van der Waals molecular interactions within a bacterium and a substratum surface gives rise to a long-range, attractive adhesion force between a bacterium and a substratum surface (Figure 4.2).

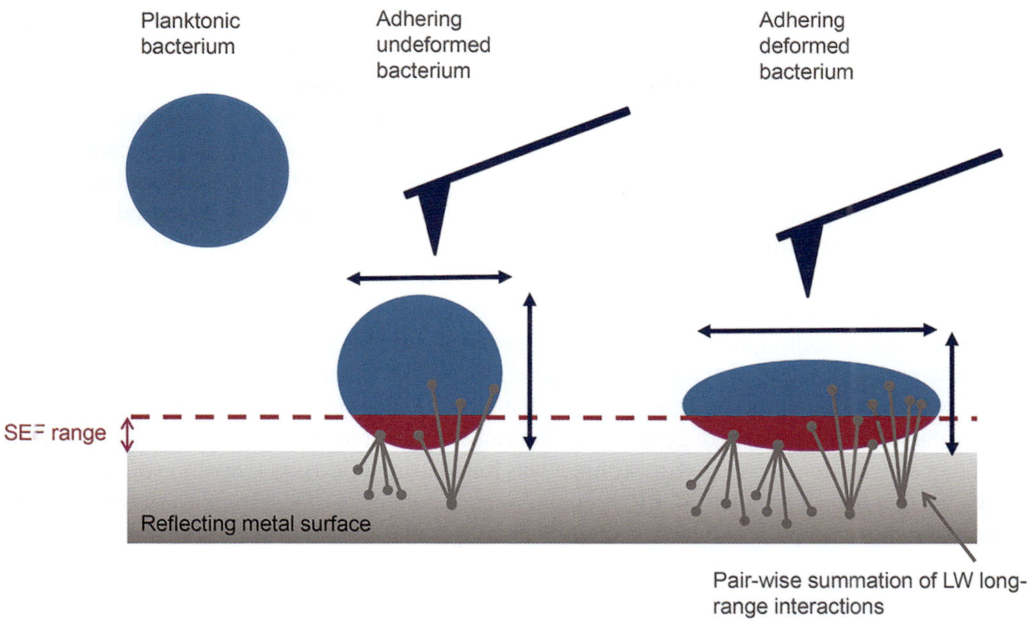

Figure 4.2 Lifshitz-Van der Waals interactions arise from the pair-wise summation of all possible molecular interactions between a bacterium and a substratum surface. Upon nanoscale cell wall deformation, more molecules inside a bacterium are brought in the close vicinity of a substratum, therewith increasing long-range Lifshitz-Van der Waals interactions (Chen *et al.* 2014). In case fluorescent bacteria are involved, more fluorophores are brought closer to a substratum surface upon cell wall deformation, i.e. within the range of SEF, and enhanced fluorescence can be measured with respect to undeformed bacteria. Deformation of individual bacteria can be measured using atomic force microscopy (AFM) imaging.

Even though bacteria have been demonstrated to adhere under conditions of unfavorable, positive values of the Gibbs free energy ΔG_{adh}, the long-range Lifshitz-Van der Waals interactions ΔG_{adh}^{LW} are nearly always favorable for adhesion to occur ($\Delta G_{adh}^{LW} < 0$) and therewith generally considered responsible for adhesion (Busscher *et al.* 2010b). Moreover, it can be argued that since bacteria of different strains and species have a highly similar, chemical bulk composition, Lifshitz-Van der Waals components ΔG_{adh}^{LW} should be largely the same for bacteria adhering in an undeformed state. Since in a deformed state more molecules are brought in closer vicinity of a substratum surface, it has been speculated that differences in the Lifshitz-Van der Waals component ΔG_{adh}^{LW} are indicative of more or less extensive bacterial cell wall deformation (Chen *et al.* 2014), as explained in Figure 4.2.

4.5 BACTERIAL SURFACE SENSING AND THE ROLE OF EPS IN BIOFILMS

4.5.1 Bacterial surface sensing and EPS production

EPS production after bacterial adhesion is one of the main hallmarks of the transition of adhering bacteria from a planktonic to a sessile phenotype and the formation of a biofilm. The analyses of surface sensing and subsequent EPS production have been mostly performed either from the perspectives of gene expression or through direct evaluation of the production of extracellular matrix components.

In adhering *S. aureus*, both *icaA* expression as well as production of poly-N-acetylglucosamine (PNAG), facilitating bacterial adhesion to surfaces and to other bacteria in multilayered biofilms (Arciola *et al.* 2015), and eDNA decreased with increasing adhesion forces between adhering bacteria and substratum surfaces (Harapanahalli *et al.* 2015a). This suggests that surface sensing through cell wall deformation is triggered by adhesion force sensing. The rigid peptidoglycan layer as part of the bacterial cell wall plays a pivotal role in counteracting adhesion force induced deformation, as in mutants lacking a rigid, cross-linked peptidoglycan layer no relation was found between adhesion forces and gene expression or matrix production (Harapanahalli *et al.* 2015a, 2015b).

Another interesting technique to analyze gene expression and EPS production is β-galactosidase assay, a colorimetric assay that detects the enzyme β-galactosidase. This enzyme is encoded in *E. coli* by the gene *lacZ*, which can be fused to another gene of interest. Using a colorimetric assay that detects the enzyme β-galactosidase, the level of gene expression can be measured from the amount of β-galactosidase produced. In a *Pseudomonas aeruginosa* strain, it was found using this assay (Davies *et al.* 1993; Davies & Geesey, 1995), that fifteen minutes after Pseudomonas adhesion to a surface, bacteria showed upregulated *algC* expression, governing the production of alginate as a matrix component.

4.5.2 Role of EPS in biofilms

There are three distinctly different operational types of EPS, each with their own function: (1) soluble EPS (sEPS), (2) tightly bound EPS (TB-EPS) and (3) loosely bound EPS (LB-EPS) (Chen *et al.* 2013).

sEPS molecules are much smaller than bacteria and once released, they will diffuse faster and adsorb to a substratum surface before initial bacterial adhesion (Tansel & Tansel, 2013). Adsorbed sEPS molecules can self-assemble and form hydrophilic patches that affect the hydrophobicity and charge properties of a substratum surface and promote bacterial adhesion. After the formation of a single, initial layer of adhering bacteria, the flexibility of sEPS patches helps to keep a favorable surface free energy state and thus allows new biofilm clusters to adhere (Tansel & Tansel, 2013). In mature biofilms, sEPS reduces its density by forming chain-like structures (Janissen *et al.* 2015; Pen *et al.* 2014). TB-EPS provides coherence to a biofilm bridging bacteria to each other to form clusters, while LB-EPS connects bacterial clusters into microcolonies (Pellicer-Nacher *et al.* 2013). eDNA is a pivotal component of TB-EPS not only promoting bacterial adhesion to surfaces (Pellicer-Nacher *et al.* 2013) but also co-adhesion between bacteria (Vilain *et al.* 2009).

In a biofilm mode of growth, adhering bacteria tend to increase their EPS production and decrease their cell size to raise the probability of biofilm survival in case of nutrient depletion during aging of a biofilm (Lin *et al.* 2014). Also, during periods of low humidity, hydrophilic EPS enables water storage needed for survival. These survival strategies allow to increase the bacterial number by cell division while conserving the total amount of biomass (James *et al.* 1995). While during the early stages of biofilm formation (12–24 hours), large amounts of eDNA and polysaccharides are required to initiate biofilm formation (Jackson *et al.* 2004; Whitchurch *et al.* 2002), during biofilm aging the major EPS components, such as polysaccharides, proteins and eDNA are accumulated and possibly become a "back up" nutrition source (Lin *et al.* 2014). In this phase of biofilm growth, eDNA helps to maintain the structural integrity of the biofilm through binding with other biofilm components and to protect its inhabitants against chemical and mechanical stress (Das *et al.* 2013).

The chemical composition of EPS, and accordingly its physical properties, differs from place to place in a biofilm of the same strain. Growth medium, temperature, relative humidity, nutrition, pH and ionic strength have a major influence on EPS composition and physical properties and therewith functions. EPS production is stimulated by low concentrations of nutrients because the matrix components can be used by the biofilm as a source of carbon, nitrogen and phosphorous in case of nutrient depletion (Jung *et al.* 2013). Low pH, high temperature and osmotic stress increase the EPS production, especially its protein content. This increases the resistance of *P. aeruginosa* in their biofilm mode of growth against heat stress and temperature-induced lysis (Lin *et al.* 2014).

4.5.2.1 EPS and resistance of biofilms against mechanical attack

In most cases, biofilms do not grow in a static environment, but under conditions of fluid flow. Fluid flow and resulting shear forces influence bacterial adhesion and biofilm formation (Brink & Nicol, 2014; Skovager *et al.* 2012), affecting deposition kinetics, architecture, dynamic behavior, physiological reaction and metabolic activities of a biofilm (Wang *et al.* 2014). EPS shows a different distribution and composition under different fluid shears in order to allow a bacterial community to persist under the environmental conditions of its formation. Biofilms growing under high flow rates appear to be thinner and more homogeneous (Kim *et al.* 2012) than biofilms formed under low flow rates. Also the prevalence of polysaccharides, proteins and TB-EPS components is increased in biofilms grown under fluid shear compared to its prevalence in biofilms grown under static conditions in order to provide more cohesion to the biofilm.

EPS provides viscoelastic properties to a biofilm (Klapper *et al.* 2002; Guelon *et al.* 2011), that are pivotal for its survival under mechanical attack, such as experienced by oral biofilms during powered toothbrushing. Powered toothbrushing, particularly in absence of direct contact between the bristles and a biofilm, can be considered as a form of transmitting energy into the biofilm (Busscher *et al.* 2010a), upon which a biofilm expands. As for all viscoelastic materials, when expansion extends above the elastic limit of the biofilm, plastic deformation commences. Biofilm detachment subsequently results when expansion and plastic deformation continue, but when brushing is arrested while in the plastic range, a "fluffed-up", very open biofilm will remain behind, i.e. the biofilm survives due to its viscoelastic properties (Rmaile *et al.* 2014).

4.5.2.2 EPS and resistance of biofilms against chemical attack

Bacteria in their biofilm mode of growth have higher resistance to antimicrobials compared to their planktonic counterparts. EPS producing strains exhibit generally higher resistance against antibiotics than non-slime producing ones (Davenport *et al.* 2014), as EPS reduces the penetration of antibiotics into the deeper layers of a biofilm and limits absorption of antibiotics to the outer layer of a biofilm. For *Acinetobacter baumannii* and *Staphylococcus aureus* this protection has been largely attributed to polysaccharides and lipids, because the biofilm resistance to antibiotics was not affected by selective removal of proteins or DNA from an EPS matrix (Davenport *et al.* 2014). The chemical resistance of biofilms is usually illustrated as resulting from a combination of structure and composition, but evidence is based on cartoons and at best highly subjective microscopic images. Recently, the viscoelastic properties of biofilms have been advocated for use to reflect structure and composition of biofilm in a more quantitative way (Peterson *et al.* 2014). The kinetics of stress relaxation of a biofilm after deformation showed that three processes mediate relaxation: (1) fast flow of water in a deformed biofilm, (2) intermediately flow of more viscous EPS and (3) slow repositioning of bacteria provided sufficiently open space is available. Interestingly, biofilms that were demonstrated to be more open by stress relaxation analysis facilitated better penetration of antimicrobials (He *et al.* 2013).

4.5.2.3 Lubricating properties of EPS in biofilms

Biofilms are slippery due to their EPS matrix (Flemming, 2011). Whether or not the slipperiness of biofilms serves any physiological role or not is unknown. Equally unknown is whether the lubricating properties of biofilm EPS, or for that matter bacteria themselves as ball-bearers, will ever find application. Yet, the lubricating properties of biofilms are another reflection of their structure and composition. Tribology measurements with AFM showed that biofilms on stainless steel of a slime producing *S. aureus* strain (*S. aureus* ATCC 12600) decreased the coefficient of friction (COF) of stainless steel to a much larger extent from 0.6 ± 0.3 to 0.1 ± 0.1 than the non-slime producing strain *S. aureus* 5298 (COF equals 0.3 ± 0.1).

4.6 METHODS TO STUDY BIOFILM COMPOSITION AND STRUCTURE

Considering the importance of biofilm structure and composition in resisting mechanical and chemical attacks, we will now briefly summarize different methods to investigate these properties.

4.6.1 Microscopic structure of biofilms

Using confocal laser scanning microscopy (CLSM), structural information can be acquired in fully hydrated biofilms, while also the distribution, shape, heterogeneity and structure of EPS can be observed (Alrhmoun *et al.* 2014). In order to provide an image, appropriate fluorescent dyes for EPS and bacteria are necessary. The staining procedure, however, causes two major limitations of CLSM imaging: (1) A thick biofilm cannot be imaged due to the insufficient penetration of dyes and (2) The biofilm cannot be reused for other studies after CLSM imaging because staining procedures cause irreversible harm to the bacteria.

As a powerful and convenient imaging tool, CLSM can be combined with for example, 3D-finite element analysis which can be constructed based on stacks of CLSM images (Bol *et al.* 2009). With the help of computer modeling, the signal pixels can be connected to each other to form a net structure, and subsequently a 3D biofilm model can be reconstructed from a series of 2D signals. The distribution and structure of EPS can be studied with fluorescence lectin-binding analysis (FLBA) (Zhang *et al.* 2014; Zippel & Neu, 2011). FLBA is a non-destructive and *in situ* EPS analysis method since the chemical dyes used in CLSM imaging are replaced by fluorescence labeled lectin which has negligible influence on properties and functions of biofilms.

Biofilm structure can also be analyzed by optical coherence tomography (OCT) which is a completely non-destructive *in situ* technique (Dreszer *et al.* 2014).

4.6.2 Composition of biofilms

Fourier transform infrared spectroscopy (FTIR) and Raman spectroscopy, amongst other methods, allow determination of the chemical composition of biofilms grown on internal reflection elements e.g. a germanium crystal placed in a flow chamber. Both techniques provide spectra showing the quantitative prevalence of specific molecular groups in a hydrated biofilm in a non-destructive way (Elzinga *et al.* 2012; Quilès & Humbert, 2014; Zalewska-Piatek *et al.* 2013). Extracellular substances such as polysaccharides, proteins, DNA and water, can be identified from the peaks in FTIR or Raman spectra, based on their specific peak wavenumber. Peak areas provide quantitative information on the amount of each extracellular substance. Spectra acquired at different stages of biofilm formation can be illustrated in a stack mode to show the variation of biofilm chemical composition during biofilm development.

Raman spectroscopy gives more or less similar information about biofilms as FTIR. Compared to FTIR spectroscopy however, Raman spectroscopy has several advantages: (1) a weaker signal of water and CO_2,

(2) wider spectral range especially for wavenumbers below 400 cm^{-1} and (3) stronger signal for symmetric bonds (e.g. —S—S—, —C—S—, —C≡C—) (Choo-Smith *et al.* 2001; Krafft & Sergo, 2006; Tang *et al.* 2013).

4.6.3 Viscoelastic properties of biofilms

Viscoelastic properties of biofilms can be determined under shear, tensile or compressive conditions. Low load compression testing (LLCT) is one of the methods to determine stress relaxation after compression (Peterson *et al.* 2014). Stress relaxation of biofilms can be modeled using three Maxwell elements representing the flow of water, more viscous EPS and bacteria repositioning in a deformed biofilm. Bacterial repositioning is more prominent in "fluffed up" biofilms than in more condensed ones. The main advantages of stress relaxation to determine biofilm structure and composition are that (1) it yields quantitative data and (2) the plunger applied e.g. in LLCT covers an area of several square millimeters, while microscopic imaging can cover only a small fraction of this, often user-selected. A cone/plate rheometer is another method to determine the viscoelastic properties of a biofilm. It basically consists of a rotatable cone on top and a fixed plate at the bottom, with viscoelastic samples, e.g. biofilms in the middle. The cone can be set into a rotating or oscillating mode while sensors at the bottom plate measure the viscous or elastic property of the samples, respectively. However, when using rheological methods, biofilms are regarded as homogeneous, soft materials, and output parameters represent average values of the biomass as a whole (Wloka *et al.* 2006). Information on possible micro-scaled structural heterogeneities in a biofilm is neglected. To address this drawback, Hohne *et al.* (2009) developed a flexible microfluidic device in order to study the micro-scaled elastic moduli and the stress relaxation of specified parts in a biofilm. Biofilm grown on a silicon wafer were covered with a thin poly(dimethyl siloxane) film and deformation of the poly(dimethyl siloxane) film was induced with a controlled air pressure. The strain and stress relaxation time of biofilms were evaluated based on the deformation of the poly(dimethyl siloxane) film.

4.6.4 Lubricating properties of biofilms with and without EPS

Biofilm tribology is still in its infancy (Ahimou *et al.* 2007; Souza *et al.* 2010). Coefficients of friction of biofilms can be measured by using a sliding wear tester. In a sliding wear test, a fixed stainless steel roller with a typical diameter of 4 mm is slided over a biofilm and the friction force generated is detected by a sensor connected to the roller-holder. On a more microscopic level, colloidal-probe AFM can be applied and such COF measurements are based on the torsion of the cantilever upon the friction between a biofilm and a silica ball, with a typical diameter of 30 μm. Often however, macroscopic and microscopic friction measurements of the COF of biofilms yield different results in terms of standard deviations and values found, as shown in Figure 4.3. The larger standard deviations observed with AFM are due to the microscopic nature of this technique, which is highly sensitive to small heterogeneities of both biofilms and the underlying surface, in contrast to the sliding wear tester which measures over a much larger surface area. The EPS producing strain (*S. aureus* ATCC 12600) shows in both cases the lowest coefficient of friction compared to the non-EPS producing strain (*S. aureus* 5298).

4.7 CONCLUDING COMMENTS

In this chapter, we reviewed the main steps in biofilm formation, starting from bacterial adhesion to surfaces. Upon initial adhesion of bacteria to a surface, the forces between a bacterium and the surface play an essential role in inducing nanoscale deformation of the bacterial cell wall including the lipid membrane. Membrane deformation eventually allows communication between an adhering bacterium and its external environment, making it aware of its adhering state. One of the main bacterial responses to surface sensing, and hallmark of the transition between planktonic and sessile phenotypes, is the production of EPS that form the biofilm matrix. The importance of EPS is ubiquitously present throughout the entire process of biofilm development, from facilitating initial bacterial adhesion to maintaining biofilm structural integrity during growth and offering "back up" resources in case of nutrient depletion during biofilm aging. During the entire biofilm life cycle, EPS protects the biofilm against chemical attacks such as from antibiotic treatment and through its viscoelastic properties, against mechanical external attacks.

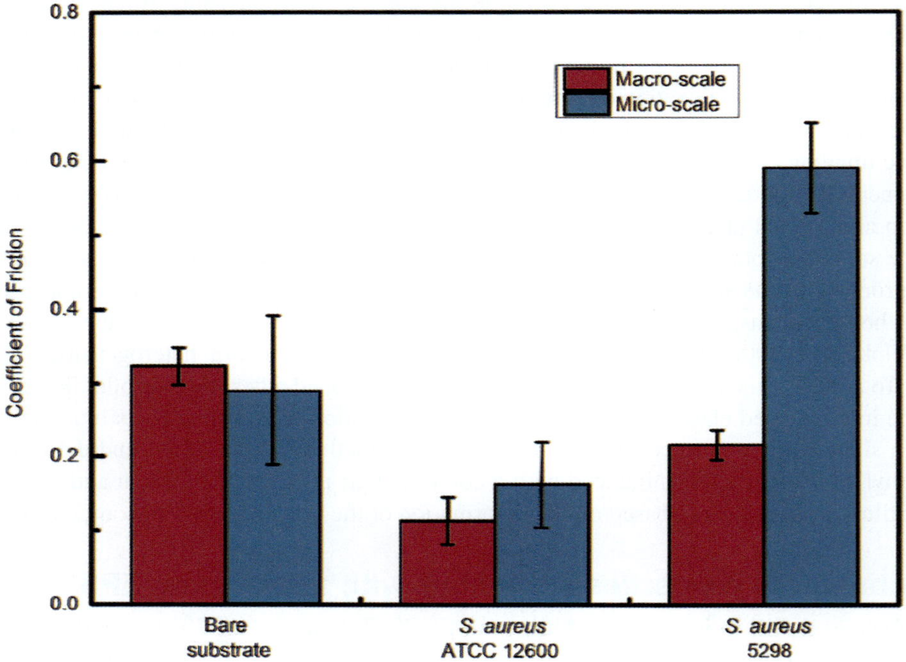

Figure 4.3 Coefficients of friction of staphylococcal biofilms on polyethylene of a EPS producing (*S. aureus* ATCC 12600) and non-EPS producing staphylococcal strain (*S. aureus* 5298), measured by a macroscopic sliding wear tester and microscopic AFM.

REFERENCES

Abdallah M., Benoliel C., Drider D., Dhulster P. and Chihib N. E. (2014). Biofilm formation and persistence on abiotic surfaces in the context of food and medical environments. *Archives of Microbiology*, **196**(7), 453–472.

Absolom D. R., Lamberti F. V., Policova Z., Zingg W., Van Oss C. J. and Neumann A. W. (1983). Surface thermodynamics of bacterial adhesion. *Applied and Environmental Microbiology*, **46**(1), 90–97.

Ahimou F., Semmens M. J., Novak P. J. and Haugstad G. (2007). Biofilm cohesiveness measurement using a novel atomic force microscopy methodology. *Applied and Environmental Microbiology*, **73**(9), 2897–2904.

Alrhmoun M., Carrion C., Casellas M. and Dagot C. (2014). Evaluation of the extracellular polymeric substances by confocal laser scanning microscopy in conventional activated sludge and advanced membrane bioreactors treating hospital wastewater. *Water Science and Technology: A Journal of the International Association on Water Pollution Research*, **69**(11), 2287–2294.

Aoki S. K., Webb J. S., Braaten B. A. and Low D. A. (2009). Contact-dependent growth inhibition causes reversible metabolic downregulation in *Escherichia coli*. *Journal of Bacteriology*, **191**(6), 1777–1786.

Arciola C. R., Campoccia D., Ravaioli S. and Montanaro L. (2015). Polysaccharide intercellular adhesin in biofilm: structural and regulatory aspects. *Frontiers in Cellular and Infection Microbiology*, **5**(7), 1–10.

Belas R. (2013). When the swimming gets tough, the tough form a biofilm. *Molecular Microbiology*, **90**(1), 1–5.

Belas R. (2014). Biofilms, flagella, and mechanosensing of surfaces by bacteria. *Trends in Microbiology*, **22**(9), 517–527.

Beloin C., Renard S., Ghigo J. M. and Lebeaux D. (2014). Novel approaches to combat bacterial biofilms. *Current Opinion in Pharmacology*, **18**, 61–68.

Bol M., Mohle R. B., Haesner M., Neu T. R., Horn H. and Krull R. (2009). 3D finite element model of biofilm detachment using real biofilm structures from CLSM data. *Biotechnology and Bioengineering*, **103**(1), 177–186.

Bolshakova A. V., Kiselyova O. I., Filonov A. S., Frolova O. Y., Lyubchenko Y. L. and Yaminsky I. V. (2001). Comparative studies of bacteria with an atomic force microscopy operating in different modes. *Ultramicroscopy*, **86**(1–2), 121–128.

Brink H. G. and Nicol W. (2014). The influence of shear on the metabolite yield of *Lactobacillus rhamnosus* biofilms. *New Biotechnology*, **31**(5), 460–467.

Busscher H. J. and Van der Mei H. C. (2012). How do bacteria know they are on a surface and regulate their response to an adhering state? *PLoS Pathogens*, **8**(8), e1002821.

Busscher H. J., Jager D., Finger G., Schaefer N. and Van der Mei H. C. (2010a). Energy transfer, volumetric expansion, and removal of oral biofilms by non-contact brushing. *European Journal of Oral Sciences*, **118**(2), 177–182.

Busscher H. J., Norde W., Sharma P. K. and Van der Mei H. C. (2010b). Interfacial re-arrangement in initial microbial adhesion to surfaces. *Current Opinion in Colloid & Interface Science*, **15**(6), 510–517.

Cairns L. S., Marlow V. L., Bissett E., Ostrowski A. and Stanley-Wall N. R. (2013). A mechanical signal transmitted by the flagellum controls signalling in *Bacillus subtilis*. *Molecular Microbiology*, **90**(1), 6–21.

Chauhan A., Bernardin A., Mussard W., Kriegel I., Esteve M., Ghigo J. M., Beloin C. and Semetey V. (2014). Preventing biofilm formation and associated occlusion by biomimetic glycocalyxlike polymer in central venous catheters. *The Journal of Infectious Diseases*, **210**(9), 1347–1356.

Chen Y., Busscher H. J., Van der Mei H. C. and Norde W. (2011). Statistical analysis of long- and short-range forces involved in bacterial adhesion to substratum surfaces as measured using atomic force microscopy. *Applied and Environmental Microbiology*, **77**(15), 5065–5070.

Chen Y. P. Zhang P., Guo J. S., Fang F., Gao X. and Li C. (2013). Functional groups characteristics of EPS in biofilm growing on different carriers. *Chemosphere*, **92**(6), 633–638.

Chen Y., Harapanahalli A. K., Busscher H. J., Norde W. and Van der Mei H. C. (2014). Nanoscale cell wall deformation impacts long-range bacterial adhesion forces on surfaces. *Applied and Environmental Microbiology*, **80**(2), 637–643.

Choo-Smith L.-P., Maquelin K., Van Vreeswijk T., Bruining H. A., Puppels G. J., Ngo Thi N. A., Kirschner C., Naumann D., Ami D., Villa A. M., Orsini F., Doglia S. M., Lamfarraj H., Sockalingum G. D., Manfait M., Allouch P. and Endtz H. P. (2001). Investigating microbial (micro)colony heterogeneity by vibrational spectroscopy. *Applied and Environmental Microbiology*, **67**(4), 1461–1469.

Costerton J. W. and Stewart P. S. (2001). Battling biofilms. *Scientific American*, **285**(1), 74–81.

Dang H. and Lovell C. R. (1999). Bacterial primary colonization and early succession on surfaces in marine waters as determined by amplified rRNA gene restriction analysis and sequence analysis of 16S rRNA genes. *Applied and Environmental Microbiology*, **66**(2), 467–475.

Das T., Sehar S. and Manefield M. (2013). The roles of extracellular DNA in the structural integrity of extracellular polymeric substance and bacterial biofilm development. *Environmental Microbiology Reports*, **5**(6), 778–786.

Davenport E. K., Call D. R. and Beyenal H. (2014). Differential protection from tobramycin by extracellular polymeric substances from *Acinetobacter baumannii* and *Staphylococcus aureus* biofilms. *Antimicrobial Agents and Chemotherapy*, **58**(8), 4755–4761.

Davies D. G. and Geesey G. G. (1995). Regulation of the alginate biosynthesis gene algC in *Pseudomonas aeruginosa* during biofilm development in continuous culture. *Applied and Environmental Microbiology*, **61**(3), 860–867.

Davies D. G., Chakrabarty A. M. and Geesey G. G. (1993). Exopolysaccharide production in biofilms: substratum activation of alginate gene expression by *Pseudomonas aeruginosa*. *Applied and Environmental Microbiology*, **59**(4), 1181–1186.

Dreszer C., Wexler A. D., Drusová S., Overdijk T., Zwijnenburg A., Flemming H.-C., Kruithof J. C. and Vrouwenvelder J. S. (2014). In-situ biofilm characterization in membrane systems using optical coherence tomography: formation, structure, detachment and impact of flux change. *Water Research*, **67**, 243–254.

Dufrene Y. F. (2003). Recent progress in the application of atomic force microscopy imaging and force spectroscopy to microbiology. *Current Opinion in Microbiology*, **6**(3), 317–323.

Dufrene Y. F. (2008). Towards nanomicrobiology using atomic force microscopy. *Nature Reviews Microbiology*, **6**(9), 674–680.

Elzinga E. J., Huang J. H., Chorover J. and Kretzschmar R. (2012). ATR-FTIR spectroscopy study of the influence of pH and contact time on the adhesion of *Shewanella putrefaciens* bacterial cells to the surface of hematite. *Environmental Science & Technology*, **46**(23), 12848–12855.

Flemming H. C. (2011). The perfect slime. *Colloids and Surfaces. B, Biointerfaces*, **86**(2), 251–259.

Flemming H. C. and Wingender J. (2010). The biofilm matrix. *Nature Reviews Microbiology*, **8**(9), 623–633.

Geng J., Beloin C., Ghigo J. M. and Henry N. (2014). Bacteria hold their breath upon surface contact as shown in a strain of *Escherichia coli*, using dispersed surfaces and flow cytometry analysis. *PloS One*, **9**(7), e102049.

Gomes L. C., Silva L. N., Simoes M., Melo L. F. and Mergulhao F. J. (2015). *Escherichia coli* adhesion, biofilm development and antibiotic susceptibility on biomedical materials. *Journal of Biomedical Materials Research Part A*, **103**(4), 1414–1423.

Guelon T., Mathias J.-D. and Stoodley P. (2011). Advances in biofilm mechanics. In: Biofilm Highlights, H.-C. Flemming, J. Wingender and U. Szewzyk (eds.), Springer, Heidelberg, Germany, pp. 111–139.

Harapanahalli A. K., Chen Y., Li J., Busscher H. J. and Van der Mei H. C. (2015a). Influence of adhesion force on *icaA* and *cidA* gene expression and production of matrix components in *Staphylococcus aureus* biofilms. *Applied and Environmental Microbiology*, **81**(10), 3369–3378.

Harapanahalli A. K., Younes J. A., Allan E., Van der Mei H. C. and Busscher H. J. (2015b). Chemical signals and mechanosensing in bacterial responses to their environment. *PLoS Pathogens*, **11**(8), e1005057.

He Y., Peterson B. W., Jongsma M. A., Ren Y., Sharma P. K., Busscher H. J. and Van der Mei H. C. (2013). Stress relaxation analysis facilitates a quantitative approach towards antimicrobial penetration into biofilms. *Plos One*, **8**(5), e63750.

Heinisch J. J., Lipke P. N., Beaussart A., El Kirat Chatel S., Dupres V., Alsteens D. and Dufrene Y. F. (2012). Atomic force microscopy - looking at mechanosensors on the cell surface. *Journal of Cell Science*, **125**(Pt 18), 4189–4195.

Hermansson M. (1999). The DLVO theory in microbial adhesion. *Colloids and Surfaces B: Biointerfaces*, **14**(1–4), 105–119.

Hohne D. N., Younger J. G. and Solomon M. J. (2009). Flexible microfluidic device for mechanical property characterization of soft viscoelastic solids such as bacterial biofilms. *Langmuir*, **25**(13), 7743–7751.

Hook A. L., Chang C. Y., Yang J., Luckett J., Cockayne A., Atkinson S., Mei Y., Bayston R., Irvine D. J., Langer R., Anderson D. G., Williams P., Davies M. C. and Alexander M. R. (2012). Combinatorial discovery of polymers resistant to bacterial attachment. *Nature Biotechnology*, **30**(9), 868–875.

Iscla I. and Blount P. (2012). Sensing and responding to membrane tension: the bacterial MscL channel as a model system. *Biophysical Journal*, **103**(2), 169–174.

Jackson K. D., Starkey M., Kremer S., Parsek M. R. and Wozniak D. J. (2004). Identification of *psl*, a locus encoding a potential exopolysaccharide that is essential for *Pseudomonas aeruginosa* PAO1 biofilm formation. *Journal of Bacteriology*, **186**(14), 4466–4475.

James G. A., Korber D. R., Caldwell D. E. and Costerton J. W. (1995). Digital image analysis of growth and starvation responses of a surface-colonizing *Acinetobacter sp. Journal of Bacteriology*, **177**(4), 907–915.

Janissen R., Murillo D. M., Niza B., Sahoo P. K., Nobrega M. M., Cesar C. L., Temperini M. L., Carvalho H. F., de Souza A. A. and Cotta M. A. (2015). Spatiotemporal distribution of different extracellular polymeric substances and filamentation mediate *Xylella fastidiosa* adhesion and biofilm formation. *Scientific Reports*, **5**, 9856.

Jucker B. A., Harms H. and Zehnder A. J. (1996). Adhesion of the positively charged bacterium *Stenotrophomonas (xanthomonas) maltophilia* 70401 to glass and teflon. *Journal of Bacteriology*, **178**(18), 5472–5479.

Jung J. H., Choi N. Y. and Lee S. Y. (2013). Biofilm formation and exopolysaccharide (EPS) production by *Cronobacter sakazakii* depending on environmental conditions. *Food Microbiology*, **34**(1), 70–80.

Kailas L., Ratcliffe E. C., Hayhurst E. J., Walker M. G., Foster S. J. and Hobbs J. K. (2009). Immobilizing live bacteria for AFM imaging of cellular processes. *Ultramicroscopy*, **109**(7), 775–780.

Karatan E. and Watnick P. (2009). Signals, regulatory networks, and materials that build and break bacterial biofilms. *Microbiology and Molecular Biology Reviews*, **73**(2), 310–347.

Keller R., Arioz C., Hansmeier N., Stenberg-Bruzell F., Burstedt M., Vikstrom D., Kelly A., Wieslander A., Daley D. O. and Hunke S. (2015). The *Escherichia coli* envelope stress sensor CpxA responds to changes in lipid bilayer properties. *Biochemistry*, **54**(23), 3670–3676.

Kim J., Park H. and Chung S. (2012). Microfluidic approaches to bacterial biofilm formation. *Molecules*, **17**(8), 9818–9834.

Klapper I., Rupp C. J., Cargo R., Purvedorj B. and Stoodley P. (2002). Viscoelastic fluid description of bacterial biofilm material properties. *Biotechnology and Bioengineering*, **80**(3), 289–296.

Krafft C. and Sergo V. (2006). Biomedical applications of Raman and infrared spectroscopy to diagnose tissues. *Spectroscopy*, **20**(5–6), 195–218.

Kwok D. Y., Lin R., Mui M. and Neumann A. W. (1996). Low-rate dynamic and static contact angles and the determination of solid surface tensions. *Colloids and Surfaces A: Physicochemical and Engineering Aspects*, **116**(1), 63–77.

Le Moal E., Fort E., Lévêque-Fort S., Cordelières F. P., Fontaine-Aupart M. P. and Ricolleau C. (2007). Enhanced fluorescence cell imaging with metal-coated slides. *Biophysical Journal*, **92**(6), 2150–2161.

Lee K., Hahn L. D., Yuen W. W., Vlamakis H., Kolter R. and Mooney D. J. (2011). Metal-enhanced fluorescence to quantify bacterial adhesion. *Advanced Materials*, **23**(12), H101–H104.

Lee K. W. K., Periasamy S., Mukherjee M, Xie C, Kjelleberg S. and Rice S. A. (2014). Biofilm development and enhanced stress resistance of a model, mixed species community biofilm. *International Society for Microbial Ecology Journal*, **8**, 894–907.

Li G., Brown P. J., Tang J. X., Xu J., Quardokus E. M., Fuqua C. and Brun Y. V. (2012). Surface contact stimulates the just-in-time deployment of bacterial adhesins. *Molecular Microbiology*, **83**(1), 41–51.

Li J., Busscher H. J., Swartjes J. J. T. M., Chen Y., Harapanahalli A. K., Norde W., Van der Mei H. C. and Sjollema J. (2014). Residence-time dependent cell wall deformation of different *Staphylococcus aureus* strains on gold measured using surface-enhanced-fluorescence. *Soft Matter*, **10**(38), 7638–7646.

Li M., Gan C., Shao W., Yu C., Wang X. and Chen Y. (2015). Effects of membrane lipid composition and antibacterial drugs on the rigidity of *Escherichia coli*: different contributions of various bacterial substructures. *Scanning*, **38**(1), 70–79.

Lin H., Chen G., Long D. and Chen X. (2014). Responses of unsaturated *Pseudomonas putida* CZ1 biofilms to environmental stresses in relation to the EPS composition and surface morphology. *World Journal of Microbiology & Biotechnology*, **30**, 3081–3090.

Liu Y., Strauss J. and Camesano T. A. (2008). Adhesion forces between *Staphylococcus epidermidis* and surfaces bearing self-assembled monolayers in the presence of model proteins. *Biomaterials*, **29**(33), 4374–4382.

Lower S. K., Yongsunthon R., Casillas-Ituarte N. N., Taylor E. S., DiBartola A. C., Lower B. H., Beveridge T. J., Buck A. W. and Fowler V. G. J. (2010). A tactile response in *Staphylococcus aureus*. *Biophysical Journal*, **99**(9), 2803–2811.

Ludecke C., Jandt K. D., Siegismund D., Kujau M. J., Zang E., Rettenmayr M., Bossert J. and Roth M. (2014). Reproducible biofilm cultivation of chemostat-grown *Escherichia coli* and investigation of bacterial adhesion on biomaterials using a non-constant-depth film fermenter. *PloS One*, **9**(1), e84837.

Mah, T.-F. (2012). Biofilm-specific antibiotic resistance. *Future Microbiology*, **7**(9), 1061–1072.

Nejadnik M. R., Van der Mei H. C., Norde W. and Busscher H. J. (2008). Bacterial adhesion and growth on a polymer brush-coating. *Biomaterials*, **29**(30), 4117–4121.

Ntziachristos V., Ripoll J., Wang L. V. and Weissleder R. (2005). Looking and listening to light: the evolution of whole-body photonic imaging. *Nature Biotechnology*, **23**(3), 313–320.

Olsen, I. (2015). Biofilm-specific antibiotic tolerance and resistance. *European Journal of Clinical Microbiology & Infectious Diseases*, **34**(5), 877–886.

Otto K. and Silhavy T. J. (2002). Surface sensing and adhesion of *Escherichia coli* controlled by the Cpx-signaling pathway. *Proceedings of the National Academy of Sciences of the United States of America*, **99**(4), 2287–2292.

Owens D. K. and Wendt R. C. (1969). Estimation of the surface free energy of polymers. *Journal of Applied Polymer Science*, **13**, 1741–1747.

Pellicer-Nacher C., Domingo-Felez C., Mutlu A. G. and Smets B. F. (2013). Critical assessment of extracellular polymeric substances extraction methods from mixed culture biomass. *Water Research*, **47**(15), 5564–5574.

Pen Y., Zhang Z. J., Morales-Garcia A. L., Mears M., Tarmey D. S., Edyvean R. G., Banwart S. A. and Geoghegan M. (2014). Effect of extracellular polymeric substances on the mechanical properties of *Rhodococcus*. *Biochimica et Biophysica Acta*, **1848**(2), 518–526.

Perozo E. and Rees D. C. (2003). Structure and mechanism in prokaryotic mechanosensitive channels. *Current Opinion in Structural Biology*, **13**(4), 432–442.

Perozo E., Kloda A., Cortes D. M. and Martinac B. (2002). Physical principles underlying the transduction of bilayer deformation forces during mechanosensitive channel gating. *Nature Structural Biology*, **9**(9), 696–703.

Peterson B. W., Busscher H. J., Sharma P. K. and Van der Mei H. C. (2014). Visualization of microbiological processes underlying stress relaxation in *Pseudomonas aeruginosa* biofilms. *Microscopy and Microanalysis*, **20**(3), 912–915.

Pillet F., Chopinet L., Formosa C. and Dague E. (2014). Atomic force microscopy and pharmacology: from microbiology to cancerology. *Biochimica et Biophysica Acta*, **1840**(3), 1028–1050.

Powl A. M., East J. M. and Lee A. G. (2008). Anionic phospholipids affect the rate and extent of flux through the mechanosensitive channel of large conductance MscL. *Biochemistry*, **47**(14), 4317–4328.

Quilès F. and Humbert F. (2014). On the production of glycogen by *Pseudomonas fluorescens* during biofilm development: an *in situ* study by attenuated total reflection-infrared with chemometrics. *Biofouling*, **30**(6), 709–718.

Rmaile A., Carugo D., Capretto L., Aspiras M., De Jager M., Ward M. and Stoodley P. (2014). Removal of interproximal dental biofilms by high-velocity water microdrops. *Journal of Dental Research*, **93**(1), 68–73.

Rodriguez-Martinez J. M. and Pascual A. (2006). Antimicrobial resistance in bacterial biofilms. *Reviews in Medical Microbiology*, **17**(3), 65–75.

Sahoo S., Rao K. K. and Suraishkumar G. K. (2006). Reactive oxygen species induced by shear stress mediate cell death in *Bacillus subtilis*. *Biotechnology and Bioengineering*, **94**(1), 118–127.

Skovager A., Whitehead K., Siegumfeldt H., Ingmer H., Verran J. and Arneborg N. (2012). Influence of flow direction and flow rate on the initial adhesion of seven *Listeria monocytogenes* strains to fine polished stainless steel. *International Journal of Food Microbiology*, **157**(2), 174–181.

Souza J. C., Henriques M., Oliveira R., Teughels W., Celis J. P. and Rocha L. A. (2010). Biofilms inducing ultra-low friction on titanium. *Journal of Dental Research*, **89**(12), 1470–1475.

Stoodley P., Sauer K., Davies D. and Costerton J. W. (2002). Biofilms as complex differentiated communities. *Annual Reviews in Microbiology*, **56**(1), 187–209.

Tang M., McEwen G. D., Wu Y., Miller C. D. and Zhou A. (2013). Characterization and analysis of mycobacteria and Gram-negative bacteria and co-culture mixtures by Raman microspectroscopy, FTIR, and atomic force microscopy. *Analytical and Bioanalytical Chemistry*, **405**(5), 1577–1591.

Tansel B. and Tansel D. Z. (2013). Adhesion strength and spreading characteristics of EPS on membrane surfaces during lateral and central growth. *Colloids and Surfaces B: Biointerfaces*, **111**, 594–599.

Torrent-Burgues J. and Sanz F. (2014). AFM in mode peak force applied to the study of un-worn contact lenses. *Colloids and Surfaces B:Biointerfaces*, **121**, 388–394.

Touhami A., Jericho M. H. and Beveridge T. J. (2004). Atomic force microscopy of cell growth and division in *Staphylococcus aureus*. *Journal of Bacteriology*, **186**(11), 3286–3295.

Turner R. D., Hurd A. F., Cadby A., Hobbs J. K. and Foster S. J. (2013). Cell wall elongation mode in gram-negative bacteria is determined by peptidoglycan architecture. *Nature Communications*, **4**, 1496.

Tuson H. H. and Weibel D. B. (2013). Bacteria-surface interactions. *Soft Matter*, **9**(18), 4368–4380.

Van Loosdrecht M. C., Lyklema J., Norde W., Zehnder A. J. (1990). Influence of interfaces on microbial activity. *Microbiological Reviews*, **54**(1), 75–87.

Van Oss C. J. (1993). Acid—base interfacial interactions in aqueous media. *Colloids and Surfaces A: Physicochemical and Engineering Aspects*, **78**, 1–49.

Van Oss C. J. (1995). Hydrophobicity of biosurfaces—origin, quantitative determination and interaction energies. *Colloids and Surfaces B: Biointerfaces*, **5**(3–4), 91–110.

Vilain S., Pretorius J. M., Theron J. and Brozel V. S. (2009). DNA as an adhesin: *Bacillus cereus* requires extracellular DNA to form biofilms. *Applied and Environmental Microbiology*, **75**(9), 2861–2868.

Vogt S. L. and Raivio T. L. (2012). Just scratching the surface: an expanding view of the cpx envelope stress response. *FEMS Microbiology Letters*, **326**(1), 2–11.

Wang C., Miao L., Hou J., Wang P., Qian J. and Dai S. (2014). The effect of flow velocity on the distribution and composition of extracellular polymeric substances in biofilms and the detachment mechanism of biofilms. *Water Science and Technology: A Journal of the International Association on Water Pollution Research*, **69**(4), 825–832.

Whitchurch C. B., Tolker-Nielsen T., Ragas P. C. and Mattick J. S. (2002). Extracellular DNA required for bacterial biofilm formation. *Science*, **295**(5559), 1487.

Wloka M., Rehage H., Flemming H.-C. and Wingender J. (2006). Structure and rheological behaviour of the extracellular polymeric substance network of mucoid *Pseudomonas aeruginosa* biofilms. *Biofilms*, **2**, 275–283.

Yeroslavsky G., Girshevitz O., Foster-Frey J., Donovan D. M. and Rahimipour S. (2015). Antibacterial and antibiofilm surfaces through polydopamine-assisted immobilization of lysostaphin as an antibacterial enzyme. *Langmuir*, **31**(3), 1064–1073.

Zalewska-Piatek B., Wilkanowicz S., Bruzdziak P., Piatek R. and Kur J. (2013). Biochemical characteristic of biofilm of uropathogenic *Escherichia coli* dr(+) strains. *Microbiological Research*, **168**(6), 367–378.

Zhang W., Gao Z., Shao D., Zhang L., Wang C. and Zhang Y. (2013). Atomic force microscopy analysis of progenitor corneal epithelial cells fractionated by a rapid centrifugation isolation technique. *PloS One*, **8**(3), e59282.

Zhang R. Y., Neu T. R., Bellenberg S., Kuhlicke U., Sand W. and Vera M. (2014). Use of lectins to in situ visualize glycoconjugates of extracellular polymeric substances in acidophilic archaeal biofilms. *Microbial Biotechnology*.

Zippel B. and Neu T. R. (2011). Characterization of glycoconjugates of extracellular polymeric substances in tufa-associated biofilms by using fluorescence lectin-binding analysis. *Applied and Environmental Microbiology*, **77**(2), 505–516.

Zobell C. E. (1943). The effect of solid surfaces upon bacterial activity. *Journal of Bacteriology*, **46**(1), 39–56.

Chapter 5

Genetics and regulation of EPS formation in *Pseudomonas aeruginosa*

Matthew J. Pestrak, Heather C. Eggleston and Daniel J. Wozniak

Dep. Microbial Infection and Immunity, Microbiology, Center for Microbial Interface Biology, The Ohio State University, 704 BRT, 460 West 12th Avenue. Columbus, OH, 43210, USA

ABSTRACT

The Gram-negative opportunistic pathogen *Pseudomonas aeruginosa* is of increasing clinical importance due to its involvement in a variety of acute and chronic infections. One of the key characteristics that allow *P. aeruginosa* to colonize and persist within the host is its ability to form a biofilm. The biofilm extracellular matrix is composed of exopolysaccharides, extracellular DNA, and proteins. *P. aeruginosa* is considered a model organism of study for biofilm development and this organism possesses a complex regulatory network to control biofilm formation. The regulation of the exopolysaccharides in particular has been studied in detail, as they play critical roles in biofilm development, structure, and resistance to antimicrobials and environmental stress. Due to the clinical and scientific importance of exopolysaccharides in biofilm formation, this chapter reviews the current knowledge concerning the structure, synthesis, and regulation of three *P. aeruginosa* exopolysaccharides, Pel, Psl, and alginate.

Keywords: *Pseudomonas aeruginosa*, biofilm, Psl synthesis, Pellicle formation, alginate

5.1 INTRODUCTION

Pseudomonas aeruginosa is a Gram-negative opportunistic pathogen that is highly capable of adapting to a range of environments due to its diverse genetic tool set. Sequencing of the *P. aeruginosa* genome revealed that it contains one of the highest proportions of transcriptional regulators yet observed in a bacterial species (Stover *et al.* 2000; Galán-Vásquez *et al.* 2011). Consequently *P. aeruginosa* has long served as a model organism for studying bacterial genetics and regulation. In addition to genetic regulatory mechanisms, *P. aeruginosa* has been a topic of research interest due to its prominence in several clinical conditions.

Nosocomial infections cost an estimated $9.8 billion annually in the United States (Zimlichman *et al.* 2013). These infections have rapidly become a critical issue as the frequency of antibiotic resistance increases. The appearance of multidrug-resistant *P. aeruginosa* strains is increasing globally; it has been estimated that multidrug resistance appears in 27–72% of patients originally infected with an antibiotic susceptible strain (Obritsch *et al.* 2012). *P. aeruginosa* is one of the most common pathogens isolated from nosocomial infections and is consistently linked to high rates of mortality (Magill *et al.* 2014). Much of the persistence and antibiotic resistance associated with *P. aeruginosa* can be attributed to the protection provided by biofilm formation. Once colonized, *P. aeruginosa* is capable of persisting within certain hosts for decades. Biofilm formation enhances persistence as bacteria encases itself in a self-produced protective matrix of exopolysaccharides (ePS), extracellular DNA (eDNA), and proteins (Ramsey et al. 2005). The presence of ePS, in particular, provides enhanced tolerance to antimicrobials, desiccation, oxidative stress, and host immune clearance (Berry *et al.* 1989; DeVault *et al.* 1990; Govan & Deretic, 1996; Billings *et al.* 2013; Tseng *et al.* 2013). Biofilm formation occurs on abiotic surfaces and medical devices, such as pulmonary ventilators, contact lenses, and catheters (Costerton *et al.* 1999). Implantation or improper cleaning of a contaminated device allows the biofilm to transition from the device to the host. The resulting infections initiated by surface associated biofilms can have devastating effects on immunocompromised patients. These medical device-related biofilms are particularly challenging to treat due to their antibiotic resistant nature, and treatment often requires replacement of the device. However, even this intervention is not 100% effective, and bacteria remaining within the host are often able to recolonize the newly implanted device or spread systemically (Bodey *et al.* 1983; Costerton *et al.* 1999). Ultimately, these infections result in significant burden on the patient and healthcare system due to additional surgeries and longer hospitalization. As antibiotic resistance becomes more prevalent, understanding how *P. aeruginosa* survives in the host and identifying alternative treatment targets will become vital for successful patient care. Given the importance of biofilms in *P. aeruginosa* resistance and adaptation to the host environment, understanding regulation of biofilm formation has never been more important.

P. aeruginosa biofilms are also found in burn and surgical site wounds, but have been primarily characterized for their role in chronic lung infections in patients suffering from the genetic disease cystic fibrosis (CF). CF is the most common genetic disease in Caucasians of Northern European descent and occurs in an estimated 1 out of every 2,500 Caucasian births (Davies *et al.* 2007). A mutation in the cystic fibrosis transmembrane conductance regulator (CFTR) leads to an imbalance in chloride ions, resulting in the buildup of dehydrated mucus in the lungs, as well as other mucosal surfaces (Rowe *et al.* 2005). The thick mucus layer traps bacteria and prevents mucociliary clearance, increasing susceptibility to infection. The pathology associated with CF can be attributed to the increased susceptibility to bacterial infection (Rowe *et al.* 2005). Although many different microbial species can be isolated from CF patients, *P. aeruginosa* is observed in almost 80% of CF individuals and correlated with impairment of lung function and eventual respiratory failure (Høiby, 2011).

To reiterate, the prevalence of *P. aeruginosa* during chronic infections is due to its ability to adapt to several environments, including the unique CF lung niche. This bacterium is highly resistant to antimicrobial treatment, and persistence in the lung has been attributed to multiple factors including type IV pili, exotoxins, proteases, lipopolysaccharide (LPS), flagella, and type III secretion proteins (Feltman *et al.* 2001; Darling *et al.* 2004).

5.1.1 Biofilm formation by *Pseudomonas aeruginosa*

The formation of the biofilm structure is determined by extracellular and intracellular signals that trigger the transition from a motile to sessile lifestyle. Biofilm formation is typically depicted as a flow of events that progresses from surface attachment to the formation of the matrix and eventual dispersal of planktonic cells from the biofilm. However, it should be emphasized that the biofilm community is a dynamic environment containing subpopulations in various stages of biofilm growth, which aids its ability to survive environmental and host mediated pressures (Boles *et al.* 2004; Parsek & Tolker-Nielsen, 2008).

Although *P. aeruginosa* utilizes a wide variety of components during biofilm production, ePS contribute significantly to the overall structure and function of the biofilm matrix. The secreted and cell-associated ePS produced by *P. aeruginosa* are key contributors to the formation, structure, and maintenance of the biofilm matrix. Understanding the process of biofilm formation is vital to resolving the function and expression of ePS. *P. aeruginosa* produces three different ePS molecules, Psl, Pel, and alginate, with all three types of ePS contributing to biofilm formation, pathogenesis, and host survival. The impact of individual ePS expression on strain properties and biofilm structure is striking and is exemplified by visible changes in colony morphology depending on the type of ePS produced. In addition to colony morphology, the primary matrix polysaccharide influences the physiological properties and structure of the biofilm. For example, strain PAO1 produces a Psl dominated matrix, while strain PA14 produces a matrix composed solely of Pel due to a deletion of *pslABCD* (Wozniak *et al.* 2003; Friedman & Kolter, 2004). Colvin *et al.* (2012) examined a panel of environmental and clinical isolates and found that the colony morphology, which is indicative of ePS production, was variable and that Psl and Pel can serve redundant roles in the biofilm matrix. From those observations, four classes of strains were proposed: Class 1 produces a Pel dominated matrix (i.e., PA14); Class 2 produces a Psl dominated matrix (i.e., PAO1); Class 3 relies on either Pel or Psl; and Class 4 overproduces Pel and Psl (i.e., small-colony variants). In addition to these classifications, mucoid isolates from chronic CF infections overproduce alginate and also require Psl and Pel for biofilm formation (Govan & Deretic *et al.* 1996; Ghafoor *et al.* 2011; Yang *et al.* 2011; Ma *et al.* 2012). Therefore, ePS provides critical contributions to the diversification and pathogenesis of *P. aeruginosa* in chronic infection, and the regulation and synthesis of each ePS has been an important focus of research.

Biofilm formation begins with the attachment of a planktonic cell to a biotic or abiotic surface, which initiates a cascade of gene expression changes. In *P. aeruginosa*, attachment involves type IV pili, flagella, and the production of Psl and Pel (O'Toole & Kolter, 1998). Biofilm initiation is a complex process that involves the simultaneous down regulation of motility genes and upregulation of biofilm related genes by a multitude of regulatory factors (O'Toole & Kolter, 1998; Merritt *et al.* 2007; Caiazza *et al.* 2007; Petrova *et al.* 2014). The initiation of biofilm formation has also been attributed to production of the bacterial second messenger cyclic dimeric (3′5′) GMP (c-di-GMP). Since its discovery, c-di-GMP has been found to control a wide variety of bacterial processes, including biofilm formation, motility, virulence, and ePS production (Simm *et al.* 2004; Hickman *et al.* 2005; Fazli *et al.* 2014; Römling *et al.* 2013). Intracellular levels of c-di-GMP are controlled by the competing activities of diguanylate cyclases (DGCs) and phosphodiesterases (PDEs), which produce and breakdown c-di-GMP, respectively. In the context of biofilms, high c-di-GMP levels are associated with the transition from motile to sessile growth. In *P. aeruginosa* several DGCs have been implicated in the promotion of biofilm formation through differing mechanisms (Table 5.1). GcbA promotes biofilm formation and initial attachment by suppressing flagellar reversal rates independent of ePS production (Petrova *et al.* 2014), whereas the Wsp system is a two-component signal transduction system that activates ePS production in response to growth on a surface (Hickman *et al.* 2005; Güvener & Harwood, 2007). Key components of this system include WspA, a membrane-bound methyl-accepting chemotaxis like (MCP) receptor, the DGC WspR, and WspF, a methylesterase that regulates WspA. Upon receiving a signal, WspA phosphorylates WspR leading to its activation and c-di-GMP production. WspF regulates the activation state of WspA through modification via methylation. A mutation in *wspF* therefore leads to constitutive activation of WspR and high levels of intracellular c-di-GMP. Thus, production of c-di-GMP by DGCs, such as WspR and GcbA, promotes the transition from planktonic to biofilm mode of growth, through the concurrent upregulation of biofilm and adhesion molecules while simultaneously down regulating motility (Römling *et al.* 2005; Jenal & Malone, 2006; Römling & Amikam, 2006; Ryan *et al.* 2006; Cotter & Stibitz, 2007; Tamayo *et al.* 2007).

Table 5.1 Regulators of ePS in *P. aeruginosa*.

Name	Alternative Name	Function	References
		Psl	
Cyclic-di-guanylate	C-di-GMP	Binds to FleQ causing conformation change and conversation to transcriptional activator	Hickman *et al.* 2008; Irie *et al.* 2012; Baraquet *et al.* 2012
FleQ	–	Transcriptional repressor in low c-di-GMP conditions; Transcriptional activator at high c-di-GMP conditions	Hickman *et al.* 2008; Baraquet *et al.* 2012
AmrZ	AlgZ	In mucoid strains: binds and inhibits transcription of the DGC, GcbA; binds to *psl* promoter to block transcription	Jones *et al.* 2013; Jones *et al.* 2014; Petrova *et al.* 2014
RpoS	–	Alternative sigma factor that promotes *psl* transcription	Irie *et al.* 2010
RsmA	–	Binds to 5′ UTR of *psl* mRNA blocking translation	Irie *et al.* 2010
SiaD	–	DGC that produces c-di-GMP in response to an exogenous Psl signal	Irie *et al.* 2012
SadC	–	DGC that produces c-di-GMP in response to an exogenous Psl signal	Irie *et al.* 2012
GacS/GacA	–	Activation of GacA leads to production of RsmY and RsmZ sRNAs to relieve RsmA repression of Psl translation	Goodman *et al.* 2004; Ventre *et al.* 2006; Workentine *et al.* 2009
LadS	–	Promotes GacS phosphorylation/activation of GacA	Goodman *et al.* 2004; Ventre *et al.* 2006; Workentine *et al.* 2009; Jing *et al.* 2010
RetS	–	Inhibits GacS phosphorylation/activation of GacA	Goodman *et al.* 2004; Ventre *et al.* 2006; Workentine *et al.* 2009; Jing *et al.* 2010
PhaF	–	Stimulates Psl translation through an unknown mechanism	Huse *et al.* 2013
		Pel	
Cyclic-di-guanylate	C-di-GMP	Binds to FleQ causing conformation change and conversion to transcriptional activator; Binds to PelD to activate Pel synthesis	Hickman *et al.* 2008; Baraquet *et al.* 2012; Lee *et al.* 2007.
FleQ	–	Transcriptional repressor in low c-di-GMP conditions; Transcriptional activator at high c-di-GMP conditions	Baraquet *et al.* 2012
TbpB	–	DGC: Activation through modulation of c-di-GMP	

TbpA	–	Phosphorylates TbpB repressing DGC activity	Ueda & Wood, 2009
Las quorum sensing	–	Activator of Pel expression through modulation of c-di-GMP	Ueda & Wood, 2009; Sakuragi & Kolter, 2007
Rhl quorum sensing	–	Activator of Pel expression through modulation of c-di-GMP	Sakuragi & Kolter, 2007
GacS/GacA	–	Activation of GacA leads to production of sRNAs, RsmY and RsmZ, to relieve RsmA repression of Pel translation	Goodman *et al.* 2004; Ventre *et al.* 2006; Workentine *et al.* 2009
SadB	–	Activates pelA transcription through unknown mechanism	Caiazza *et al.* 2007
SadC	–	DGC: Activation through modulation of c-di-GMP	Merritt *et al.* 2007
BifA	–	PDE: Repression through modulation of c-di-GMP	Kuchma *et al.* 2007
PelD	–	Pel biosynthesis protein that contains c-di-GMP binding site that regulates synthesis activity	Lee *et al.* 2007
		Alginate	
AlgT	AlgU, σ^{22}	Alternative sigma factor required for *algD* promoter expression	Martin *et al.* 1993a; Hershberger *et al.* 1995; Schurr *et al.* 1995; Schulz *et al.* 2015
MucA	–	Anti-sigma factor that binds and sequesters AlgT preventing *algD* expression	Mathee *et al.* 1999; Martin *et al.* 1993b; Xie *et al.* 1996; Schurr *et al.* 1996; Wood *et al.* 2006; Qui *et al.* 2007; Limoli *et al.* 2014
MucB	AlgN	Periplasmic protein that protects MucA from protease degradation and enhances MucA sequestration of AlgT	Goldberg *et al.* 1993; Martin *et al.* 1993c; Mathee *et al.* 1997; Rowen & Deretic, 2000; Cezairliyan & Sauer, 2009
Prc	–	Protease that degrades truncated periplasmic domain of MucA releasing AlgT in *muc*-22 strains	Reiling *et al.* 2005
AlgW	–	Protease that degrades truncated periplasmic domain of MucA releasing AlgT in response to a build of MucE	Boucher *et al.* 1996; Qiu *et al.* 2007; Wood & Ohman, 2009
MucP	–	Protease that degrades truncated periplasmic domain of MucA releasing AlgT in response to a build of MucE	Boucher *et al.* 1996; Qiu *et al.* 2007; Damron & Yu, 2011

(*Continued*)

Table 5.1 Regulators of ePS in *P. aeruginosa* (*Continued*).

Name	Alternative Name	Function	References
		Alginate	
MucC	–	Inhibits alginate production during temperature and osmotic stress conditions	Boucher *et al.* 1997; Martinez-Salazar *et al.* 1996; Nunez *et al.* 2000
MucD	AlgY	Breaks down MucE to prevent MucE buildup and subsequent activation of AlgW and MucP	Boucher *et al.* 1996; Qiu *et al.* 2007; Wood & Ohman, 2009; Damron & Yu, 2011
mucD-AS	–	Natural anti-sense transcript of MucD that blocks translation	Yang *et al.* 2011
ClpX/ClpP	–	Cytoplasmic proteases that break down remaining MucA on AlgT after release from inner membrane	Qui *et al.* 2008
RpoN	NtrA, σ^{54}	Sigma factor that represses *algD* transcription in *muc-22* strains but promotes *algD* transcription in *muc-23* mucoid strains	Boucher *et al.* 2000
AlgR	AlgR1	Binds to RB1, RB2, and RB3 sites upstream of *algD* to facilitate transcription.	Deretic *et al.* 1988; Mohr *et al.* 1992
FimS	AlgZ	Predicted cognate sensor kinase of AlgR and negative regulator of alginate production	Whitchurch *et al.* 1996; Yu *et al.* 1997
AlgQ	AlgR2	Anti-sigma factor that promotes *algD* transcription by sequestering RpoD	Kato *et al.* 1989; Konyecsni & Deretic *et al.* 1990; Dove & Hochschild, 2001
H_p-1	AlgP, AlgR3	Activates *algD* transcription	Kato *et al.* 1990; Deretic & Konyecsni, 1990; Deretic *et al.* 1992
AlgB	–	Activates *algD* transcription	Leech *et al.* 2007
KinB	–	Negatively regulates alginate production by inhibiting AlgW protease activity	Damron *et al.* 2009; Damron *et al.* 2012; Withers *et al.* 2013
CysB	–	Activates algD transcription	Delic-Atttree *et al.* 1997
Cyclic-di-guanylate	C-di-GMP	Binds to cytoplasmic region of Alg44 to activate synthesis activity	Merighi *et al.* 2007; Hay *et al.* 2009

Elevated levels of c-di-GMP produced by the Wsp system enhance the production of the two ePS molecules, Psl and Pel, which play a crucial role in the biofilm initiation process. The ePS Psl facilitates cell-cell and cell-surface interactions and is necessary for the initiation of attachment (Ma *et al.* 2009; Colvin *et al.* 2010). However, the specific role of Pel in attachment remains uncertain. Colvin *et al.* (2012) found that *pel* mutants are proficient in surface attachment. However, recent work by Cooley *et al.* (2013) has shown that Pel does facilitate surface attachment in cooperation with Psl. The combined effects of Pel and Psl increased surface attachment strength, as Pel shortened the range of interaction between the cell and surface relative to Psl alone (Cooley *et al.* 2013). It has also been demonstrated that the mediators of attachment and adherence differ in response to biotic versus abiotic surfaces. For example, the type III secretion system translocon is necessary for the formation of cell-associated aggregates on epithelial cells, but not abiotic surfaces such as in flow cells or static microtiter assays (Tran *et al.* 2014). Therefore, it can be concluded from the current body of work that the ePS Psl and Pel, in addition to other mediators, function to facilitate surface attachment in a surface-dependent manner.

The process of biofilm formation continues by the transition from a monolayer of adherent cells to the formation of ordered structures termed microcolonies or "mushrooms." The differences in the physical properties of Psl, Pel, and alginate have structural implications on biofilms produced by different *P. aeruginosa* strains. Colvin *et al.* (2011) showed in the strain PA14 that Pel is required for the maintenance of intercellular interactions and therefore acts as the primary scaffold for biofilm structure. Subsequent deletion of Pel resulted in a severe reduction in biofilm formation due to the bacterium's inability to transition from a monolayer to microcolony mode of growth. However, the deletion of Pel in the strain PAO1 did not recapitulate these findings. Psl producing strains, such as PAO1, form mature microcolonies that consist of a cavity containing planktonic cells and lysed cells, as well as eDNA located closer to the substratum, or "stalk," of the microcolony. A peripheral Psl "shell" surrounds the cavity and stalk structure (Ma *et al.* 2009). The presence of Psl is necessary for maintenance of biofilm biomass, as measured by biofilm thickness post-attachment (Ma *et al.* 2006). The biofilm also contains an additional protein adhesion component, CdrA, which binds Psl and promotes Psl-dependent aggregation of cells (Borlee *et al.* 2010).

New evidence suggests that Psl is important for the establishment and maintenance of infection. A positive selection for Psl producing strains occurs during the early stages of CF infection (Huse *et al.* 2013). However, in late stages of infection, alginate-producing strains become more prevalent (Lam *et al.* 1980; Deretic *et al.* 1995; Singh *et al.* 2000). Alginate overproduction leads to the formation of complex biofilm architecture with large microcolonies separated by water channels with increased thickness and heterogeneity relative to isogenic PAO1 (Hentzer *et al.* 2001). O-acetylation of alginate is responsible for the thickness and viscosity of the mucoid biofilm structure and is critical for microcolony maintenance and cell aggregation (Nivens *et al.* 2001; Tielen *et al.* 2004). Overproduction of alginate provides a physical and chemical barrier from antibiotic treatment, reactive oxygen species, and both non-opsonic and opsonic phagocytic killing (Aaron *et al.* 2002; Govan & Deretic, 1996; Pier, 1998).

As microcolonies mature, cell density and colony thickness increase to a threshold that promotes the phenomenon of "seeding dispersal." Seeding dispersal occurs when motile cells from the interior of the microcolony disperse from "local break points" and adhere to surrounding surfaces (Purevdorj-Gage *et al.* 2005; Ma *et al.* 2009). Though rhamnolipid molecules are not necessary for dispersal, the Las and rhamnolipid-associated quorum sensing systems are required for efficient dispersal. Dispersion can be initiated by a variety of factors including reactive oxygen or nitrogen species, cis-2-decenoic acid, cell lysis, and decreased intracellular levels of c-di-GMP due to induction of PDE activity by extracellular carbon availability, starvation, and nitric oxide signaling (Sauer *et al.* 2004; Barraud *et al.* 2006; Morgan *et al.* 2006; Davies & Marques, 2009; Harmsen *et al.* 2010). Seeding dispersal has been observed in the non-mucoid strain PAO1, but not in the mucoid clinical isolate FRD1. FRD1 relies upon the dispersal method termed "erosion," which occurs when the shear force of fluid causes passive removal of cells from the biofilm matrix. In PAO1, dispersion of the biofilm may also be aided by Pel, as recent findings by Chew *et al.* (2014) demonstrate that Pel production reduced cross-linking within the biofilm, promoting dissemination and spreading of the biofilm. The opposing result was found for Psl production, as it increased elasticity and cross-linking within the biofilm matrix, which promoted stability. Overall, it is clear that ePS perform critical roles in biofilm formation, from the facilitation of initial attachment to microcolony formation, maturation, and dispersion for further colonization.

The production of ePS with diverse functional roles at distinct time points during biofilm formation requires complex regulatory mechanisms. Transcriptional, post-transcriptional, and environmental factors work in concert to ensure the correct ePS is being produced both temporally and spatially within the biofilm. The following sections will cover each of the *P. aeruginosa* ePS structures and synthesis while highlighting the various regulatory mechanisms utilized by *P. aeruginosa* to control their expression.

5.2 PSL (POLYSACCHARIDE SYNTHESIS LOCUS)

Although Psl was discovered relatively recently (Friedman & Kolter, 2004; Jackson *et al.* 2004; Matsukawa *et al.* 2004), a critical role for this ePS has been identified in biofilm formation on biotic and abiotic surfaces during the early stages of development (Ma *et al.* 2006) as well as during pathogenic infections (Huse *et al.* 2013). Psl serves a functional role in biofilm initiation, cell-cell and cell-surface interactions, biofilm matrix structure and stability, and antibiotic resistance (Ma *et al.* 2006; Billings *et al.* 2014). As mentioned previously, Psl cross-linking in the biofilm matrix promotes elasticity and structure formation resulting in greater resistance to shear forces (Chew *et al.* 2014). The physiological characteristics of Psl suggest production is crucial during the early stages of biofilm initiation and development for surface attachment and microcolony formation, while in late-stage biofilm development the down regulation of Psl promotes dissemination (Chew *et al.* 2014).

A role for Psl in persistence during infection has also been established. In the host, there is a positive selective pressure for Psl overproducing strains during CF lung infection (Huse *et al.* 2013; Malone *et al.* 2012). A persistent variant of *P. aeruginosa* termed small-colony variants (SCV) is caused by the overproduction of Psl and Pel (Starkey *et al.* 2009). SCV prevalence increases after long-term antibiotic treatment of chronic lung infections, possibly due to overproduction of Psl, which provides resistance to antibiotics and host immune clearance (Haussler *et al.* 1999; Schneider *et al.* 2008; Billings *et al.* 2014; Mishra *et al.* 2012). Although significant progress has been made in the years since the discovery of Psl, much about the ePS structure, synthesis, and role in biofilm development remains to be investigated. The following section will summarize the current state of knowledge regarding Psl structure, synthesis and regulation.

5.2.1 Psl composition and structure

A number of recent comprehensive reviews are available that discuss the composition and structure of Psl in detail (Franklin *et al.* 2011; Ma *et al.* 2013; Mann *et al.* 2011). Psl structure and synthesis appear to closely resemble the *E. coli* group 1 capsular polysaccharide (Franklin *et al.* 2011). It is composed of mannose, rhamnose, and glucose in a 3:1:1 ratio, and it appears to exist in at least two forms: a high molecular weight cell-associated form and a cell free low molecular weight form (Byrd *et al.* 2009). Cell free Psl is primarily composed of a pentasaccharide consisting of D-mannose, D-glucose, and L-rhamnose

(Byrd *et al.* 2009; Figure 5.1). The structure of the cell-associated form is unknown but believed to consist of repeats of this pentasaccharide unit. Surface bound Psl is predicted to function in attachment and cell-cell interactions with adjacent Psl-producing bacteria (Ma *et al.* 2009).

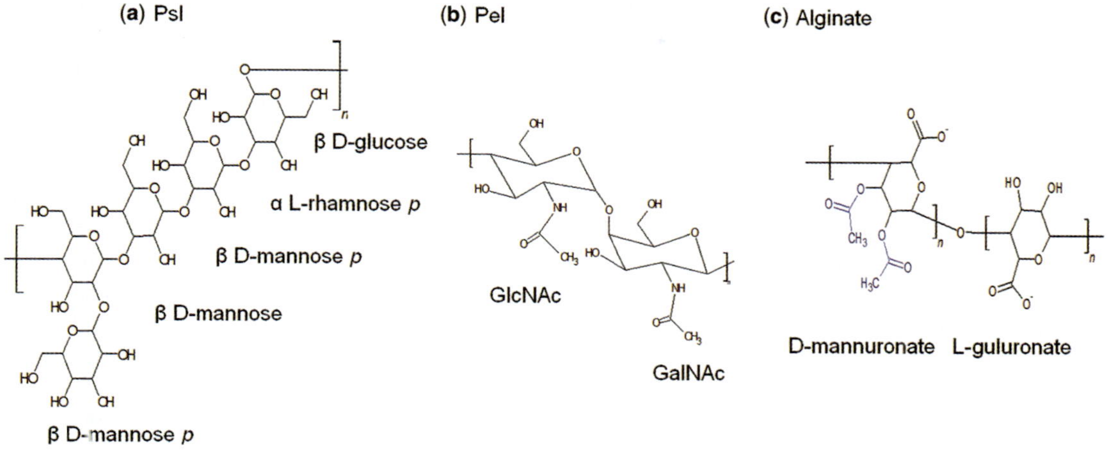

Figure 5.1 Structures of *P. aeruginosa* exopolysaccharides. Predicted carbohydrate repeating units of Psl, Pel, and alginate polymers. Blue labeled regions of alginate indicate where O-acetylation sometimes occurs. Structures were generated using ChemSketch, Version 14.0, Advanced Chemistry Development, Inc., Toronto, On, Canada, www.acdlabs.com, 2015.

5.2.2 Psl synthesis

Psl is synthesized by the enzymes in the polysaccharide synthesis locus, which consists of fifteen co-transcribed genes *pslA-pslO* (PA2231-PA2245) (Friedman & Kolter, 2004; Jackson *et al.* 2004; Matsukawa *et al.* 2004). Ten of these genes, *pslACDEFHIJKL*, are required for Psl production, while *pslGMNO* mutant strains still produce Psl (Byrd *et al.* 2009; Baker *et al.* 2015). The activity of *pslB* is also essential for Psl production but shares functional redundancy with *wbpW* (Byrd *et al.* 2009). Sequence and structure homology analysis has been used to predict the function of the Psl synthesis proteins and has been described in detail in a review by Franklin *et al.* (2011). This section will summarize the predicted biosynthesis mechanisms for each of the biosynthetic proteins.

As mentioned previously, the Psl biosynthetic components resemble the machinery used by *E. coli* for group 1 capsular synthesis. Psl polymers are likely built on an isoprenoid lipid carrier, and then assembled and exported through the action of polysaccharide polymerases and outer membrane polysaccharide export proteins (Franklin *et al.* 2011; Cuthbertson *et al.* 2009). The sugar nucleotide precursors GDP-D-mannose, UDP-D-glucose, and dTDP-L-rhamnose are essential for Psl production (Byrd *et al.* 2009). PslB is the only Psl synthesis protein involved in precursor formation and is predicted to function as a phosphomannose isomerase and GDP-mannose pyrophosphorylase that produces GDP-D-mannose (Franklin *et al.* 2011). PslB is functionally redundant with the LPS A-band protein, WbpW (Byrd *et al.* 2009). Thus, deletion of both *pslB* and *wbpW* is necessary to disrupt Psl production (Byrd *et al.* 2009). Similarly, PslB and WbpW likely share functional redundancy with AlgA (in alginate producing strains) since AlgA also produces GDP-D-mannose (Darzins *et al.* 1986; Franklin *et al.* 2011). Pools of the Psl precursor UDP-D-glucose are generated by AlgC and GalU, while dTDP-L-rhamnose is generated by

RmlC (Byrd *et al.* 2009). Thus, unlike previously identified *P. aeruginosa* ePS synthesis systems, AlgC, GalU, and RmlC are encoded by genes outside of the *psl* operon.

Following precursor formation, glycosyltransferase (GT) enzymes typically function to transfer activated sugar subunits to the inner membrane polymerization machinery. The remaining predicted cytoplasmic Psl synthesis proteins, PslC, PslF, PslH, and PslI, contain GT domains. Based on structural homology predictions, PslF, PslH, and PslI contain CAZy GT-4 domains while PslC contains a CAZy GT-2 domain (Franklin *et al.* 2011). Thus, these proteins are predicted to transfer activated sugar subunits from their nucleotide carrier to form the repeating polysaccharide unit that makes up Psl.

The polymerization machinery for ePS is typically found embedded within the inner membrane. PslA, PslE, PslJ, PslK, PslG, and PslL have predicted transmembrane topology and likely localize to the inner membrane (Franklin *et al.* 2011; Baker *et al.* 2015.) PslA and PslE also contain large domains that extend into the cytoplasm and share structural similarities to proteins in the Wzy-capsule synthesis pathway, suggesting a possible role in the transport and polymerization of Psl (Franklin *et al.* 2011). The cytoplasmic portion of PslA contains a predicted ePS biosynthesis polyprenyl glycosylphosphotransferase and a Rossmann fold, which may suggest that it plays a similar role to WbaP in capsule synthesis (Franklin *et al.* 2011). WbaP functions to accept activated precursor subunits from a GT and transfers it to an isoprenoid carrier at the cytoplasmic face of the inner membrane. In addition, WbaP functions as a flippase to transfer the subunit-isoprenoid across the inner membrane (Wang *et al.* 1996). PslE contains two transmembrane domains and both a cytoplasmic and periplasmic portion. PslE shows structural similarity to the Wzz component of the Wzy-capsule synthesis system (Franklin *et al.* 2011); Wzz regulates the chain length of the growing LPS polysaccharide during capsule synthesis (Franco *et al.* 1996). The periplasmic portion of PslE may interact with the outer membrane protein PslD to coordinate Psl polymerization and ePS secretion out of the cell (Franklin *et al.* 2011). The function of the remaining inner membrane proteins, PslJ, PslK, and PslL, is difficult to predict due to a lack of sequence and structure homology with known proteins. However, considering the similarities of PslA and PslE to the Wzy-capsule synthesis proteins, PslJKL functions are thought to aid PslA and PslE in polymerization and flipping activity (Franklin *et al.* 2011). PslG is primarily located within the periplasm with a single transmembrane domain that anchors the protein to the inner membrane (Baker *et al.* 2015). Crystal structure and functional analysis of PslG revealed its similarity in both structure and function to members of the glycoside hydrolase family 39 (Baker *et al.* 2015). PslG hydrolase activity may function in processing the growing polysaccharide chain during export out of the cell, as a mechanism to prevent the buildup of Psl within the periplasm, or in modulating ratios of cell-associated to released Psl (Franklin *et al.* 2011).

5.2.3 Transcriptional regulation of Psl

The transcriptional regulators of Psl are summarized in Figure 5.2. The role of the secondary messenger molecule c-di-GMP is becoming increasingly apparent in the transcriptional and post-transcriptional regulation of *P. aeruginosa* biofilms. As mentioned previously, c-di-GMP mediated regulation has been found in a wide variety of bacterial processes, and high intracellular levels are associated with the transition from a motile to a sessile or biofilm lifestyle in a variety of bacteria including *P. aeruginosa* (Simm *et al.* 2004). Deletion of *wspF*, a DGC repressor, and constitutive production of c-di-GMP results in elevated *psl* transcription and biofilm formation (Hickman *et al.* 2005). However, it is not yet fully understood how c-di-GMP regulates Psl production. The transcriptional regulator FleQ, described in further detail in the following section (11.3.3), has been implicated in Psl regulation. ChIP-seq analysis observed FleQ binding to the *psl* promoter (Hickman *et al.* 2008). Although not directly observed in the context of Psl production, it is likely that FleQ functions as a transcriptional regulator playing a similar role for Psl regulation as it does for Pel (see below). Irie *et al.* (2012) suggests c-di-GMP may have an additional role in regulating Psl production by a self-signaling mechanism that stimulates two DGCs, SiaD and SadC to increase c-di-GMP levels. This mechanism would function as a positive feedback system where Psl stimulates its own production by elevating c-di-GMP levels; the receptor that recognizes Psl in this system has not yet been identified.

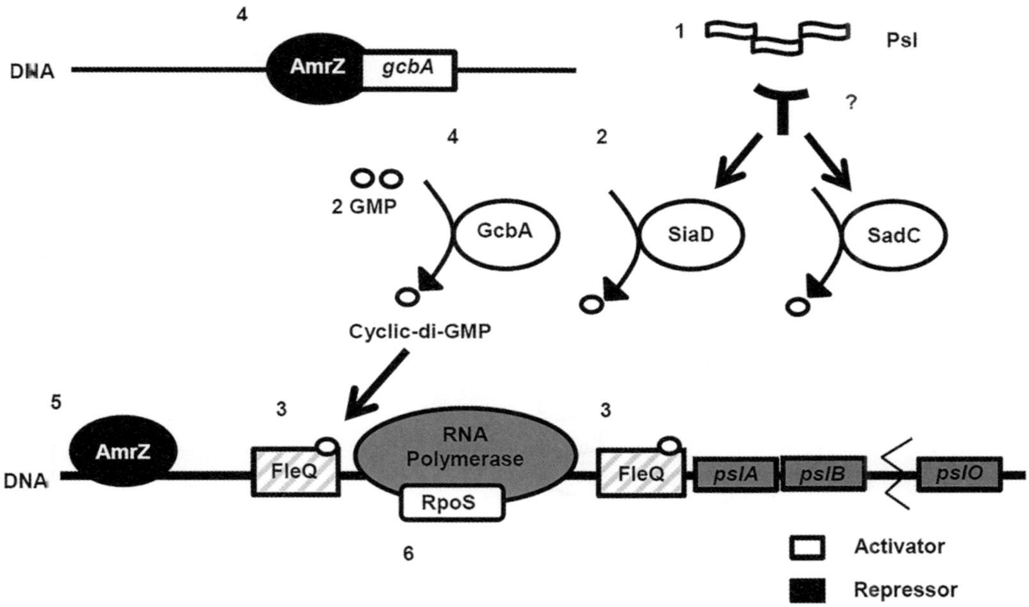

Figure 5.2 Transcriptional Regulation of Psl. 1) Psl functions as a signal to activate an unknown receptor stimulating SiaD and SadC activity. 2) SiaD and SadC diguanylate cyclase activity elevates c-di-GMP levels to enhance Psl production via FleQ. 3) In low c-di-GMP conditions, FleQ represses Psl production by blocking RNA polymerase from binding to the *psl* promoter, but enhances RNA polymerase binding following c-di-GMP binding. 4) In alginate producing mucoid strains, AmrZ inhibits *gcbA* transcription reducing c-di-GMP production and subsequent FleQ activation. 5) AmrZ also binds directly to the *psl* promoter inhibiting transcription of the *psl* operon. 6) The stationary phase σ factor, RpoS, promotes psl transcription.

In alginate producing mucoid strains, Psl production is down regulated compared to non-mucoid strains (Ghafoor *et al.* 2011; Jones *et al.* 2014). In these strains, the ribbon-helix-helix DNA binding protein, AmrZ, inhibits Psl production by down regulating c-di-GMP levels and *psl* transcription. AmrZ binds and inhibits transcription of the DGC-encoding gene *gcbA* (*adcA*), resulting in reduced intracellular c-di-GMP levels (Jones *et al.* 2014; Petrova *et al.* 2014). In addition, AmrZ represses Psl production by directly binding to the *psl* promoter and repressing transcription (Jones *et al.* 2013). Given the role that AmrZ plays in alginate production (described in 11.4.3) it is likely AmrZ plays an important role in the transition from a Psl to an alginate dominant biofilm by concurrently repressing Psl production and activating alginate production.

Another Psl transcriptional regulator has been identified that is not influenced by c-di-GMP levels. The stationary-phase σ factor RpoS is an alternate subunit of RNA polymerase. Environmental stress activates RpoS translation, which leads to RpoS-mediated transcription of genes associated with stationary phase growth and virulence (Battesti *et al.* 2011). RpoS serves as a critical σ factor that promotes RNA polymerase *psl* promoter interactions. Deletion of *rpoS* reduced *psl* transcription while overproduction stimulated it (Irie *et al.* 2010). ePS regulation in this manner is resourceful in that it may utilize a universal σ factor to activate biofilm formation in stressful environments.

5.2.4 Post-transcriptional regulation of Psl

In addition to transcriptional regulation, multiple Psl post-transcriptional regulators have been discovered in recent years (Figure 5.3). RsmA binds to the 5′ untranslated region (UTR) of *psl* mRNA and stabilizes an mRNA stem-loop secondary structure preventing ribosome access to the ribosome binding site. Deletion of *rsmA* had no effect on transcription, but resulted in three- to four-fold greater translation, suggesting that RsmA functions as a translational repressor of Psl (Irie *et al.* 2010). In a similar manner, RsmA represses translation of the Psl inducing DGC, SadC (Moscoso *et al.* 2014; Irie *et al.* 2012).

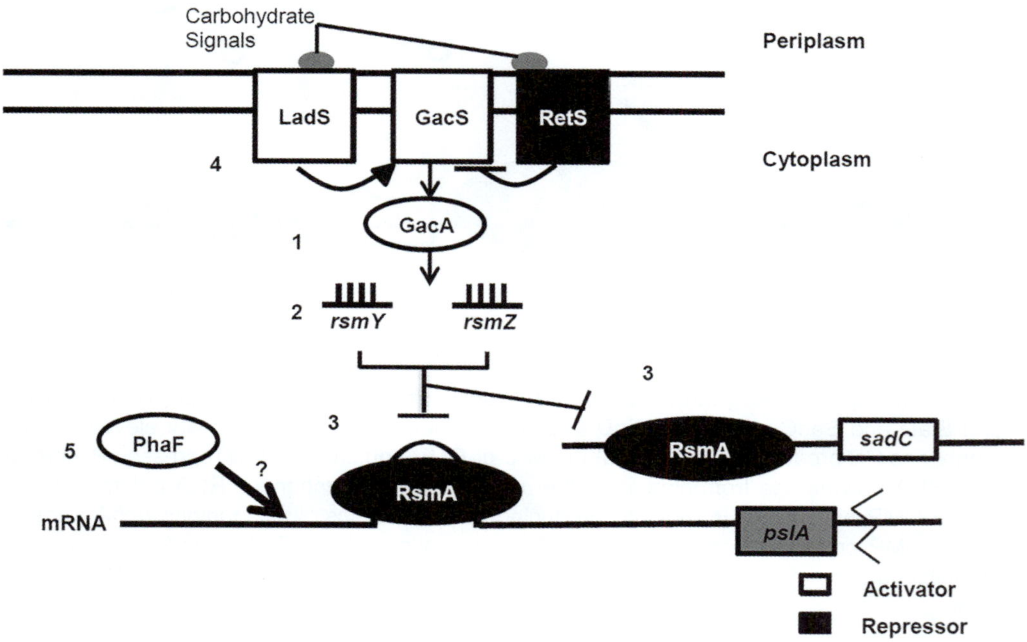

Figure 5.3 Post-Transcriptional Regulation of Psl. 1) GacS/GacA function as a two component system that promotes Psl synthesis by inducing the production of two sRNAs, *rsmY* and *rsmZ* 2) The sRNAs *rsmY* and *rsmZ* bind and inhibit RsmA preventing translation inhibition of the *psl* operon and the DGC *sadC* 3) RsmA binds to the 5′ UTR of *psl* and *sadC* mRNA stabilizing a secondary structure that inhibits translation 4) Binding of a carbohydrate ligand to the periplasmic domains of LadS or RetS result in modification of GacS sensor activity either enhancing or repressing *rsmY* and *rsmZ* production 5) PhaF stimulates translation of Psl synthesis proteins through an unknown mechanism.

Psl production also appears to be regulated by a two-component system sensor kinase and response regulator pair. For these systems, an environmental stimulus is recognized by the sensor kinase, resulting in the donation of a phosphoryl group to the cognate response regulator, and activation of a downstream effector(s) (Stock *et al.* 2000). Typically these effects are associated with changes in gene expression because many response regulators bind directly to specific DNA sites to influence transcription. The GacS family of response regulators is an important regulatory mechanism in a variety of bacterial species, and it functions primarily to alter virulence and stress response factors including extracellular enzyme production, toxin secretion, quorum sensing molecules, and motility (Heeb & Haas, 2001). GacS/GacA (response regulator and sensor kinase, respectively) functions as a two-component system in *P. aeruginosa* to control virulence factor production, as well as Pel and Psl production (Ventre *et al.* 2006; Goodman *et al.* 2004; Goodman *et al.* 2009). Activation of the GacS/GacA system results in the transcription of two small regulatory RNAs (sRNAs), *rsmY* and *rsmZ* (Goodman *et al.* 2004; Ventre *et al.* 2006; Workentine *et al.* 2009; Brencic *et al.* 2009a, Brencic *et al.* 2009b). These sRNAs bind RsmA directly and prevent RsmA mediated translation inhibition (Valverde *et al.* 2003; Brencic *et al.* 2009a). Thus, the GacS/GacA system promotes Psl production in *P. aeruginosa* through the activation of *rsmY* and *rsmZ* sRNA expression. An additional level of GacS/GacA control involves the action of two regulatory sensor kinase proteins, LadS and RetS (Ventre *et al.* 2006; Goodman *et al.* 2009). Based on their lectin-like conformations, structural analysis of these proteins predict that their periplasmic domains receive similar but distinct carbohydrate signals (Ventre *et al.* 2006; Goodman *et al.* 2009; Jing *et al.* 2010). Upon signal recognition, LadS stimulates GacA phosphorylation activity, while RetS functions in an opposite manner to repress GacA activity. The resulting impact of their activity is that LadS stimulates Psl- while RetS represses Psl-production.

Biofilm forming ability and expression changes in clonal *P. aeruginosa* isolates from CF patients over time led to the identification of six genes/operons that enhance Psl production when overexpressed (Huse *et al.* 2013). Expression of one of these genes, PhaF, increase translation of *pslA* 5-fold, but had no effect on transcription (Huse *et al.* 2013). PhaF appears to function as an activator of *psl* translation independently of *rsmZ* and *rsmY* expression (Huse *et al.* 2013). The remaining five genes/operons, PA1106, PA1323–24, PA1592, PA3691–92, and PA5178, also induce Psl production when overexpressed (Huse *et al.* 2013). However, these genes did not alter *psl* transcription or translation, indicating they may act post-translationally.

It has also been proposed that competition for key sugar precursors, including glucose-6-phosphate and fructose-6-phosphate, may function indirectly in post-translational regulation of Psl biosynthesis (Ma *et al.* 2012). Considering Psl precursors are also used to produce other ePS molecules, such as alginate, it is reasonable that a reduced precursor pool in alginate producing cells would result in reduced Psl production.

5.3 PEL (PELLICLE FORMATION)

The original identification of the *pel* operon and the Pel polysaccharide was performed in strain PA14. Screening of a PA14 transposon library identified mutants unable to form pellicles, which are mat-like biofilms that form at the air-liquid interface during static growth. These mutants contained various transposon insertions within an operon of seven adjacent genes, each with predicted functions for polysaccharide synthesis and transport. The identified polysaccharide, named "Pel" due its role in pellicle formation, exhibited cellulase sensitive binding of Congo red, thus providing a method to visualize the relative polysaccharide concentration within the matrix (Friedman & Kolter, 2004). It was later shown that PA14 produces only the Pel polysaccharide due to a four-gene deletion within the *psl* operon (Δ*pslABCD*). Early characterization showed that the production of Pel was necessary for formation of solid surface biofilms in strains PA14 and PAK (Friedman & Kolter, 2004). This early work laid the foundation for the involvement of Pel in initial adherence, while later studies elucidated several additional roles, including cell-cell signaling, interspecies interactions, aminoglycoside antibiotic protection, biofilm dissemination, and SCV formation (Kirisits *et al.* 2005; Starkey *et al.* 2009; Khan *et al.* 2010; Yang *et al.* 2011; Chew *et al.* 2014; Colvin *et al.* 2011).

5.3.1 Pel composition and structure

Though various roles for Pel have been identified in detail, the structure of the polysaccharide was only recently determined. Originally hypothesized to be a glucose-rich polymer, current work has identified Pel as a linear ePS consisting of $1 \rightarrow 4$ glycosidic linkages of *N*-acetylglucosamine and *N*-acetylgalactosamine (Friedman & Kolter, 2004; Urch *et al.* 2009; Jennings *et al.* 2015; Figure 5.1). Whereas alginate is anionic and Psl neutrally charged, Pel is cationic due to the presence of amino sugars. This cationic nature allows Pel to interact with anionic eDNA to form cross-links within the biofilm matrix (Jennings *et al.* 2015).

5.3.2 Pel synthesis

The Pel biosynthesis components are encoded by *pelA-G* (PA3064-PA3058). Current knowledge of Pel biosynthesis consists of experimental models of PelC and PelD, and putative functions proposed for the remaining proteins (Vasseur *et al.* 2007; Bonsor *et al.* 2007; Lee *et al.* 2007; Ueda & Wood, 2009; Keiski *et al.* 2010; Kowalska *et al.* 2010). A comprehensive overview is provided in this section, and an in-depth description of the biosynthetic enzymes that produce Pel can be found in an excellent review by Franklin *et al.* (2011).

The *pel* operon does not encode any sugar-nucleotide precursor synthesis enzymes, suggesting that the production of Pel components relies on enzymes located outside of the *pel* operon (Byrd *et al.* 2009). These precursors are most likely bound and polymerized by PelF, the single predicted glycosyltransferase, which remains localized to the cytoplasm (Coutinho *et al.* 2003). The lack of an additional glycosyltransferase located within the inner membrane suggests that Pel does not require an additional lipid carrier for transport. Instead, PelD, PelE, and PelG, each of which are located in the inner membrane and contain multiple transmembrane (TM) domains, may function separately or in complex to transport Pel from the cytoplasm to the periplasm (Krogh *et al.* 2001). PelD specifically is essential for the production of the Pel polysaccharide; this function requires the binding of c-di-GMP to the C-terminus of PelD within the cytoplasm (Lee *et al.* 2007; Ueda & Wood, 2009). The roles of PelE and PelG are less clear. It has been proposed that PelE functions to transport Pel through participation in a scaffold complex with PelA, PelB, or PelD, while PelG, in association with PelA, aids in the transport of Pel or small cofactors (Krogh *et al.* 2001). Once transported to the periplasm, PelA, whose *in vitro* deacetylation activity is necessary for Pel synthesis, is proposed to modify polymerized Pel before transport by PelB (Colvin *et al.* 2013). PelB, a fusion protein containing periplasmic and outer membrane domains, may aid in protection of Pel during periplasmic transport in addition to its TPR domain that may participate as a scaffold within the secretion complex (Keiski *et al.* 2010). PelB also contains a beta-barrel structure within the outer membrane domain that is structurally similar to AlgE, a protein involved in the export of alginate across the outer membrane. The structural similarity between these two proteins suggests that PelB may also be involved in the transport of Pel across the outer membrane. An additional outer membrane protein, PelC, possesses a C-terminus essential for polymer production, though the configuration and insertion into the outer membrane remains to be determined (Vasseur *et al.* 2007; Jennings *et al.* 2015).

Though much of the knowledge concerning Pel structure and synthesis remains unknown, past work has provided a deeper understanding of the role and regulation of the Pel polysaccharide during biofilm formation; these topics will be covered in-depth in the following sections.

5.3.3 Transcriptional regulation of Pel

The following sections covering the transcriptional and post-transcriptional regulation of Pel synthesis are summarized in Figure 5.4. A key regulator of Pel expression is the ubiquitous bacterial second messenger, c-di-GMP, which contributes both directly and indirectly to the regulation of Pel and Psl production (Hickman *et al.* 2005; Irie *et al.* 2011). The expression of Pel is dependent on the direct binding of c-di-GMP to FleQ, a transcription factor previously characterized for its role as a master regulator of flagella expression (Arora *et al.* 1997; Hickman *et al.* 2008). FleQ is a DNA binding protein that is involved in the regulation of Pel and Psl, as well as CdrA, a high molecular weight adhesin that binds Psl in vitro (Baraquet *et al.* 2012; Borlee *et al.* 2010). FleQ binds two separate sites upstream and downstream of the *pel* operon transcription start site (Baraquet *et al.* 2012). These sites are occupied simultaneously by one molecule of FleQ per site, with no preference demonstrated in the order or amount of binding. When FleQ is bound to both sites, FleN forms a complex to bridge the two bound FleQ proteins, which then induces DNA bending of the *pel* promoter. Transcriptional repression by FleQ occurs with low levels of c-di-GMP; when c-di-GMP levels are elevated, c-di-GMP binds both molecules of FleQ, leading to a cascade of conformational changes that allow the DNA to relax and RNA polymerase to bind and initiate *pel* transcription.

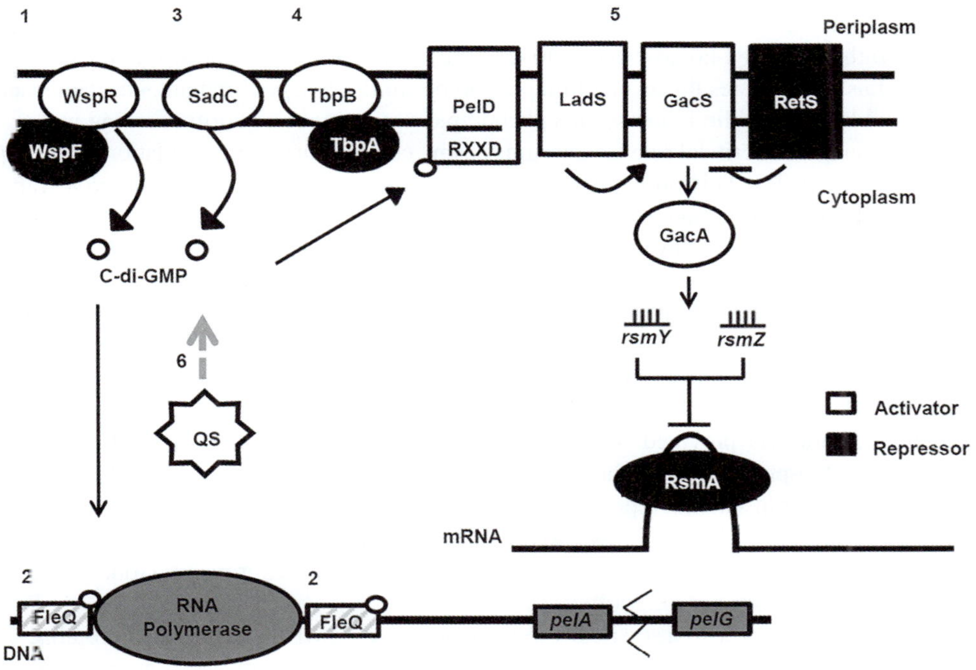

Figure 5.4 Regulation of Pel. 1) WspF binds and inhibits WspR DGC activity preventing c-di-GMP production. 2) C-di-GMP activates *pel* transcription by binding to FleQ causing a conformational change that promotes RNA polymerase binding and transcription. 3) The DGCs SadC and TbpB produce c-di-GMP that activates Pel production by binding to the C-terminus degenerate RXXD motif of PelD. 4) The DGC TbpB is inhibited via phosphorylation by TbpA. 5) C-di-GMP binding to the is required for the synthesis and export of Pel. 5) The sRNAs *rsmY* and *rsmZ* bind and inhibit RsmA preventing translation inhibition of the *psl* operon and the DGC *sadC*. 6) Quorum sensing (QS) systems Las and Rhl are hypothesized to affect the intracellular levels of c-di-GMP.

Additional transcriptional regulation is provided by the Wsp chemosensory system, mentioned previously for its role in the regulation of the Psl polysaccharide. WspF acts to repress the DGC WspR to maintain low levels of intracellular c-di-GMP. The mutation of *wspF* or *wspA*, which has been shown to occur in SCVs, results in elevated levels of c-di-GMP due to constitutive activation of WspR. The increased levels of c-di-GMP then lead to increased synthesis of the Pel and Psl polysaccharides (Hickman *et al.* 2005).

5.3.4 Post-transcriptional regulation of Pel

Extracellular signals, modulated through *tbpAB* and the *las* and *rhl* quorum sensing systems provide an additional level of regulation for Pel production. TbpA is proposed to reduce intracellular levels of c-di-GMP by dephosphorylating the DGC TbpB (Ueda & Wood, 2009). Loss of TbpA results in a hyper-biofilm forming phenotype characterized by enhanced aggregation and pellicle formation, indicating a likely role for TbpA mediated repression of Pel production through modulation of c-di-GMP levels. TbpA and c-di-GMP levels are further regulated by the *las* quorum sensing system (Hengee, 2009; Ueda & Wood, 2009), as suggested by the identification of a conserved LasR binding site upstream of the *tbpA* transcription start site (Ueda & Wood, 2009).

As one of the primary components of the *las* system, LasR acts to activate the expression of *lasI*, which mediates the synthesis of acyl-homoserine lactone (AHL) and induces the expression of several virulence genes. Loss of LasI or LasR results in a biofilm phenotype similar to a *pelA* mutant, i.e., decreased Congo Red binding and loss of biofilm biomass; this phenotypic similarity of a quorum sensing mutant to a *pel* deficient mutant suggests a critical role for quorum sensing components in the regulation of *pel* expression (Sakuragi & Kolter, 2007). This observation is further supported by a lack of *pel* operon transcription within *las* mutants. The restoration of the wrinkly colony morphology phenotype, which is characteristic of *pel* expression, can be achieved by the addition of synthetic AHL. The loss of *rhlI* or *rhlR*, two activators required for rhamnolipid synthesis, result in reduced transcription of the *pel* operon. This suggests that quorum sensing *las* and *rhl* systems primarily act at an indirect level of regulation, most likely through alteration of intracellular c-di-GMP. Alternatively, it has been shown that the development of biofilm structure is dependent upon quorum sensing mediated regulation of swarming motility and carbon source availability, not ePS production (Shrout *et al.* 2006).

Additional regulation is provided by the GacSA system, mentioned previously for its role in Psl regulation (11.2.4). Phosphorylation of GacA increases the expression of the RsmY and RsmZ sRNAs that bind and inactivate the translational repressor RsmA. RsmA acts to repress *pel* and *psl* expression, as well as the DGC, SadC (described below) (Moscoso *et al.* 2014; Merritt *et al.* 2007). Therefore, elevated levels of RsmY and RsmZ relieve RsmA mediated repression to enhance *pel* expression (Goodman *et al.* 2004; Ventre *et al.* 2006; Workentine *et al.* 2009).

The transition from a motile to a sessile lifestyle involves c-di-GMP and associated proteins, including SadB, SadC, and BifA. SadC is an inner membrane-localized protein with DGC activity; loss of SadC results in delayed irreversible attachment, hyper-swarming, and a decrease in Congo red binding (Merritt *et al.* 2007). Overexpression of SadC restored Congo red binding to greater than WT levels, but not in a *pelA* deletion background. However, multi-copy expression of SadC did not alter the levels of the *pelA* or *pelG* transcripts, suggesting that SadC indirectly regulates the production of Pel through its production of c-di-GMP (Merritt *et al.* 2007). SadB also functions in the transition from reversible to irreversible attachment, control of swarming motility, and production of Pel. Overexpression of SadB in a *sadC* mutant background restored biofilm formation (Caiazza *et al.* 2007). This is likely due to SadB mediated activation of *pelA;* however, the biochemical activity of SadB has yet to be determined. BifA functions as a PDE and therefore negatively regulates biofilm formation through degradation of c-di-GMP (Kuchma *et al.* 2007). Finally, the expression of Pel is dependent on the direct binding of c-di-GMP to the cytoplasmic C-terminus of PelD via a RXXD motif. Although RXXD motifs typically function as sites of allosteric inhibition for a DGC, c-di-GMP binding is required for the synthesis and export of the Pel polysaccharide (Lee *et al.* 2007).

The current understanding of the regulation of Pel polysaccharide synthesis is centered on the role of c-di-GMP. However, it is to be expected that the delineation of the structure and synthesis of Pel will generate a more refined understanding of the role of Pel in the production and maintenance of the biofilm structure, as well as identification of additional contributors to the regulatory framework of Pel biosynthesis.

5.4 ALGINATE

Alginate is currently the most well defined ePS due to its contribution to *P. aeruginosa* persistence in CF infections (Hay *et al.* 2009). It's designation as the first ePS identified in *P. aeruginosa* has allowed for extensive investigation into the complexity of alginate biosynthesis and regulation (Carlson *et al.* 1966; Doggett 1969). Overproduction of alginate is responsible for the mucoid phenotype, with the conversion to mucoidy often associated with late stages of infection and a worsening prognosis (Martin *et al.* 1993; Derry, 2005). Mucoid conversion provides protection from environmental and host stressors by enhancing viscosity and stability of the biofilm matrix while limiting host immune clearance (Devault *et al.* 1990; Govan & Deretic, 1996; Pier, 1998; Pier *et al.* 2001; Aaron *et al.* 2002). Mucoid strains of *P. aeruginosa* are solely isolated from clinical infections and have not been found in the environment outside of a host, emphasizing the specialized role the mucoid phenotype plays during infection (Bennett, 1974; Suman *et al.* 1993; Murphy *et al.* 2008; Cullen & McLean, 2015). The following sections will summarize the current knowledge of alginate synthesis and the complex regulatory system that controls its expression.

5.4.1 Alginate composition and structure

Alginate was originally identified in *P. aeruginosa* as polyuronic acids (Carlson & Matthews, 1966). A homologous polymer was additionally identified in the bacterium *Azotobacter vinelandii*. Therefore, *A. vinelandii* has often been utilized as a model system for the investigation of the mechanism and regulation of alginate biosynthesis (Pindar & Bucke, 1975; Pigott *et al.* 1981). Alginate is a heteropolymer with a high molecular weight, composed of non-repetitive monomers of β-1, 4- linked L-guluronic and D-mannuronic acid, which are O-acetylated at the 2,3 positions (Linker & Jones, 1964; Evans & Linker, 1973; Figure 5.1). Evans and Linker (1973) additionally demonstrated that the composition of alginate produced by clinical isolates demonstrates wide variation in O-acetylation (0–13.4%), guluronate content (0–75%), and molecular weight (1×10^4–4×10^5 Daltons). Pier *et al.* (2001) found that this O-acetylation is essential for the resistance of mucoid *P. aeruginosa* opsonic killing. The amount of O-acetylation also contributes to alginate composition in two key ways by determining: 1) the degree of mannuronate to guluronate epimerization and 2) the viscosity of the biofilm matrix (Nivens *et al.* 2001).

5.4.2 Alginate synthesis

An in-depth review of alginate biosynthesis is covered by Franklin *et al.* (2011); summarized below are the key details of the current knowledge. The alginate biosynthesis operon consists of *algD-algA*, encoded by PA3540-PA3551, which includes all required alginate synthesis components except *algC* (Govan *et al.* 1996; Chitnis & Ohman, 1993). Alginate is synthesized by polymerization of the activated sugar precursor GDP-mannuronic acid. The biosynthesis of GDP-mannuronic acid begins with the conversion of fructose-6-phosphate to mannose-6-phosphate by AlgA, a bifunctional activity as a phosphomannose isomerase and GDP-mannose pyrophosphorylase (Pelissier *et al.* 2010). AlgC, a phosphomannomutase and phosphoglucomutase converts mannose-6-phosphate to mannose-1-phosphate, which can then be activated with the addition of GDP by AlgA (Zielinski *et al.* 1991; Regni *et al.* 2002). AlgD converts activated mannose to activated mannuronate. The conversion of mannose to mannuronate is considered the rate-limiting step of alginate synthesis, and requires NADH to bind with GDP-mannose in the AlgD active site (Snook *et al.* 2003). Additionally, AlgD is the first enzyme within the alginate biosynthetic pathway whose expression is up regulated in mucoid cells, suggesting that it functions as a key enzymatic checkpoint (Deretic *et al.* 1987).

Once GDP-mannuronic acid is synthesized, it is polymerized and transported across the inner membrane by Alg8 and Alg44 (Maharaj *et al.* 1993; Remminghorst & Rehm, 2006; Oglesby *et al.* 2008). The alginate polymer is protected from degradation in the periplasmic space by the interaction of AlgG, AlgK, AlgX, and AlgL (Jain & Ohman, 1998; Jain *et al.* 2003; Robles-Price *et al.* 2004; Keiski *et al.* 2010; Weadge *et al.* 2010). AlgK provides structural support as a scaffold component and AlgL acts as an alginate lyase to degrade free alginate within the periplasmic space to prevent cell lysis (Schiller *et al.* 1993; Jain & Ohman, 1998; Jain & Ohman, 2005). AlgG, a C-5 epimerase, converts mannuronate to guluronate residues, which are then O-acetylated at the -2, -3, or -2,3 position by a complex of AlgF, AlgJ, and AlgI (Chitnis & Ohman, 1990; Franklin & Ohman, 1993; Franklin *et al.* 1994; Franklin & Ohman, 2002). AlgE, an outer membrane protein, then transports the copolymer of alginate to the surface of the cell (Rehm *et al.* 1994; Whitney *et al.* 2009; Hay *et al.* 2010; Whitney *et al.* 2011). AlgK, which contains a lipid moiety that anchors the protein to the outer membrane, is proposed to interact with AlgE to ensure the proper localization and insertion of AlgE, and together, to function as a novel secretion complex (Keiski *et al.* 2010).

P. aeruginosa environmental isolates produce little detectable alginate. However, non-mucoid strains can exhibit the mucoid phenotype when grown in conditions mimicking the CF lung (Terry *et al.* 1991). In fact, mucoid isolates from CF are clonal variants of previously isolated non-mucoid isolates (Ogle *et al.* 1987; Martin *et al.* 1993). There are several factors that trigger the transition to mucoidy when present during growth, these include surfactant, sub-inhibitory concentrations of antibiotics and antimicrobial peptides, nutrient limitation, reactive oxygen species, and ethanol (Govan, 1975; Govan & Fyfe, 1978; Devault *et al.* 1990; Mathee *et al.* 1999; Limoli *et al.* 2014). Though spontaneous reversion of mucoid variants to non-mucoid commonly occurs once a selective environmental pressure is removed, acquisition of stable mutations in the *muc* operon, *mucABCD*, initiates alginate overproduction. MucA is the primary regulator of alginate synthesis, and disruption of this regulator, either by physiological stimuli or genetic mutation, results in the conversion to mucoidy (Fyfe & Govan, 1981; Martin *et al.* 1993; Damron *et al.* 2012). In addition, a hyper-mutator phenotype frequently observed during CF infection may promote the generation of *muc* mutations (Warren *et al.* 2011).

5.4.3 Transcriptional regulation of alginate

As mentioned previously, the majority of genes responsible for alginate synthesis are located in an operon (*algD-algA* [PA3540-PA3551]) and are transcriptionally controlled by the *algD* promoter. Regulation and activation of the *algD* promoter is summarized in Figure 5.5 and 5.6. The network of transcriptional regulators that influence alginate production at this site is complex, and many regulatory components function by altering promoter activity. The sigma factor AlgT (also known as AlgU and σ^{22}) closely resembles members of the σ^{E} sigma factor family and is required to activate alginate synthesis gene expression. AlgT functions by binding to the *algD* promoter and associating with RNA polymerase to induce transcription (Martin *et al.* 1993). Hershberger *et al.* (1995) and Schurr *et al.* (1995) showed that AlgT binds to the σ^{22} consensus site on both the *algD* and *algR* promoters (discussed in further detail later in this section). AlgT has a broader role as an alternative σ factor, and CHiP-seq and RNA-seq analysis has revealed that 194 genes are uniquely controlled by AlgT (Schulz *et al.* 2015). AlgT likely functions during stress, which is demonstrated by reduced survival of *algT* mutant strains in high temperature conditions and in the presence of reactive oxygen species (Martin *et al.* 1994; Yu *et al.* 1995).

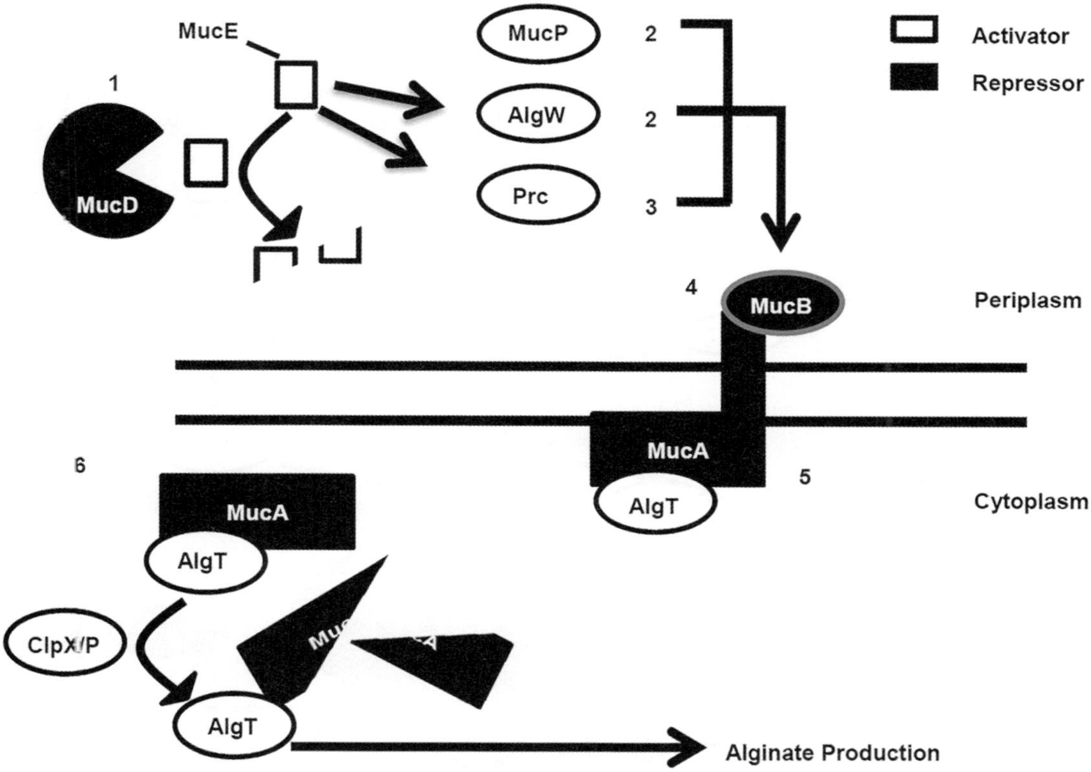

Figure 5.5 Activation of the alginate synthesis sigma factor AlgT 1) MucD degradation of MucE prevents buildup within the periplasm 2) Accumulation of MucE results in activation of the proteases AlgW and MucP which degrade the periplasmic portion of MucA resulting in release of AlgT. 3) The protease Prc degrades truncated MucA, which results from muc-2 and muc-22 mutations, and leads to release of AlgT. 4) MucB binds to the periplasmic portion of MucA and protects it from protease degradation. 5) MucA functions as an anti-sigma factor that binds AlgT and prevents σ factor activity 6) Degredation of the periplasmic portion of MucA, ClpX and ClpP proteases degrade and the remaining portion of MucA releasing active AlgT.

In wild type nonmucoid *P. aeruginosa* strains, MucA is anchored to the inner membrane by a single transmembrane domain, and its cytoplasmic portion acts as an anti-sigma factor that directly binds AlgT, preventing σ factor activity and transcriptional activation (Xie *et al.* 1996; Schurr *et al.* 1996). The *muc-2* or *muc-22* mutation results in loss of a functional MucA anti-sigma factor, leading to mucoid conversion due to constitutive AlgT activity (Govan & Deretic, 1996). Mucoid strains containing either of these mutations are frequently isolated from chronic CF lung infections, and up to 84% of mucoid *P. aeruginosa* isolates contain a mutation in *mucA* (MacGeorge *et al.* 1986; Flynn & Ohman, 1988; Boucher *et al.* 1997). The mechanisms leading to the generation of the *muc*-2 and *muc*-22 mutations are still being investigated. However, considering the prevalence of mucoid development during CF lung infection, the *mucA* gene appears to be highly susceptible to mutation within the host lung environment. The inflammatory environment within the CF lung contains an abundance of host produced mutagenic factors such as reactive oxygen species. *In vitro* studies have observed increased rates of mucoid conversion after exposure to sub-lethal concentrations of polymorphonucleocyte products such as hydrogen peroxide and antimicrobial peptides (Mathee *et al.* 1999; Limoli *et al.* 2014). The resulting loss of functional MucA by these mutations leads to constitutive alginate production (Martin *et al.* 1993; Xie *et al.* 1996; Schurr *et al.* 1996). In addition to mutagenic factors, proteolytic degradation of MucA triggered by cell wall stress and overexpression of an envelope protein (discussed below) stimulate AlgT activity (Wood *et al.* 2006; Qui *et al.* 2007).

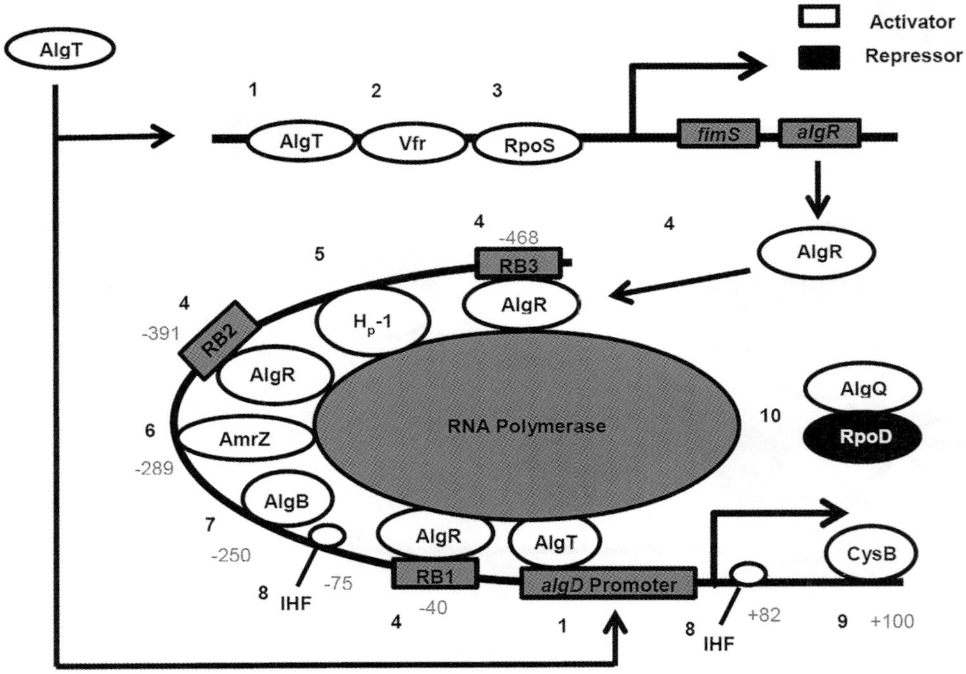

Figure 5.6 Transcriptional Regulation of Alginate. 1,2,3) AlgT binds upstream of the *algD* and *algR* promoter and Vfr and RpoS bind upstream of the *algR* promoter to activate transcription. 4) AlgR binds at three sites (RB1, RB2, and RB3) upstream of the *algD* promoter facilitate RNA polymerase binding and transcription. 5) H_p-1 and AlgQ are functionally redundant and promote *algD* transcription at the same site. 6,7,8,9) AmrZ, AlgB, IHF, and CysB bind at various sites around the *algD* promoter to enhance transcription of Alginate biosynthesis proteins. 10) AlgQ acts as an anti-sigma factor to sequester RpoD (σ^{70}) allowing AlgT to interact with RNA polymerase.

MucB (AlgN) localizes to the periplasm and is able to interact with the carboxy-terminus of MucA (Mathee *et al.* 1997; Rowen & Deretic, 2000). MucB is predicted to act as a receptor for environmental stress signals, and it regulates *algT* expression by binding to a periplasmic domain of MucA and enhances sequestration of AlgT (Mathee *et al.* 1997). The *muc-2* and *muc-22* mutations result in a truncated MucA C-terminal periplasmic domain (Boucher *et al.* 1997). Truncated MucA is more susceptible to protease degradation, while MucB binding blocks and protects MucA from proteolytic cleavage (Cezairliyan & Sauer, 2009). The activity of multiple proteases impact mucoid conversion. The protease Prc degrades truncated MucA, while AlgW and MucP degrade MucA in response to high concentrations of a membrane stress signaling protein, MucE (Reiling *et al.* 2005; Boucher *et al.* 1996; Qiu *et al.* 2007; Wood & Ohman, 2009; Damron & Yu, 2011). AlgW and MucP protease activity is regulated by the serine protease MucD, which degrades MucE that would otherwise accumulate and activate degradation of MucA (Boucher *et al.* 1996; Damron & Yu, 2011; Qiu *et al.* 2007; Wood & Ohman, 2009). While overexpressing MucD suppresses a MucE induced mucoid phenotype (Qiu *et al.* 2007), the accumulation of other factors in the periplasm are still capable of activating AlgW. Finally, two cytoplasmic proteases, ClpX and ClpP, were identified that activate AlgT by removing the residual portion of MucA that remains attached to AlgT following degradation by the membrane associated proteases (Qui *et al.* 2008). In addition to MucABD, there is evidence that MucC works synergistically with MucA and MucB to repress alginate production in conditions of high temperature and osmolarity (Boucher *et al.* 1997). While the mechanism of MucC regulation in *P. aeruginosa* is still not fully understood, MucC may directly inhibit AlgT activity during stress as observed in *Azotobacter vinelandii* (Martínez-Salazar *et al.* 1996; Núñez *et al.* 2000).

A third class of mucoid strains, *muc-23*, contains a mutation in the *pilA* gene resulting in truncated PilA, the major structural subunit of the type IV pilus (Withers *et al.* 2013). The truncated PilA contains an exposed phenylalanine-threonine-phenylalanine (FTF) sequence in the C-terminal region that activates AlgW-dependent proteolytic breakdown of MucA and constitutive AlgT activity (Withers *et al.* 2013). Additionally, the mechanism for *muc-23* mucoid conversion differs in that it requires the sigma factor RpoN (σ54) for activation of *algD* transcription (Boucher *et al.* 2000; Withers *et al.* 2013). While this mutation results in conversion to mucoidy, *muc-23* mutants have not yet been isolated from a CF lung infection (Boucher *et al.* 1997).

In addition to AlgT, a multitude of other transcription factors are also involved in regulating *algD* promoter activity (Fig. 5.6). AlgR is the response regulator of a two-component system, and is co-transcribed with *fimS* (*algZ*), which was originally predicted to act as a histidine sensor kinase (Whitchurch *et al.* 1996; Yu *et al.* 1997). The interaction between these two components remains to be determined, and direct phosphotransfer between the two proteins has not been observed. However, since the discovery of FimS it has been determined that phosphorylation of AlgR is not required for alginate synthesis, and removal of the phosphorylation site on AlgR had no effect on alginate production (Ma *et al.* 1998). Although FimS may not function as a histidine kinase, a deletion mutant showed increased production of alginate indicating it may have a minor inhibitory role (Yu *et al.* 1997). AlgR is required to activate *algD* expression, and it functions by binding directly to three different sites on the *algD* promoter deemed RB1, RB2, and RB3 located at −468, −391 and −40 relative to the transcription start site (Deretic *et al.* 1991; Mohr *et al.* 1992). The abnormal distance upstream of the AlgR binding sites relative to the *algD* promoter suggests that DNA looping may be required for regulatory interactions to occur between AlgR and RNA polymerase. Although the majority of the alginate biosynthesis proteins are located within the *algD* operon, the *algC* gene is the exception and is required for alginate synthesis and is located outside of the *algD* operon (Zielinski *et al.* 1991; Ye *et al.* 1994). In addition to *algD* activation, AlgR also serves as an activator of *algC* transcription (Fujiwara *et al.* 1993)). As an additional level of control, AlgT, RpoS, and Vfr promote alginate production by enhancing *algR* and *fimS* transcription. AlgT and Vfr activate *algR* and *fimS* expression while RpoS activates only *algR* expression by binding upstream of the transcription start site (Pritchett *et al.* 2015).

The *psl* transcription factor AmrZ (AlgZ) was first discovered for its role in alginate regulation. The ribbon-helix-helix DNA binding protein binds to a sequence −280 bp upstream of the *algD* promoter and promotes *algD* transcription (Baynham & Wozniak, 1996). Similar to previously mentioned *algD* activation factors, *amrZ* transcription requires the AlgT sigma factor (Baynham & Wozniak, 1996). Recent reports suggest that AmrZ may have an additional auto-regulatory role in repressing its own synthesis. Two AmrZ binding sites upstream of the *amrZ* promoter were identified and binding resulted in repression of *amrZ* expression (Ramsey *et al.* 2005).

A number of other DNA binding proteins regulate alginate production by binding on or near the *algD* promoter. ChIP analysis revealed AlgB binds upstream of the *algD* promoter (–250) *in vivo*, and an R442E mutation at the predicted AlgB-DNA binding site completely abolished DNA binding (Leech *et al.* 2008). AlgB activates alginate production, and its deletion results in loss of the mucoid phenotype and significantly reduced alginate production (Goldberg & Ohman, 1987). AlgB belongs to the NtrC family of two-component response regulators (Wozniak & Ohman, 1991; Goldberg & Dahnke, 1992); however, in mucoid strains its function appears to be independent of its sensor kinase, KinB, and its phosphorylation state (Ma *et al.* 1998). While KinB may not regulate AlgB activity, it negatively regulates alginate production by controlling AlgW protease activity in mucoid strains (Damron *et al.* 2009; Damron *et al.* 2012). Consistent with this finding, deletion of *kinB* in non-mucoid strains results in alginate production (Damron *et al.* 2009; Withers *et al.* 2014). The histone-like protein integration host factor (IHF) also regulates alginate production by binding to two different regions of the *algD* promoter. The first IHF binding site is between −80 bp and −68 bp and the second is between +79 bp and +105 bp relative to the *algD* transcription start site. Mutations at either of these sites reduced *algD* transcription, revealing that IHF binding is required for alginate production (Wozniak, 1994). An additional DNA binding protein, CysB, binds downstream of the *ihf2* binding site (+100) and promotes *algD* promoter activity (Delic-Attree *et al.* 1997).

Another regulator of alginate production is AlgQ (AlgR2), which was identified as a transcriptional activator that functions cooperatively with AlgR to promote *algD* expression (Kato *et al.* 1989). In the presence of AlgR, overexpression of *algQ* resulted in alginate production (Kato *et al.* 1989). More recent evidence has shown that AlgQ is an Rsd family protein that acts as an anti-sigma factor to sequester RpoD allowing AlgT to initiate transcription (Dove & Hochschild, 2001; Yuan *et al.* 2008). AlgP (AlgR3) is a histone-like regulatory protein that was renamed to H_p-1 when its broader role in structure and regulation was identified in systems outside of alginate production. Removal of H_p-1from *P. aeruginosa* results in loss of alginate production (Deretic & Konyecsni, 1990; Deretic *et al.* 1992; Kato *et al.* 1990). Complementation studies identified partial functional redundancy between H_p-1 and AlgQ, and both transcription factors were able to restore alginate production in an *alg52* mutant strain (Kato *et al.* 1990). The regulatory mechanism of these proteins has yet to be determined. However, previous reports on H_p-1 DNA binding affinity suggests it may play a structural role in the formation of a DNA loop (Medvedkin *et al.* 1995).

5.4.4 Post-transcriptional regulation of alginate

Alg44 is essential for the polymerization of alginate, though the exact role is not fully understood (Remminghorst & Rehm, 2006). C-di-GMP binds the cytoplasmic PilZ domain, inducing conformational changes; however, the effects of these changes on the polymerization and/or export of alginate have yet to be identified (Merighi *et al.* 2007; Hay *et al.* 2009). It has been proposed that the binding of c-di-GMP exposes sites involved in protein-protein interactions, specifically for Alg8; Alg8 is a glycosyltransferase that is proposed to form a co-polymerase complex with Alg44, as evidenced by co-stability experiments (Benach *et al.* 2007; Habazettl *et al.* 2011; Ryan *et al.* 2012; Rehman *et al.* 2013). A local pool of c-di-GMP is provided to Alg44 by MucR, an inner membrane protein containing cytoplasmic domains necessary for DGC and PDE activity; this production of c-di-GMP by MucR occurs primarily in the biofilm mode of growth (Hay *et al.* 2009; Li *et al.* 2013).

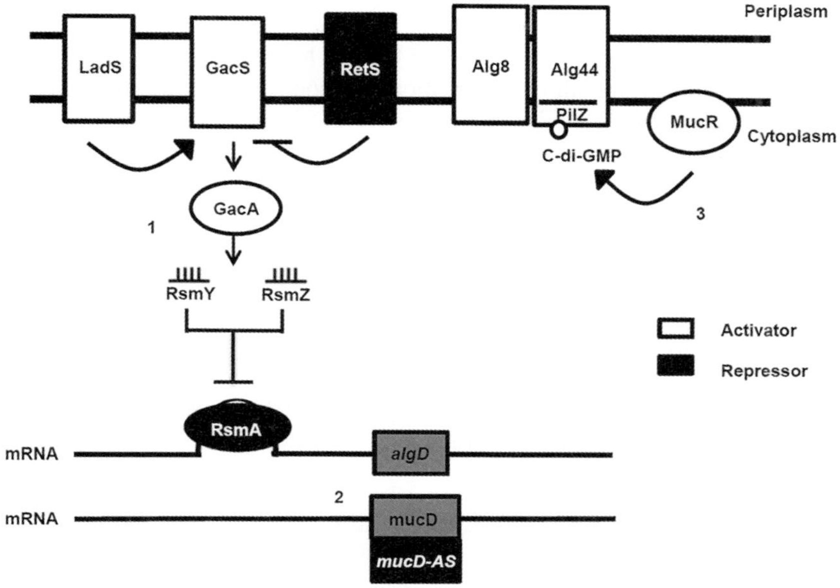

Figure 5.7 Post-transcriptional regulation of Alginate. 1) RsmA binds to the 5' UTR of *algD* mRNA stabilizing a secondary structure that prevents ribosome binding and translation; RsmA repression is relieved by the production of the sRNAs, rsmY and rsmZ, which are produced via the GacS/GacA two component signaling system. 2) The translation of *mucD* is repressed by the anti-sense transcript mucD-AS, which prevents alginate production. 3) A local pool of c-di-GMP provided by MucR binds the PilZ domain of the C-terminus of Alg44, which in complex with Alg8, allows for the polymerization and export of alginate.

The GacSA two-component system, mentioned previously for its role in the regulation of the Psl and Pel polysaccharides, is also likely to be involved in the regulation of alginate biosynthesis. RsmA binds the 5′ leader sequence of the *algD* mRNA in *A. vinelandii*. (Rehm, 2010; Manzo *et al.* 2011). Though the GacSA system has not been shown to directly control alginate biosynthesis in *P. aeruginosa*, the expression of *algD*, as well as *alg8*, were upregulated in an *rsmA* mutant; similarly, GacA induction of *algD* is required for the production of alginate (Castañeda *et al.* 2001; Burrowes *et al.* 2006). Together, these findings suggest a role for the GacSA system in the regulation of alginate biosynthesis by *P. aeruginosa*.

Post-translational regulation also occurs through the action of a natural anti-sense transcript, designated mucD-AS, which acts to block translation of MucD resulting in subsequent alginate production (Yang *et al.* 2011). Similar to other ePS, alginate shares common sugar precursors with other biosynthetic pathways. For example, glucose-6-phosphate, fructose-6-phosphate, and mannose-6-phosphateare are utilized in both Psl and LPS synthesis. Limited availability of precursor components may serve as a form of indirect post-translational regulation of alginate biosynthesis (Ma *et al.* 2012).

It is evident from the preceding paragraphs that the early discovery of alginate allowed for extensive investigation into the complex, multi-level regulation of the expression and biosynthesis of alginate. It is only to be expected that future work will not only continue to provide additional knowledge of alginate, but will shift focus to similar in-depth explorations of Pel and Psl polysaccharide production.

5.5 CONCLUSIONS AND PERSPECTIVES

The study of *P. aeruginosa* as a model organism for biofilm formation, polysaccharide production, and genetic regulation has yielded a wealth of information that contributes to advancing our understanding of the biofilm as a complex, multi-cellular community. The contributions of ePS to biofilm formation, structure, and maintenance have been shown to be indispensable. This importance is further demonstrated by the critical roles of ePS during biofilm formation by other microbial species, including organisms such as *Staphylococcus aureus*, *Vibrio cholerae*, and *Escherichia coli*. Identification of the temporal and spatial contribution of secreted ePS under conditions that mimic the host environment will significantly enhance our understanding of biofilm development, and therefore our ability to modulate biofilm formation and reduce the burden of chronic infection. It is important to note that additional components, such as eDNA, proteins, and lipids also provide notable contributions to the biofilm matrix. Future work within the *P. aeruginosa* field, as well as within the biofilm community at large, will focus on determining the interaction of such components with secreted ePS, as well as investigating more thoroughly the contribution of multiple polymeric substances within polymicrobial biofilms.

REFERENCES

Aaron S. D., Ferris W., Ramotar K., Vandemheen K., Chan F. and Saginur R. (2002). Single and combination antibiotic susceptibilities of planktonic adherent, and biofilm-grown *Pseudomonas aeruginosa* isolates cultured from sputa of adults with cystic fibrosis. *J. Clin. Microbiol.*, **40**, 4172–4179.

Arora S. K., Ritchings B. W., Almira E. C., Lory S. and Ramphal R. (1997). A transcriptional activator, FleQ, regulates mucin adhesion and flagellar gene expression in *Pseudomonas aeruginosa* in a cascade manner. *J. Bacteriol.*, **179**, 5574–5581.

Baker P., Whitfield G. B., Hill P. J., Little D. J., Pestrak M. J., Robinson H., Wozniak D. J. and Howell P. L. (2015). Characterization of the *Pseudomonas aeruginosa* glycoside hydrolase PslG reveals that its levels are critical for Psl polysaccharide biosynthesis and biofilm formation. *J. Biol. Chem.*, **290**(47), 28374–28387. doi: 10.1074/jbc.M115.674929.

Baraquet C., Murakami K., Parsek M. R. and Harwood C. S. (2012). The FleQ protein from *Pseudomonas aeruginosa* functions as both a repressor and an activator to control gene expression from the Pel operon promoter in response to c-di-GMP. *Nucl Acids Res.*, **40**, 7207–7218.

Barraud N., Schleheck D., Klebensberger J., Webb J. S., Hassett D. J., Rice S. A. and Kjelleberg S. (2009). Nitric oxide signaling in Pseudomonas aeruginosa biofilms mediates phosphodiesterase activity, decreased cyclic di-GMP levels, and enhanced dispersal. *J. Bacteriol.*, **191**, 7333–7342.

Battesti A., Majdalani N., and Gottesman S. (2011). The RpoS-mediated general stress response in *Escherichia coli*. *Annu Rev Microbiol*, **65**, 189–213.

Baynham P. J. and Wozniak D. J. (1996). Identification and characterization of AlgZ, an AlgT-dependent DNA-binding protein required for Pseudomonas aeruginosa algD transcription. *Mol Microbiol.*, **22**, 97–108.

Benach J., Swaminathan S. S., Tamayo R., Handelman S. K., Folta-Stogniew E., Ramos J. E., Forouhar F., Neely H., Seetharaman J., Camilli A. and Hunt J. F. (2007). The structural basis of cyclic diguanylate signal transduction by PilZ domains. *EMBO J.*, **26**, 5153–5166.

Bennett J. V. (1974). Nosocomial infections due to *Pseudomonas*. *J. Infect. Dis.*, **130**.

Berry A., DeVault J. D., and Chakrabarty A. M. (1989). High osmolarity is a signal for enhanced algD transcription in mucoid and nonmucoid Pseudomonas aeruginosa strains. *J Bacteriol.*, **171**(5), 2312–2317.

Billings N., Millan M., Caldara M., Rusconi R., Tarasova Y., Stocker R. and Ribbeck K. (2013). The extracellular matrix component Psl provides fast-acting antibiotic defense in *Pseudomonas aeruginosa* biofilms. *PLoS Pathog*, **9**(8), 1003526.

Bodey G. P., Bolivar R., Fainstein V., Jadeja L., Reviews S. and Apr N. M. (1983). Infections caused by *Pseudomonas aeruginosa*. *Rev Infect Dis.*, **5**(2), 279–313.

Boles B. R., Thoendel M. and Singh P. K. (2004). Self-generated diversity produces 'insurance effects' in biofilm communities. *Proc. Natl. Acad. Sci. U.S A.*, **101**, 16630–16635.

Bonsor D. A., Grishkovskaya I., Dodson E. J. and Kleanthous C. (2007). Molecular mimicry enables competitive recruitment by a natively disordered protein. *J. Am. Chem. Soc.*, **129**, 4800–4807.

Borlee B. R., Goldman A. D., Murakami K., Samudrala R., Wozniak D. J. and Parsek M. R. (2010). *Pseudomonas aeruginosa* uses a cyclic-di-GMP-regulated adhesin to reinforce the biofilm extracellular matrix. *Mol Microbiol.*, **75**(4), 827–842.

Boucher J. C., Schurr M. J., Mudd M. H., and Yu H. (1996). Two distinct loci affecting conversion to mucoidy in *Pseudomonas aeruginosa* in Cystic Fibrosis encode homologs of the serine protease HtrA. *J Bacteriol.*, **178**(2), 511–523.

Boucher J. C., Yu H., Mudd M. H. and Deretic V. (1997). Mucoid *Pseudomonas aeruginosa* in Cystic Fibrosis: characterization of *muc* mutations in clinical isolates and analysis of clearance in a mouse model of respiratory infection. *Infect Immun*, **65**, 3838–3846.

Boucher J. C., Schurr M. J., and Deretic V. (2000). Dual regulation of mucoidy in *Pseudomonas aeruginosa* and sigma factor antagonism, Mol. *Microbiol.*, **36**, 341–351.

Boucher J. C., Schurr M. J., Yu H., Rowen D. W., and Deretic V. (1997). *Pseudomonas aeruginosa* in Cystic Fibrosis: role of *mucC* in the regulation of alginate production and stress sensitivity. *Microbiology*, **143**(1997), 3473–3480.

Brencic A., and Lory S. (2009). Determination of the regulon and identification of novel mRNA targets of *Pseudomonas aeruginosa* RsmA. *Mol Microbiol*, **72**(3), 612–632.

Brencic A., Mcfarland K. A., Mcmanus H. R., Castang S., Dove S. L., and Lory S. (2009). The GacS/GacA signal transduction system of *Pseudomonas aeruginosa* acts exclusively through its control over the transcription of the RsmY and RsmZ regulatory small RNAs. *Mol Microbiol*, **73**(3), 434–445.

Burrowes E., Baysse C., Adams C. and O'Gara F. (2006). Influence of the regulatory protein RsmA on cellular functions in *Pseudomonas aeruginosa* PAO1, as revealed by transcriptome analysis. *Microbiology*, **152**, 405–418.

Byrd M. S., Sadovskaya I., Vinogradov E., Lu H., Sprinkle A. B., Richardson S. H., Ma L., Ralston B., Parsek M. R., Anderson E. M., Lam J. S. and Wozniak D. J. (2009). Genetic and biochemical analyses of the *Pseudomonas aeruginosa* Psl exopolysaccharide reveal overlapping roles for polysaccharide synthesis enzymes in Psl and LPS production. *Mol Microbiol.*, **73**(4), 622–638.

Caiazza N. C. and O'Toole G. A. (2004). *SadB* is required for the transition from reversible to irreversible attachment during biofilm formation by Pseudomonas aeruginosa PA14. *J. Bacteriol.*, **186**, 4476–4485.

Carlson D. M. and Matthews L. W. (1966). Polyuronic acids produced by *Pseudomonas aeruginosa*. *Biochemistry*, **5**, 2817–2822.

Castañeda M., Sánchez J., Moreno S., Núñez C. and Espín G. (2001). The global regulators GacA and σS form part of a cascade that controls alginate production in *Azotobacter vinelandii*. *J. Bacteriol.*, **183**, 6787–6793.

Cezairliyan B. O. and Sauer R. T. (2009). Control of *P. aeruginosa* AlgW protease cleavage of MucA by peptide signals and MucB. *Molecular Microbiology*, **72**(2), 368–379.

Chew S. C., Kundukad B. and Seviour T. (2014). Dynamic remodeling of microbial biofilms by functionally distinct. *mBio*, **5**(4), 1–11.

Chitnis C. E. and Ohman D. E. (1990). Cloning of *Pseudomonas aeruginosa algG*, which controls alginate structure. *J. Bacteriol.*, **172**, 2894–2900.

Colvin K. M., Alnabelseya N., Baker P., Whitney J. C., Howell P. L. and Parsek M. R. (2013). PelA deacetylase activity is required for Pel polysaccharide synthesis in *Pseudomonas aeruginosa*. *J. Bacteriol.*, **195**, 2329–2339.

Colvin K. M., Irie Y., Tart C. S., Urbano R., Whitney J. C., Ryder C., Howell P. L., Wozniak D. J. and Parsek M. R. (2012). The Pel and Psl polysaccharides provide *Pseudomonas aeruginosa* structural redundancy within the biofilm matrix. *Environ. Microbiol.*, **14**, 1913–1928.

Colvin K. M., Gordon V. D., Murakami K., Borlee B. R., Wozniak D. J., Wong G. C. and Parsek M. R. (2011). The Pel polysaccharide can serve a structural and protective role in the biofilm matrix of *Pseudomonas aeruginosa. PLoS Pathog.*, **7**.

Cooley B., Thatcher T., Hashmi S. and L'Her G. (2013). The extracellular polysaccharide Pel makes the attachment of *P. aeruginosa* to surfaces symmetric and short-ranged. *Soft Matter*, **9**, 3871–3876.

Costerton J. W., Stewart P. S. and Greenberg E. P. (1999). Bacterial biofilms: a common cause of persistent infections. *Science*, **284**, 1318–1323.

Coutinho P. M., Deleury E., Davies G. J. and Henrissat B. (2003). An evolving hierarchical family classification for glycosyltransferases. *J. Mol. Biol.*, **328**, 307–317.

Cullen L. *et al.* (2015). Phenotypic characterization of an international *Pseudomonas aeruginosa* reference panel: strains of cystic fibrosis origin show less *in vivo* virulence than non-CF strains. *Microbiology.*, **161**, 1961–1977.

Cuthbertson L., Mainprize I. L., Naismith J. H. and Whitfield C. (2009). Pivotal roles of the outer membrane polysaccharide export and polysaccharide copolymerase protein families in export of extracellular polysaccharides in gram-negative bacteria. *Microbiol Mol Biol Rev.*, **73**(1), 155–177.

Damron F. H. and Goldberg J. B. (2012). Proteolytic regulation of alginate overproduction in *Pseudomonas aeruginosa. Mol Microbiol*, **84**, 595–607.

Damron F. H., Qiu D. and Yu H. D. (2009). The *Pseudomonas aeruginosa* sensor kinase KinB negatively controls alginate production through AlgW-Dependent MucA Proteolysis. *J Bacteriol.*, **191**(7), 2285–2295.

Damron F. H., Owings J. P., Okkotsu Y., Varga J. J., Schurr J. R., Goldberg J. B. and Schurr M. J. (2012). Analysis of the *Pseudomonas aeruginosa* Regulon controlled by the sensor kinase KinB and sigma factor RpoN. *J Bacteriol.*, **194**(6), 1317–1330.

Damron F. H. and Yu H. D. (2011). *Pseudomonas aeruginosa* MucD regulates the alginate pathway through activation of MucA degradation via MucP proteolytic activity. *Journal of Bacteriology*, **193**(1), 286–291.

Darling K. E., Dewar A. and Evans T. J. (2004). Role of the cystic fibrosis transmembrane conductance regulator in internalization of *Pseudomonas aeruginosa* by polarized respiratory epithelial cells. *Cell Microbiol.*, **6**(6), 521–533.

Darzins A., Frantz B., Vanags R. I. and Chakrabarty A. M. (1986). Nucleotide sequence analysis of the phosphomannose isomerase gene (*pmi*) of *Pseudomonas aeruginosa* and comparison with the corresponding *Escherichia coli gene manA. Gene*, **42**(3), 293–302.

Davies J. C., Alton E. W. and Bush A. (2007). Cystic fibrosis. *BMJ*, **335**(7632), 1255–1259.

Davies D. G. and Marques C. N. H. (2009). A fatty acid messenger is responsible for inducing dispersion in microbial biofilms. *J. Bacteriol.*, **191**, 1393–1403.

Delic-Attree I., Toussaint B., Garin J. and Vignais P. M. (1997). Cloning, sequence and mutagenesis of the structural gene of *Pseudomonas aeruginosa CysB*, which can activate *algD* transcription. *Mol Microbiol.*, **24**(6), 1275–1284.

Deretic V. and Konyecsni W. M. (1990). A prokaryotic regulatory factor with a histone HI-Like carboxy-terminal domain: clonal variation of repeats within *algP*, a gene involved in regulation of mucoidy in *Pseudomonas aeruginosa. J. Bacteriol.*, **172**(10), 5544–5554.

Deretic V., Dikshit R., Konyecsni W. M., Chakrabarty A. M. and Misra T. K. (1989). The *algR* Gene, which regulates mucoidy in *Pseudomonas aeruginosa*, belongs to a class of environmentally responsive genes. *J. Bacteriol.*, **171**(3), 1278–1283.

Deretic V., Gill J. F. and Chakrabarty A. M. (1987). Gene *algD* coding for GDP mannose dehydrogenase is transcriptionally activated in mucoid *Pseudomonas aeruginosa. J. Bacteriol.*, **169**, 351–358.

Deretic V., Hibler N. S. and Holt S. C. (1992). Immunocytochemical analysis of AlgP (Hp1), ahHistone-like element participating in control of mucoidy in *Pseudomonas aeruginosa. J. Bacteriol.*, **174**(3), 824–831.

DeVault J. D., Berry A., Misra T. K., Darzins A., and Chakrabarty A. M. (1989). Environmental sensory signals and microbial pathogenesis: *Pseudomonas aeruginosa* infection in cystic fibrosis. *Biotechnology*, **7**, 352–357.

DeVault J. D., Kimbara K. and Chakrabarty A. M. (1990). Pulmonary dehydration and infection in Cystic Fibrosis: evidence that ethanol activates alginate gene expression and induction of mucoidy in *Pseudomonas aeruginosa. Mol. Microbiol.*, **4**, 737–745.

Doggett R. G. (1969). Incidence of mucoid *Pseudomonas aeruginosa* from clinical sources. *Appl. Microbiol.*, **18**, 936–937.

Dove S. L. and Hochschild A. (2001). Bacterial two-hybrid analysis of interactions between region 4 of the σ^{70} subunit of RNA polymerase and the transcriptional regulators Rsd from *Escherichia coli* and AlgQ from *Pseudomonas aeruginosa*. *Genes Dev.*, **183**(21), 6413–6421.

Evans L. R. and Linker A. (1973). Production and characterization of the slime polysaccharide of *Pseudomonas aeruginosa*. *J. Bacteriol.*, **116**, 915–924.

Feltman H., Schulert G., Khan S., Jain M., Peterson L. and Hauser A. R. (2001). Prevalence of type III secretion genes in clinical and environmental isolates of *Pseudomonas aeruginosa*. *Microbiology*, **147**, 2659–2669.

Flynn J. and Ohman D. (1988). Cloning of genes from mucoid *Pseudomonas aeruginosa*, which control spontaneous conversion to the alginate production phenotype. *J. Bacteriol.*, **170**, 1452–1460.

Franco A. V., Liu D., and Reeves P. R. A. (1996). Wzz (Cld) protein determines the chain length of K lipopolysaccharide in *Escherichia coli* O8 and O9 strains. *J. Bacteriol.*, **178**(7), 1903–1907.

Franklin M. J., Nivens D. E., Weadge J. T., and Howell P. L. (2011). Biosynthesis of the *Pseudomonas aeruginosa* extracellular polysaccharides, Alginate, Pel, and Psl. *Front Microbiol.*, **2**, 167.

Franklin M. J. and Ohman D. E. (1993). Identification of *algF* in the alginate biosynthetic gene cluster of *Pseudomonas aeruginosa*, which is required for alginate acetylation. *J. Bacteriol.*, **175**, 5057–5065.

Franklin M. J. and Ohman D. E. (2002). Mutant analysis and cellular localization of the AlgI, AlgJ, and AlgF proteins required for O acetylation of alginate in *Pseudomonas aeruginosa*. *J. Bacteriol.*, **184**, 3000–3007.

Franklin M. J., Chitnis C. E., Gacesa P., Sonesson A., White D. C. and Ohman D. E. (1994). *Pseudomonas aeruginosa* AlgG is a polymer level alginate C5-mannuronan epimerase. *J. Bacteriol.*, **176**, 1821–1830.

Friedman L. and Kolter R. (2004). Two genetic Loci Produce distinct carbohydrate-rich structural components of the *Pseudomonas aeruginosa* biofilm matrix. *J. Bacteriol.*, **186**(14), 4457–4465.

Friedman L. and Kolter R. (2004). Two genetic loci produce distinct carbohydrate-rich structural components of the Pseudomonas aeruginosa biofilm matrix. *J. Bacteriol.*, **186**, 4457–4465.

Fujiwara S., Zielinski N. A. and Chakrabarty A. M. (1993). Enhancer-Like activity of AlgR1-binding site in alginate gene activation: positional, orientational, and sequence specificity. *J. Bacteriol.*, **175**(17), 5452–5459.

Fyfe J. A. and Govan J. R. (1980). Alginate synthesis in mucoid *Pseudomonas aeruginosa*: a chromosomal locus involved in control. *J. Gen. Microbiol.*, **119**, 443–450.

Galán-vásquez E., Luna B. and Martínez-antonio A. (2011). The regulatory network of *Pseudomonas aeruginosa*. *Microbial Informatics and Experimentation*, **1**(1), 3.

Ghafoor A., Hay I. D. and Rehm B. H. A. (2011). Role of exopolysaccharides in *Pseudomonas aeruginosa* biofilm formation and architecture. *Appl. Environ. Microb.*, **77**(15), 5238–5246.

Goldberg J. B. and Dahnke T. (1992). *Pseudomonas aeruginosa* AlgB, which modulates the expression of alginate, is a member of the NtrC subclass of prokaryotic regulators. *Mol Microbiol.*, **6**, 59–66.

Goldberg J. B. and Ohman D. E. (1987). Construction and characterization of *algB* mutants of *Pseudomonas aeruginosa*: role of *algB* in high-level production of alginate. *J. Bacteriol.*, **169**, 1593–1602.

Goodman A. L., Merighi M., Hyodo M., Ventre I., Filloux A. and Lory S. (2009). Direct interaction between sensor kinase proteins mediates acute and chronic disease phenotypes in a bacterial pathogen. *Genes Dev.*, **23**(2), 249–259.

Goodman A. L., Kulasekara B., Rietsch A., Boyd D., Smith R. S. and Lory S. (2004). A signaling network reciprocally regulates genes associated with acute infection and chronic persistence in *Pseudomonas aeruginosa*. *Dev Cell*, **7**(5), 745–754.

Goodman A. L., Merighi M., Hyodo M., Ventre I., Filloux A. and Lory S. (2009). Direct interaction between sensor kinase proteins mediates acute and chronic disease phenotypes in a bacterial pathogen. *Genes and Development*, **23**(2), 249–259.

Govan J. R. and Deretic V. (1996). Microbial pathogenesis in Cystic Fibrosis: mucoid *Pseudomonas aeruginosa* and *Burkholderia cepacia*. *Microbiol. Rev.*, **60**, 539–574.

Govan J. R. W. (1975). Mucoid strains of *Pseudomonas aeruginosa*: the influence of culture medium on the stability of mucus production. *J. Med. Microbiol.*, **8**, 513–522.

Habazettl J., Allan M. G., Jenal U. and Grzesiek S. (2011). Solution structure of the PilZ domain protein *PA4608* complex with cyclic di-GMP identifies charge clustering as molecular readout. *J. Biol. Chem.*, **286**, 14304–14314.

Harmsen M., Yang L., Pamp S. J. and Tolker-Nielsen T. (2010). An update on *Pseudomonas aeruginosa* biofilm formation, tolerance, and dispersal. *FEMS Immunol Med Mic.*, **59**, 253–268.

Häussler S., Tummler B., Weissbrodt H., Rohde M. and Steinmetz I. (1999). Small-colony variants of *Pseudomonas aeruginosa* in Cystic Fibrosis. *Clin Infect Dis.*, **29**, 621–625.

Hay I. D., Gatland K., Campisano A., Jordens J. Z. and Rehm B. H. A. (2009). Impact of alginate overproduction on attachment and biofilm architecture of a supermucoid *Pseudomonas aeruginosa* strain. *Appl. Environ. Microbiol.*, **75**, 6022–6025.

Hay I. D., Rehman Z. U., Ghafoor A. and Rehm B. H. A. (2010). Bacterial biosynthesis of alginates. *J Chem Technol Biot*, **85**, 752–759.

Heeb S. and Haas D. (2001). Regulatory roles of the GacS/GacA two-component system in plant-associated and other gram-negative bacteria. *Mol Plant Microbe In.*, **14**(12), 1351–1363.

Hengge R. (2009). Principles of c-di-GMP signaling in bacteria. *Nat. Rev. Microbiol.*, **7**, 263–273.

Hentzer M. *et al.* (2001). Alginate overproduction affects *Pseudomonas aeruginosa* biofilm structure and function. *J. Bacteriol.*, **183**, 5395–5401.

Hershberger C. D., Ye R. W., Parsek M. R., Xie Z.-D. and Chakrabarty A. M. (1995). The *algT* (*algU*) gene of *Pseudomonas aeruginosa*, a key regulator involved in alginate biosynthesis, encodes an alternative sigma factor (σE). *Proc. Natl. Acad. Sci. USA*, **92**, 7941–7945.

Hickman J. W. and Harwood C. S. (2008). Identification of FleQ from Pseudomonas aeruginosa as a c-di-GMP-responsive transcription factor. *Mol Microbiol.*, **69**(2), 206–221.

Hickman J. W., Tifrea D. F. and Harwood C. S. (2005). A chemosensory system that regulates biofilm formation through modulation of cyclic diguanylate levels. *Proc. Natl. Acad. Sci. USA*, **102**(40), 14422–14427.

Høiby N. (2011). Recent advances in the treatment of *Pseudomonas aeruginosa* infections in cystic fibrosis. *BMC*, **9**(1), 32.

Huse H. K., Kwon T., Zlosnik J. E., Speert D. P., Marcotte E. M. and Whiteley M. (2013). *Pseudomonas aeruginosa* enhances production of a non-alginate exopolysaccharide during long-term colonization of the Cystic Fibrosis lung. *PLoS One*, **8**(12), e82621.

Irie Y., Borlee B. R., O'Connor J. R., Hill P. J., Harwood C. S., Wozniak D. J. and Parsek M. R. (2012). Self-produced exopolysaccharide is a signal that stimulates biofilm formation in *Pseudomonas aeruginosa*. *Proc. Natl. Acad. Sci. USA*, **109**(50), 20632–20636.

Irie Y., Starkey M., Edwards A. N., Wozniak D. J., Romeo T. and Parsek M. R. (2010). *Pseudomonas aeruginosa* biofilm matrix polysaccharide Psl is regulated transcriptionally by RpoS and post-transcriptionally by RsmA. *Mol. Micro.*, **78**(1), 158–172.

Irvin R. T., Govan J. W., Fyfe J. A. and Costerton J. W. (1981). Heterogeneity of antibiotic resistance in mucoid isolates of *Pseudomonas aeruginosa* obtained from Cystic Fibrosis patients: role of outer membrane proteins. *Antimicrob. Agents Chemother.*, **19**, 1056–1063.

Jackson K. D., Starkey M., Kremer S., Parsek M. R. and Wozniak D. J. (2004). Identification of *psl*, a locus encoding a potential exopolysaccharide that is essential for Pseudomonas aeruginosa PAO1 biofilm formation. *J. Bacteriol.*, **186**(14), 4466–4475.

Jain S. and Ohman D. E. (1998). Deletion of *algK* in mucoid *Pseudomonas aeruginosa* blocks alginate polymer formation and results in uronic acid secretion. *J. Bacteriol.*, **180**, 634–641.

Jain S. and Ohman D. E. (2005). Role of an alginate lyase for alginate transport in mucoid *Pseudomonas aeruginosa*. *Infect. Immun.*, **73**, 6429–6436.

Jain S., Franklin M. J., Ertesvåg H., Valla S. and Ohman D. E. (2003). The dual roles of AlgG in C-5-epimerization and secretion of alginate polymers in *Pseudomonas aeruginosa*. *Mol. Microbiol.*, **47**, 1123–1133.

Jennings L. K., Storek K. M., Ledvina H. E., Coulon C., Marmont L. S., Sadovskaya I., Secor P. R., Tseng B. S., Scian M., Filloux A., Wozniak D. J., Howell L. P. and Parsek M. R. (2015). Pel is a cationic exopolysaccharide that cross-links extracellular DNA in the Pseudomonas aeruginosa biofilm matrix. *Proc Natl Acad. Sci USA*, **112**(36), 11353–11358.

Jones C. J., Newsom D., Kelly B., Irie Y., Jennings L. K., Xu B. and Wozniak D. J. (2014). ChIP-Seq and RNA-Seq reveal an AmrZ-mediated mechanism for cyclic di-GMP synthesis and biofilm development by *Pseudomonas aeruginosa*. *PLoS Path*, **10**(3), e1003984.

Jones C. J., Ryder C. R., Mann E. E. and Wozniak D. J. (2013). AmrZ modulates *Pseudomonas aeruginosa* biofilm architecture by directly repressing transcription of the psl operon. *J. Bacteriol.*, **195**(8), 1637–1644.

Kato J., Chu L., Kitano K., DeVault J. D., Kimbara K., Chakrabarty A. M. and Misra T. K. (1989). Nucleotide sequence of a regulatory region controlling alginate synthesis in *Pseudomonas aeruginosa*: characterization of the *algR2* gene. *Gene*, **84**, 31–38.

Kato, J., Misra T. K. and Chakrabarty A. M. (1990). AlgR3, a protein resembling eukaryotic histone HI, regulates alginate synthesis in Pseudomonas aeruginosa. *Proc. Natl. Acad. Sci. USA*, **87**, 2887–2891.

Keiski C. L., Harwich M., Jain S., Neculai A. M., Yip P., Robinson H., Whitney J. C., Riley L., Burrows L. L., Ohman D. E. and Howell L. P. (2010). AlgK Is a TPR-Containing Protein and the Periplasmic Component of a Novel Exopolysaccharide Secretin. *Structure*, **18**, 265–273.

Kowalska K., Soscia C., Combe H., Vasseur P., Voulhoux R. and Filloux A. (2010). The C-terminal amphipathic alpha-helix of *Pseudomonas aeruginosa* PelC outer membrane protein is required for its function. *Biochimie*, **92**, 33–40.

Krogh A., Larsson B., von Heijne G. and Sonnhammeer E. L. (2001). Predicting transmembrane protein topology with a hidden Markov model: application to complete genomes. *Mol. Biol.*, **305**, 567–580.

Kuchma S. L., Brothers K. M., Merritt J. H., Liberati N. T., Ausubel F. M. and O'Toole G. A. (2007). BifA, a cyclic-di-GMP phosphodiesterase, inversely regulates biofilm formation and swarming motility by *Pseudomonas aeruginosa* PA14. *J. Bacteriol.*, **189**, 8165–8178.

Lam J., Chan R., Lam K. and Costerton J. W. (1980). Production of mucoid microcolonies by *Pseudomonas aeruginosa* within infected lungs in cystic fibrosis. *Infect. Immun.*, **28**, 546–556.

Lee V. T., Matewish J. M., Kessler J. L., Hyodo M., Hayakawa Y. and Lory S. (2007). A cyclic-di-GMP receptor required for bacterial exopolysaccharide production. *Mol. Microbiol.*, **65**, 1474–1484.

Leech A. J., Sprinkle A., Wood L., Wozniak D. J. and Ohman D. E. (2008). The NtrC family regulator AlgB, which controls alginate biosynthesis in mucoid *Pseudomonas aeruginosa*, binds directly to the *algD* promoter. *J. Bacteriol.*, **190**(2), 581–589.

Li Y., Heine S., Entian M., Sauer K. and Frankenberg-Dinkel N. (2013). NO-induced biofilm dispersion in *Pseudomonas aeruginosa* is mediated by an MHYT domain-coupled phosphodiesterase. *J. Bacteriol.*, **195**, 3531–3542.

Limoli D. H., Rockel A. B., Host K. M., Jha A., Kopp B. T., Hollis T. and Wozniak D. J. (2014). Cationic antimicrobial peptides promote microbial mutagenesis and pathoadaptation in chronic infections. *PLoS Pathog.*, **10**(4).

Linker A. and Jones R. S. (1964). A polysaccharide resembling alginic acid from a *Pseudomonas* microorganism. *Nature*, **204**, 187–188.

Ma L., Conover M., Lu H., Parsek M. R., Bayles K. and Wozniak D. J. (2009). Assembly and development of the *Pseudomonas aeruginosa* biofilm matrix. *PLoS Pathog.*, **5**(3).

Ma L., Jackson K. D., Landry R. M., Parsek M. R. and Wozniak D. J. (2006). Analysis of *Pseudomonas aeruginosa* conditional psl variants reveals roles for the Psl polysaccharide in adhesion and maintaining biofilm structure pos-tattachment. *J. Bacteriol.*, **188**(23), 8213–8221.

Ma L., Wang S., Wang D., Parsek M. R. and Wozniak D. J. (2012). The roles of biofilm matrix polysaccharide Psl in mucoid *Pseudomonas aeruginosa* biofilms. *FEMS Immunol. Med. Microbiol.*, **65**(2), 377–380.

Ma L., Wang J., Wang S., Anderson E. M., Lam J. S., Parsek M. R. and Wozniak D. J. (2012). Synthesis of multiple *Pseudomonas aeruginosa* biofilm matrix exopolysaccharides is post-transcriptionally regulated. *Environ. Microbiol.*, **14**, 1995–2005.

Ma S., Selvaraj U. M. A., Ohman D. E., Quarless R., Hassett D. J. and Wozniak D. J. (1998). Phosphorylation-independent activity of the response regulators AlgB and AlgR in promoting alginate biosynthesis in mucoid *Pseudomonas aeruginosa*. *J. Bacteriol.*, **180**(4), 956–968.

Magill S. S., Edwards J. R., Bamberg W., Beldavs Z. G., Dumyati G., Kainer M. A., *et al.* (2014). Multistate point-prevalence survey of health care-associated infections. *N Engl J Med.*, **370**(13), 1198–1208.

Maharaj R., May T. B., Wang S. K. and Chakrabarty A. M. (1993). Sequence of the alg8 and alg44 genes involved in the synthesis of alginate by *Pseudomonas aeruginosa*. *Gene*, **136**, 267–269.

Malone J. G., Jaeger T., Manfredi P., Dötsch A., Blanka A., Bos R., *et al.* (2012). The YfiBNR signal transduction mechanism reveals novel targets for the evolution of persistent *Pseudomonas aeruginosa* in cystic fibrosis airways. *PLoS Pathog.*, **8**(6), e1002760.

Mann E. E. and Wozniak D. J. (2015). Pseudomonas biofilm matrix composition and niche biology. *FEMS Microbiol Rev.*, **36**(4), 893–916.

Manzo J., Cocotl-Yañez M., Tzontecomani T., Martínez V., Bustillos R., Velásquez C., *et al.* (2012). Post-transcriptional regulation of the alginate biosynthetic gene algD by the Gac/Rsm system in *azotobacter vinelandii*. *J. Mol. Microbiol. Biotechnol.*, **21**, 147–159.

Martin D. W., Holloway B. W. and Deretic V. (1993). Characterization of a locus determining the mucoid status of *Pseudomonas aeruginosa*: AlgU shows sequence similarities with a Bacillus sigma factor. *J. Bacteriol.*, **175**(4), 1153–1164.

Martin D. W., Schurr M. J., Mudd M. H., Govan J. R., Holloway B. W. and Deretic V. (1993). Mechanism of conversion to mucoidy in *Pseudomonas aeruginosa* infecting cystic fibrosis patients. *Proc Natl Acad Sci USA*, **90**, 8377–8381.

Martin D. W., Schurr M. J., Mudd M. H. and Deretic V. (1993). Differentiation of *Pseudomonas aeruginosa* into the alginate-producing form: inactivation of *mucB* causes conversion to mucoidy. *Mol Microbiol*, **9**(3), 497–506.

Martin D. W., Schurr M. J., Yu H. and Deretic V. (1994). Analysis of promoters controlled by the putative sigma factor AlgU regulating conversion to mucoidy in *Pseudomonas aeruginosa*: relationship to stress response. *J. Bacteriol.*, **176**, 6688–6696.

Martinez-Salazar J., Moreno S., Najera R., Boucher J., Espin G., Soberon-Chavez G. and Dereticc V. (1996). Characterization of the genes coding for the putative sigma factor AlgU and its regulators MucA, MucB, MucC, and MucD in *Azotobacter vinelandii* and evaluation of their roles in alginate biosynthesis. *J. Bacteriol.*, **178**(7), 1800–1808.

Mathee K., Ciofu O., Sternberg C., Lindum P. W., Campbell J. I., Jensen P., *et al.* (1999). Mucoid conversion of *Pseudomonas aeruginosa* by hydrogen peroxide: a mechanism for virulence activation in the cystic fibrosis lung. *Microbiology*, **145**, 1349–1357.

Mathee K., Pherson C. J. and Ohman D. E. (1997). Posttranslational control of the *algT* (*algU*) -encoded σ^{22} for expression of the alginate regulon in *Pseudomonas aeruginosa* and localization of its antagonist proteins MucA and MucB (AlgN). *J. Bacteriol.*, **179**(11), 3711–3720.

Matsukawa M. and Greenberg E. P. (2004). Putative exopolysaccharide synthesis genes influence *Pseudomonas aeruginosa* biofilm development putative exopolysaccharide synthesis genes influence *Pseudomonas aeruginosa* biofilm development. **186**(14), 4449–4456.

Medvedkin V. N., Permyakov E. A., Klimenko L. V., Mitin Y. V., Matsushima N., Nakayama S. and Kretsinger R. H. (1995). Interactions of (Ala*Ala*Lys*Pro)n and (Lys*Lys*Ser*Pro)n with DNA. Proposed coiled-coil structure of AlgR3 and AlgP from *Pseudomonas aeruginosa*. *Protein Eng.*, **8**(1), 63–70.

Merighi M., Lee V. T., Hyodo M., Hayakawa Y. and Lory S. (2007). The second messenger bis-(3′-5′)-cyclic-GMP and its PilZ domain-containing receptor Alg44 are required for alginate biosynthesis in *Pseudomonas aeruginosa*. *Mol. Microbiol.*, **65**, 876–895.

Merritt J. H., Brothers K. M., Kuchma S. L. and O'Toole G. A. (2007). SadC reciprocally influences biofilm formation and swarming motility via modulation of exopolysaccharide production and flagellar function. *J. Bacteriol.*, **189**, 8154–8164.

Mian F. A., Jarman T. R. and Righelato R. C. (1978). Biosynthesis of exopolysaccharide by *Pseudomonas aeruginosa*. *J. Bacteriol.*, **134**, 418–422.

Mishra M., Byrd M. S., Sergeant S., Azad A. K., Parsek M. R., McPhail L., *et al.* (2012). *Pseudomonas aeruginosa* Psl polysaccharide reduces neutrophil phagocytosis and the oxidative response by limiting complement-mediated opsonization. *Cell Microbiol.*, **14**(1), 95–106.

Mohr C. D., Leveau J. H. J., Krieg D. P., Hibler N. S. and Deretic V. (1992). AlgR-binding sites within the *algD* promoter make up a set of inverted repeats separated by a large intervening segment of DNA. **174**(20), 6624–6633.

Morgan R., Kohn S., Hwang S. H., Hassett D. J. and Sauer K. (2006). BdlA, a chemotaxis regulator essential for biofilm dispersion in *Pseudomonas aeruginosa*. *J. Bacteriol.*, **188**, 7335–7343.

Moscoso J., Jaeger T., Valentini M., Hui K., Jenal U. and Filloux A. (2014). The diguanylate cyclase SadC is a central player in Gac/Rsm-mediated biofilm formation in *Pseudomonas aeruginosa*. *J. Bacteriol.*, **196**(23), 4081–4088. doi: 10.1128/JB.01850-14.

Murphy T. F., Brauer A. L., Eschberger K., Lobbins P., Grove L., Cai X. and Sethi S. (2008). *Pseudomonas aeruginosa* in chronic obstructive pulmonary disease. *Am. J. Respir. Crit. Care Med.*, **177**, 853–860.

Nivens D. E., Ohman D. E., Williams J. and Franklin M. J. (2001). Role of alginate and its O acetylation in formation of *Pseudomonas aeruginosa* microcolonies and biofilms. *J. Bacteriol.*, **183**, 1047–1057.

Núñez C., León R., Guzmán J., Espín G. and Soberón-Chávez G. (2000). Role of *Azotobacter vinelandii mucA* and *mucC* gene products in algiante production. *Journal of Bacteriology*, **182**(23), 6550–6556.

O'Toole G. A. and Kolter R. (1998). Flagellar and twitching motility are necessary for *Pseudomonas aeruginosa* biofilm development. *Mol. Microbiol.*, **30**, 295–304.

O'Toole G. A. and Kolter R. (1998). Initiation of biofilm formation in *Pseudomonas fluorescens* WCS365 proceeds via multiple, convergent signaling pathways: a genetic analysis. *Mol. Microbiol.*, **28**, 449–461.

Obritsch M. D., Pharm D., Fish D. N., Maclaren R. and Jung R. (2005). Nosocomial Infections Due to Multidrug-Resistant *Pseudomonas aeruginosa. Epidemiology and Treatment Options.*, **25**(10), 1353–1364.

Ogle J. W., Janda J. M., Woods D. E. and Vasil M. L. (1987). Characterization and use of a DNA probe as an epidemiological marker for *Pseudomonas aeruginosa. J Infect Dis.*, **155**, 119–126.

Oglesby L. L., Jain S. and Ohman D. E. (2008). Membrane topology and roles of *Pseudomonas aeruginosa* Alg8 and Alg44 in alginate polymerization. *Microbiology*, **154**, 1605–1615.

Parsek M. R. and Tolker-Nielsen T. (2008). Pattern formation in *Pseudomonas aeruginosa* biofilms. *Curr Opin Microbiol.*, **11**, 560–566.

Pelissier M. C., Lesley S. A., Kuhn P. and Bourne Y. (2010). Structural insights into the catalytic mechanism of bacterial guanosine-diphospho-D-mannose pyrophosphorylase and its regulation by divalent ions. *J. Biol. Chem.*, **285**, 27468–27476.

Petrova O. E., Cherny K. E. and Sauer K. (2014). The *Pseudomonas aeruginosa* diguanylate cyclase GcbA, a homolog of *P. Fluorescens* GcbA, promotes initial attachment to surfaces, but not biofilm formation, via regulation of motility. *J. Bacteriol.*, **196**(15), 2827–2841.

Pindar D. F. and Bucke C. (1975). The biosynthesis of alginic acid by *Azotobacter vinelandii. Biochem. J*, **152**, 617–622.

Poschet J. F., Boucher J. C., Firoved A. M. and Deretic V. (2001). Conversion to mucoidy in *Pseudomonas aeruginosa* infecting cystic fibrosis patients. *Method Enzymol*, **336**, 65–76.

Pritchett C. L., Little A. S., Okkotsu Y., Frisk A., Cody W. L., Covey C. R. and Schurr M. J. (2015). Expression Analysis of the *Pseudomonas aeruginosa* AlgZR Two-Component Regulatory System. *J. Bacteriol.*, **197**(4), 736–748. http://doi.org/10.1128/JB.02290-14.

Pritt B., O'Brien L. and Winn W. (2007). Mucoid *Pseudomonas* in cystic fibrosis. *Am. J. Clin. Pathol.*, **128**, 32–34.

Purevdorj-Gage B., Costerton W. J. and Stoodley P. (2005). Phenotypic differentiation and seeding dispersal in non-mucoid and mucoid *Pseudomonas aeruginosa* biofilms. *Microbiology*, **151**, 1569–1576.

Qiu D., Eisinger V. M., Head N. E., Pier G. B. and Yu H. D. (2008). ClpXP proteases positively regulate alginate overexpression and mucoid conversion in *Pseudomonas aeruginosa. Microbiology*, **154**, 2119–2130. http://doi.org/10.1099/mic.0.2008/017368-0.

Qiu D., Eisinger V. M., Rowen D. W. and Yu H. D. (2007). Regulated proteolysis controls mucoid conversion in *Pseudomonas aeruginosa. Proceedings of the National Academy of Sciences of the United States of America*, **104**(19), 8107–8112. http://doi.org/10.1073/pnas.0702660104.

Ramsey D. M. and Wozniak D. J. (2005). Understanding the control of *Pseudomonas aeruginosa* alginate synthesis and the prospects for management of chronic infections in cystic fibrosis. *Mol Microbiol*, **56**(2), 309–322. http://doi.org/10.1111/j.1365-2958.2005.04552.x.

Regni C., Tipton P. A. and Beamer L. J. (2002). Crystal structure of PMM/PGM: an enzyme in the biosynthetic pathway of *P. aeruginosa* virulence factors. *Structure*, **10**, 269–279.

Rehm B. H. A., Boheim G., Tommassen J. and Winkler U. K. (1994). Overexpression of *algE* in *Escherichia coli*: subcellular localization, purification, and ion channel properties. *J. Bacteriol.*, **176**, 5639–5647.

Rehman Z. U., Wang Y., Moradali M. F., Hay I. D. and Rehm B. H. A. (2013). Insights into the assembly of the alginate biosynthesis machinery in *Pseudomonas aeruginosa. Appl. Environ. Microbiol.*, **79**, 3264–3272.

Reiling S. A., Jansen J. A., Henley B. J., Singh S., Chattin C., Chandler M. and Rowen D. W. (2005). Prc protease promotes mucoidy in *mucA* mutants of *Pseudomonas aeruginosa*. *Microbiology*, **151**, 2251–2261.

Reinhardt A., Köhler T., Wood P., Rohner P., Dumas J. L., Ricou B., *et al.* (2007). Development and persistence of antimicrobial resistance in *Pseudomonas aeruginosa*: a longitudinal observation in mechanically ventilated patients. *Antimicrob Agents Chemother.*, **51**(4), 1341–1350.

Remminghorst U. and Rehm B. H. A. (2006). In vitro alginate polymerization and the functional role of Alg8 in alginate production by *Pseudomonas aeruginosa*. *Appl. Environ. Microbiol.*, **72**, 298–305.

Robles-Price A., Wong T. Y., Sletta H., Valla S. and Schiller N. L. (2004). AlgX is a periplasmic protein required for alginate biosynthesis in *Pseudomonas aeruginosa*. *J. Bacteriol.*, **186**, 7369–7377.

Rowe S. M. and Miller S. E. J. (2005). Cystic fibrosis. *N Engl J Med.*, **352**, 1992–2001.

Rowen D. W. and Deretic V. (2000). Membrane-to-cytosol redistribution of ECF sigma factor AlgU and conversion to mucoidy in *Pseudomonas aeruginosa* isolates from cystic fibrosis patients. *Mol Microbiol*, **36**(2), 314–327. http://doi.org/10.1046/j.1365-2958.2000.01830.x.

Ryan R. P., Tolker-Nielsen T. and Dow J. M. (2012). When the PilZ don't work: effectors for cyclic di-GMP action in bacteria. *Trends Microbiol*, **20**, 235–242.

Sakuragi Y. and Kolter R. (2007). Quorum-sensing regulation of the biofilm matrix genes (pel) of *Pseudomonas aeruginosa*. *J. Bacteriol.*, **189**, 5383–5386.

Sauer K., Cullen M. C., Rickard A. H., Zeef L. A., Davies D. G. and Gilbert P. (2004). Characterization of nutrient-induced dispersion in *Pseudomonas aeruginosa* PAO1 biofilm. *J. Bacteriol.*, **186**, 7312–7326.

Sauer R. T. and Cezairliyan B. O. (2009). Control of *Pseudomonas aeruginosa* AlgW protease cleavage of MucA by peptide signals and MucB. *Mol Microbiol*, **72**(2), 368–379. http://doi.org/10.1111/j.1365-2958.2009.06654.x.Control

Schiller N. L., Monday S. R., Boyd C. M., Keen N. T. and Ohman D. E. (1993). Characterization of the *Pseudomonas aeruginosa* alginate lyase gene (algL): cloning, sequencing, and expression in *Escherichia coli*. *J. Bacteriol.*, **175**, 4780–4789.

Schulz S., Eckweiler D., Bielecka A. and Nicolai T. (2015). Elucidation of sigma factor-associated networks in *Pseudomonas aeruginosa* reveals a modular architecture with limited and function-specific crosstalk. *PLoS Pathogens*, **11**(3), 1–21. http://doi.org/10.1371/journal.ppat.1004744

Schurr M. J., Yu H., Boucher J. C., Hibler N. S. and Deretic V. (1995). Multiple promoters and induction by heat shock of the gene encoding the alternative sigma factor AlgU (σE) which controls mucoidy in cystic fibrosis isolates of *Pseudomonas aeruginosa*. *J Bacteriol*, **177**, 5670–5679.

Schurr M. J., Yu H., Martinez-Salazar J. M., Boucher J. C. and Deretic V. (1996). Control of AlgU, a member of the σ^E-like family of stress sigma factors, by the negative regulators MucA and MucB and *Pseudomonas aeruginosa* conversion to mucoidy in Cystic Fibrosis. *J. Bacteriol.*, **178**, 4997–5004.

Schneider M., Mühlemann K., Droz S., Couzinet S., Casaulta C. and Zimmerli S. (2008). Clinical characteristics associated with isolation of small-colony variants of *Staphylococcus aureus* and *Pseudomonas aeruginosa* from respiratory secretions of patients with cystic fibrosis. *J. Clin. Microbiol.*, **46**, 1832–1834.

Shrout J. D., Chopp D. L., Just C. L., Hentzer M., Givskov M. and Parsek M. R. (2006). The impact of quorum sensing and swarming motility on *Pseudomonas aeruginosa* biofilm formation is nutritionally conditional. *Mol Microbiol.*, **62**(5), 1264–1277.

Simm R., Morr M., Kader A., Nimtz M. and Römling U. (2004). GGDEF and EAL domains inversely regulate cyclic di-GMP levels and transition from sessibility to motility. *Mol Microbiol.*, **53**(4), 1123–1134.

Singh P. K., Schaefer A. L., Parsek M. R., Moninger T. O., Welsh M. J. and Greenberg E. P. (2000). Quorum-sensing signals indicate that cystic fibrosis lungs are infected with bacterial biofilms. *Nature*, **407**, 762–764.

Snook C. F., Tipton P. A. and Beamer L. J. (2003). Crystal structure of GDP-mannose dehydrogenase: a key enzyme of alginate biosynthesis in *P. aeruginosa*. *Biochemistry*, **42**, 4658–4668.

Starkey M., Hickman J. H., Ma L., Zhang N., De Long S., Hinz A., *et al.* (2009). *Pseudomonas aeruginosa* rugose small-colony variants have adaptations that likely promote persistence in the cystic fibrosis lung. *J. Bacteriol.*, **191**(11), 3492–3503.

Stock A. M., Robinson V. L. and Goudreau P. N. (2000). Two-component signal transduction. *Annu. Rev. Biochem.*, **69**, 183–215.

Stover C. K., Pham X. Q., Erwin A. L., Mizoguchi S. D., Warrener P., Hickey M. J., *et al.* (2000). Complete genome sequence of *Pseudomonas aeruginosa* PAO1, an opportunistic pathogen. **406**, 959–964.

Suman E., Blat K. G. and Hegde B. K. (1993). Mucoid *Pseudomonas aeruginosa* in urinary tract infection. *Trop Doct*, **23**, 180–181.

Terry J. M., Piña S. E. and Mattingly S. J. (1991). Environmental conditions which influence mucoid conversion *Pseudomonas aeruginosa* PAO1. *Infect Immun*, **59**, 471–477.

Tran C. S., Rangel S. M., Almblad H., Kierbel A., Givskov M., Tolker-Nielsen T., *et al.* (2014). The *Pseudomonas aeruginosa* type III translocon is required for biofilm formation at the epithelial barrier. *PLoS Pathog.*, **10**, e1004479.

Tseng B. S., Zhang W., Harrison J. J., Quach T. P., Song J. L., Penterman J., Singh P. K., Chopp D. L., Packman A. I. and Parsek M. R. (2013). The extracellular matrix protects *Pseudomonas aeruginosa* biofilms by limiting the penetration of tobramycin. *Environ Microbiol*, **15**, 2865–2878. doi: 10.1111/1462-2920.12155.

Ueda A. and Wood T. K. (2009). Connecting quorum sensing, c-di-GMP, pel polysaccharide, and biofilm formation in *Pseudomonas aeruginosa* through tyrosine phosphatase TpbA (PA3885). *PLoS Pathog.*, **5**(6). doi: 10.1371/journal.ppat.1000483.

Valverde C., Heeb S., Keel C. and Haas D. (2003). RsmY, a small regulatory RNA, is required in concert with RsmZ for GacA-dependent expression of biocontrol traits in *Pseudomonas fluorescens* CHA0. *Mol Microbiol*, **50**(4), 1361–1379. doi: 10.1046/j.1365-2958.2003.03774.x.

Vasseur P., Soscia C., Voulhoux R. and Filloux A. (2007). PelC is a *Pseudomonas aeruginosa* outer membrane lipoprotein of the OMA family of proteins involved in exopolysaccharide transport. *Biochimie.*, **89**, 903–915.

Ventre I., Goodman A. L., Vallet-Gely I., Vasseur P., Soscia C., Molin S., *et al.* (2006). Multiple sensors control reciprocal expression of *Pseudomonas aeruginosa* regulatory RNA and virulence genes. *Proc Natl Acad Sci USA*, **103**(1), 171–176.

Wang L., Liu D. and Reeves P. R. (1996). C-terminal half of *Salmonella enterica* WbaP (RfbP) is the galactosyl-1-phosphate transferase domain catalyzing the first step of O-antigen synthesis. *J. Bacteriol.*, **178**(9), 2598–2604.

Warren A. E., Boulianne-Larsen C. M., Chandler C. B., Chiotti K., Kroll E., Miller S. R. and Rosenzweig F. (2011). Genotypic and phenotypic variation in *Pseudomonas aeruginosa* reveals signatures of secondary infection and mutator activity in certain cystic fibrosis patients with chronic lung infections. **79**(12), 4802–4818. http://doi.org/10.1128/IAI.05282-11.

Weadge J. T., Yip P. P., Robinson H., Arnett K., Tipton P. A. and Howell P. L. (2010). Expression, purification, crystallization and preliminary X-ray analysis of *Pseudomonas aeruginosa* AlgX. *Acta Crystallogr. Sect. F Struct. Biol. Cryst. Commun.*, **66**, 588–591.

Wei Q. and Ma L. Z. (2013). Biofilm matrix and its regulation in *Pseudomonas aeruginosa*. *Int. J. Mol. Sci.*, **14**(10), 20983–21005.

Whitchurch, C. B., Alm, R. A. and Matrick, J. S. (1996). The alginate regulator AlgR and an associated sensor FimS are required for twitching motility in *Pseudomonas aeruginosa*. *Proc. Natl. Acad. Sci. USA*, **93**, 9839–9843.

Whitchurch, C. B., Erova, T. E., Emery, J. A., Sargent, J. L., Harris, J. M., Semmler, A. B. T., *et al.* (2002). Phosphorylation of the *Pseudomonas aeruginosa* response regulator AlgR Is essential for Type IV fimbria-mediated twitching motility. **184**(16), 4544–4554. http://doi.org/10.1128/JB.184.16.4544.

Whitney, J. C. C., Neculai, A. M., Ohman, D. E. and Howell, P. L. (2009). Expression, refolding, crystallization and preliminary X-ray analysis of *Pseudomonas aeruginosa* AlgE. *Acta Crystallogr. Sect. F Struct. Biol. Cryst. Commun.*, **65**, 463–466.

Whitney, J. C., Hay, I. D., Li, C., Eckford, P. D., Robinson, H., Amaya, M. F., *et al.* (2011). Structural basis for alginate secretion across the bacterial outer membrane. *Proc. Natl. Acad. Sci. USA*, **108**, 13083–13088.

Whitney, J. C., Colvin, K. M., Marmont, L. S., Robinson, H., Parsek, M. R. and Howell, P. L. (2012). Structure of the cytoplasmic region of PelD, a degenerate diguanylate cyclase receptor that regulates exopolysaccharide production in *Pseudomonas aeruginosa*. *J. Biol. Chem.*, **287**, 23582–23593. doi: 10.1074/jbc.M112.375378.

Withers, R. T., Damron, H. F., Yin, Y. and Yu, H. D. (2013). Truncation of type IV pilin induces mucoidy in *Pseudomonas aeruginosa* strain PAO579. *Microbiology Open*, **2**(3), 459–470.

Withers, T. R., Yin, Y. and Yu, H. D. (2014). Identification of novel genes associated with alginate production in *Pseudomonas aeruginosa* using mini- himar1 mariner transposon-mediated mutagenesis. *J Vis Exp.*, **85**, 2–7. 10.3791/51346.

Wood L. F., Leech A. J. and Ohman D. E. (2006). Cell wall-inhibitory antibiotics activate the alginate biosynthesis operon in *Pseudomonas aeruginosa*: roles of sigma (AlgT) and the AlgW and Prc proteases. *Mol Microbiol*, **62**, 412–426. doi: 10.1111/j.1365-2958.2006.05390.x.

Wood L. F. and Ohman D. E. (2009). Use of cell wall stress to characterize sigma 22 (AlgT/U) activation by regulated proteolysis and its regulon in *Pseudomonas aeruginosa*. *Mol Microbiol*, **72**(1), 183–201. http://doi.org/10.1111/j.1365-2958.2009.06635.x.

Workentine M. L., Chang L., Ceri H. and Turner R. J. (2009). The GacS-GacA two-component regulatory system of *Pseudomonas fluorescens*: a bacterial two-hybrid analysis. *FEMS Microbiol Lett*, **292**(1), 50–56. doi: 10.1111/j.1574-6968.2008.01445.x.

Wozniak D. J., Wyckoff T. J., Starkey M., Keyser R., Azadi P., O'Toole G. and Parsek M. R. (2003). Alginate is not a significant component of the extracellular polysaccharide matrix of PA14 and PAO1 *Pseudomonas aeruginosa* biofilms. *Proc. Natl. Acad. Sci. USA*, **100**, 7907–7912.

Wozniak D. J. (1994). Integration host factor and sequences downstream of the *Pseudomonas aeruginosa algD* transcription start site are required for expression. *J. Bacteriol.*, **176**(16), 5068–5076.

Wozniak D. J. and Ohman D. E. (1991). *Pseudomonas aeruginosa* AlgB, a two-component response regulator of the NtrC family, is required for *algD* Transcription. *J. Bacteriol.*, **173**(4), 1406–1413.

Xie Z., Hershberger C. D., Shankar S., Ye R. and Chakrabarty A. (1996). Sigma factor – anti-sigma factor interaction in alginate synthesis: inhibition of AlgT by MucA. *J. Bacteriol.*, **178**(16), 4990–4996.

Jing X., Jaw J., Robinson H. and Schubot F. (2010). Crystal structure and oligomeric state of the RetS signaling kinase sensory domain. *Proteins*, **78**(7), 1631–1640. doi: 10.1002/prot.22679.

Yang L., Hengzhuang W., Wu H., Damkiær S., Jochumsen N., Song Z., *et al.* (2012). Polysaccharides serve as scaffold of biofilms formed by mucoid *Pseudomonas aeruginosa*. *FEMS Immunol Med Microbiol.*, **65**(2), 366–376. http://dx.doi.org/10.1111/j.1574-695X.2012.00936.x.

Yang Z., Jin X., Rao X. and Hu F. (2011). A natural antisense transcript regulates *mucD* gene expression and biofilm biosynthesis in *Pseudomonas aeruginosa*. *Mikrobiologiia*, **80**, 756–762.

Ye R., Zielinski N. A. and Chakrabarty A. M. (1994). Purification and characterization of phosphomannomutase / phosphoglucomutase from *Pseudomonas aeruginosa* involved in biosynthesis of both alginate and lipopolysaccharide. *J. Bacteriol.*, **176**(16), 4851–4857.

Yu H., Mudd M., Boucher J. C., Schurr M. J. and Deretic V. (1997). Identification of the *algZ* gene upstream of the response regulator *algR* and its participation in control of alginate production in *Pseudomonas aeruginosa*. *J. Bacteriol.*, **179**, 187–193.

Yu H., Schurr M. and Deretic V. (1995). Functional equivalence of *Escherichia coli* σ^E and *Pseudomonas aeruginosa* AlgU: *E. coli rpoE* restores mucoidy and reduces sensitivity to reactive oxygen intermediates in algU mutants of *P. aeruginosa*. *J. Bacteriol.*, **177**(11), 3259–3268.

Yuan A., Gregory B., Sharp J., McCleary K., Dove S. and Hochschild A. (2008). Rsd family proteins make simultaneous interactions with regions 2 and 4 of the primary sigma factor. *Mol Micro.*, **70**(5), 1136–1151.

Zielinski N. A., Chakrabarty A. M. and Berry A. (1991). Characterization and regulation of the *Pseudomonas aeruginosa algC* gene encoding phosphomannomutase. *J. Biol. Chem.*, **266**(15), 9754–9763.

Zimlichman E., Henderson D., Tamir O., Franz C., Song P., Yamin C., *et al.* (2013). Health care–associated infections a meta-analysis of costs and financial impact on the US health care system. *JAMA Intern Med*, **173**(22), 2039–2046. http://doi.org/10.1001/jamainternmed.2013.9763.

Chapter 6

Amyloids – a neglected child of the slime

Morten S. Dueholm and Per Halkjær Nielsen

Center for Microbial Communities, Department of Chemistry and Bioscience, Aalborg University, Fredrik Bajers Vej 7H, 9220 Aalborg, Denmark, e-mail: md@bio.aau.dk; phn@bio.aau.dk

ABSTRACT

Functional amyloids define a group of highly ordered fibrillar protein polymers, which are characterized by a cross-β quaternary structure, an ability to self-assemble from their monomeric counterparts, and an extreme resistance towards thermal and chemical denaturants. These properties make them excellent nanomaterials. Organisms from all domains of life make use of functional amyloids, and biofilm-associated bacteria are not an exception. Histological staining of biofilms with amyloid-specific dyes or amyloid conformation-specific antibodies has revealed that functional amyloids are widespread among microorganisms in biofilms that originate from nature as well as in engineered systems. They are therefore believed to play important roles in biofilm ecology. Despite their omnipresence in biofilms, few amyloids have been purified from biofilm-associated bacteria and investigated in depth. However, the few amyloids which have been studied so far have already provided an astonishing demonstration of how amyloids are exploited in biofilms with roles ranging from fimbriae and other cell appendages for adhesion and biofilm formation to structural components of cell envelopes and spore coats. They may even be used as cytotoxins and as reservoirs for quorum-sensing signaling molecules.

In this chapter, we provide an overview of the structural and physiochemical properties of amyloids and highlight the histological studies that provide evidence for the omnipresence of amyloids within biofilms. We discuss the few functional amyloid systems that have recently been characterized. We also discuss why so few amyloid systems have been characterized with focus on strategies of identifying novel functional amyloid systems.

Keywords: Functional amyloids, extracellular polymeric substance, adhesion, biofilm formation, cell envelope components, cytotoxins, quorum sensing

6.1 INTRODUCTION

Amyloids are fascinating structures. Biophysically, they can be defined as any fibrillar polypeptide aggregate with a cross-β quaternary structure (Fandrich, 2007). In the cross-β structure, the polypeptide folds back on itself in a regular manner so that adjacent chain segments are laterally arranged perpendicularly to the fibril axis (Figure 6.1) (Jimenez *et al.* 1999; Nelson *et al.* 2005; Sunde *et al.* 1997). Multiple amyloid filaments (protofilaments) often twine about one another to make up the mature amyloid fibril when formed *in vitro* (Figure 6.1c) (Fandrich, 2007).

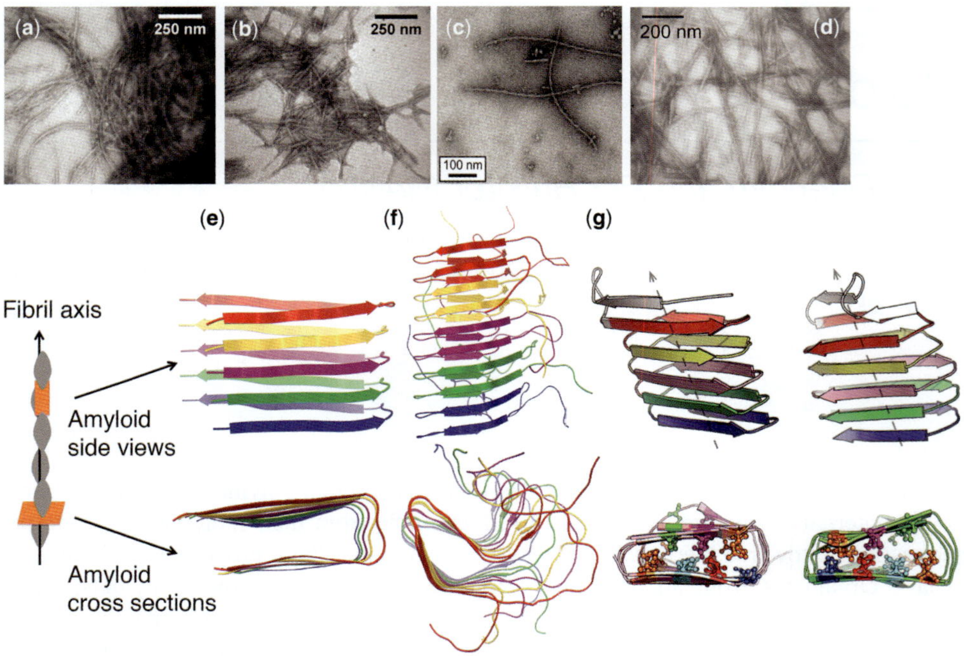

Figure 6.1 Fibrillar morphology and molecular structure of selected amyloids. (a–d) Transmission electron microscopy of amyloids purified from *Pseudomonas* sp. UK4 (a) and *Escherichia coli* SM2258 (b) or formed *in vitro* by chaplins from *Streptomyces coelicolor* (c) (Sawyer *et al.* 2011) or the human protein glucagon (d) (Pedersen *et al.* 2006). (e) Molecular structure of the Alzheimer's disease-related $A\beta_{1-42}$ amyloid (Lührs *et al.* 2005). The structure was created using solid-state NMR (ssNMR) data [Protein Data Bank (PDB): 2BEG]. (f) Molecular structure of the HET-s(218–289) prion (Wasmer *et al.* 2008). HET-s(218–289) forms a β-solenoid with a triangular hydrophobic core. The structure was obtained from ssNMR data [Protein Data Bank (PDB): 2RNM]. (g) Molecular structure of the curli amyloid from *E. coli*. The structure was created based on the covariation of amino acids in sequences of curli homologs and validated using ssNMR, electron microscopy and x-ray diffraction data. The data was consistent with both a left-handed (left) and a right-handed (right) β-roll structure (Tian *et al.* 2015).

One of the most striking features of the amyloids is their ability to self-assemble in a nucleation-dependent process from their monomeric constituents given the right environmental conditions (Figure 6.2a) (Booth *et al.* 1997; Dueholm *et al.* 2011; Linding *et al.* 2004). The first step of amyloid formation involves the formation of an amyloidogenic precursor that can act as the initial seed for amyloid polymerization (Figure 6.2b) (Jahn *et al.* 2006). Subsequently, the amyloid fibrils formed act as seed for further amyloid polymerization.

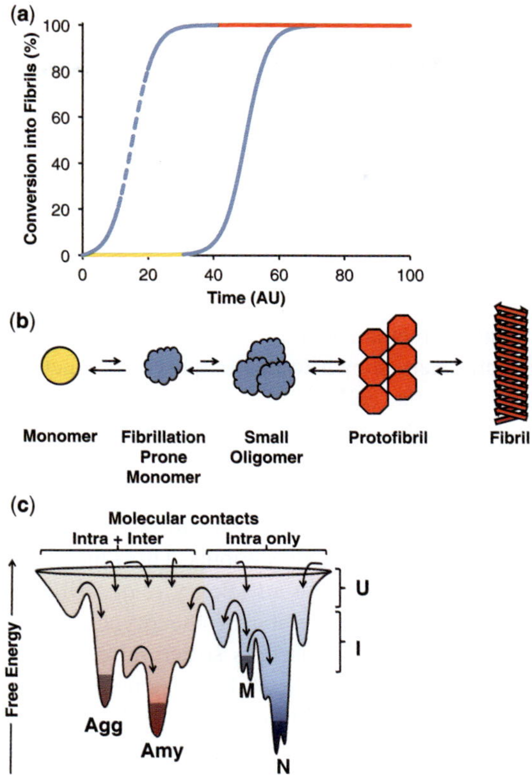

Figure 6.2 Conversion from monomeric precursor to amyloid fibrils. (a) Classical fibrillation kinetics in the absence (solid line) and presence (dotted line) of fibrillation prone-seeds. The fibrillation process involves a lag phase, where the fibrillation-prone intermediates are formed (yellow). The lag phase is followed by an elongation phase, in which the protofibrils form and incorporate additional monomers (blue). When all monomers are depleted, the fibrillation process enters the stationary phase (red). During the elongation phase and the stationary phase, the amyloids form their mature structure. (b) Molecular species involved in the fibrillation process. The native monomers are assumed to be in equilibrium with fibrillation-prone monomer structures. The latter may associate to form small oligomers, which in turn aggregate to produce structured protofibrils. The protofibrils act as nuclei for further growth by incorporation of additional precursors and associate to form the mature amyloid fibrils. (c) The folding funnel as described by Clark (2004) explains how the amyloid conformation could represent a universal, energetic minimum for aggregated proteins. The following abbreviations are used: M = molten globule, N = native state, Agg = amorphous aggregates, Amy = amyloid fibrils, U = unfolded structures, I = partly folded intermediates. Stable conformations are highlighted in dark colors.

Many proteins, which are structurally unrelated in their native monomeric forms, are able to fold into mature amyloids of similar overall structure (Ban *et al.* 2004; Dueholm *et al.* 2011; Makin & Serpell, 2005). Consequently, the amyloid conformation has been postulated to represent a universal, energetic minimum for aggregated proteins (Figure 6.2c) (Dobson, 1999, 2003). Extreme resistance toward thermal and chemical denaturants consequently characterizes most amyloids (Makin *et al.* 2005).

Amyloid formation is commonly associated with devastating human conditions, including Alzheimer's and Parkinson's disease (Dobson, 1999; Dovidchenko *et al.* 2014). Here, amyloid formation is the result of protein misfolding or general cellular proteostasis defects (Chiti & Dobson, 2006; Frokjaer & Otzen, 2005; Uversky & Fink, 2004). During disease-related amyloid formation, small oligomers often accumulate, and it is these folding intermediates, rather than the mature amyloid fibrils, that are cytotoxic and lead to the disease conditions (Figure 6.2b) (Andreasen *et al.* 2015; Stefani, 2012). Amyloids are, however, much more than just malicious entities. Throughout all domains of life, nature has harnessed the amyloid structure for beneficial use (Dueholm *et al.* 2013a; Fowler *et al.* 2007; Hammer *et al.* 2008; Maury, 2009). These amyloids are named functional amyloids to distinguish them from their disease-related counterparts.

Among archaea, the filamentous methanogenic *Methanosaeta* grow and divide within an extremely stable tubular sheath that recently were shown to be composed of functional amyloids (Dueholm *et al.* 2015). The amyloid sheaths protect the cells from protozoan grazing and cell turgor pressure, while allowing the uptake of substrates and the release of methane. Filamentous fungi use functional amyloids, known as class-I hydrophobins, to form amphipathic monolayers at hydrophobic/hydrophilic interfaces. This reduces the surface tension of the growth medium and allows the fungi to break the water/air interface and produce aerial hyphae for the dispersal of spores (Kwan *et al.* 2006; Ren *et al.* 2013; Wosten & de Vocht, 2000). The spores are also coated with hydrophobins. This renders them hydrophobic and provides resistance to wetting and eases their dispersal in air (Bayry *et al.* 2012). Yeast cells use amyloid prion proteins as a non-genetic, but transmissible regulator of phenotypes (Bayry *et al.* 2012; Wickner *et al.* 2007). In arthropods, functional amyloids are used as structural components of the silk moth chorion and in the form of spider silk (Hamodrakas *et al.* 2004; Iconomidou *et al.* 2006; Kenney *et al.* 2002). Functional amyloids also have beneficial usage in the human body. They form the scaffolds upon which melanin, a core component of our skin pigmentation, is synthesized (McGlinchey *et al.* 2009). They are also used as a storage form for peptide hormones (Maji *et al.* 2009). However, the largest structural and functional diversity of beneficial amyloids is probably found within the matrix of extracellular polymeric substances (EPS) of microbial biofilms.

6.2 VISUALIZATION OF AMYLOIDS AND THEIR ABUNDANCE IN BIOFILMS

Histological stains have been used to identify disease-related amyloids since their discovery more than 150 years ago, and these stains also target the functional amyloids (Kyle, 2001; Sipe & Cohen, 2000). The most popular amyloid stains are Congo Red (CR) and Thioflavin T (ThT). These dyes alter their spectral properties upon binding to amyloid fibrils, which allows easy detection without the need of washing to remove excess dye (Groenning, 2010; Westermark *et al.* 1999). Although CR and ThT are commonly used to detect amyloid aggregates, they are not completely specific to amyloids, and false positive and false negative results may occur (Groenning, 2010; Khurana *et al.* 2001). To improve the specificity of amyloid detection, amyloid conformational-specific antibodies can be used (O'Nuallain & Wetzel, 2002). The monoclonal antibodies WO1 and WO2 recognize generic epitopes within the amyloid fold that are not present within unfolded or natively folded protein monomers (Larsen *et al.* 2007; O'Nuallain & Wetzel, 2002).

CR, ThT and WO1/WO2 can be combined with fluorescence *in situ* hybridization (FISH), a microscopy-based technique that uses rRNA-targeting oligonucleotides for the identification of specific bacterial taxa in complex systems (Amann & Fuchs, 2008). This allows quantification of the amyloid abundance within biofilms in terms of a biovolume fraction (Larsen *et al.* 2007, 2008). However, it also allows amyloid expression to be linked with specific bacteria taxa *in situ* in microbial communities (Larsen *et al.* 2007, 2008).

The combination of FISH with the amyloid probes has shown that amyloids are abundant in a wide range of environmental and engineered biofilms, occupying 5–40% of the biovolume, depending on the source of the biofilm (Larsen *et al.* 2007, 2008). Amyloid expression has been observed for bacteria belonging to several bacterial phyla, including Proteobacteria (Alpha-, Beta-, Gamma-, and Deltaproteobacteria), Bacteriodetes, Chloroflexi, and Actinobacteria (Figure 6.3) (Jordal *et al.* 2009; Larsen *et al.* 2007, 2008). CR and ThT also provide evidence for functional amyloids within archaeal and fungal biofilms (Chimileski *et al.* 2014; Garcia *et al.* 2011).

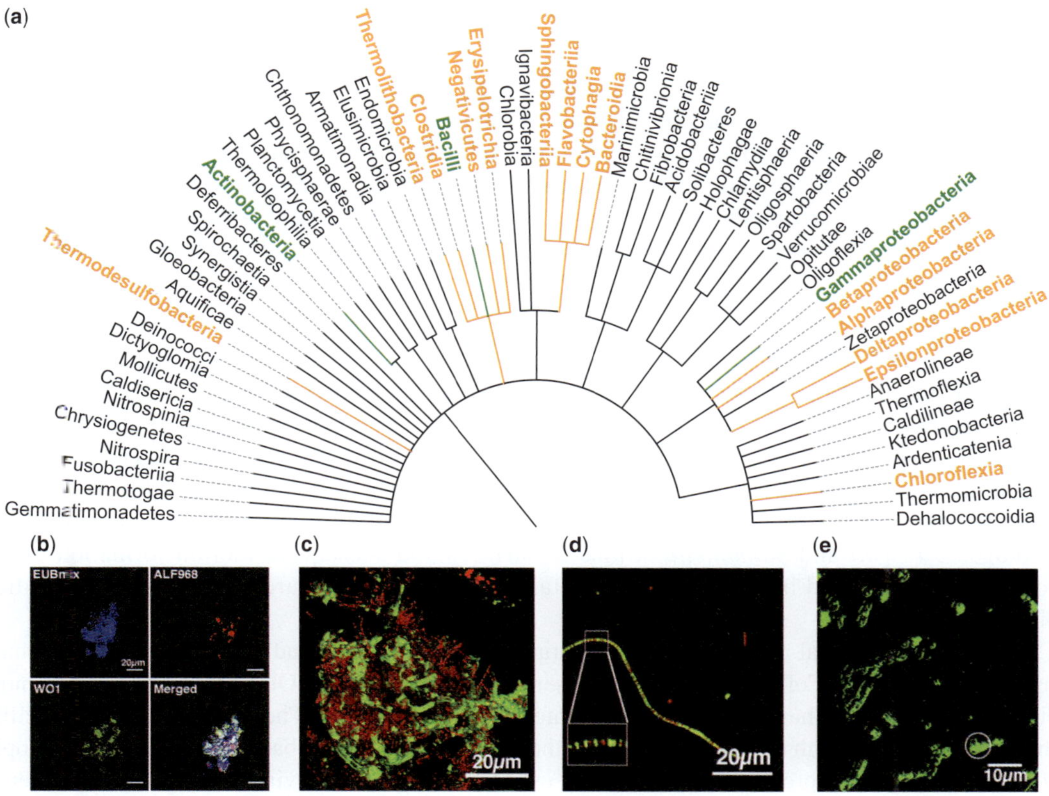

Figure 6.3 Diversity of functional bacterial amyloids. (a) Phylogenetic distribution of functional amyloid systems within the Bacterial domain. Bacterial classes, where amyloids have been purified and characterized biophysically, are shown in green. Bacterial classes where amyloids have been identified based on histological staining with amyloid-specific dyes or antibodies or by the presence of genes encoding amyloid systems are shown in orange. The phylogenetic tree was generated based on the NCBI taxonomy using PhyloT (http://phylot.biobyte.de/index.html) and visualized using iTOL (Letunic & Bork, 2007). (b) Simultaneous labeling of microcolonies in a biofilm from a freshwater lake with antibody (WO1) and oligonucleotide probes by FISH. EUBmix is an oligonucleotide probe that targets almost all bacteria. The ALF968 probe targets the class Alphaproteobacteria (Larsen *et al.* 2007). (c) Staining of amyloids in an activated sludge floc with ThT (green). Bacteria were counter strained with propidium iodide (red) (Larsen *et al.* 2008). (d) Simultaneous labeling of filaments with the WO1 antibody (green) and a Chloroflexi specific FISH probe (red) (Larsen *et al.* 2008). (e) Staining of amyloids on *Bacillus mycoides* cells and spores with the WO2 antibody (Jordal *et al.* 2009).

6.3 ONLY A FEW FUNCTIONAL AMYLOIDS HAVE BEEN CHARACTERIZED

Although the histological studies have revealed an astonishing diversity of functional amyloids within biofilms, very few of these functional amyloids have been isolated and have had their amyloid structure confirmed by biophysical methods (Table 6.1). Furthermore, only three of these have been studied in depth. These are the curli amyloids produced by *Escherichia coli* and *Salmonella* spp. (Chapman *et al.* 2002; Collinson *et al.* 1991), the functional amyloid of *Pseudomonas* (Fap) (Dueholm *et al.* 2010, 2013c), and the TasA amyloids produced by *Bacillus subtilis* (Romero et *al.* 2010).

6.3.1 Curli fimbriae

Curli fimbriae of the enteric bacteria *E. coli* and *Salmonella* (Figure 6.1a) were the first functional amyloids to be discovered within the bacterial domain of life (Chapman *et al.* 2002; Romling *et al.* 1998), and they remain the best characterized functional amyloid system by far. Although the curli system was originally believed to be restricted to the Enterobacteriales, a recent evolutionary study demonstrated that the genetic elements required for curli production are phylogenetic wide-spread and can be found in at least four bacterial phyla (Bacteriodetes, Firmicutes, Proteobacteria and Thermodesulfobacteria) (Dueholm *et al.* 2012).

6.3.1.1 Biological role

Multiple functions have been assigned to the curli fimbriae. In *E. coli* and *Salmonella*, they serve as structural components within the EPS matrix of biofilms and are essential for the development of mature mushroom-shaped biofilms (Kikuchi *et al.* 2005; Vidal *et al.* 1998). Curli also act as adhesins for the initial attachment of the bacteria to abiotic surfaces such as Teflon and stainless steel (Austin *et al.* 1998; Pawar *et al.* 2005; Ryu *et al.* 2004; Ryu & Beuchat, 2005). The curli-mediated attachment of pathogenic *E. coli* and *Salmonella* strains to surfaces and subsequent biofilm formation poses a serious threat in the food industry and in hospital settings as biofilms are often difficult to eradicate (Høiby *et al.* 2010).

Curli mediate bacterial interaction with multiple host proteins, including the extracellular matrix components fibronectin (Collinson *et al.* 1993; Olsen *et al.* 1989), laminin (Olsen *et al.* 1993), plasminogen (Sjöbring *et al.* 1994), and human contact phase proteins (Nasr *et al.* 1996). These interactions may facilitate the bacterial spread through the host and promote the internalization of the bacteria into host cells (Gophna *et al.* 2001, 2002). These properties were initially used to classify curli as virulence factors. However, the involvement of curli in infections is also supported by many other observations. Whereas most laboratory strains of *E. coli* and *Salmonella* only produce curli below 30°C (Arnqvist *et al.* 1992), a significant fraction of *E. coli* isolates obtained from human blood is able to express curli at 37°C (Bian *et al.* 2000). Sera from sepsis patients have, furthermore, been shown to contain significantly more antibodies directed against the curli, compared to sera from healthy individuals (Bian *et al.* 2000). Finally, it has been shown that curli-expressing strains are able to induce significantly higher levels of proinflammatory cytokines in human macrophages, compared to non-curlinated strains (Bian *et al.* 2000). These observations suggest a direct implication of curli in human sepsis related to *E. coli* infections. The implication in sepsis is furthermore supported by studies showing that curli-proficient *Salmonella* strains and purified curli are able to stimulate the inert pathway of inflammation through the interaction with Toll-like receptor 2 (Tukel *et al.* 2005, 2009). Curli were consequently described as pathogen-associated molecular patterns (PAMP) (Tukel *et al.* 2005).

Table 6.1 Functional amyloids related to biofilms that have been purified and characterized by biophysical methods.

Microorganism	Major Amyloid Subunit	Function	Experimental Evidence[a]	References
Bacillus subtilis	TasA	Biofilm component.	EM, Congo red, and self-polymerization of depolymerized fibrils. ThT binding shown for *in vitro* polymerized fibrils.	(Romero *et al.* 2010, 2011, 2014)
Escherichia coli and *Salmonella* sp.	CsgA	Biofilm formation, virulence factor.	EM, CR, ThT, CD, x-ray fibre diffraction, ssNMR, AB, and self-polymerization.	(Barnhart & Chapman, 2006; Chapman *et al.* 2002; Dueholm *et al.* 2012; Shewmaker *et al.* 2009)
Klebsiella pneumoniae	Microcin E492	Cytotoxin that kills closely related Enterobacteriaceae by forming pores in their cytoplasmic membranes.	EM, CD, CR, ThT and self-polymerization.	(Bieler *et al.* 2005; Marcoleta *et al.* 2013; Shahnawaz & Soto, 2012)
*Mycobacterium tuberculosis**	MTP	Adhesive pili used for host invasion.	EM and resistance to depolymerization in boiling SDS-PAGE loading buffer.	(Alteri *et al.* 2007)
Pseudomonas	FapC	Adhesion, biofilm formation, and putative virulence factor.	EM, ThT, WO1, CD, FTIR, and self-polymerization.	(Dueholm *et al.* 2010, 2013b, 2013c)
*Streptococcus mutans**	Adhesin P1 (antigen I/II, PAc)	Forms a fibrous layer on the bacterial cell surface that contributes to cariogenicity. Provide interactions with host matrix proteins and other oral bacteria.	EM, ThT, CR, and self-polymerization.	(Oli *et al.* 2012)
Staphylococcus aureus	Phenol soluble modulins	Promote biofilm integrity	EM, ThT, CR, CD, and self-polymerization. All experiments were done on *in vitro* polymerized fibrils.	(Schwartz *et al.* 2012)
Xanthomonas spp. and other plant pathogens	HpaG	Virulence factor that induces a hypersensitive response in plant cells, resulting in cell death.	EM, CR, CD, self-polymerization, and protease K resistance. All experiments were done on *in vitro* polymerized fibrils.	(Oh *et al.* 2007)

**Structural analyses of native fibrils are required to confirm the amyloid nature.*

[a]*Abbreviations:* CD, circular dichroism; CR, Congo red; EM, electron microscopy; FTIR, Fourier transform infrared spectroscopy; ssNMR, solid state nuclear magnetic resonance; ThT, Thioflavin T; WO1, binding of the amyloid conformational antibody WO1.

Colonization of fresh products such as alfalfa sprouts, lettuce, melons, tomatoes, and spinach by pathogenic *Salmonella* and *E. coli* strains can lead to human infections and, in the worst cases, to foodborne outbreaks (Yaron & Römling, 2014). Laboratory strains of *E. coli* such as K-12 and DH5α are unable to colonize plants. However, when these strains are transformed with plasmids encoding the curli system, they become efficient plant colonizers (Jeter & Matthysse, 2005; Macarisin *et al.* 2012; Torres *et al.* 2005). This implies that curli play a direct role in plant colonization. However, studies carried out on shiga toxin-producing *E. coli* O157:H7 wild type strains and curli-deficient derivatives have shown that, although curli is important for plant colonization by some strains, curli are dispensable in others (Boyer *et al.* 2007; Macarisin *et al.* 2012; Patel *et al.* 2011; Torres *et al.* 2005). This suggests that some strains have redundant systems for plant colonization.

6.3.1.2 Biogenesis

The controlled biogenesis of curli fimbriae has been thoroughly studied. In *E. coli* and *Salmonella*, two divergently transcribed operons are required for expression of the functional amyloid (Figure 6.4a). The *csgBAC* operon encodes the major curli subunit, CsgA, and its homolog CsgB. The latter acts as a nucleus for polymerization of soluble CsgA after translocation to the cell surface (Chapman *et al.* 2002; Hammar *et al.* 1995). The *csgBAC* operon also encodes a third protein, CsgC, which is required for correct assembly of the mature curli fimbriae (Gibson *et al.* 2007; Taylor *et al.* 2011). Interestingly, CsgC contains a functional oxido-reductase motif (CXXC) that has been suggested to target a conserved cysteine in CsgG, the only other Cys-containing Csg protein (Taylor *et al.* 2011). CsgC has furthermore been shown to function as an efficient and selective inhibitor of amyloid formation, preventing the formation of intracellular aggregates or cytotoxic oligomers of CsgA during curli biogenesis (Evans *et al.* 2015). The *csgDEFG* operon encodes CsgD, a transcriptional activator of the *csgBAC* operon, CsgE, and CsgF, which act as chaperones and are required for effective curli assembly, and finally CsgG, a lipoprotein that forms a trans-membrane β-barrel outer-membrane macromolecular exporter used for the secretion of the curli subunits to the extracellular environment (Chapman *et al.* 2002; Hammar *et al.* 1995; Nenninger *et al.* 2009, 2011; Robinson *et al.* 2006). CsgG and CsgE apparently form a complex and work in concert with one another to facilitate the secretion of the curli subunits (Goyal *et al.* 2014; Nenninger *et al.* 2011; Robinson *et al.* 2006). CsgF represents the third component of the curli secretion complex. CsgF is transported to the extracellular environment through CsgG, where it forms a constitutive complex with CsgG (Nenninger *et al.* 2009; Robinson *et al.* 2006). CsgF is required for anchoring of CsgB to the outer membrane and is therefore a prerequisite of the nucleating function of CsgB (Nenninger *et al.* 2009).

6.3.1.3 Biotechnological applications

The high structural stability displayed by curli fibrils combined with their biocompatibility makes them ideal choices for next generation nanomaterials. With a detailed knowledge of the curli biogenesis machinery at hand, researches now have the tools to manipulate the curli systems for biotechnological applications. Van Gerven *et al.* (2014) demonstrated that the fusion of small heterologous sequences to *csgA* often resulted in the secretion of functional fusion proteins that folded into amyloid fibrils with the fusion peptides surface-exposed. This finding was subsequently applied to create functionalized biofilm materials with various valuable properties (Nguyen *et al.* 2014). Direct practical use was demonstrated by Zhong *et al.* (2014), who created an extremely strong underwater adhesive by fusing CsgA with fusing mussel foot proteins. Later Botyanszki *et al.* (2015) showed that curli fibrils, functionalized with the SpyTag capture domain, could be used to immobilize enzymes that were fused with the complementary SpyCatcher attachment domain, and that this immobilization had strong positive effect on the stability of the enzymes.

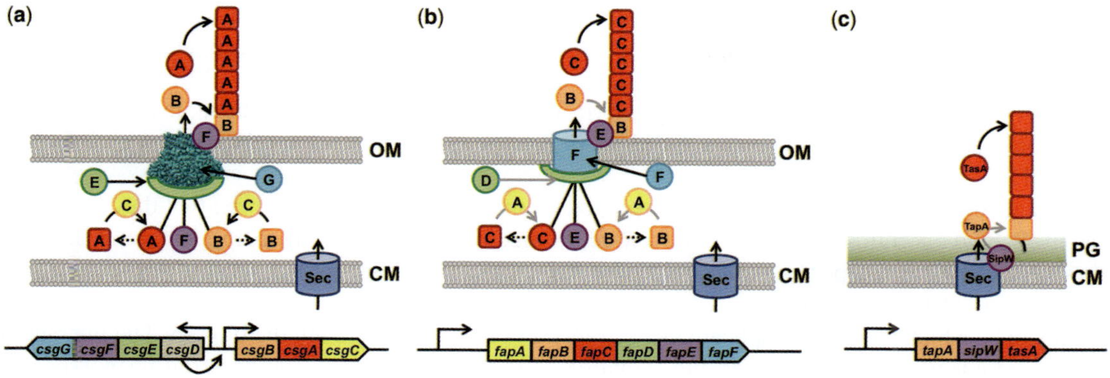

Figure 6.4 Schematic representation of curli, Fap and TasA biogenesis. (a) The curli biosynthesis requires the products of the *csgDEFG* and *csgBAC* operons. All Csg proteins, except the transcriptional regulator CsgD, are secreted to the periplasm via the Sec translocon, based on the presence of N-terminal signal sequences. CsgG forms a multimeric membrane pore together with CsgE, which is used for the secretion of the curli subunits and CsgF. Outside the cell, CsgF anchors secreted CsgB subunits to the cell surface where they act as the initial nucleators of CsgA polymerization. CsgC works as a chaperon in the periplasm by preventing the formation of cytotoxic CsgA oligomers. The structure of CsgG was obtained from PDB: 4UV3 (Goyal *et al.* 2014) (b) The Fap system is encoded in the *fapABCDEF* operon. All Fap proteins are secreted to the periplasm in a Sec dependent manner. FapF is proposed to form a membrane pore likely associated with FapD used for the secretion of the Fap subunits (FapB and FapC) as well as FapE. FapE is proposed to bind FapB to the cell surface, where it may act as the initial nucleator for FapC polymerization. FapA is believed to act as a chaperon, preventing the formation of cytotoxic FapB and FapC oligomers inside the periplasm. (c) The TasA system requires the products of the *tapA-sipW-tasA* operon. SipW is a signal peptidase that is believed to associate with a Sec translocon. This membrane transporter complex is then used for the secretion of TapA and TasA. TapA is covalently attached to the peptidoglycan, where it is required for the polymerization of the major TasA amyloid subunit, TasA. The following abbreviations are used: CM: cytoplasmic membrane, OM: outer membrane, PG: peptidoglycan.

6.3.2 Functional amyloids of *Pseudomonas* (Fap)

Many *Pseudomonas* species, including the opportunistic pathogen *P. aeruginosa*, the plant growth promoting *P. protogenes*, and the bioremediation-related *P. putida* are able to produce functional amyloids known as Functional amyloids in *Pseudomonas* (Fap) (Dueholm *et al.* 2010, 2013c). However, an evolutionary study has shown that the genetic elements encoding the Fap system are not restricted to the pseudomonads, but can be found in at least 15 other genera within the Proteobacteria, suggesting a more common use of this amyloid system (Dueholm *et al.* 2013b).

6.3.2.1 Biological role

The biological roles of Fap are still poorly understood, however, it is clear that Fap is associated with biofilm structure and function. Overexpression of the Fap system in several *Pseudomonas* species results in highly aggregative phenotypes that form extraordinarily strong biofilms (Dueholm *et al.* 2013c). A recent study based on atomic force microscopy (AFM) imaging and force spectroscopy showed that the amyloids render individual cells more resistant to drying and increase their hydrophobicity (Zeng *et al.* 2015). These properties could be important for the ecology of the bacteria in environments that occasionally encounter periods of low water activity such as soils. The amyloids also increase the overall hydrophobicity of biofilms and increase their stiffness a striking 20-fold (Zeng *et al.* 2015). The increased stiffness is likely caused by the rigid amyloid fibrils acting as a reinforcement of the more flexible biofilm matrix provided by eDNA and polysaccharides, similar to the role of steel in reinforced concrete (Zeng *et al.* 2015).

The Fap amyloids may also have more specialized functions. We recently showed that they are able to selectively bind quorum sensing molecules and redox mediators (Seviour *et al.* 2015). Amyloids therefore allow the bacteria to keep these small molecules in the vicinity of the cells, even in turbulent systems (Seviour *et al.* 2015). Binding of redox mediators such as pyocyanin changes the conductive properties of the amyloids, which may have implications for transport of electrons through the biofilm. The interaction of the amyloids with the quorum sensing signaling system strongly affects the bacterial phenotype. We recently documented major changes in the proteome of *P. aeruginosa* PAO1 upon expression of amyloid fibrils (Herbst *et al.* 2014). These changes included the down-regulation of classical virulence factors, including elastase B and the secretion system of alkaline protease. Amyloid also induced an unexpected activation of the alginate and pyoverdine biogenesis machinery, which turns wild type PAO1 into a mucoid phenotype with properties similar to those observed in *P. aeruginosa* isolates from chronic infections (Herbst *et al.* 2014; Smith *et al.* 2006).

Although a direct role of Fap in relation to chronic infections has never been established, there are several indications for such functions. Wiehlmann *et al.* (2007) identified a Fap-deficient mutant as one of the most attenuated in a *Caenorhabditis elegans* infection model and in a polymorphonuclear neutrophil leukocyte phagocytosis assay among 480 random transposon *P. aeruginosa* deletion mutants. Recently, it was also shown that transcription of the *fap* operon was highly up-regulated in murine models of acute burn and chronic surgical wound infections when compared to traditional laboratory conditions (Turner *et al.* 2014). Fap amyloids were finally observed in a murine based corneal infection model and in a murine based implant model with both Fap-specific antibodies and a novel Fap-specific dye (CDy11) (Kim *et al.* 2016).

6.3.2.2 Biogenesis

The molecular machinery required for Fap biosynthesis has many similarities to that used for the biogenesis of curli in *E. coli,* although there is no indication of a common evolutionary history of the two (Dueholm *et al.* 2012, 2013b). The genetic elements required for Fap biosynthesis are encoded in the *fapABCDEF* operon (Figure 6.4b). All Fap proteins are produced with N-terminal signal sequences that target them for export to the periplasmic space via the Sec translocon (Dueholm *et al.* 2013c). FapF integrates with the outer membrane, where, according to structural predictions, it forms a multimeric β-barrel membrane pore in analogy to CsgG of the curli system (Dueholm *et al.* 2013c; Herbst *et al.* 2014). FapB, FapC, and FapE are transported outside the cells (Herbst *et al.* 2014), probably via FapF. FapC and FapB represent the major and minor subunits of the mature Fap fibril, respectively (Dueholm *et al.* 2010). FapB shares sequence homology with FapC, and accordingly it has been suggested that FapB plays a nucleating role in the Fap biosynthesis, similar to that played by CsgB in the curli system (Dueholm *et al.* 2010, 2013c). This is supported by the fact that FapB is required for Fap biogenesis (Zeng *et al.* 2015). FapE is also required, and it is present in preparations of purified Fap (Herbst *et al.* 2014; Zeng *et al.* 2015). This supports a direct interaction of FapE with the growing amyloids, and it is likely that FapE plays a role similar to that of CsgF of the curli system in the organization of the extracellular Fap biogenesis apparatus. FapA is important for correct assembly of the mature Fap fibrils and deletion of the *fapA* gene results in the production of fibrils composed mainly of FapB (Dueholm *et al.* 2013c). Accordingly, it has been hypothesized that FapA acts as a periplasmic chaperon that keeps FapB and FapC in an unfolded conformation, until they are secreted in analogy to CsgC of the curli system (Dueholm *et al.* 2013c).

6.3.3 TasA from *Bacillus*

The Gram-positive model organism *Bacillus subtilis*, the pathogenic *B. cereus*, and the plant beneficial rhizobacterium *B. amyloliquefaciens* all produce functional amyloids composed of the protein TasA (Caro-Astorga *et al.* 2015; Romero *et al.* 2010; Zhang *et al.* 2015). However, identification of TasA homologs within the RefSeq database with the blastP algorithm suggests that TasA amyloids may be produced by many other *Bacillus* and some *Pontibacillus* species.

6.3.3.1 Biological role

TasA was originally identified as a secreted spore coat protein required for correct assembly of *B. subtilis* spores, although it was shown that TasA was not required for the viability of the spores (Serrano *et al.* 1999; Stöver & Driks, 1999a). When overexpressed in *E. coli,* TasA furthermore displays broad-spectrum antimicrobial activity (Stöver & Driks, 1999a). Accordingly, it was suggested that TasA could be used to inhibit the growth of other organisms and thereby provide a competitive advantage for the developing spores. The initial studies on TasA suggested that transcription of *tasA* was associated with the early onset of sporulation (Stöver & Driks, 1999a). However, it was revealed that *tasA* was also transcribed under sporulation-repressing conditions, suggesting that TasA is a transition-phase protein, rather than a sporulation-specific protein (Stöver & Driks, 1999b). Later, TasA was shown to be a major EPS component of *B. subtilis* biofilms and pellicles, together with exopolysaccharides (Branda *et al.* 2004, 2006). Within the biofilms, TasA acts as a structural component that provides rigidity to the biofilm (Chu *et al.* 2006; Romero *et al.* 2010).

6.3.3.2 Biogenesis

The genes required for TasA amyloid biogenesis are located within the *tapA-sipW-tasA* operon (Figure 6.4c) (Serrano *et al.* 1999; Stöver & Driks, 1999b). SipW is a type I signal peptidase that is believed to associate with a Sec translocon (Stöver & Driks, 1999c; Tjalsma *et al.* 2000). This complex likely represents the export system for the major amyloid subunit TasA and the peptidoglycan-anchored protein TapA (Branda *et al.* 2006; Romero *et al.* 2011, 2014). TapA facilitates the assembly of TasA on the cell surface (Romero *et al.* 2014). Although the exact mechanism is unknown, it is known that a sequence of 8 amino acid residues in the N-terminal region of TapA is required for amyloid biogenesis, whereas the C-terminal part is dispensable (Romero *et al.* 2014). Pairwise alignment of TasA and TapA reveals homology between the two proteins, especially in the N-terminal region. Consequently, it may be speculated that TapA acts a peptidoglycan-bound nucleator of TasA amyloid formation in analogy to CsgB of the curli system.

6.3.4 Other EPS-associated functional amyloids

Functional amyloids have also been identified and characterized for a few other biofilm-forming bacteria and archaea (Table 6.1), although these systems are still not very well described.

Klebsiella pneumoniae produce an antimicrobial peptide, known as Microcin E492 (MccE492), which is able to transverse the outer membrane of Enterobacteriaceae in a TonB-dependent manner.

When inside the periplasm, MccE492 forms oligomers that interact with the cytoplasmic membrane and forms pores which destabilize the cytoplasmic membrane and kill the bacterium (Destoumieux-Garzón *et al.* 2003). Interestingly, the activity of MccE492 peaks during the exponential growth phase, but is almost undetectable in the stationary phase, despite the fact that the peptide accumulates in both phases (de Lorenzo *et al.* 1984). This suggests that the activity of MccE492 is regulated by a conformational change of the peptide. Accordingly, Bieler *et al.* (2005) demonstrated that MccE492 was able to form biologically inactive amyloid fibrils *in vitro* and that such fibrils were also produced *in vivo*. The amyloid state may consequently be used *in vivo* to regulate the biological activity of the antimicrobial peptide. Although amyloids are often considered an end point of protein folding, Shahnawaz and Soto, (2012) demonstrated that the amyloids of MccE492 could depolymerize to release cytotoxic oligomers upon changing environmental conditions. MccE492 amyloids are therefore proposed to be the first example of an offensive EPS component.

Alteri *et al.* (2007) describe the first discovery of *Mycobacterium tuberculosis* pili (MTP), and it was later shown that these pili were involved in biofilm formation (Ramsugit *et al.* 2013). Electron microscopy of MTP reveals an ultrastructure similar to that of *E. coli* curli, and they also bind Congo red, which supports an amyloid structure (Alteri *et al.* 2007). However, currently, no direct structural data supports the amyloid fold of MTP. MTP is likely involved in pathogenicity as the pili are recognized by IgG antibodies contained in sera originating from patients suffering from tuberculosis and are required for bacterial binding to the host extracellular matrix protein laminin (Alteri *et al.* 2007).

Oli *et al.* (2012) reported that *Streptococcus mutans*, a bacterium associated with the formation of dental caries, produces functional amyloids composed of the cell surface adhesin P1 (antigen I/II, PAc). However, the amyloid nature of P1 was concluded based mainly on histological staining with Congo red and ThT, which are known to bind unspecifically to some non-amyloid EPS components (Groenning, 2010; Khurana *et al.* 2001). The study did not involve any structural characterization (X-ray fiber diffraction, CD, or FTIR), and the lack of direct evidence of the amyloid-defining cross-β-sheet structure questions the amyloid properties of P1. Indeed, a recent crystal structure of the adhesin P1 did not provide support for an amyloid structure (Heim *et al.* 2013).

The human pathogen *Staphylococcus aureus* and other Staphylococci secrete small peptides (20–40 amino acid residues) known as phenol-soluble modulins (PSM) (Mehlin *et al.* 1999; Wang *et al.* 2007). The PSMs are secreted to the extracellular environment via a dedicated ATP-binding cassette transporter called the PSM transporter (Chatterjee *et al.* 2013). Outside the cell, the PSMs form amphipathic α-helixes that are able to penetrate and form pores in the membranes of competing microbes and host cells (Wang *et al.* 2007). This provides *S. aureus* with a selective advantage in the environment and allows the bacterium to invade host tissue and evade the immune cells during infection. The implication of PSMs in infections is further supported from animal infection studies showing more severe infections with wild type strains compared to derivatives that lack the genes encoding the PSMs (Wang *et al.* 2007). PSMs are also important for *S. aureus* biofilm architecture. The amphipathic PSM α-helixes display surfactant-like properties and are involved in the development of channels within biofilms and also used for the dispersal of cells from mature biofilms, facilitating the spread of *S. aureus* (Periasamy *et al.* 2012). However, PSMs may also function as structural components of biofilms in its amyloid state. Schwartz *et al.* (2012, 2014) demonstrated that PSMs are able to form amyloid-like fibrils *in vivo* and that these extracellular fibrils enhance the stability of *S. aureus* biofilms in relation to chemical and mechanical stresses. The conformational change into amyloid fibrils is therefore clearly associated with a change in biological activity and function (Schwartz *et al.* 2012). The biological role of PSMs resembles that of the *K. pneumonia* bacteriocin MccE492, and it would be interesting to investigate if PSM amyloids can also be depolymerized by environmental factors. If this is the case, the amyloid might act as an inert storage form of PSMs, which can be stimulated to release active PSMs to promote biofilm disassembly, antimicrobial activity, or virulence when needed (Schwartz *et al.* 2012).

Xanthomonas axonopodia and many other plant pathogenic bacteria produce harpins, which are heat stable, glycine-rich, type III-secreted proteins that trigger a hypersensitive response (HR) when infiltrated into the intercellular space of host plant leaves (Oh *et al.* 2007; Wei *et al.* 1992). When harpins are subjected to plant apoplast-like conditions, they structurally rearrange from an initial α-helical state, first into spherical oligomers and prefibrillar species, and subsequently into mature amyloid-like fibrils (Oh *et al.* 2007). The transition to the amyloid state is a requirement for the HR-activity of the harpins as is evident from the inability of an HR-inactive harpin mutant to form amyloids (Oh *et al.* 2007). Although mature amyloids are the clinical hallmark of amyloid diseases, it is generally accepted that it is prefibrillar oligomers or protofibrillar species that lead to the actual cell death (Breydo & Uversky, 2015; Stefani, 2012). It is therefore tempting to speculate that it is the oligomers and other prefibrils that trigger the HR and that the mature amyloids act as reservoirs that can release active species when required, similarly to the proposed function of MccE492 and PSM amyloids.

6.4 ISOLATION AND CHARACTERIZATION OF FUNCTIONAL AMYLOIDS

Amyloids are very common in the microbial world as demonstrated by their detection in almost all environmental biofilms, independent of their origin, using amyloid-specific dyes or antibodies. However, functional amyloids have still only been purified and confirmed for a handful of common model organisms. There are four major reasons why so few amyloids have been characterized. The first relates to the insolubility and extreme stability of most functional amyloids, which exclude them from traditional protein analyses such as SDS-PAGE. The second is that many functional amyloids are highly adhesive and therefore bind to pipette tips and other consumables. Large sample volumes are therefore required for purification. The third is that the amyloid structure has to be preserved throughout the purification, as structural rearrangement will make it impossible to confirm the amyloid nature of the protein. The final reason is that confirmation of the amyloid structure is challenging and requires the combined use of high-resolution microscopic techniques (TEM or AFM to confirm the fibrillar morphology), structural analysis (FTIR or x-ray fiber diffraction to confirm the β-sheet structure), and specific binding of amyloid indicator dyes (Congo red or ThT) or antibodies (Fandrich, 2007). Thus, there is a lot of novel functional amyloids to be discovered, and there are currently two purification strategies that have proved successful (Figure 6.5).

In the first strategy, amyloids are liberated from the bacterial cells by mechanical shearing without rupturing the cells (Figure 6.5a). This requires that the amyloids are relatively weakly attached to the cell surface and that the cells are fairly rigid. The bacteria are subsequently separated from the EPS, containing the amyloids, by centrifugation. The crude amyloid preparation contains non-amyloid EPS components, lipids, and bound, extracellular proteins. These are removed by extractions with organic solvents or detergents. This strategy has been employed for the purification of *Mycobacterium tuberculosis* pili (MTP) (Alteri *et al.* 2007), the P1 adhesin of *Streptococcus mutans* (Oli *et al.* 2012), and *E. coli* curli (Chapman *et al.* 2002). The major advantages are the simplicity and the fact that it does not require explicit information of the cell wall architecture of the bacteria.

The second strategy relies on the unusual chemical and mechanical strength of the functional amyloids as it involves the selective degradation and solubilization of all other cell macromolecules by hydrolytic enzymes and detergents (Figure 6.5b). The first step in this strategy is cell lysis, which can be achieved using cell wall degrading enzymes, such as lysozyme and mutanolysin, together with freeze-thaw cycles. The lysis buffer should also contain RNases and DNases that can degrade released nucleic acids as these would otherwise cause an increase in viscosity and hamper subsequent sample handling. After the cell lysis and enzymatic digestion, the sample is heated in SDS to unfold and dissolve contaminating proteins. These are subsequently removed from the insoluble amyloids, which are pelleted by centrifugation and washed with buffer. The samples are now ready for biophysical analyses. This strategy has been used to isolate curli fibrils from *E. coli* (Dueholm *et al.* 2011) and *Salmonella* (Collinson *et al.* 1991), the Fap amyloids from *Pseudomonas* (Dueholm *et al.* 2010), chaplins from *Streptomyces coelicolor* (Claessen *et al.* 2003), phenol-soluble modulins from *Staphylococcus aureus* (Schwartz *et al.* 2012), and the amyloid cell wall sheaths from *Methanosaeta* (Dueholm *et al.* 2015). The major advantage of this strategy is that it can be used to isolate amyloids that are tightly associated with the bacterial cells or integrated in the cell wall. It can also be used when the amyloids are associated with other EPS components.

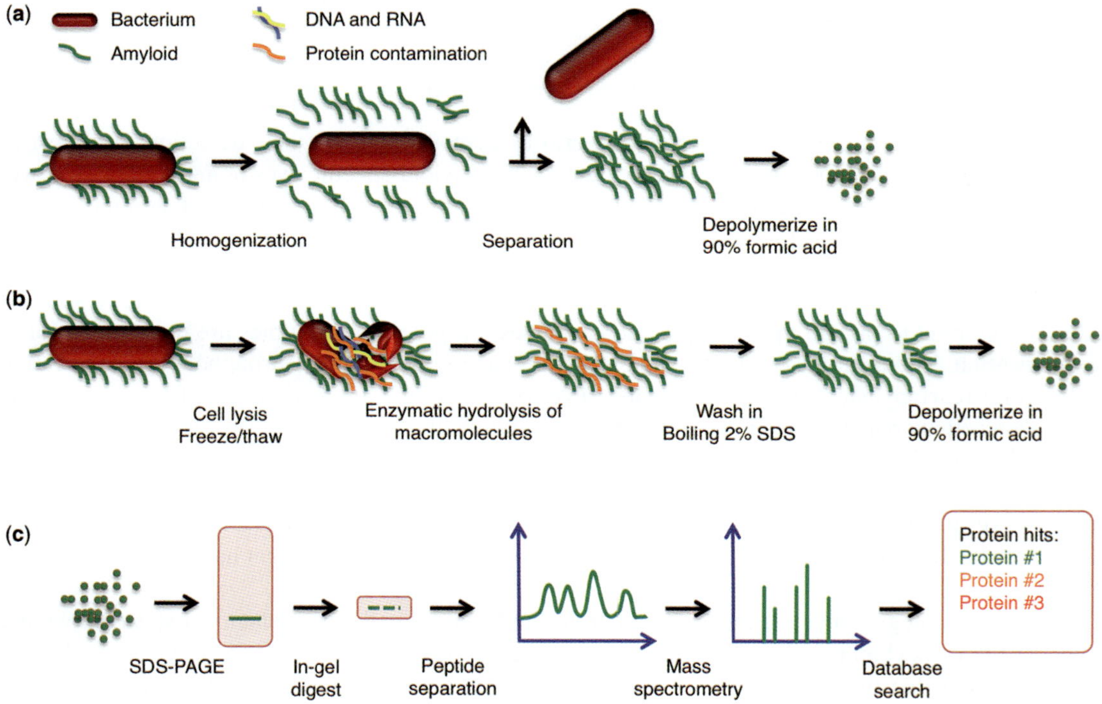

Figure 6.5 General strategies used for the purification of functional amyloids. (a) The amyloids are released from the bacterial cells by homogenization with a tissue grinder or a sonication bath. The amyloids are separated from the cells by differential or gradient centrifugation. The purified amyloid can be depolymerized by treatment with concentrated formic acid. (b) The bacterial cells are first lysed by freeze-thaw cycles in a buffer containing cell wall and nucleic acid degrading enzymes. The enzymatic hydrolysis is then continued under optimal conditions. Non-amyloid protein contaminants are solubilized with hot SDS and subsequently removed by centrifugation and buffer washes. The purified amyloid can be depolymerized by treatment with concentrated formic acid. (c) Formic acid depolymerized monomers are dissolved in SDS-PAGE loading buffer containing 8M urea and separated by SDS-PAGE. Protein bands are hereafter cut out, digested with trypsin, and analyzed by traditional MS/MS. Identified peptides are compared to a protein database in order to identify the amyloid protein.

Due to their large size and proteolytic resistance, functional amyloids are not directly applicable for identification by mass spectrometry. However, they can be depolymerized in aggressive solvents such as concentrated formic acid (Chapman *et al.* 2002; Collinson *et al.* 1991), trifluoroacetic acid (TFA) (Claessen *et al.* 2003; Jordal *et al.* 2009), and hexafluoroisopropanol (HFIP) (Cegelski *et al.* 2009). The depolymerized monomers can be lyophilized and separated by SDS-PAGE and identified by MS/MS. Urea (8M) may be included in the SDS-PAGE sample-loading buffer as some amyloid monomers aggregate in traditional loading buffers (Dueholm *et al.* 2013c).

6.5 CONCLUDING REMARK

Amyloids are common in biofilms, yet we know little about diversity and biological roles of these fascinating structures. The abundance of amyloids in nature raises some fundamental questions regarding structure and function. In order to address these questions, new functional microbial amyloids should be isolated and their function determined. The combined knowledge of many functional amyloid systems can then be used to generate a generic understanding of functional microbial amyloids. This knowledge may allow us to control amyloid formation in the future. This may have implications for the treatment of amyloid diseases and problematic biofilms. The functional amyloids display some very interesting properties: They are able to self-assemble from their monomeric constituent, they are remarkably stable, and they are biocompatible. These properties make them ideal choices for next generation manipulative nanomaterial.

REFERENCES

Alteri C. J., Xicohtencatl-Cortes J., Hess S., Caballero-Olin G., Giron J. A. and Friedman R. L. (2007). *Mycobacterium tuberculosis* produces pili during human infection. *Proc. Natl. Acad. Sci. U.S.A.*, **104**, 5145–5150.

Amann R. and Fuchs B. M. (2008). Single-cell identification in microbial communities by improved fluorescence *in situ* hybridization techniques. *Nat. Rev. Microbiol.*, **6**, 339–348.

Andreasen M., Lorenzen N. and Otzen D. (2015). Interactions between misfolded protein oligomers and membranes: A central topic in neurodegenerative diseases? *Biochim. Biophys. Acta.*, **1848**, 1897–1907.

Arnqvist A., Olsen A., Pfeifer J., Russell D. G. and Normark S. (1992). The Crl protein activates cryptic genes for curli formation and fibronectin binding in *Escherichia coli* HB101. *Mol. Microbiol.*, **6**, 2443–2452.

Austin J. W., Sanders G., Kay W. W. and Collinson S. K. (1998). Thin aggregative fimbriae enhance *Salmonella enteritidis* biofilm formation. *FEMS Microbiol. Lett.*, **162**, 295–301.

Ban T., Hoshino M., Takahashi S., Hamada D., Hasegawa K., Naiki H. and Goto Y. (2004). Direct observation of Aβ amyloid fibril growth and inhibition. *J. Mol. Biol.*, **344**, 757–767.

Barnhart M. M. and Chapman M. R. (2006). Curli biogenesis and function. *Annu. Rev. Microbiol.*, **60**, 131–147.

Bayry J., Aimanianda V., Guijarro, J. I., Sunde M. and Latgé J.-P. (2012). Hydrophobins—unique fungal proteins. *PLoS Pathog.*, **8**, e1002700.

Bian Z., Brauner A., Li Y. and Normark S. (2000). Expression of and cytokine activation by *Escherichia coli* curli fibers in human sepsis. *J. Infect. Dis.*, **181**, 602–612.

Bieler S., Estrada L., Lagos R., Baeza M., Castilla J. and Soto C. (2005). Amyloid formation modulates the biological activity of a bacterial protein. *J. Biol. Chem.*, **280**, 26880–26885.

Booth D. R., Sunde M., Bellotti V., Robinson C. V., Hutchinson W. L., Fraser P. E., Hawkins P. N., Dobson C. M., Radford S. E., Blake C. C. and Pepys M. B. (1997). Instability, unfolding and aggregation of human lysozyme variants underlying amyloid fibrillogenesis. *Nature*, **385**, 787–793.

Botyanszki Z., Tay P. K. R., Nguyen P. Q., Nussbaumer M. G. and Joshi N. S. (2015). Engineered catalytic biofilms: Site-specific enzyme immobilization onto E. coli curli nanofibers. *Biotechnology and Bioengineering*, **112**, 2016–2024. doi:10.1002/bit.25638

Boyer R. R., Sumner S. S., Williams R. C., Pierson M. D., Popham D. L., Kniel and K. E. (2007). Influence of curli expression by *Escherichia coli* 0157:H7 on the cell's overall hydrophobicity, charge, and ability to attach to lettuce. *J. Food Prot.*, **70**, 1339–1345.

Branda S. S., Chu F., Kearns D. B., Losick R. and Kolter R. (2006). A major protein component of the *Bacillus subtilis* biofilm matrix. *Mol. Microbiol.*, **59**, 1229–1238.

Branda S. S., González-Pastor J. E., Dervyn E., Ehrlich S. D., Losick R. and Kolter R. (2004). Genes involved in formation of structured multicellular communities by *Bacillus subtilis*. *J. Bacteriol.*, **186**, 3970–3979.

Breydo L. and Uversky V. N. (2015). Structural, morphological, and functional diversity of amyloid oligomers. *FEBS Lett., Dynamics, flexibility, and intrinsic disorder in protein assemblies*, **589**, 2640–2648.

Caro-Astorga J., Pérez-García A., de Vicente A. and Romero D. (2015). A genomic region involved in the formation of adhesin fibers in *Bacillus cereus* biofilms. *Front. Microbiol.*, **5**, 1–11. doi:10.3389/fmicb.2014.00745

Cegelski L., Pinkner J. S., Hammer N. D., Cusumano C. K., Hung C. S., Chorell E., Åberg V., Walker J. N., Seed P. C., Almqvist F., Chapman M. R. and Hultgren S. J. (2009). Small-molecule inhibitors target *Escherichia coli* amyloid biogenesis and biofilm formation. *Nat. Chem. Biol.*, **5**, 913–919.

Chapman M. R., Robinson L. S., Pinkner J. S., Roth R., Heuser J., Hammar M., Normark S. and Hultgren S. J. (2002). Role of *Escherichia coli* curli operons in directing amyloid fiber formation. *Science*, **295**, 851–855.

Chatterjee S. S., Joo H. -S., Duong A. C., Dieringer T. D., Tan V. Y., Song Y., Fischer E. R., Cheung, G. Y. C., Li M. and Otto M. (2013). Essential *Staphylococcus aureus* toxin export system. *Nat. Med.*, **19**, 364–367.

Chimileski S., Franklin M. J. and Papke R. T. (2014). Biofilms formed by the archaeon *Haloferax volcanii* exhibit cellular differentiation and social motility, and facilitate horizontal gene transfer. *BMC Biol.*, **12**. doi:10.1186/s12915-014-0065-5

Chiti F. and Dobson C. M. (2006). Protein misfolding, functional amyloid, and human disease. *Annu. Rev. Biotech.*, **75**, 333–366.

Chu F., Kearns D. B., Branda S. S., Kolter R. and Losick R. (2006). Targets of the master regulator of biofilm formation in *Bacillus subtilis*. *Mol. Microbiol.*, **59**, 1216–1228.

Claessen D., Rink R., de Jong W., Siebring J., de Vreugd P., Boersma F. G., Dijkhuizen L. and Wosten H. A. (2003). A novel class of secreted hydrophobic proteins is involved in aerial hyphae formation in *Streptomyces coelicolor* by forming amyloid-like fibrils. *Genes Dev.*, **17**, 1714–1726.

Clark P. L. (2004). Protein folding in the cell: Reshaping the folding funnel. *Trends Biochem. Sci.*, **29**, 527–534.

Collinson S. K., Doig P. C., Doran J. L., Clouthier S., Trust T. J. and Kay W. W. (1993). Thin, aggregative fimbriae mediate binding of *Salmonella enteritidis* to fibronectin. *J. Bacteriol.*, **175**, 12–18.

Collinson S. K., Emody L., Muller K. H., Trust T. J. and Kay W. W. (1991). Purification and characterization of thin, aggregative fimbriae from *Salmonella enteritidis*. *J. Bacteriol.*, **173**, 4773–4781.

De Lorenzo V., Martínez J. L. and Asensio C. (1984). Microcin-mediated interactions between *Klebsiella pneumoniae* and *Escherichia coli* strains. *J. Gen. Microbiol.*, **130**, 391–400.

Destoumieux-Garzón D., Thomas X., Santamaria M., Goulard C., Barthélémy M., Boscher B., Bessin Y., Molle G., Pons A.-M., Letellier L., Peduzzi J. and Rebuffat S. (2003). Microcin E492 antibacterial activity: Evidence for a TonB-dependent inner membrane permeabilization on *Escherichia coli*. *Mol. Microbiol.*, **49**, 1031–1041.

Dobson C. M. (2003). Protein folding and misfolding. *Nature*, **426**, 884–890.

Dobson C. M. (1999). Protein misfolding, evolution and disease. *Trends Biochem. Sci.*, **24**, 329–332.

Dovidchenko N. V., Leonova E. I. and Galzitskaya O. V. (2014). Mechanisms of amyloid fibril formation. *Biochemistry (Moscow)*, **79**, 1515–1527.

Dueholm M. S., Albertsen M., Otzen D. and Nielsen P. H. (2012). Curli functional amyloid systems are phylogenetically widespread and display large diversity in operon and protein structure. *PLoS ONE*, **7**, e51274.

Dueholm M. S., Larsen P., Finster K., Stenvang M. R., Christiansen G., Vad B. S., Bøggild A., Otzen D. E. and Nielsen P. H. (2015). The tubular sheaths encasing *Methanosaeta thermophila* filaments are functional amyloids. *J. Biol. Chem.*, **290**, 20590–20600.

Dueholm M. S., Nielsen P. H., Chapman M. and Otzen D. (2013a). Functional amyloids in bacteria. In: Amyloid Fibrils and Prefibrillar Aggregates, D. E. Otzen (ed.), Wiley-VCH Verlag GmbH & Co. KGaA, pp. 411–438.

Dueholm M. S., Nielsen S. B., Hein K. L., Nissen P., Chapman M., Christiansen G., Nielsen P. H. and Otzen D. E. (2011). Fibrillation of the major curli subunit CsgA under a wide range of conditions implies a robust design of aggregation. *Biochemistry*, **50**, 8281–8290. doi:10.1021/bi200967c

Dueholm M. S., Otzen D. and Nielsen P. H. (2013b). Evolutionary insight into the functional amyloids of the pseudomonads. *PLoS ONE*, **8**, e76630. doi:10.1371/journal.pone.0076630

Dueholm M. S., Petersen S. V., Sønderkaer M., Larsen P., Christiansen G., Hein K. L., Enghild J. J., Nielsen J. L., Nielsen K. L., Nielsen P. H. and Otzen D. E. (2010). Functional amyloid in *Pseudomonas. Mol. Microbiol.*, **77**, 1009–1020.

Dueholm M. S., Søndergaard M. T., Nilsson M., Christiansen G., Stensballe A., Overgaard M. T., Givskov M., Tolker-Nielsen T., Otzen D. E. and Nielsen P. H. (2013c). Expression of Fap amyloids in *Pseudomonas aeruginosa, P. fluorescens*, and *P. putida* results in aggregation and increased biofilm formation. *MicrobiologyOpen*, **2**, 365–382. doi:10.1002/mbo3.81

Evans M. L., Chorell E., Taylor J. D., Åden J., Götheson A., Li F., Koch M., Sefer L., Matthews S. J., Wittung-Stafshede P., Almqvist F. and Chapman M. R. (2015). The bacterial curli system possesses a potent and selective inhibitor of amyloid formation. *Mol. Cell*, **57**, 445–455. doi:10.1016/j.molcel.2014.12.025

Fandrich M. (2007). On the structural definition of amyloid fibrils and other polypeptide aggregates. *Cell. Mol. Life Sci.*, **64**, 2066–2078.

Fowler D. M., Koulov A. V., Balch W. E. and Kelly J. W. (2007). Functional amyloid–from bacteria to humans. *Trends Biochem. Sci.*, **32**, 217–224.

Frokjaer S. and Otzen D. E. (2005). Protein drug stability: a formulation challenge. *Nat. Rev.*, **4**, 298–306.

Garcia M. C., Lee J. T., Ramsook C. B., Alsteens D., Dufrêne Y. F. and Lipke P. N. (2011). A role for amyloid in cell aggregation and biofilm formation. *PLoS ONE*. **6**, e17632. doi:10.1371/journal.pone.0017632

Gibson D. L., White A. P., Rajotte C. M. and Kay W. W. (2007). AgfC and AgfE facilitate extracellular thin aggregative fimbriae synthesis in *Salmonella enteritidis. Microbiology*, **153**, 1131–1140.

Gophna U., Barlev M., Seijffers R., Oelschlager T. A., Hacker J. and Ron E. Z. (2001). Curli fibers mediate internalization of *Escherichia coli* by eukaryotic cells. Infect. *Immun.*, **69**, 2659–2665.

Gophna U., Oelschlaeger T. A., Hacker J. and Ron E. Z. (2002). Role of fibronectin in curli-mediated internalization. *FEMS Microbiol. Lett.*, **212**, 55–58.

Goyal P., Krasteva P. V., Van Gerven N., Gubellini F., Van den Broeck I., Troupiotis-Tsaïlaki A., Jonckheere W., Péhau-Arnaudet G., Pinkner J. S., Chapman M. R., Hultgren S. J., Howorka S., Fronzes R. and Remaut H. (2014). Structural and mechanistic insights into the bacterial amyloid secretion channel CsgG. *Nature*. doi:10.1038/nature13768

Groenning M. (2010). Binding mode of Thioflavin T and other molecular probes in the context of amyloid fibrils–current status. *J. Chem. Biol.*, **3**, 1–18. doi:10.1007/s12154-009-0027-5

Hammar M., Arnqvist A., Bian Z., Olsen A. and Normark S. (1995). Expression of two *csg* operons is required for production of fibronectin- and Congo red-binding curli polymers in *Escherichia coli* K-12. *Mol. Microbiol.*, **18**, 661–670.

Hammer N. D., Wang X., McGuffie B. A. and Chapman M. R. (2008). Amyloids: friend or foe? *J. Alzheimers Dis.*, **13**, 407–419.

Hamodrakas S. J., Hoenger A. and Iconomidou V. A. (2004). Amyloid fibrillogenesis of silkmoth chorion protein peptide-analogues via a liquid-crystalline intermediate phase. *J. Struct. Biol.*, **145**, 226–235.

Heim K. P., Crowley P. J. and Brady L. J. (2013). An intramolecular interaction involving the N terminus of a *Streptococcal* adhesin affects its conformation and adhesive function. *J. Biol. Chem.*, **288**, 13762–13774. doi:10.1074/jbc.M113.459974

Herbst F.-A., Søndergaard M. T., Kjeldal H., Stensballe A., Nielsen P. H. and Dueholm M. S. (2014). Major proteomic changes associated with amyloid-induced biofilm formation in *Pseudomonas aeruginosa* PAO1. *J Proteome Res.*, doi:10.1021/pr500938x

Høiby N., Bjarnsholt T., Givskov M., Molin S. and Ciofu O. (2010). Antibiotic resistance of bacterial biofilms. *Int. J. Antimicrob. Agents*, **35**, 322–332. doi:10.1016/j. ijantimicag.2009.12.011

Iconomidou V. A., Chryssikos G. D., Gionis V., Galanis A. S., Cordopatis P., Hoenger A. and Hamodrakas S. J. (2006). Amyloid fibril formation propensity is inherent into the hexapeptide tandemly repeating sequence of the central domain of silkmoth chorion proteins of the A-family. *J. Struct. Biol.*, **156**, 480–488.

Jahn T. R., Parker M. J., Homans S. W. and Radford S. E. (2006). Amyloid formation under physiological conditions proceeds via a native-like folding intermediate. *Nat. Struct. Mol. Biol.*, **13**, 195–201. doi:10.1038/nsmb1058

Jeter C. and Matthysse A. G. (2005). Characterization of the binding of diarrheagenic strains of *E. coli* to plant surfaces and the role of curli in the interaction of the bacteria with alfalfa sprouts. *Mol. Plant Microbe Interact.*, **18**, 1235–1242. doi:10.1094/MPMI-18-1235

Jimenez J. L., Guijarro J. I., Orlova E., Zurdo J., Dobson C. M., Sunde M. and Saibil H. R. (1999). Cryo-electron microscopy structure of an SH3 amyloid fibril and model of the molecular packing. *EMBO J.*, **18**, 815–821.

Jordal P. B., Dueholm M. S., Larsen P., Petersen S. V., Enghild J. J., Christiansen G., Hojrup P., Nielsen P. H. and Otzen D. E. (2009). Widespread abundance of functional bacterial amyloid in mycolata and other gram-positive bacteria. *Appl. Environ. Microbiol.*, **75**, 4101–4110.

Kenney J. M., Knight D. and Wise M. J., Vollrath F. (2002). Amyloidogenic nature of spider silk. *Eur. J. Biochem.*, **269**, 4159–4163.

Khurana R., Uversky V. N., Nielsen L. and Fink A. L. (2001). Is Congo red an amyloid-specific dye? *J. Biol. Chem.*, **276**, 22715–22721.

Kikuchi T., Mizunoe Y., Takade A., Naito S. and Yoshida S. (2005). Curli fibers are required for development of biofilm architecture in *Escherichia coli* K-12 and enhance bacterial adherence to human uroepithelial cells. *Microbiol. Immunol.*, **49**, 875–884.

Kim J.-Y., Sahu S., Yau Y.-H., Wang X., Shochat S. G., Nielsen P. H., Dueholm M. S., Otzen D. E., Lee J., Delos Santos M. M. S., Yam J. K. H., Kang N.-Y., Park S.-J., Kwon H., Seviour T., Yang L., Givskov M. and Chang Y.-T. (2016). Detection of pathogenic biofilms with bacterial amyloid targeting fluorescent probe, CDy11. *J. Am. Chem. Soc.*, **138**, 402–407. doi:10.1021/jacs.5b11357

Kwan A. H., Winefield R. D., Sunde M., Matthews J. M., Haverkamp R. G., Templeton M. D. and Mackay J. P. (2006). Structural basis for rodlet assembly in fungal hydrophobins. *Proc. Natl. Acad. Sci. U.S.A.*, **103**, 3621–3626.

Kyle R. A. (2001). Amyloidosis: a convoluted story. *Brit. J. Haemat.*, **114**, 529–538.

Larsen P., Nielsen J. L., Dueholm M. S., Wetzel R., Otzen D. and Nielsen P. H. (2007). Amyloid adhesins are abundant in natural biofilms. *Environ. Microbiol.*, **9**, 3077–3090.

Larsen P., Nielsen J. L., Otzen D. and Nielsen P. H. (2008). Amyloid-like adhesins produced by floc-forming and filamentous bacteria in activated sludge. *Appl. Environ. Microbiol.*, **74**, 1517–1526.

Letunic I. and Bork P. (2007). Interactive tree of life (iTOL): an online tool for phylogenetic tree display and annotation. *Bioinformatics*, **23**, 127–128. doi:10.1093/bioinformatics/btl529

Linding R., Schymkowitz J., Rousseau F., Diella F. and Serrano L. (2004). A comparative study of the relationship between protein structure and β-aggregation in globular and intrinsically disordered proteins. *J. Mol. Biol.*, **342**, 345–353.

Lührs T., Ritter C., Adrian M., Riek-Loher D., Bohrmann B., Döbeli H., Schubert D. and Riek R. (2005). 3D structure of Alzheimer's amyloid-β(1–42) fibrils. *Proc. Natl. Acad. Sci. U.S.A.*, **102**, 17342–17347. doi:10.1073/pnas.0506723102

Macarisin D., Patel J., Bauchan G., Giron J. A. and Sharma V. K. (2012). Role of curli and cellulose expression in adherence of *Escherichia coli* O157:H7 to spinach leaves. *Foodborne Pathog. Dis.*, **9**, 160–167. doi:10.1089/fpd.2011.1020

Maji S. K., Perrin M. H., Sawaya M. R., Jessberger S., Vadodaria K., Rissman R. A., Singru P. S., Nilsson K. P. R., Simon R., Schubert D., Eisenberg D., Rivier J., Sawchenko P., Vale W. and Riek R. (2009). Functional amyloids as natural storage of peptide hormones in pituitary secretory granules. *Science*, **325**, 328–332. doi:10.1126/science.1173155

Makin O. S., Atkins E., Sikorski P., Johansson J. and Serpell L. C. (2005). Molecular basis for amyloid fibril formation and stability. *Proc. Natl. Acad. Sci. U.S.A.*, **102**, 315–320. doi:10.1073/pnas.0406847102

Makin O. S. and Serpell L. C. (2005). Structures for amyloid fibrils. *FEMS J.*, **272**, 5950–5961.

Marcoleta A., Marín M., Mercado G., Valpuesta J. M., Monasterio O. and Lagos R. (2013). Microcin E492 amyloid formation is retarded by posttranslational modification. *J. Bacteriol.*, **195**, 3995–4004. doi:10.1128/JB.00564-13

Maury C. P. J. (2009). The emerging concept of functional amyloid. *J. Intern. Med.*, **265**, 329–334. doi:10.1111/j.1365-2796.2008.02068.x

McGlinchey R. P., Shewmaker F., McPhie P., Monterroso B., Thurber K. and Wickner R. B. (2009). The repeat domain of the melanosome fibril protein Pmel17 forms the amyloid core promoting melanin synthesis. *Proc. Natl. Acad. Sci. U.S.A.*, **106**, 13731–13736.

Mehlin C., Headley C. M. and Klebanoff S. J. (1999). An inflammatory polypeptide complex from *Staphylococcus epidermidis*: Isolation and characterization. *J. Exp. Med.*, **189**, 907–918. doi:10.1084/jem.189.6.907

Nasr A., Olsén A., Sjöbring U., Müller-Esterl W. and Björck L. (1996). Assembly of human contact phase proteins and release of bradykinin at the surface of curli-expressing *Escherichia coli*. *Mol. Microbiol.*, **20**, 927–935.

Nelson R., Sawaya M. R., Balbirnie M., Madsen A. O., Riekel C., Grothe R. and Eisenberg D. (2005). Structure of the cross-β spine of amyloid-like fibrils. *Nature*, **435**, 773–778.

Nenninger A. A., Robinson L. S., Hammer N. D., Epstein E. A., Badtke M. P., Hultgren S. J. and Chapman M. R. (2011). CsgE is a curli secretion specificity factor that prevents amyloid fibre aggregation. *Mol. Microbiol.*, **81**, 486–499.

Nenninger A. A., Robinson L. S. and Hultgren S. J. (2009). Localized and efficient curli nucleation requires the chaperone-like amyloid assembly protein CsgF. *Proc. Natl. Acad. Sci. U.S.A.*, **106**, 900–905.

Nguyen P. Q., Botyanszki Z., Tay P. K. R. and Joshi N. S. (2014). Programmable biofilm-based materials from engineered curli nanofibres. *Nature Communications*, **5**, 4945. doi:10.1038/ncomms5945

Oh J., Kim J. G., Jeon E., Yoo C. H., Moon J. S., Rhee S. and Hwang I. (2007). Amyloidogenesis of type III-dependent harpins from plant pathogenic bacteria. *J. Biol. Chem.*, **282**, 13601–13609.

Oli M. W., Otoo H. N., Crowley P. J., Heim K. P., Nascimento M. M., Ramsook C. B., Lipke P. N. and Brady L. J. (2012). Functional amyloid formation by *Streptococcus mutans*. *Microbiology*, **158**, 2903–2916. doi:10.1099/mic.0.060855-0

Olsen A., Arnqvist A., Hammar M., Sukupolvi S. and Normark S. (1993). The RpoS sigma factor relieves H-NS-mediated transcriptional repression of *csgA*, the subunit gene of fibronectin-binding curli in *Escherichia coli*. *Mol. Microbiol.*, **7**, 523–536.

Olsen A., Jonsson A. and Normark S. (1989). Fibronectin binding mediated by a novel class of surface organelles on *Escherichia coli*. *Nature*, **338**, 652–655.

O'Nuallain B. and Wetzel R. (2002). Conformational Abs recognizing a generic amyloid fibril epitope. *Proc. Natl. Acad. Sci. U.S.A.*, **99**, 1485–1490.

Patel J., Sharma M. and Ravishakar S. (2011). Effect of curli expression and hydrophobicity of Escherichia coli O157:H7 on attachment to fresh produce surfaces. *J. Appl. Microbiol.*, **110**, 737–745. doi:10.1111/j.1365-2672.2010.04933.x

Pawar D. M., Rossman M. L. and Chen J. (2005). Role of curli fimbriae in mediating the cells of enterohaemorrhagic *Escherichia coli* to attach to abiotic surfaces. *J. Appl. Microbiol.*, **99**, 418–425.

Pedersen J. S., Dikov D., Flink J. L., Hjuler H. A., Christiansen G. and Otzen D. E. (2006). The changing face of glucagon fibrillation: structural polymorphism and conformational imprinting. *J. Mol. Biol.*, **355**, 501–523.

Periasamy S., Joo, H.-S., Duong A. C., Bach, T.-H. L., Tan V. Y., Chatterjee S. S., Cheung, G. Y. C. and Otto M. (2012). How *Staphylococcus aureus* biofilms develop their characteristic structure. *Proc. Natl. Acad. Sci. U.S.A.*, **109**, 1281–1286. doi:10.1073/pnas.1115006109

Ramsugit S., Guma S., Pillay B., Jain P., Larsen M. H., Danaviah S. and Pillay M. (2013). Pili contribute to biofilm formation *in vitro* in *Mycobacterium tuberculosis*. *Antonie van Leeuwenhoek* **104**, 725–735. doi:10.1007/s10482-013-9981-6

Ren Q., Kwan A. H. and Sunde M. (2013). Two forms and two faces, multiple states and multiple uses: Properties and applications of the self-assembling fungal hydrophobins. *Biopolymers*, **100**, 601–612. doi:10.1002/bip.22259

Robinson L. S., Ashman E. M., Hultgren S. J. and Chapman M. R. (2006). Secretion of curli fibre subunits is mediated by the outer membrane-localized CsgG protein. *Mol. Microbiol.*, **59**, 870–881.

Romero D., Aguilar C., Losick R. and Kolter R. (2010). Amyloid fibers provide structural integrity to *Bacillus subtilis* biofilms. *Proc. Natl. Acad. Sci. U.S.A.*, **107**, 2230–2234.

Romero D., Vlamakis H., Losick R. and Kolter R. (2014). Functional analysis of the accessory protein TapA in *Bacillus subtilis* amyloid fiber assembly. *J. Bacteriol.*, **196**, 1505–1513. doi:10.1128/JB.01363-13

Romero D., Vlamakis H., Losick R. and Kolter R. (2011). An accessory protein required for anchoring and assembly of amyloid fibres in *B. subtilis* biofilms. *Mol. Microbiol.*, **80**, 1155–1168. doi:10.1111/j.1365-2958.2011.07653.x

Romling U., Bian Z., Hammar M., Sierralta W. D. and Normark S. (1998). Curli fibers are highly conserved between *Salmonella typhimurium* and *Escherichia coli* with respect to operon structure and regulation. *J. Bacteriol.*, **180**, 722–731.

Ryu J. H. and Beuchat L. R. (2005). Biofilm formation by *Escherichia coli* O157:H7 on stainless steel: effect of exopolysaccharide and curli production on its resistance to chlorine. *Appl. Environ. Microbiol.*, **71**, 247–254.

Ryu J. H., Kim H., Frank J. F. and Beuchat L. R. (2004). Attachment and biofilm formation on stainless steel by *Escherichia coli* O157:H7 as affected by curli production. *Lett. Appl. Microbiol.*, **39**, 359–362.

Sawyer E. B., Claessen D., Haas M., Hurgobin B. and Gras S. L. (2011). The assembly of individual chaplin peptides from *Streptomyces coelicolor* into functional amyloid fibrils. *PLoS ONE*, **6**, e18839. doi:10.1371/journal.pone.0018839

Schwartz K., Sekedat M. D., Syed A. K., O'Hara B., Payne D. E., Lamb A. and Boles B. R. (2014). The AgrD N-terminal leader peptide of *Staphylococcus aureus* has cytolytic and amyloidogenic properties. *Infect. Immun.*, **82**, 3837–3844. doi:10.1128/IAI.02111-14

Schwartz K., Syed A. K., Stephenson R. E., Rickard A. H. and Boles B. R. (2012). Functional amyloids composed of phenol soluble modulins stabilize *Staphylococcus aureus* biofilms. *PLoS Pathog.*, **8**, e1002744. doi:10.1371/journal.ppat.1002744

Serrano M., Zilhão R., Ricca E., Ozin A. J., Moran C. P. and Henriques A. O. (1999). A *Bacillus subtilis* secreted protein with a role in endospore coat assembly and function. *J. Bacteriol.*, **181**, 3632–3643.

Seviour T., Hansen S. H., Yang L., Yau Y. H., Wang V. B., Stenvang M. R., Christiansen G., Marsili E., Givskov M., Chen Y., Otzen D. E., Nielsen P. H., Geifman-Shochat S., Kjelleberg S. and Dueholm M. S. (2015). Functional amyloids keep quorum-sensing molecules in check. *J. Biol. Chem.*, **290**, 6457–6469. doi:10.1074/jbc.M114.613810

Shahnawaz M. and Soto C. (2012). Microcin amyloid fibrils A are reservoir of toxic oligomeric species. *J. Biol. Chem.*, **287**, 11665–11676. doi:10.1074/jbc.M111.282533

Shewmaker F., McGlinchey R. P., Thurber K. R., McPhie P., Dyda F., Tycko R. and Wickner R. B. (2009). The functional curli amyloid is not based on in-register parallel β-sheet structure. *J. Biol. Chem.*, **284**, 25065–25076.

Sipe J. D. and Cohen A. S. (2000). Review: history of the amyloid fibril. *J. Struct. Biol.*, **130**, 88–98.

Sjöbring U., Pohl G. and Olsén A. (1994). Plasminogen, absorbed by *Escherichia coli* expressing curli or by *Salmonella enteritidis* expressing thin aggregative fimbriae, can be activated by simultaneously captured tissue-type plasminogen activator (t-PA). *Mol. Microbiol.*, **14**, 443–452.

Smith E. E., Buckley D. G., Wu Z., Saenphimmachak C., Hoffman L. R., D'Argenio D. A., Miller S. I., Ramsey B. W., Speert D. P., Moskowitz S. M., Burns J. L., Kaul R. and Olson M. V. (2006). Genetic adaptation by *Pseudomonas aeruginosa* to the airways of cystic fibrosis patients. *Proc. Natl. Acad. Sci. U.S.A.*, **103**, 8487–8492. doi:10.1073/pnas.0602138103

Stefani M. (2012). Structural features and cytotoxicity of amyloid oligomers: Implications in Alzheimer's disease and other diseases with amyloid deposits. *Prog. Neurobiol.*, The Neurotoxicity of Mutant Proteins 20 Years after the discovery of the first mutant gene involved in neurodegeneration, **99**, 226–245. doi:10.1016/j.pneurobio.2012.03.002

Stöver A. G. and Driks A. (1999a). Secretion, localization, and antibacterial activity of TasA, a *Bacillus subtilis* spore-associated protein. *J. Bacteriol.*, **181**, 1664–1672.

Stöver A. G. and Driks A. (1999b). Regulation of synthesis of the *Bacillus subtilis* transition-phase, spore-associated antibacterial protein TasA. *J. Bacteriol.*, **181**, 5476–5481.

Stöver A. G. and Driks A. (1999c). Control of synthesis and secretion of the *Bacillus subtilis* protein YqxM. *J. Bacteriol.*, **181**, 7065–7069.

Sunde M., Serpell L. C., Bartlam M., Fraser P. E., Pepys M. B. and Blake C. C. (1997). Common core structure of amyloid fibrils by synchrotron x-ray diffraction. *J. Mol. Biol.*, **273**, 729–739.

Taylor J. D., Zhou Y., Salgado P. S., Patwardhan A., McGuffie M., Pape T., Grabe G., Ashman E., Constable S. C., Simpson P. J., Lee W., Cota E., Chapman M. R. and Matthews S. J. (2011). Atomic resolution insights into curli fiber biogenesis. *Structure*, **19**, 1307–1316.

Tian P., Boomsma W., Wang Y., Otzen D. E., Jensen M. H. and Lindorff-Larsen K. (2015). Structure of a functional amyloid protein subunit computed using sequence variation. *J. Am. Chem. Soc.*, **137**, 22–25. doi:10.1021/ja5093634

Tjalsma H., Stöver A. G., Driks A., Venema G., Bron S. and van Dijl J. M. (2000). Conserved serine and histidine residues are critical for activity of the ER-type signal peptidase SipW of *Bacillus subtilis*. *J. Biol. Chem.*, **275**, 25102–25108. doi:10.1074/jbc.M002676200

Torres A. G., Jeter C., Langley W. and Matthysse A. G. (2005). Differential binding of *Escherichia coli* O157:H7 to alfalfa, human epithelial cells, and plastic is mediated by a variety of surface structures. *Appl. Environ. Microbiol.*, **71**, 8008–8015.

Tukel C., Raffatellu M., Humphries A. D., Wilson R. P., Andrews-Polymenis H. L., Gull T., Figueiredo J. F., Wong M. H., Michelsen K. S., Akcelik M., Adams L. G. and Baumler A. J. (2005). CsgA is a pathogen-associated molecular pattern of *Salmonella enterica* serotype Typhimurium that is recognized by Toll-like receptor 2. *Mol. Microbiol.*, **58**, 289–304.

Tukel C., Wilson R. P., Nishimori J. H., Pezeshki M., Chromy B. A. and Baumler A. J. (2009). Responses to amyloids of microbial and host origin are mediated through toll-like receptor 2. *Cell Host Microbe.*, **6**, 45–53.

Turner K. H., Everett J., Trivedi U., Rumbaugh K. P. and Whiteley M. (2014). Requirements for *Pseudomonas aeruginosa* acute burn and chronic surgical wound infection. *PLoS Genet.*, **10**, e1004518. doi:10.1371/journal.pgen.1004518

Uversky V. N. and Fink A. L. (2004). Conformational constraints for amyloid fibrillation: the importance of being unfolded. *Biochim. Biophys. Acta*, **1698**, 131–153.

Van Gerven N., Goyal P., Vandenbussche G., De Kerpel M., Jonckheere W., De Greve H. and Remaut H. (2014). Secretion and functional display of fusion proteins through the curli biogenesis pathway. *Mol. Microbiol.*, **91**, 1022–1035. doi:10.1111/mmi.12515

Vidal O., Longin R., Prigent-Combaret C., Dorel C., Hooreman M. and Lejeune P. (1998). Isolation of an *Escherichia coli* K-12 mutant strain able to form biofilms on inert surfaces: involvement of a new *ompR* allele that increases curli expression. *J. Bacteriol.*, **180**, 2442–2449.

Wang R., Braughton K. R., Kretschmer D., Bach, T.-H. L., Queck S. Y., Li M., Kennedy A. D., Dorward D. W., Klebanoff S. J., Peschel A., DeLeo F. R. and Otto M. (2007). Identification of novel cytolytic peptides as key virulence determinants for community-associated MRSA. *Nat. Med.*, **13**, 1510–1514. doi:10.1038/nm1656

Wasmer C., Lange A., Melckebeke H. V., Siemer A. B., Riek R. and Meier B. H. (2008). Amyloid fibrils of the HET-s(218–289) prion form a β solenoid with a triangular hydrophobic core. *Science*, **319**, 1523–1526. doi:10.1126/science.1151839

Wei Z. M., Laby R. J., Zumoff C. H., Bauer D. W., He S. Y., Collmer A. and Beer S. V. (1992). Harpin, elicitor of the hypersensitive response produced by the plant pathogen *Erwinia amylovora*. *Science*, **257**, 85–88.

Westermark G. T., Johnson K. H. and Westermark P. (1999). Staining methods for identification of amyloid in tissue. *Methods in enzymology*, **309**, 3–25.

Wickner R. B., Edskes H. K., Shewmaker F. and Nakayashiki T. (2007). Prions of fungi: inherited structures and biological roles. *Nat. Rev. Microbiol.*, **5**, 611–618.

Wiehlmann L., Munder A., Adams T., Juhas M., Kolmar H., Salunkhe P. and Tummler B. (2007). Functional genomics of *Pseudomonas aeruginosa* to identify habitat-specific determinants of pathogenicity. *Int. J. Med. Microbiol.*, **297**, 615–623.

Wosten H. A. and de Vocht M. L. (2000). Hydrophobins, the fungal coat unravelled. *Biochim. Biophys. Acta*, **1469**, 79–86.

Yaron S. and Römling U. (2014). Biofilm formation by enteric pathogens and its role in plant colonization and persistence. *Microb. Biotechnol.*, **7**, 496–516. doi:10.1111/1751-7915.12186

Zeng G., Vad B. S., Dueholm M. S., Christiansen G., Nilsson M., Tolker-Nielsen T., Nielsen P. H., Meyer R. L. and Otzen D. E. (2015). Functional bacterial amyloid increases *Pseudomonas* biofilm hydrophobicity and stiffness. *Front. Microbiol.*, **6**, 1–14. doi:10.3389/fmicb.2015.01099

Zhang N., Yang D., Wang D., Miao Y., Shao J., Zhou X., Xu Z., Li Q., Feng H., Li S., Shen Q. and Zhang R. (2015). Whole transcriptomic analysis of the plant-beneficial rhizobacterium *Bacillus amyloliquefaciens* SQR9 during enhanced biofilm formation regulated by maize root exudates. *BMC Genomics*, **16**, 1–20. doi:10.1186/s12864-015-1825-5

Zhong C., Gurry T., Cheng A. A., Downey J., Deng Z., Stultz C. M. and Lu T. K. (2014). Strong underwater adhesives made by self-assembling multi-protein nanofibres. *Nat Nano*, **9**, 858–866. doi:10.1038/nnano.2014.199

Chapter 7

Bacterial exopolysaccharides from unusual environments and their applications

Jean Guezennec

AiMB (Advices in Marine Biotechnology), 17 Rue d'Ouessant, 29290 Plouzane, France

ABSTRACT

It is not well known that micro-organisms originating from unusual marine environments offer a valuable source of innovative metabolites such as new drugs, pigments, vitamins, enzymes and biopolymers as well. Deep sea hydrothermal vents, microbial mats and Antarctic ecosystems harbour novel microbial diversity that produces a wide variety of potentially promising exopolysaccharides. The commercial value of these biopolymers is increasingly acknowledged in medicine, food industry, cosmetics and for treating some environmental problems. Nevertheless to face new challenges there is still a need for novel exopolysaccharides and their biotechnological potential remains largely untapped.

Keywords: Exopolysaccharides, application of exopolysaccharides

7.1 INTRODUCTION

A vast number of bacterial extracellular polysaccharides (EPS) have been reported over recent decades and their composition, structure, biosynthesis and functional properties have been extensively studied. (Sutherland, 2002). Despite the great diversity of molecular structures already described for these bacterial EPS, only a few have been industrially developed (Kumar *et al.* 2007; Rehm, 2009; Ullrich, 2009). Nevertheless to face to new challenges in many industrial sectors including human health, there is a need for innovative exopolymers and novel chemical structures (Guezennec, 2002).

In the course of the discovery of novel polysaccharides of biotechnological interest, it is now widely accepted that micro-organisms originating from unusual ecosystems such as deep sea hydrothermal vents, Southern Ocean and microbial mats will provide a valuable resource not only for exploitation in novel biotechnological processes but also as models for investigating how biomolecules are stabilized when subjected to changing or extreme conditions. These ecosystems have been recently studied for their bacterial diversity and as a source of innovative metabolites such as secondary metabolites, pigments, enzymes and biopolymers as well (Guezennec *et al.* 2002; Mancuso *et al.* 2005; Guezennec *et al.* 2011).

7.2 EPS FROM DEEP SEA HYDROTHERMAL VENTS

Deep-sea hydrothermal vents were discovered in 1977 and are characterized by high pressures, high temperature gradients, and high levels of toxic elements such as sulfides and/or heavy metals (Prieur, 1997). Indeed, deep-sea hydrothermal vents now offer a new source of a variety of fascinating microorganisms well adapted to these extreme environments (Harmen *et al.* 1997) (Figure 7.1).

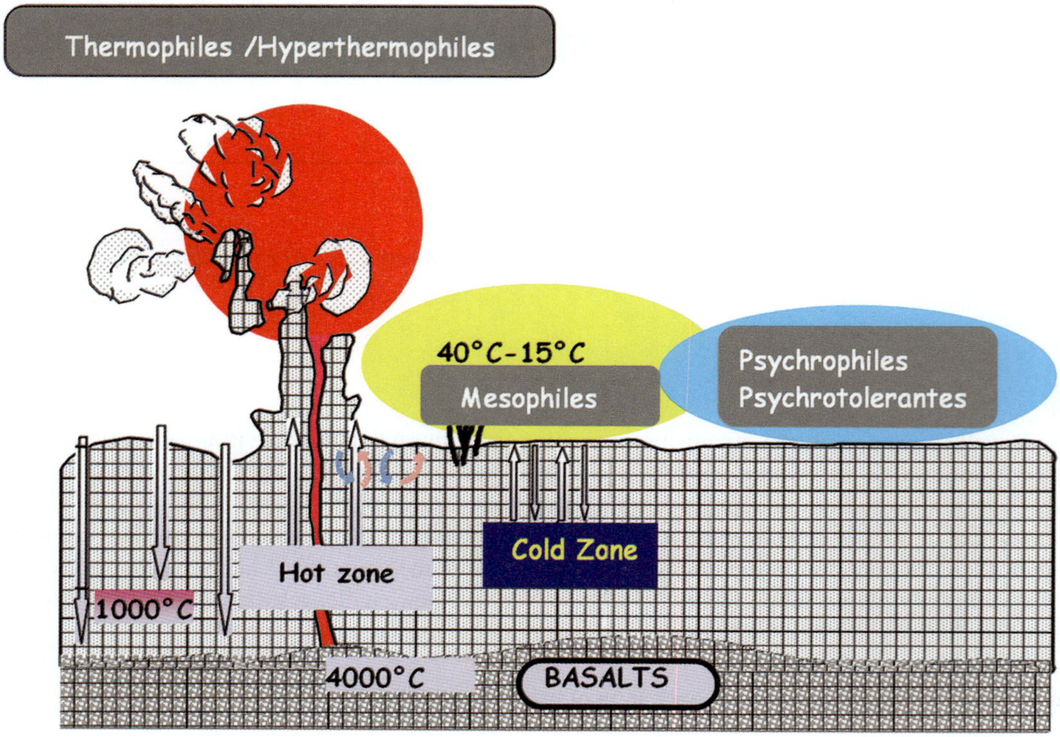

Figure 7.1 Approximate microorganisms distribution nearby deep-sea hydrothermal vents.

In most shallow water and terrestrial ecosystems energy mainly comes from sunlight, but in the deep ocean there is total darkness. Deep sea hydrothermal vents often expel nutrient rich water, containing methane and sulfur compounds, which provide electrons and energy. Vent bacteria can synthesize all the compounds they need to live from these nutrients, a process called chemosynthesis. And these bacteria sometimes grow into a thick mat, covering the hydrothermal vent and form the basis of the entire hydrothermal vent ecosystem (Figure 7.2). Biofilm formation nearby deep sea vents involves the initial attachment of cells to a solid support followed by EPS production and biofilm development. These associated exopolysaccharides play an important role in microbial adaptation, creating an environment that allows cell adhesion, concentration of nutrients and protecting them against severe environmental conditions such as high concentrations in toxic elements as present nearby deep sea vents (Table 7.1).

Figure 7.2 Colored bacterial biofilms found in the vicinity of deep-sea hydrothermal vents.

Table 7.1 Heavy metals as determined in the vicinity of some deep sea hydrothermal vents (D. Desbruyères, 2010).

	Sea Water	East Pacific Ride	Mid-Atlantic Ride	Mid-Atlantic Ride
Location		21°N	Rainbow (36°13' N)	Logatchev (14°45' N)
Deep (m)	2650	2650	2300	3000
Temperature (°C)	2	350	365	353
pH	7,8	3,5	2,8	3,3
H_2S (mM)	0	3–12	1,2	0,5–0,8
Hydrogen (mM)	0	0,05–1	12	16
Methane (μM)	0	25–100	2100	2500
Fer (μM)	**<2**	**750–6500**	**24000**	**2500**
Manganese (μM)	**<2**	**360–1140**	**2250**	**330**
Copper (μM)	**<2**	**10–40**	**140**	**27**
Zinc (μM)	**≤2**	**40–100**	**160**	**29**
Sulfate (mM)	30	0	0	0
Chloride (mM)	546	565	750	515

Bacteria associated with deep-sea hydrothermal *conditions* have demonstrated their ability to produce under laboratory conditions unusual extracellular polysaccharides in an aerobic carbohydrate-based medium. To date, screenings have been performed mainly on mesophilic heterotrophic bacteria rather than on thermophilic and hyperthermophilic microorganisms, even though the latter microorganisms are considered to be biotechnologically attractive due to their production of lipids and thermostable enzymes (Cornec *et al.* 1997; Littuati *et al.* 1998; Querellou & Guezennec, 2011). So far, four main EPS producers have been identified, i.e., *Pseudomonas* sp, *Pseudoalteromonas* sp, *Alteromonas* sp, and *Vibrio* sp. Only a few polymers have so far been fully characterized, but considerable information related to the chemical composition of those polymers, and their properties in terms of rheology or metal binding capability, has been determined (Rougeaux *et al.* 1996; Guezennec, 2002; see other refs in the text). Most of them have uronic acid contents from 10% to 40% and high molecular masses up to 10^6 g mol^{-1} (Table 7.2).

Table 7.2 Monosaccharides of EPS produced under laboratory conditions by some bacteria originating from deep-sea hydrothermal vents.

	Strain	Origin	Monosaccharides
HE 800	*Vibrio* sp	Tube worm	GlcA/GlcNAc/GalNAc
HYD 657	*Alteromonas* sp	Tube worm	Fuc/Rha/Glc/Gal/Man/ GlcA/GalA
ST 716	*Alteromonas* sp	Hydrothermal fluid	Glc / Gal / 4,6Pyr -Man /GlcA /GalA
ST 721	*Pseudoalteromonas* sp	Hydrothermal fluid	Glc / Gal / [SO_3H] -Man /Rha/GlcA
GY 785	*Alteromonas* sp	Hydrothermal fluid	Glc / Gal /Man /Rha/GlcA /GalA
HYD 1644	*Alteromonas sp*	Alvinellidae	Glc / Gal /GlcA/ 3Lac-GlcA /GalA
HYD 1545	*Alteromonas sp*	*Alvinella pompejana*	Glc / Gal / 4,6Pyr - Gal /GlcA /GalA

Abbreviations: Ara: Arabinose; Fuc: Fucose; Gal: Galactose; Glc: Glucose; Man: Mannose; Rha: Rhamnose; GlcA: Glucuronic acid: GalA: Galacturonic Acid; GlcNAc: N-Acetyl Glucosamine; GalNAc: N-Acetyl Galactosamine; Ac: Acetate; Pyr; Pyruvate; Succ: Succinate.

7.3 EPS FROM COLD ENVIRONMENTS

The Antarctic marine environment as well as other cold environments is characterized by challenging conditions for the survival of microorganisms either psychrophilic or psychrotolerant microorganisms. The necessity for all these microorganisms to adapt to the cold implies the biosynthesis of a broad range of products of biotechnological interest including cold-adapted enzymes, polyunsaturated fatty acids (PUFA) and EPS. It is well known that bacterial cells growing in cold aquatic systems are surrounded by EPS (Decho, 1990). These exopolysaccharides may be providing a cryoprotectant role in these environments of high salinities and low temperature (Marx *et al.* 2009). But these biopolymers also have a specific ecological role in trace metal binding such as Fe^{3+} where availability in the Southern Ocean is know to limit primary metabolite production as well as for the uptake of other dissolved metals. Bacterial strains isolated from Southern -Ocean particulate material or from sea ice were shown to produce under laboratory conditions and at low temperatures exopolysaccharides (Mancuso-Nichols *et al.* 2004). For some bacteria (*i.e.*, CAM 025 – see Table 7.3) EPS production was 30-fold more at –2°C and 10°C than at 20°C (Mancuso-Nichols *et al.* 2005). To date few information are available related to structural analysis of these biopolymers from the cold environment.

Table 7.3 Monosaccharides of EPS produced under laboratory conditions by some bacteria isolated from cold environments. (For Abbreviations see Table 7.2).

	Strain	Origin	Monosaccharides	Substituents
CAM 015	*Pseudoalteromonas sp*	Melted fast Ice	Ara/Rha/Man/Glu	Nd
CAM 023	"	Plankton tow	Ara/Glc	Nd
CAM 025	"	Melted fast Ice	Glc/GalA	Ac
CAM 036	"	Plankton tow	Man/Glc/GalA/GlcNAc	Sulf, Ac, Succ.
CAM 064	"	Melted fast Ice	Man/GalNAc	Ac, Pyr
CAM 090	*Shewanella sp*	Melted Fast Ice	Man/Glc/GlcA/GalNAc	Ac, Pyr
	Pseudoalteromonas strain SM9913	Deep sea sediments	Glc	
	Colwellia psyschroerythraea strain 34H	Arctic sediments	Unknown	

7.4 EPS FROM MICROBIAL MATS

In the French Polynesian atolls, microbial mats develop in water ponds exposed to changing conditions (Figure 7.3). These microbial mats, which are called "kopara" by the inhabitants of Tuamotu archipelago, are laminated communities primarily composed of phototrophic and chemotrophic prokaryotes and contain large amounts of exopolysaccharide (Rougeaux *et al.* 2001). In such unusual environments, bacteria co-exist with cyanobacteria and may form a synergistic relationship. The vertical stratification of such communities is a response to the organisms physiological requirements to the gradients of light, oxygen, redox potential, sulfide and pH (Figure 7.4). Similar microbial mats are found in many lagoons in Baja California, lakes in the Sinaï, Egypt and still waters in Spain. "Kopara" mats are characterized by varied physical and chemical parameters, with pH values ranging from 6 to 10.5, salinity levels ranging from 5 g/l to 42 g/l, temperatures ranging from 20°C during the night up to 42°C around mid-day and light intensities varying from one site to another (Mao Che *et al.* 2001).

Figure 7.3 Microbial mats present on some French Polynesian atolls.

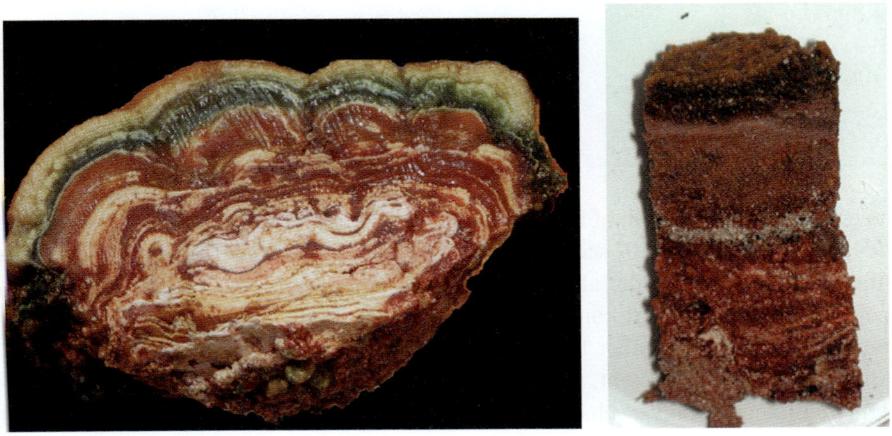

Figure 7.4 Microbial stratification within "kopara" mats.

Various macromolecules, including proteins, lipids, nucleic acids and exopolysaccharides form the architectural matrix in the intercellular space of microbial biofilms. These EPS preserve the integrity of the mat, prevent desiccation, maintain cell hydrophobicity and are also involved in the precipitation of carbonates (Defarge *et al.* 1994a, Defarge *et al.* 1994b). Basically extracellular polymers help microorganisms to compete and survive in changing environmental conditions by altering the physical and geochemical microenvironment around the cells (Figure 7.5). Since 2001, this ecosystem has been screened for innovative biomolecules, including novel microbial exopolysaccharides secreted under laboratory conditions by both bacteria and cyanobacteria (Table 7.4). (Raguenes *et al.* 2004; Richert *et al.* 2005; Guezennec *et al.* 2011).

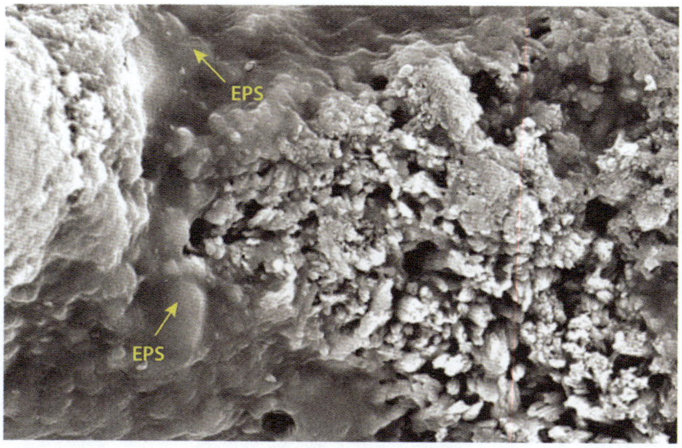

Figure 7.5 Scanning electron micrograph showing exopolysaccharides (yellow arrows) present in microbial mats (1 cm = 100 μm).

Table 7.4 Monosaccharides of EPS produced under laboratory conditions by bacteria isolated from Polynesian microbial mats. (For Abbreviations see Table 7.2).

	Strain	Origin	Monosaccharides	Substituents
RA 19	*Pyrococcus* sp	Microbial mats	Gal/Glc/Rha/Fuc/ Gal A	Sulf > 25%
Mo 203	*Alteromonas* sp	"	Glc/Gal/Rham/GalA/GluA	
Mo 245	*Vibrio* sp	"	GlcNAc/GlcA/GalNAc	Ac
Mo 169	*Alteromonas* sp	"	Man/Gal/Glc/GluA/GalA	
Mo 278	*Alteromonas sp*	"	Gal/Glc/Man/ GluA/GalA/ GluNAc	
Mo 300	*Vibrio* sp	"	Rha/Glc/GlcNAc	Sulf > 15%

So far, the most remarkable result of this screening was the discovery of *Paracoccus zeaxanthinificiens* subsp. *payriae* (designated as RA 19 see Table 7.4). This bacterium, under laboratory conditions produce a highly sulfated exopolysaccharide, along with a non-water soluble macromolecule that was assumed to be a glycoprotein. The chemical composition of the water-soluble exopolysaccharide appeared to be different from that produced by other bacteria from similar environments. The most important feature is undoubtedly its high sulfate content, which can be up to 29% w/w, in its native state. The biosynthesis of such a highly sulfated exopolysaccharide by a marine bacterium has, to the extent of our knowledge, never been reported in the literature. The natural occurrence of sulfate is known to improve the biological activity of some polysaccharides with anti-viral activity, including those with activity against the human immunodeficiency virus (HIV). In addition, certain biological activities, such as blood coagulation, are influenced by the sulfate content and position on some polysaccharides (Colliec-Joault *et al.* 2001; Colliec-Jouault *et al.* 2004; Matou *et al.* 2005; Matou *et al.* 2007).

7.5 BIOMEDIAL APPLICATIONS OF EPS

Biopolymers produced by marine bacteria can offer useful tools for regenerative medicine (Colliec Jouault *et al.* 2004; Colliec *et al.* 2012; Delbarre-Ladrat *et al.* 2014). One of the most intriguing discovery was a new facultatively anaerobic, heterotrophic, and mesophilic bacterium isolated from a Pompei worm tube (polychaete *Alvinella pompejana*) collected from a deep-sea hydrothermal field of the East Pacific Rise and named *Vibrio diabolicus* (Raguenes *et al.* 1997) (Figure 7.6a, b). During stationary phase of growth in batch cultures in the presence of glucose, this bacterium produced an EPS characterized by equal amounts of uronic acid and hexosamine (N-acetyl glucosamine and N-acetyl galactosamine), which can be regarded as the combination of both hyaluronan and non-sulfated chondroitin disaccharidic repeating units (designated as HE 800 EPS) (Rougeaux *et al.* 1999) (Figure 7.7). This polymer was found to be a strong bone-healing material in an experimental model (Zanchetta *et al.* 2003a, 2003b). Previous studies demonstrated its effectiveness to enhance *in vivo* bone regeneration and to support osteoblastic cell metabolism in culture. In order to assess the usefulness of this high-molecular weight polymer in tissue engineering and tissue repair, *in vitro* reconstructed connective tissues containing HE800 EPS were performed (Senni *et al.* 2013). This polysaccharide promotes both collagen structuring and extracellular matrix settle by dermal fibroblasts. Furthermore, from the native HE800 EPS, a low-molecular weight sulfated derivative (HE800 DROS) displaying chemical analogy with heparan-sulfate, was designed (Rederstorff *et al.* 2011a; Rederstorff *et al.* 2011b).It was demonstrated that HE800 DROS mimics some properties of heparan-sulfate, such as promotion of fibroblast proliferation and inhibition of matrix metalloproteinase (MMP) secretion. Therefore, it can suggested that the HE800EPS family can be considered as an innovative biotechnological source of glycosaminoglycan-like compounds useful to design biomaterials and drugs for tissue engineering and repair (Senni *et al.* 2008a, Senni *et al.* 2008b).

Figure 7.6 Pompei worm (*Alvinella pompejana*) hosting a GAG (Glycosy Amino Glycan) producing *Vibrio sp.*

Hyaluronan Block *Chondroitin Block*

→3)-β-D-Glc*p*NAc-(1→ 4)-β-D-Glc*p*A-(1 →4)-β-D-Glc*p*A-(1→ 4)-α-Gal*p*NAc-(1

Figure 7.7 Repeating unit of Exopolysaccharide produced by the deep sea hydrothermal vent bacteria *Vibrio diabolicus* (Rougeaux *et al.* 1999).

A bacterium named *Alteromonas infernus* was isolated from a sample of fluid collected among a dense population of *Riftia pachytila* (Raguenes *et al.* 1997). The EPS (designated as GY 785) produced is a highly branched acidic heteropolysaccharide with a high molecular weight (>1.5 × 10^6 g/mol) with a low sulphate content (≤10%). Its nonasaccharide repeating unit is composed of uronic acid (galacturonic and glucuronic acid) and neutral sugars (galactose and glucose) and substituted with one sulfate group (Figure 7.8) (Roger *et al.* 2004). A recent study demonstrated that this high molecular weight EPS associated with an injectable silylated hydroxy propyl methyl cellulose-based hydrogel (Si-HPMC) could significantly improve the mechanical properties of this new hydrogel construct. The attachment of chondrocytes and osteoblasts was induced when the cells were cultivated in two-dimensional culture on the top of the hydrogel containing 0.67% of EPS. In three-dimensional culture, this EPS increased viability and proliferation of the chondrocytes. The hydrogel supplemented with 0.67% GY785 EPS presented an interesting feature for cartilage tissue engineering applications (Rederstorff *et al.* 2011; Velasco *et al.* 2010; Ruiz Velasco *et al.* 2011).

SO_3^-
↓
2
→4)-β–D-Glc*p*-(1 → 4)-α-D-Gal*p*A-(1 → 4)-β-D-Gal*p*-(1 →
3
↑
1
α-D-Glc*p* -(1 → 3)- β-D-Glc*p*A
4
↑
1
α-D-Glcp -(1 → 2)- β-D-Glc*p*A
4
↑
1
β-D-Glcp -(1 →3)- α-D-Gal*p*

Figure 7.8 Repeating unit of the bioactive EPS GY785 produced by *Alteromonas infernus* (Roger *et al.* 2004).

The structure and chemical characteristics of bacterial EPS may also be modified to obtain desired properties. The same EPS (EPS 785) was modified to obtain new heparin-like compounds. The resulting EPS, contrary to the native one, presented anticoagulant properties similar to heparin, suggesting that by carrying out modifications in natural EPS their functionality and specificity may be improved (Matou *et al.* 2005; Matou *et al.* 2007).

7.6 EPS AS ANTIBIOFILM AGENT

Biological fouling in marine and freshwater environments creates a number of problems for engineered structures. The deleterious effects arise from microbial biofilm formation, often followed by successional colonization by a variety of macrofouling organisms. To date, effective prevention strategies have primarily relied upon antimicrobial agents that usually show toxicity against non-target organisms. While a variety of non-toxic surface modification technologies have been employed, their efficacy has been very limited in *in situ* environments.

Several exopolysaccharides (EPS) were evaluated as potential antibiofilm agents. In a first approach pure EPS from marine bacteria originating from deep-sea vents were dip-coated onto cleaned glass slides, placed into a flow cell apparatus with real-time imaging, then exposed to natural flowing seawater under laboratory conditions over a 5 day time-frame. All purified EPS formulations inhibited biofouling (primarily, bacteria) over the test period. Surface area fouling was reduced by 90% relative to a cleaned glass control substrate. None of the exopolymers evaluated showed any evidence of antimicrobial activity or cytotoxicity (Guezennec *et al.* 2012) (Figure 7.9). Additional studies performed on 316L stainless steel exposed in natural flowing seawater for 10 days gave similar data as expressed by electrochemical techniques (Figure 7.10). Bacterial EPS were supposed to mediate cell-to-cell and cell-to-surface interactions that are required for the formation, cohesion and stabilization of bacterial biofilms. These preliminary results thus suggest that bacterial EPS may be an effective inhibitor of the initial stages of biofilm formation and subsequent biofouling activity and a "green" alternative to more conventional methods (Jiang *et al.* 2011; Sayem *et al.* 2011; Bernal *et al.* 2012; Rendueles *et al.* 2012).

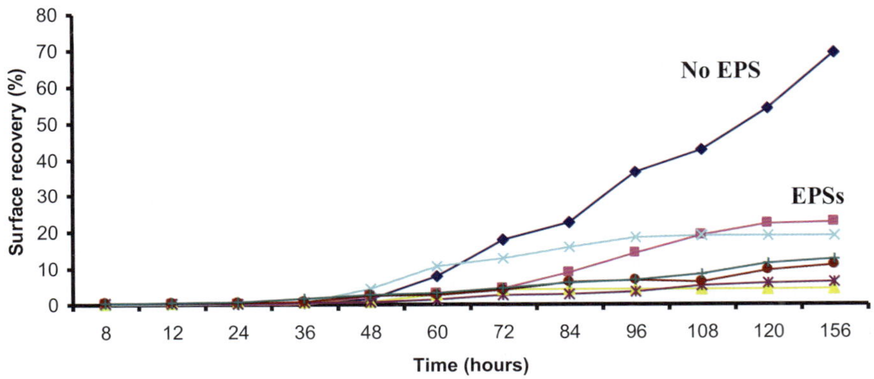

Figure 7.9 Microbial surface recovery versus time (average of three data points/time point). Standard deviation is about 5% for each measured value. All experiments were performed in triplicate.

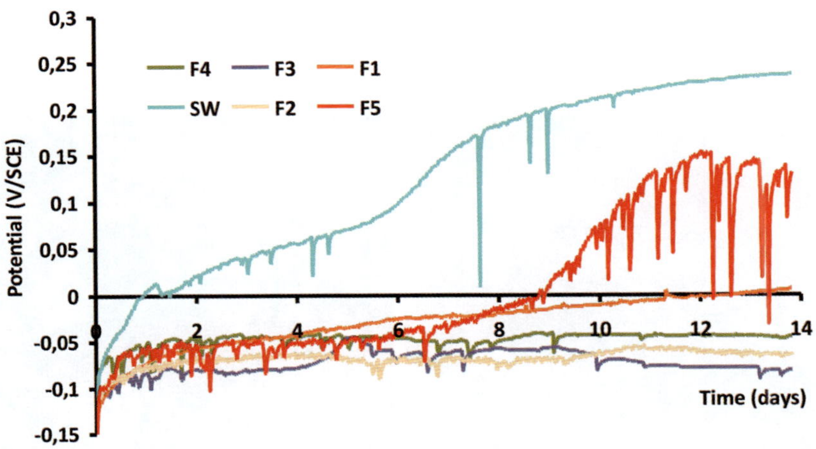

Figure 7.10 E_{Eq}/SCE versus time for 316 L SS treated with EPS (F1 to F5) or without (SW: Natural seawater). No biofilm was detected on surfaces treated at T_0 by all EPS, with the exception of EPS F1 while a significant biofilm was present on non-treated surfaces (SW days.) after 6 days.

7.7 EPS AS BIODETOXIFIERS

Water pollution due to toxic heavy metals remains a serious environmental and public health problem. Heavy metals are not biodegradable and accumulate in living organisms, often resulting in deleterious effects. As a result of public and media pressure, different states have progressively introduced stricter regulations with regard to the metal discharges and treatments associated with industrial operations. A list of the immediate heavy metals of most concern involving, aluminium, chromium, manganese cadmium, iron, mercury and nickel has been established by the World Health Organization.

Compared to more conventional technologies for metal removal from effluents (e.g., chemical precipitation, chemical oxidation, electrochemical treatment, reverse osmosis, ion exchange, etc.) biosorption may represent a more economically feasible approach. Biosorption can be defined as a process in which a metal is sequestered by chemical sites naturally present in the biosorbents (Volesky, 2001). One advantage of biosorption is that it is cost-effective, using cheap biosorbents which are either abundant in nature such as seaweeds and microbial biomass, or obtained as bioproducts of microbial growth; e.g., exopolymers. Although they compete with polysaccharides from other sources, e.g., from algae (alginates, carrageenans), crustacea (chitin) or plants bacterial EPS production is less subject to variability due to marine pollution, crop failure or climatic impact.

Bacteria isolated from different marine ecosystems, including deep sea hydrothermal vents and microbial mats, were shown to synthesize unusual exopolysaccharides (EPS) under laboratory conditions. Bacterial EPS contain ionisable functional groups such as carboxyl, amine, sulphate and to a lesser extent hydroxyl groups that enable these biopolymers to bind heavy metals. Experiments were conducted on the capability of some "exotic" EPS to bind a large number of toxic ions (Cu, Pb, Cd, Zn, Ni, Ag, Hg, U, Co, Fe).

Binding capacities up to 625 mg Cu/g EPS, 316 mg Pb/g EPS, 120 mg Cd/g EPS, 100 mg Ni/g EPS 75 mg Zn/g EPS or even 333 mg Ag/g EPS have been determined for mostly uronic rich and/or highly sulphated EPS from either deep sea vents or microbial mats (Figure 7.11a, b, 7.12) (Loaec *et al.* 1997; Ginisty *et al.* 1998; Loaec *et al.* 1998; Ginisty *et al.* 2002; Moppert *et al.* 2009; Deschatre *et al.* 2013; Deschatre *et al.* 2015). Such data raises interesting questions about biopolymers in the field of bioremediation. Isolated- or engineered-bacterial exopolysaccharides, or even crude bacterial biomass could be incorporated in biofilter technology and provide useful applications for toxic metal remediation. However such development also depends on the cost-effectiveness of these treatments.

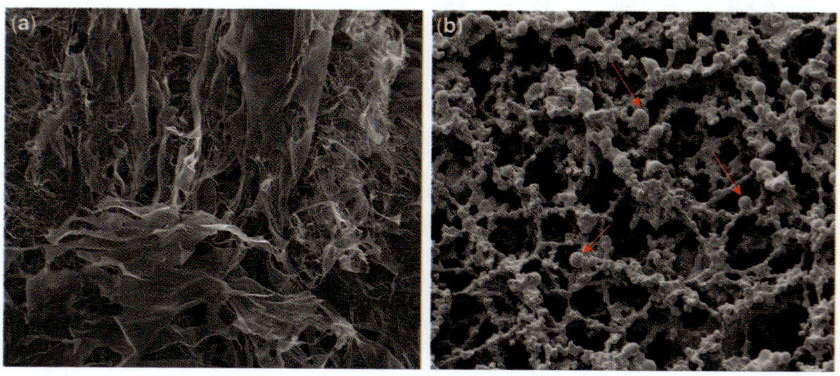

Figure 7.11 Scanning electron micrographs showing Left: Native EPS (Deep sea vents) Right: Spheres (as indicated by red arrows) showed the binding of Ni by this EPS (1 cm = 12.5 μm).

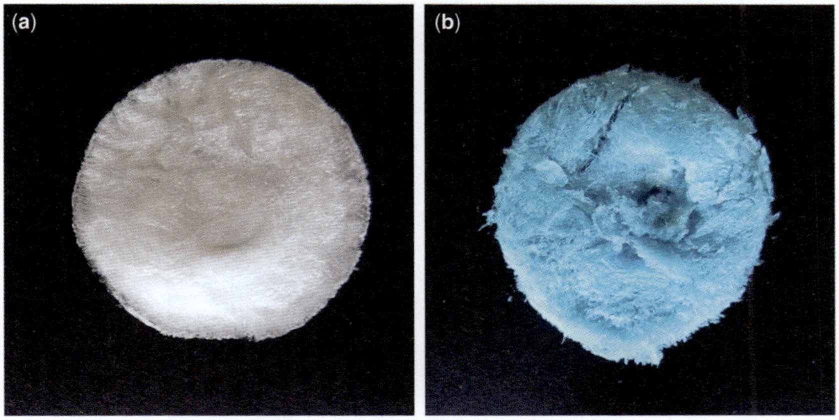

Figure 7.12 EPS and metallic ions: From left to right: Native EPS (white): EPS/Cu (II) (blue).

7.8 EPS AND..... BLACK PEARLS

In French Polynesia, cultured pearls are the product of grafting and rearing of *Pinctada margaritifera* pearl oysters in their natural environment. Pearl culture includes several stages. To start with, the pearl oysters are collected and raised to serve either as donor or receiver. The grafting process then takes place following a surgical operation during which the graft, a small piece of mantle tissue, is inserted into the "pearl pocket" of the receiving oyster together with a nacre bead, the nucleus. Once inserted into the receiving oyster, the external epithelial-cells of the graft multiply to form a pearl sac around the nucleus (Cochennec *et al.* 2010). The pearl sac then starts to deposit calcium carbonate polymorphs layers onto the nucleus. This is the starting point for the future pearl. A rearing period of about 18 months is needed to finally produce a pearl with a sufficiently thick layer of nacre (0.8 mm) (Figure 7.13).

Unfortunately, nucleus rejections and oyster mortality occur during the first 45 days following grafting. These phenomena, which typically affect 40 to 50% of oysters within three weeks following the operation, appear to result from bacterial infections or from an inappropriate grafting practice. The development of an inflammatory reaction after insertion of the nucleus and contamination by pathogenic bacteria, combined with the absence of rapid healing of the tissue incised during the graft are the most likely causes of nucleus rejection (Cochennec *et al.* 2010; Gueguen *et al.* 2013). To prevent bacterial infections, synthetic antibiotics have been applied during the grafting practice. The use of such antibiotics presents an environmental problem, due to their incomplete biodegradability, and a public health problem because of the potential for development of bacterial resistance.

Amazingly the coating of the nucleus with selected bacterial EPS from deep sea vent bacterial isolates has been found to reduce the failure rate of graft procedures. Exopolysaccharides were shown to form a stable and homogeneous film on the surface of the nucleus (Figure 7.14). The coating of the nucleus by this polymeric film appears to accelerate the healing of the recipient oyster graft site. Moreover, the combination of EPS film with natural antimicrobial peptides (AMP), such as tachyplesine, significantly reduces bacterial levels associated with the oysters. The presence of EPS alone or in combination with this AMP agent has resulted in significant reduction in the failure of grafting procedures, and has led to a reduction in overall recipient oyster mortality rates. Another point to be considered is the calcium and magnesium binding by the EPS homogeneous film, which has been found to produce a higher grade quality black pearl. A surprising and interesting application for bacterial exopolysaccharide!

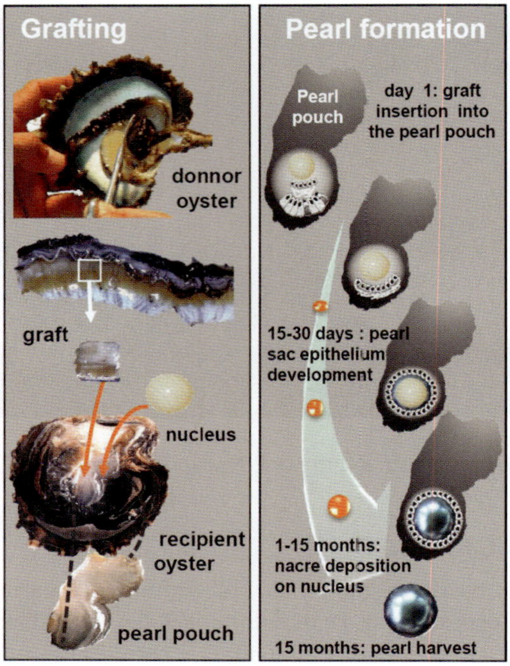

Figure 7.13 The different steps of the grafting process and pearl formation in the pearl oyster *Pinctada margaritifera*. During the grafting process, a small piece of mantle tissue from the donor oyster (the graft), is inserted into the "pearl pocket" of the receiving oyster together with a nacre bead, the nucleus. Once inserted into the receiving oyster, the external epithelial of the graft multiplies and forms a pearl sac around the nucleus. The pearl sac then starts to deposit nacre (aragonite) layers onto the nucleus. A rearing period of about 18 months is needed to finally produce a pearl with a sufficiently thick layer of nacre (illustration C. Montagnani).

Figure 7.14 Formation of a stable EPS film on the surface on nucleus. The scraping indicates the presence of a thin homogeneous polymeric film (EPS) on the surface of the nucleus.

7.9 EPS IN EOR/MEOR

Enhanced oil recovery methods a long time ago have been employed in helping to prolong the life-time of oil and to obtain the maximum amount of recoverable oil (Brown, 2010). The efficiency of oil recovery during the primary stage is relatively low, leaving the majority of oil to be removed by secondary and/or tertiary recovery methods. While the most common and the cheapest method of secondary oil recovery is water flooding or inert gas injection, tertiary or enhanced oil recovery (EOR) processes include all methods that use external sources of energy and/or materials to recover oil that cannot be produced economically by conventional methods. EOR processes can be classified as thermal methods (steam flooding, hot water drive, in situ combustion), chemical (polymer, surfactant) or miscible (hydrocarbon gas, CO_2, nitrogen). Since the last decades, petroleum industry has employed two types of polymers in the EOR projects: synthetics, mainly HPAM (partially hydrolyzed polyacrylamide) and polysaccharides like xanthan gum (Mothe *et al.* 2006). But these polymers have their rheological properties compromised by the conditions found in reservoirs like, high pressures, high temperatures, high salinities, hardness of connate water, bacterial degradation, etc.

Such extreme conditions are encountered in the vicinity of deep sea vents and in that it can be hypothesized that exopolysaccharides produced by bacteria originating from such ecosystems could be appropriate to be injected to such conditions. An EPS was secreted by *Alteromonas macleodii* subsp. *fijiensis*, which is an aerobic, mesophilic, heterotrophic bacterium isolated from a diluted hydrothermal vent fluid at a depth of 2600 m in a rift system of the North Fiji basin. The repeating unit of this polymer secreted during its stationary growth phase was characterized by means of chemical analysis and NMR studies and deduced to be a hexasaccharide with three linked uronic acids and with a side chain ended by a 4,6-O-(1-carboxyethylidene)-mannose residue (Rougeaux *et al.* 1998). This EPS exhibits a significant thickening power and a pronounced shear-thinning behaviour. More interestingly this biopolymer exhibits a higher thermostability than xanthan combined with similar rheological properties when submitted for three weeks to either high or low salinity at 90°C. Amazingly the repeating unit of this polymer shows some similarities with xanthan (Figures 7.15a, b). However despite these interesting properties and subsequent applications of this polysaccharide, other factors including the yield, price, and the markets for such new molecules will ultimately determine its commercial development in this domain.

(a) ←4)-β-D-Glc*p*-(1←4)-α-D-Gal*p*A-(1←4)-α-D-Gal*p*-(1←
3
↑
1
α-D-Glc*p*A
3
↑
1
β-D-Glc*p*A
4
↑
1
4,6-Pyr-β-D-Man*p*

(b) ←4)-β-D-Glc*p*-(1←4)-β-D-Glc*p*-(1←
3
↑
1
6 Ac-α D-Man*p*
3
↑
1
β-D-Glc*p*A
4
↑
1
4 -Pyr-β-D-Man*p*

Figure 7.15 Repeating unit of a) EPS produced by a deep sea vent bacteria, *Alteromonas macleodii* subsp. *fijiensis*, b) xanthan gum used in MEOR.

An other point to be considered could be the screening of thermophilic (heat-loving) microorganisms from deep sea vents. Biofilms formed by extremely thermophilic *Thermotoga maritima* and *Methanococcus jannaschii* were found in nearby deep-sea vents (Muralidharan *et al.* 1997).

7.10 EPS IN COSMETICS

Marine microorganisms constitute a tremendous source of structurally-diverse bioactive metabolites. Cosmetics industry is progressively turning to the sea in the search for new ingredients. Consumer concern about the unwanted signs of aging and stress-related effects has led to the development of products for skin firming, skin lightening, moisturizing, smoothing, line removing, anti-aging, skin repair, skin protecting, and more. Cosmetics are defined as articles to be applied to human body for cleansing, beautifying, promoting attractiveness, or altering the appearance without affecting body structure or functions ("Forever young!"). More recently, the cosmetic industry introduced a special class of products, the cosmeceuticals not regulated by the FDA as drugs, as a combination of cosmetics and pharmaceuticals, as bioactive ingredients are now combined with creams, lotions and ointments.

One of the most common molecular classes of compounds used in the personal care bioactive ingredients sector are polysaccharides from algae and bacterial exopolysaccharides (EPS). Deep-sea hydrothermal and microbial mats with diverse physico-chemical characteristics, encourage the development of microorganisms with chemical defence systems against hard and changing environments. Exopolysaccharides surrounding bacterial cells are one of these systems and could be of a great interest in the cosmetic industry.

The mesophilic strain *Alteromonas macleodii* subsp. *fijiensis* biovar deepsane was collected on the East Pacific Rise at 2600 m depth. (Cambon-Bonavita *et al.* 2002). It was isolated from a polychaete annelid *Alvinella pompejana* and is able to synthesize and excrete the exopolysaccharide "Deepsane". Biological activities have been screened and some protective properties have been established. "Deepsane" is commercially available in cosmetics under the name of Abyssine® (patent PCT 94907582-4) for soothing and reducing irritation of sensitive skin against chemical, mechanical and UVB aggression (Thibodeau et Takeoka, 2006). Although discovered in the late 80's studies on the chemical structure of Deepsane (Le Castaouec, 2010; Le Costaouec *et al.* 2012) were recently performed. The data shows a very complex branched repeating unit of 16–18 sugars with 7 different types of monosaccharides (neutral and acidic) and 3 types of substituents (sulphate, lactyl and pyruvyl groups) (see Table 7.2-EPS designated as HDY 657-).

In the same way, two different EPS produced by an *Alteromonas* sp and *Vibrio* sp respectively isolated from Polynesian microbial mats have found applications in the cosmetic area. The first one commercially available as Exo-H™ restores hyaluronic acid secretion in senescent skin cells, stimulates lipid synthesis in the skin, supports the dermo-epidermal junction and induces filaggrin expression. The second one produced by a *Vibrio* sp and commercially available as Exo-T™ stimulates desquamation and differentiation markers for skin regeneration and optimal skin texture.

To date only a few EPS from cold environments have been launched in the cosmetic market. A *Pseudoalteromonas* strain isolated from the coasts of Antarctic waters produce a mixture of glycoporoteins and exopolysaccharides. This combination (SeaCode®) provides complete protection of the skin but reconstruction effects as well. Further studies are in progress to determine the biotechnological interest of other cold-loving microorganisms.

7.11 CONCLUSIONS

Bacterial exopolysaccharides possess a wide variety of properties that may not be found in more traditional polymers of plant origin. Although they compete with polysaccharides from other sources, e.g., from algae (alginates, carrageenans), crustacean (chitin), or plants, their production is less subject to variability due to marine pollution, crop failure or climatic impact.

But the greatest potential of bacterial EPS is related to their use in high value market niches, such as cosmetics, pharmaceuticals and biomedicine, in which traditional polymers fail to comply with the required degree of purity or lack some specific functional properties. Such markets will provide opportunities for the development of up-and-coming bacterial EPS providing that they have unique desirable physicochemical properties In such high-value applications, product quality wholly surpasses cost production and product yield issues. Separation processes is especially challenging when polymers are considered for medical applications because a high-purity product is envisaged. For environmental applications a high degree in purity is not necessary but conversely low production costs are required.

Unusual marine ecosystems may provide the market with innovative biopolymers such as exopolysaccharides. Unusual and changing physical and chemical conditions favour the development of microorganisms well adapted through either enzymes or secondary metabolites or biopolymers to such conditions. And it can be hypothesized that novel exopolysaccharides will be identified in the very near future.

REFERENCES

Abu Sayem S.-M., Manzo E., Ciavatta L., Tramice A., Cordone A., Zabnfradino A., De Felice M. and Varcamonti M. (2011). Anti-biofilm activity of an exopolysaccharide from a sponge-associated strain of *Bacillus licheniformis*. *Microbial cell factories*, **10**, 74.

Bernal P. and Llamas M. A. (2012). Promising biotechnological applications of antibiofilm exopolysaccharides. *Microb. Biotechnol*, **5**(6), 670–673.

Brown L. R. (2010). Microbial enhanced oil recovery (MEOR). *Current Opinion in Microbiology* **13**, 316–320.

Cambon-Bonavita M. A., Raguénès G., Jean J., Vincent P. and Guézennec J. (2002). A novel polymer produced by a bacterium isolated from a deep-sea hydrothermal vent polychaete annelid. *Journal of Applied Microbiology*, **93**, 310–315.

Cochennec-Laureau N., Montagnani C., Saulnier D., Fougerouse A., Levy P. and Lo C. (2010). A histological examination of grafting success in pearl oyster *Pinctada margaritifera* in French Polynesia. *Aquat. Living Resour.*, **23**, 131–140.

Colliec-Jouault S., Chevolot L., Helley D., Ratiskol J., Bros A. and Sinquin C. (2001). Characterization, chemical modifications and in vitro anticoagulant properties of an exopolysaccharide produced by *Alteromonas infernus*. *Biochim Biophys Acta*, **1528**, 141–151.

Colliec-Jouault S., Zanchetta Ph., Helley D., Ratiskol J., Sinquin C., Fischer A. M. and Guézennec J. (2004). Exopolysaccharides produced by bacteria isolated from deep-sea hydrothermal vents: new agents with therapeutic potential. *Pathol. Biol.*, **52**, 127–130.

Colliec-Jouault S., Bavington C. and Delabarre-Ladrat C. (2012). Heparin-like Entities from marine organisms Heparin-A century of progress. R. Lever *et al.* (eds). Handbook of Experimental Pharmacology 207, Springler Verlab Berlin Heideberg.

Cornec L., Loaec M., Rougeaux H., Guezennec J. and Dietrich J. (1997). Molécules d'intérêt biotechnologique des microorganismes hydothermaux: Enzymes thermostables et exopolysaccharides. *Oceanis*, **23**(1), 11–25.

Guezennec J. (2002). Deep-sea hydrothermal vents: A new source of innovative bacterial exopolysaccharides of biotechnological interest? *Journal of Industrial Microbiology & Biotechnology*, **29**, 204–208.

Guezennec J., Moppert X., Raguenes G., Richert L., Costa B. and Simon Colin C. (2011). Microbial mats in French Polynesia and their biotechnological applications. *Process Biochemistry*, **46**, 16–22.

Guezennec J., Herry J. M., Kouzayha A., Bachere E., Mittelman M. W and Bellon Fontaine M.-N. (2012). Exopolysaccharides from unusual marine environments inhibit early stages of biofouling. *International Biodeterioration & Biodegradation*, **66**, 1–7.

Desbruyères D. Les tresors des abysses. Eds Quae Collection Carnet des Sciences. p. 184.

Decho A. W. (1990). Microbial exopolymer secretions in ocean environments: Their role(s) in food webs and marine processes. *Oceanogr. Mar. Biol. Annu. Rev.*, **28**, 73–153.

Defarge C., Trichet J. and Coute A. (1994a). On the appearance of cyanobacterial calcification in modern stromatolites. *Sediment Geol.*, **94**(2), 9–11.

Defarge C., Trichet J., Maurin A. and Hucher M. (1994b). Kopara in Polynesian atolls: early stages of formation of calcareous stromatolites. *Sediment Geol.*, **89**, 9–23.

Delabarre-Ladrat C., Sinquin C., Lebellenger L., Zykwinska A. and Colliec-Jouault S. (2014). Exopolysaccharides produced by marine bacteria and their applications as glycosyaminoglycan-like molecules. *Frontiers in Chemistry*, 2, **85**.

Deschatre M., Ghillebaert F., Guezennec J. and Simon-Colin C. (2013). Sorption of Copper(II) and Silver(I) by four bacterial exopolysaccharides. *Appl. Biochem. Biotech.*, **171**, 1313–1327.

Deschatre M., Lescop B., Simon-Colin C., Ghillebaert F., Guezennec J. and Rioual S. (2015). Characterization of exopolysaccharides after sorption of silver ions in aqueous solution. *J. Environ. Chem. Eng.*, **3**(1), 210–216.

Ginisty P., Besnainou B., Sahut C. and Guezennec J. (1998). Biosorption of cobalt by *Pseudomonas halodenitrificans*: influence of cell wall treatment by alkali and alkaline-earth metals and ion-exchange mechanisms. *Biotechnology Letters*, **20**(11), 1035–1039.

Ginisty P., Sahut C. and Guézennec J. (2002). Biosorption of Cs^+, Sr^{2+}, Co^{2+}, Pb^{2+}, Eu^{3+} by *Pseudomonas halodenitrificans* et *Alteromonas infernus*. *Biotechnology Letters*, **3**, 10–14.

Gueguen Y., Montagnani C., Joubert B., Marie C., Belliard A., Tayale J., Fievet P., Levy D., Piquemal F., Marin G., Moullac Le., Ky C. L., Garen P., Lo C. and Saulnier D.(2013). Characterization of molecular processes involved in the pearl formation in *Pinctada margaritifera* for the sustainable development of Pearl Farming industry in French Polynesia, In 'Recent Advances in Pearl Research – *Proceedings of the International Symposium on Pearl Research 2011*' Watabe S., Maeyama K. and Nagasawa H. eds., Terrapub, Tokyo, 183–193.

Harmsen H., Prieur D. and Jeanthon C. (1997). Distribution of microorganisms in deep-sea hydrothermal vent chimneys investigated by whole-cell hybridization and enrichment culture of thermophilic subpopulations. *Appl. Environ. Microbiol.*, **63**(7), 2876–2883.

Jiang P., Jingbao L., Han F., Duan G., Lu X., Gu Y. and Yu W. (2011). Antibiofilm activity of an exopolysaccharide from marine bacterium *Vibrio* sp. QY101. *Plos One*, **6**(4), e–18514.

Kumar S., Mody K. and Jha B. (2007). Bacterial exopolysaccharides-a perception. *J. Basic. Microbiol.*, **47**, 103–117.

Le Coataouëc T. (2010). Elucidation Structurale et Modifications d'un Exopolysaccharide Bactérien D'origine Hydrothermale (Structural elucidation and modifications of an exopolysaccharide produced by a bacteria from hydrothermal origin). Ph D thesis. Université de Bretagne Occidentale, Brest, France.

Le Costaouec T., Cerantola S., Ropartz D., Ratiskol J., Sinquin C., Colliec Jouault S. and Boisset C. (2012). Structural data on a bacterial exopolysaccharides produced by a deep-sea *Alteromonas macleodii* strain. *Carbohydr Polym.* **90**(1), 49–59.

Lattuati A., Guezennec J., Metzger P. and Largeau C. (1998). Lipids of *Thermococcus hydrothermalis*, an archaea isolated from a deep-sea hydrothermal vent. *Lipids*, **33**(3), 319–326.

Loaec M., Olier R. and Guezennec J. (1998). Chelating properties of bacterial exopolysaccharides from deep-sea hydrothermal vents. *Carbohydr. Polym*, **35**, 65–70.

Loaec M., Olier R. and Guézennec J. (1997). Uptake of lead, cadmiun and zinc by a novel bacterial exopolysaccharide. *Wat Res.*, **31**(5), 1171–1179.

Mancuso Nichols C.-A., Garon S., Bowman J. P., Raguénès G. and Guézennec J. (2004). Production of exopolysaccharides by Antarctic marine bacterial isolates. *J. Appl. Microb.*, **96**, 1057–1066.

Mancuso Nichols C., Guézennec J., and Bowman J. P. (2005). Bacterial exopolysaccharides from extreme marine environments with special consideration of the southern ocean, sea ice, and deep-sea hydrothermal vents: A review. *Mar Biotechnol,* **7**, 253–71.

Mancuso Nichols C., Bowman J.-P. and Guézennec J. (2005). Effects of incubation temperature on growth and production of exopolysaccharides by an antarctic sea ice bacterium grown in batch culture. *Appl. Environ Microbiol*, **71**, 3519–323.

Mao Che L., Andrefouet S., Bothorel V., Guezennec M., Rougeaux H. and Guezennec J. (2001). Physical, chemical and microbiological characteristics of microbial mats (Kopara) in the south pacific atolls of French Polynesia. *Can J Microbiol*, **47**, 994–1012.

Matou S., Colliec-Jouault S., Galy-Fouroux I., Ratiskol J., Sinquin C, Guezennec J., Fisher A.-M. and Helley D. (2005). Effect of an oversulfated exopolysaccharide on angiogenesis induced by fibroblast growth factor-2 or vascular endothelial growth factor in vitro. *Biochem. Pharmacol.*, **69**, 751–759.

Matou S., Colliec-Jouault S., Helley D., Ratiskol J., Sinquin C., Boisset C., Guezennec J. and Fischer A.-M. (2007). Use of low-molecular weight highly sulfated polysaccharide derivatives for modulating angiogenesis. *U.S. Patent 0,259,833.*

Marx J. G., Carpenter S. D. and Deming J. W. (2009). Production of cryoprotectant extracellular polysaccharide substances (EPS) by the marine psychrophilic bacterium *Colwellia psychrerythraea* strain 34H under extreme conditions. *Can. J. Microbiol.*, **55**(1), 63–72.

Matou S., Colliec-Jouault S., Galy6fouroux I., Ratiskol J., Sinquin C, Guezennec J., Fisher A.-M. and Helley D. (2005). Effect of an oversulfated exopolysaccharide on angiogenesis induced by fibroblast growth factor-2 or vascular endothelial growth factor in vitro. *Biochem. Pharmacol.*, **69**, 751–759.

Matou S., Colliec-Jouault S., Helley D., Ratiskol J., Sinquin C., Boisset C., Guezennec J. and Fischer A.-M. (2007). Use of low-molecular weight highly sulfated polysaccharide derivatives for modulating angiogenesis. *U.S. Patent 0,259,833.*

Moppert X., Le Costaouec T., Raguénès G., Simon-Colin C., Crassous P., Costa B. and Guezennec J. (2009). Investigation into the uptake of copper, iron and selenium by a highly sulphated bacterial exopolysaccharide isolated from microbial mats. *J. Ind Microbiol Biotechnol.*, **36**, 599–604.

Mothe C. G., Correia D. Z., De França F. P. and Riga A. T. (2006). Thermal and rheological study of polysaccharides for enhanced oil recovery. *J. of Thermal Analysis and Calorimetry*, **85**(1), 31–36.

Muralidharan V., Rinker K. D., Hirsch I. S., Bower E. J. and Kelly R. M. (1997). Hydrogen transfer between methanogens and fermentative heterotrophs in hyperthermophilic cocultures. *Biotechnol. Bioeng*, **56**, 268–278.

Prieur D. (1997). Microbiology of deep-sea hydrothermal vents. *Trends in Biotechnology*, **15**(7), 242–244.

Raguénès G., Peres A., Ruimy R., Pignet P., Christen R., Loaec M., Rougeaux H., Barbier G. and Guezennec J. (1997). *Alteromonas infernus* sp.nov, a new polysaccharide producing bacterium isolated from a deep-sea hydrothermal vent. *J. Appl. Bacteriol*, **82**, 422–430.

Raguénès G., Christen R., Guezennec J., Pignet P. and Barbier G. (1997). *Vibrio diabolicus* sp.nov., a new polysaccharide-secreting organism isolated from a deep-sea vent polychaete annelid, *Alvinella pompejana. Int. J. Syst. Bacteriol*, **47**, 989–995.

Raguenes G., Moppert X., Richert L., Costa, Payri C et Guézennec J. (2004). A novel exopolymer-producing bacterium, *Paracoccus zeaxanthinifaciens* subsp *payriae*, isolated from a "kopara" mat located in Rangiroa an atoll of French Polynesia. *Current Microbiology*, **49**, 3145–3151.

Rederstorff E., Fatimi A., Sinquin C., Ratiskol J., Merceron C., Vinatier C., Weiss P. and Colliec-Jouault S. (2011a). Sterilization of Exopolysaccharides Produced by Deep-Sea Bacteria: Impact on Their Stability and Degradation. *Mar. Drugs*, **9**, 224–241.

Rederstorff E., Weiss P., Sourice S., Pilet P., Xie F., Sinquin C., Colliec-jouault S., Guicheux J. and Laïb S. (2011b). An in vitro study of two GAG-like marine polysaccharides incorporated into injectable hydrogels for bone and cartilage tissue engineering. *Acta Biomaterialia*, **7**, 2119–2130.

Rehm B. H. A. (ed.) (2009). Microbial production of biopolymers and polymer precursors: applications and perspectives. Caister Academic Press.

Rendueles O., Kaplan J.-B. and Ghigo J.-M. (2012). Antibiofilm polysaccharides. *Environmental Microbiology* **15**(2), 334–346.

Richert L., Golubic S., Le Guedes R., and Guezennec J. (2005). Characterization of exopolysaccharides produced by cyanobacyteria isolated from Polynesian microbial mats. *Current Microbiology*, **51**(6), 379–384.

Roger O., Kervarec N., Ratiskol J., Colliec-Jouault S. and Chevolot L. (2004). Structural studies of the main exopolysaccharide produced by the deep-sea bacterium *Alteromonas infernus. Carbohydr. Res.*, **339**, 2371–2380.

Rougeaux H., Kervarec N., Pichon R., Raguenes G. and Guezennec J. (1996). Novel bacterial exopolysaccharides from deep-sea hydrothermal vents. *Carbohydr. Polym.*, **31**, 237–242.

Rougeaux H., Kervarec N., Pichon R. and Guezennec J. (1999). Structure of the exopolysaccharide of *Vibrio diabolicus* isolated from a deep-sea hydrothermal vent. *Carbohydr Res*, **322**, 40–45.

Rougeaux H., Talaga Ph., Carlson R. W. and Guezennec J. (1998). Structural studies of an exopolysaccharide produced by *Alteromonas macleodii* subsp *fijiensis* originating from a deep-sea hydrothermal vent. *Carbohydr. Res*, **312**, 53–59.

Rougeaux H., Guezennec M., Mao Che L., Payri C., Deslandes E. and Guezennec J. (2001). Microbial communities and exopolysaccharides from polynesian mats. *Mar Biotechnol.*, **3**, 181–187.

Ruiz Velasco C., Baudhuin M., Sinquin C., Maillasson M., Heymann D., Colliec-Jouault S. and Padrines M. (2011). Effects of a sulfated exopolysaccharide produced by *Altermonas infernus* on bone biology. *Glycobiology* **21**, 781–795.

Senni K., Sinquin C., Colliec-Jouault S., Godeau G. and Guezennec J. (2008a) Use of a polysaccharide which is excreted by the *Vibrio diabolicus* species for the engineering of non-mineralized connective-tissue. *U.S. Patent 0,317,860.*

Senni K., Gueniche F., Yousfi M., Fioretti F., Godeau G., Colliec-Jouault S., Ratiskol J., Sinquin C., Raguenes G., Courtois A. and Guezennec J. (2008b) Sulfated depolymerized derivatives of exopolysaccharides (EPS) from mesophilic marine bacteria, method for preparing same, and use thereof in tissue regeneration. *U.S. Patent 0,131,472.*

Senni K., Guencihe F., Changatage S., Septier D., Sinquin C., Ratiskol J., Lutomski D., Godeau G., Guezennec J. and Colliec-Jouault S. (2013). Unusual Glycosaminoglycans from a Deep Sea Hydrothermal Bacterium Improve Fibrillar Collagen Structuring and Fibroblast Activities in Engineered Connective Tissues. *Mar. Drugs*, **13**, 1351–1369.

Sutherland I. (2002). A sticky business: Microbial polysaccharides; Current products and future trends. *Microbiol. Today*, **29**, 70–71.

Thibodeau A. and Takeoka A. (2006). The applications and functions of new exopolysaccharide "Deepsane" from the deepest oceans. *Fragr. J*, **34**, 61–68.

Ullrich M. (ed.) (2009). Bacterial polysaccharides: Current innovations and future trends. Casiter Academic Press.

Velasco C. R., Colliec-Jouault S., Redini F., Heymann D. and Padrines M. (2010). Proteoglycans on bone tumor development. *Drug Discov. Today*, **15**, 553–560.

Volesky B. (2001). Detoxification of metal-bearing effluents: biosorption for the next century. *Hydrometallurgy*, **59**(2–3), 203–216.

Zanchetta Ph., Lagarde M. and Guézennec J. (2003a). A new bone-healing material: A hyaluronic acid-like bacterial exopolysaccharide. *Calcif Tissue Int.*, **72**, 74–79.

Zanchetta P., Lagarde N. and Guézennec J. (2003b). Systematic effects on bone healing of a new hyaluronic acid-like bacterial exopolysaccharide. *Calcif Tissue Int.*, **73**, 232–236.

Chapter 8

Mechanical properties of biofilms

Stefania Fabbri[1] and Paul Stoodley[1,2]

[1]National Centre for Advanced Tribology at Southampton (nCATS), Engineering Sciences, University of Southampton, Southampton SO17 1BJ, United Kingdom.

[2]Department of Microbial Infection and Immunity and the Department of Orthopaedics, Centre for Microbial Interface Biology, The Ohio State University, Columbus, Ohio 43210, USA.

ABSTRACT

An extensive knowledge of biofilm mechanical behaviour in response to different physical forces is crucial for assessing biofilm survival in ancient and modern natural environments as well as relating biofilm structure and function. During the last two decades, various mechanical properties of different types of biofilms have been investigated borrowing techniques for studying polymers and biological tissues. Despite a diversity of different types of biofilms and a variety of methods applied by independent research groups the mechanical behaviour of microbial biofilms is consistently described as viscoelastic. The extracellular polymeric substance (EPS) matrix is itself a multi-component material which gives flexibility to the biofilm structure by changing shape in response to an applied force. Recently, both the advances in microbiology and in the mechanical testing techniques have shown that is not only the biology of EPS that contributes to viscoelasticity and structure but also the biophysical factors under which the biofilm is grown. Herewith, mechanical properties are determined by an interaction between a structure's composition and external environmental changes which may provide possible survival fitness. Biofilm adaptation to persist on surfaces involves structural and physiological changes which are represented in the viscoelasticity of a biofilm. Understanding the role of biofilm mechanics in survival opens up new research paths which might help the desig of new strategies for controlling and preventing biofilms, as well as revealing new insights into the role of biofilm formation in prokaryotic biology. In this chapter we review various mechanical testing methodologies applied to biofilms and discuss the findings in the context of survival strategies and implications for infection and fouling. We also provide some basic but necessary background on the material properties of elastic solid, viscous liquids and viscoelastic materials.

Keywords: EPS, matrix, physical properties, biofilm mechanics, detachment, modelling

8.1 INTRODUCTION

In medical and dental fields, biofilms play an important role in the development and persistence of infectious diseases (Costerton *et al.* 1999). Biofilm formation is thought to be the reason behind about 60% of the bacterial infections reported in developing countries (Fux *et al.* 2005). Urinary tract, catheters, dental plaque, gingivitis, endocarditis, cystic fibrosis, permanent indwelling devices (joint prostheses, heart valves) are some examples of infections that are biofilm-based in the medical field (Costerton *et al.* 1999; Fux *et al.* 2005). Biofilms are also involved in biofouling issues such as water treatment systems (Hagadorn & Bottjer, 1997; Salta *et al.* 2013) energy losses on ship hulls and in fluid pipelines (Schultz & Swain, 2000; Towler *et al.* 2003) causing serious complications in the industry sectors. In nature, biofilms are ubiquitous: not only appearing as benthic biofilms in lakes and streams (microbial mats), but they have also been identified in fossils (Hall-Stoodley *et al.* 2004; Cuadrado *et al.* 2014; Noffke *et al.* 2013; Goehring, 2013; Noffke *et al.* 2001).

Once attached to the surface, microorganisms excrete extracellular polymeric substances (EPS) from the nutrients they metabolize. The EPS conveys to the biofilm a three-dimensional structure (matrix) which allows bacteria to withstand stressful environmental events such as extreme temperatures, high-shear flows or environmental pH changes (Flemming & Wingender 2010). Additionally, the high-cell concentration within a biofilm generates a spectrum of chemical communications between microbes which allow the biofilm to adapt quicker to evolutionary changes than freely suspended bacteria populations (Konovalova & Sogaard-Andersen, 2011).

Although biofilm mechanical properties have been studied for decades and represent an important part for the understanding of biofilm systems it is only recently that their implications for survival on surfaces and mechanical disruption has begun to be appreciated. An extensive knowledge of biofilm mechanical behaviour in response to different physical forces is beneficial for understanding biofilm resistance in ancient and modern natural environments as well as biofilm structure and functions, such as cell dispersal and mass transfer, under various hydrodynamic conditions.

During the last two decades, the mechanical properties of different types of biofilms have been investigated borrowing techniques for studying polymers or biological tissues. However, special considerations of biofilm with respect to materials testing (including small volumes and thickness, structural heterogeneity and their attachment to surfaces) has prompted the development of many creative testing methodologies. Despite a diversity of methods, research groups and types of biofilms tested their mechanical behaviour is consistently described as viscoelastic. The EPS components, which are kept together by physicochemical interactions (electrostatic interactions, van der Waals forces and hydrogen bonding) appear to be the main contributors to biofilm viscoelasticity and mechanical stability (Korstgens, 2001). A viscoelastic material behaves in different ways depending on the external conditions applied, e.g., time-scale of interest, the magnitude and frequency of the external force and the spatial scale of testing. Energy dissipation through viscoelasticity allows biofilms to tolerate rapidly-changing shear stresses without completely detaching from the surface. Biofilms have been shown to be able to migrate in ripple- or stream-like structures when exposed to high-velocity flows and may even grow in a streamlined profile in order to resist shearing forces (Stoodley *et al.* 1998).

Recently, both the advances in microbiology and in the mechanical testing techniques have shown that is not only the biology of EPS that contributes to viscoelasticity and structure but also the biophysical factors (e.g., strain, fluid flow and pressure) under which the biofilm is grown. Shear forces can also influence biofilm structure and cause bacteria to undergo phenotypic adaptation (Stoodley *et al.* 2002). Fluid/structure interactions in biofilms can modulate the dynamics of biofilm growth, tolerance to antibiotics, and virulence (Peterson *et al.* 2015). However, the interaction between biofilm structure and chemical composition with their viscoelastic properties remains one of the fundamental questions concerning biofilm research. This question has stimulated modellers to generate mathematical models with the aim of expanding our basic science understanding of the physics of biofilms and leading to new insight between the structure/function interactions (De Leenheer & Cogan, 2009). In addition, mechanical modes will help the design of more effective strategies to remove biofilms as well as to better control and predict their attachment and growth in natural, industrial and medical environments.

In the first part of the review we provide a brief theoretical background of material science required to discuss and characterize bacterial biofilms from a mechanical point of view. Then, we detail the main findings regarding biofilm mechanics, from the macroscale to the microscale level, and then we go on to discuss detachment. Mathematical models that have been developed to characterize biofilm structure and mechanical responses are also examined. Finally, we explained these findings in terms of biofilm versatility and strength with the aim of enlightening the link between biofilm physiology (such as viscoelasticity) and biofilm biological functions.

8.2 MECHANICAL BACKGROUND

Investigating the mechanical properties of a generic material requires describing how the material deforms when an external force is applied. The main goal is to find the relationship (known as the "constitutive law") between the state of stress inside the material and the corresponding deformation. Each material has its own behaviour which is explored through standardized techniques and instruments. Depending on the way the deforming force is applied (Figure 8.1a), two stresses and relative deformations (strains) are considered:

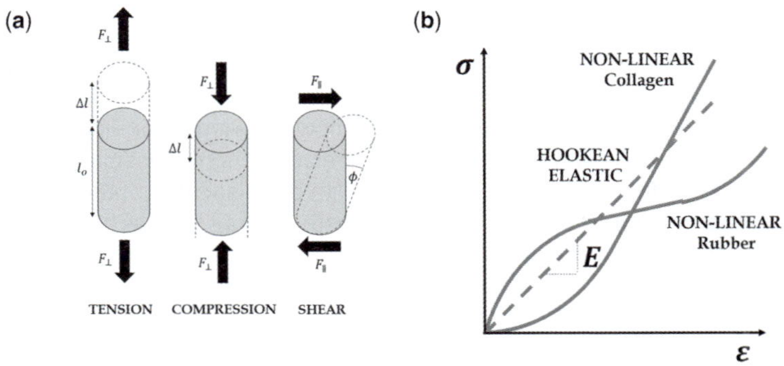

Figure 8.1 a) Schematic representation of a sample subjected to a normal stress, during compression or stretching, and to a shear stress. During a normal stress the force acts perpendicularly ($F_{\perp}$) to the surface area (A_0) generating a compression (or stretching) of the sample (Δl). For a shear stress the force acts parallel ($F_{\parallel}$) to the surface area (A_0) creating a deformation angle Φ. **b)** Representation of stress-strain curves of Hookean-elastic and non-linear elastic materials for a compression (or tension) experiment. For a pure elastic material, the stress applied is proportional to the strain by the constant E (Eq. 8.5): the increase in its length (Δl) by an applied force doubles each time the force is doubled. Many biomaterials (such as collagen) display non-linear J-shaped curves: small increases in stress give large deformations, then at large deformations the material becomes stiffer. Many lightly cross-linked polymers (such as rubber) display an S-shaped curve. The initial part of the curve shows a decreasing in the stiffness with increasing load, than it shows the same behaviour as a J-shaped curve after (Cambridge, 2006).

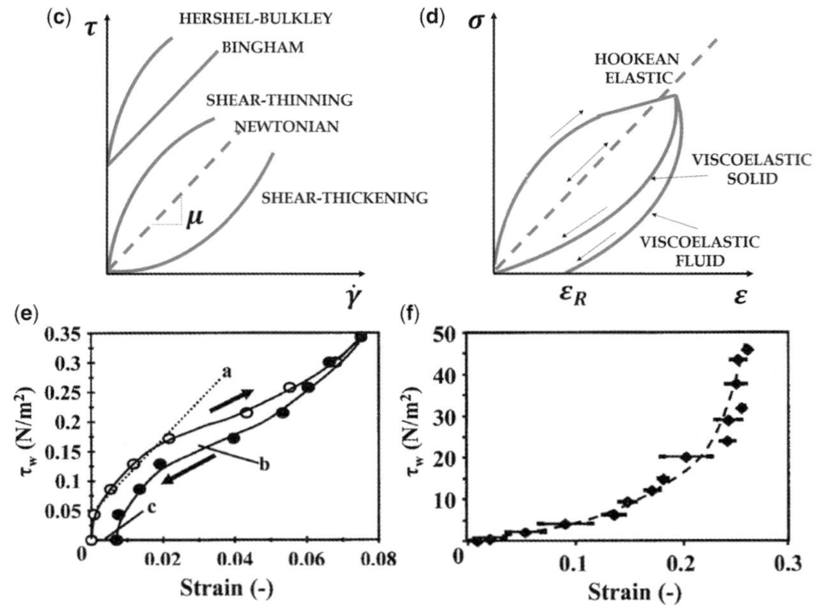

Figure 8.1 (*Continued*) c) Representation of stress-strain curves of Newtonian and non-Newtonian fluids. For a pure viscous material, the viscosity μ doesn't vary with the shear rate as it happens for a non-Newtonian material. The "apparent" fluid viscosity (μ_{app}) can decrease (shear-thinning) or increase (shear-thickening) with increasing $\dot{\gamma}$. Instead a visco-plastic fluid starts to flow only after a certain yield stress τ^* is exceeded. The fluid then can show either linear (Bingham plastic) or non-Newtonian behaviour (Hershel-Bulkley) after (Chhabra, 2010). **d)** Illustration of a pure elastic behaviour with full energy recovery compared to a viscoelastic behaviour. The energy dissipation generates hysteresis in a viscoelastic solid or a permanent deformation (ε_R) in a viscoelastic fluid after (Meyers & Chawla 2009). **e)** Stress-strain curve of *P. aeruginosa* FRD1 strain biofilm grown under laminar flow showing a non-linear S-shaped curve when exposed to low-shear stresses. Biofilm also exhibits viscoelastic energy dissipation in the loading (a) and unloading (b) curves and residual strain (c) (image used with permission of author (Klapper *et al.* 2002). **f)** Stress-strain curve of the mucoid *P. aeruginosa* FRD1 cystic fibrosis isolate biofilm grown under laminar flow showing a non-linear elastic J-shaped curve when exposed to high-shear stresses (image used with permission of author (Klapper *et al.* 2002). Panels *E* and *F* used with authors permission.

- *Normal stress* (σ), when the sample is either compressed or stretched. It is defined as the ratio between the perpendicular component of the applied force and the initial area of the sample:

$$\sigma = \frac{F_{\perp}}{A_0} \tag{8.1}$$

- *Shear stress* (τ), when the sample is twisted. It is defined as the ratio between the parallel component of the applied force and the initial area of the sample:

$$\tau = \frac{F_{\parallel}}{A_0} \tag{8.2}$$

- *Strain* (ε), defined as the ratio between the change in length and the initial length of the sample:

$$\varepsilon = \frac{\Delta l}{l_0} \tag{8.3}$$

- *Shear strain* (γ), defined as the tangent of the angle (Φ) between the loaded and unloaded configurations:

$$\gamma = \tan(\phi) \tag{8.4}$$

8.2.1 Hookean solids and newtonian fluids

Generally the behaviour of many solids can be approximated as pure elastic (Hookean solids) for infinitesimal deformations and, many liquids as pure viscous (Newtonian fluids) for low strain rates (Banks *et al.* 2010). If a deforming force is applied to an elastic solid, it deforms and relax back to its original shape when the force ceases (stores energy). In contrast, if a stress is applied to a viscous fluid, it won't resist to the stress applied and it will flow (deform) indefinitely without recovering the initial shape (dissipates energy). For Hookean solids the measured strain is always directly proportional to the stress applied (Hosford, 2010)(Figure 8.1b) according to:

$$\sigma = E\varepsilon \tag{8.5}$$

$$\tau = G\gamma \tag{8.6}$$

where E and G are the Young's (or elastic) and shear modulus respectively. The units of E and G in the metric system are Newton per square metre (N/m^2) or Pascal (Pa).

The elastic constants G and E are closely related to each other:

$$G = \frac{E}{2(1 + \upsilon)} \tag{8.7}$$

where ν is the Poisson's ratio of the material considered. The intuitive interpretation of the elastic parameters is simple: the higher their value, the stiffer the material will be (to achieve the same relative deformation it is necessary to supply the specimen with a greater force). E and G are commonly used to characterize the stiffness of an elastic material. Ceramic and metals are usually considered elastic linear materials and the Young's moduli generally are in the range 200–500 GPa and 50–210 GPa respectively (Cotterell & Williams 2010).

On the other hand, when a fluid is subjected to an applied shear and starts to flow, frictional forces always take place in the fluid as a result of cohesion movements between fluid particles. These forces give rise to a fluid property called viscosity (μ) and is a measure of the liquid resistance to flow (Gresham, 2008). It has been found experimentally that the velocity at which the fluid is deformed (deformation rate, $\dot{\gamma} = \gamma/t$) is directly proportional to the shear stress applied by the viscosity (Batchelor, 2000):

$$\tau = \mu\dot{\gamma} \tag{8.8}$$

For a Newtonian fluid the viscosity is independent of the stress or strain rate, therefore Eq. (8.8) is a straight line (Figure 8.1c). Most low molecular weight substances and gases exhibit Newtonian flow characteristics. Commonly viscosity measurement of linear viscous fluids are performed with a viscometer. The viscosity in the metric system is expressed as Pa · s. Typical values of viscosity for scores of common fluids (at 25°C) are: air (18.6 μPa · s), water (8.90×10^{-4} Pa · s), ethanol (1.074 mPa · s), olive oil (81 mPa · s) and pitch (2.3×10^{8} Pa · s).

8.2.2 Non-linear behaviour

Hookean and Newtonian behaviour are idealizations. In reality, most solid and fluids, mainly for high stress or strains, present a non-linear behaviour, i.e., σ-ε or τ-$\dot{\gamma}$ relationships are not linear. Rubber and various living tissues are examples of non-linear elastic solids (Figure 8.1B). For small stresses or strains, the response will be linear and therefore linear elastic theory can be applied. "Apparent" elastic parameters (E_{app} and G_{app}) are estimated from the liner region of the stress-strain curves. For instance, E_{app} of polymers and rubbers range 1–3 GPa and 3–100 MPa respectively. For comparison, living tissues have the following "apparent" elastic moduli: pig brain (260–490 Pa), human liver (640 Pa), rat skeletal muscle (100 kPa) and bovine cartilage (950 kPa). On the other hand, biological fluids (blood, saliva), polymeric melts and solutions and fluid and semisolid foods (honey, peanut butter) are examples of non-linear viscous fluids (Figure 8.1c). This type of fluids show an "apparent" viscosity (μ_{app}), which varies with the shear rate. The study of complex non-linear fluids is called rheology. Rheometers are the common instruments used to measure non-linear fluids behaviour and viscosity dependence on the applied stress (or strain). According to the η-γ relationship, we can classify this type of fluid into two classes (Chhabra, 2010): 1) *shear-thinning* fluids, where the fluid viscosity falls with increasing $\dot{\gamma}$, and *shear-thickening* fluids, which exhibit an apparent viscosity which progressively rises with increasing $\dot{\gamma}$. Another peculiar non-linear behaviour is the so called *visco-plastic*, where a fluid behaves like an elastic solid below a certain threshold stress value (yield stress), but when the yield stress is exceeded, the fluid starts to behave like a fluid. For stresses beyond the yield stress, if the dependence between σ and $\dot{\gamma}$ is linear (Newtonian), the fluid is called *Bingham plastic*, instead if the dependence displays a shear-thinning behaviour, the fluid is called *Hershel-Bulkley*.

8.2.3 Viscoelasticity

Biofilms, as well as polymers, soft tissues and biological fluids present not only a non-linear behaviour but they are also viscoelastic. Viscoelasticity means the material behaves either as an elastic solid or a viscous fluid depending how fast it is deformed and how quickly it recovers (Banks *et al.* 2010). In other words, viscoelastic materials respond to an instantaneous stress with a deformation which varies in time. In literature, a viscoelastic material can be classified as viscoelastic fluid or viscoelastic solid. Despite the similarity of the two names, these material behaviours are quite different. A viscoelastic solid is a true solid that show a retarded response to stress (in this case "viscoelasticity" refers to viscous retardation), whereas a viscoelastic fluid is a material that act like a solid on short time scales and a fluid on longer time scales (in this case "viscoelasticity" refers to adding elastic behaviour to a viscous fluid). Because of the viscous component, viscoelastic materials do not store 100% of the energy under deformation. For a viscoelastic solid the forward and backward deformation curves are different but initial and final deformation are the same (hysteresis) (Figure 8.1d). Instead, for a viscoelastic fluid the initial and final deformation don't coincide and the material present a residual deformation.

If the material is subjected to a small deformation (or stresses), the material response is considered to be linear viscoelastic or in the linear viscoelastic range (Ponnamma & Thomas, 2014). Linear viscoelasticity is the simplest response of a viscoelastic material and is based on the Boltzmann superposition principle. According to this theory the material response is a function of the entire loading history and each loading makes independent contribution to the final behaviour. Following these principles, the stress-strain relationship for a linear viscoelastic material is expressed by the so called Boltzmann-Volterra equation (Liu, 2013):

$$\sigma(t) = \int_{-\infty}^{t} G(t - t')\varepsilon(t')dt' \tag{8.9}$$

where $G(t\text{–}t')$ is the defined as the relaxation modulus. The stress is linked to the history of deformation: ε is a function of t' and therefore represents the entire deformation from the past ($-\infty$) to the actual time (t). The way in which σ is linked to ε is determined by the function G which must obey the condition of "vanishing memory": i.e., the more the deformation is in the past, less the material remembers (Banyavichyus *et al.* 1981).

8.2.4 Rheology of viscoelastic materials

The mechanical characterization of viscoelastic materials consists in performing specific tests in order to enable the observation of the time dependency and non-linearity of the material response. Three main rheological tests are commonly performed: creep, stress-relaxation and dynamic testing, and all of these tests have been performed on biofilms. Creep and stress-relaxation are used for studying material viscoelastic response over long time scales. For testing the short time response behaviour, dynamic tests which use oscillating loads are performed.

8.2.4.1 Creep and relaxation tests

The stress-relaxation test is achieved by producing a constant deformation to the material and measuring the relaxation of the internal stress which was built up during the initial deformation (Figure 8.2a). The resulting stress for a viscoelastic solids slowly relax and reach an equilibrium value (residual stress, σ_R), while for viscoelastic fluids the stress goes to zero (Irgens, 2008). Instead the creep test consists in measuring the time dependent strain resulting from the application of a steady uniaxial stress (Figure 8.2b). In this case, the strain increases until approaching an asymptotic level. If the load is removed, the resulting strain slowly decreases until zero for a viscoelastic solid, while for viscoelastic fluid the resulting ε decreases until an equilibrium value (residual strain, ε_R) (Irgens, 2008).

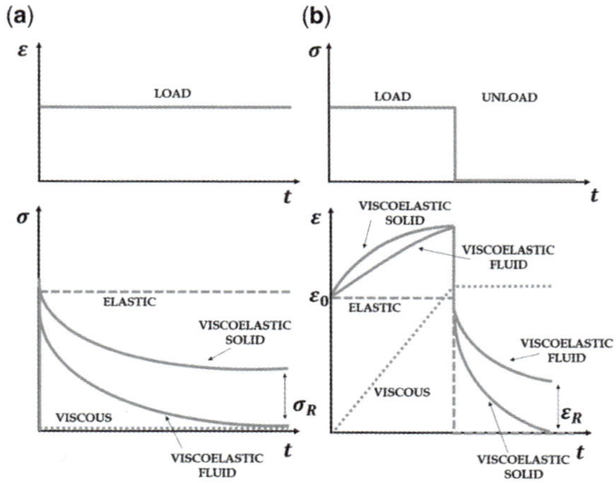

Figure 8.2 a) Schematic representation example of a stress relaxation test for a pure elastic, pure viscous and viscoelastic material. When a constant strain is applied (LOAD) over a period of time to a viscoelastic sample, the resulting stress decreases over time until it reaches an equilibrium value σ_R (viscoelastic solid), or it goes to zero (viscoelastic fluid) after (Lakes, 1998) and (Irgens, 2008). **b)** Schematic representation example of a creep and recovery test for a pure elastic, pure viscous and viscoelastic material. When a constant stress is applied to a viscoelastic sample (LOAD), after an immediate elastic response (ε_0) the resulting strain increases over time (viscoelastic fluids) or approaches an asymptotic level (viscoelastic solid). When the stress is removed (UNLOAD), the strain immediately decreases (elastic behaviour) and then gradually decreases to a residual strain ε_R (viscoelastic fluid) or to zero (viscoelastic solid) after (Lakes, 1998) and (Irgens, 2008).

8.2.4.2 Dynamic test

Dynamic tests consist in subjecting the sample to a controlled small, sinusoidal oscillatory strain and measuring the resulting stress. In purely elastic solids the stress and strain are in phase (store energy) (Figure 8.3a). On the other hand, in purely viscous fluids, the stress and strain have a 90° phase lag (dissipate energy). The reason for this 90° lag (known as the phase angle (δ)) is that, as we saw in Eq. (8.5) and (8.6), for an ideal elastic solid the maximum strain is proportional to the maximum stress. However, as we see in Eq. (8.8), for an ideal Newtonian liquid the maximum stress is proportional to the rate of strain. Since the rate of strain (i.e., the change in strain as a function of time) is effectively the first derivative of a sinusoidal oscillation (i.e., a cosine) then the rate of strain will be zero when the absolute value of strain is at a maximum. Viscoelastic materials behave somewhere in the middle and therefore display an intermediate phase angle (Schramm, 1994). We can represent oscillatory stress and strain as:

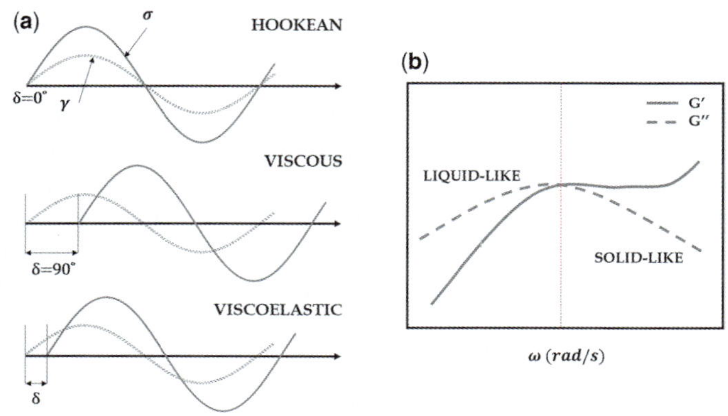

Figure 8.3 a) Schematic representation of the measured stress in response to an oscillatory strain for a Hookean solid, a Newtonian fluid and a viscoelastic material. For a pure elastic solid, the stress is in phase with the applied strain ($\delta = 0°$). For a pure viscous fluid, the strain and the stress are out of phase ($\delta = 90°$). For a viscoelastic materials the stress-strain response is between 0° and 90 after (Schramm, 1994). **b)** Frequency (ω) dependence of G' and G'' 'for a generic polymer solution. For low strain rates, the material behaves more like a liquid ($G'' > G'$)' instead for high strain rates the material response becomes solid-like ($G' > G''$) after (Mezger, 2006).

$$\varepsilon(t) = \varepsilon_0 \sin(\omega t) \tag{8.10}$$

$$\sigma(t) = \sigma_0 \sin(\omega t + \delta) \tag{8.11}$$

By applying the Fourier Transform to Eq. (8.9) we obtain:

$$\hat{\sigma}(\omega) = \hat{G}(\omega) \cdot \hat{\varepsilon}(\omega) \tag{8.12}$$

All transformed functions are complex numbers. The term $\hat{G}(\omega)$ is the Complex modulus and can be rewritten as:

$$\hat{G}(\omega) = G' + i \cdot G'' \tag{8.13}$$

where the real part (Storage modulus) indicates the material ability to accumulate elastic energy while the Imaginary part (Loss modulus), describes the ability to dissipate energy.

Substituting Eq. (8.9) in Eq. (8.12) we obtain:

$$\sigma(t) = \varepsilon_0[G'\sin(\omega t) + G''\cos(\omega t)] \tag{8.14}$$

Using the formula for the sine of a sum Eq. (8.14) yields the following for the Storage and Loss moduli:

$$G' = \frac{\sigma_0}{\varepsilon_0}\cos(\delta) \tag{8.15}$$

$$G'' = \frac{\sigma_0}{\varepsilon_0}\sin(\delta) \tag{8.16}$$

by determining G' and G'', we can understand if a viscoelastic material acts in a more viscous or more elastic manner in response to a given applied stress (Figure 8.3b).

8.2.4.3 Modelling viscoelasticity

A mathematical model describing the response of a viscoelastic material (such as biofilms) during creep and relaxation experiments is essential for predicting deformation, failure and detachment. Such models, called rheological models, are linear networks of two fundamental elements: elastic solid "Hookean" springs (-ww-) and viscous liquid "Newtonian" dashpots (-⊏-) (Barnes *et al.* 1989). The first represents the purely elastic response, expressed by Eq. (8.5), the second the dissipative and delayed response, expressed by Eq. (8.8). The simplest viscoelastic models consist of one single spring and one single dashpot in series (Maxwell model) or in parallel (Kelvin-Voigt model) (Fig. 8.4a–b). The Kelvin-Voigt model displays an exponential strain creep (viscoelastic solids) (Figure 8.4a), whereas the Maxwell model exhibits an exponential relaxation (viscoelastic liquids) (Figure 8.4b). Maxwell and Kelvin models are adequate for qualitative and conceptual analyses. Complex viscoelastic materials like biofilms are modelled combining multiple elements of Kelvin-Voigt and Maxwell. The Burger's model and the generalized Maxwell model are the most used in literature. The Burger model, which results in a series combination of a Maxwell element with a Voigt element, is used to model the creep response (Figure 8.4c-d). The generalized Maxwell model, which is a combination of multiple Maxwell elements in parallel, is used to model the relaxation response (Figure 8.4e-f). As a result, a final stress relaxation time coefficient (τ_r), which estimates the time necessary to pass from the initial stress at the beginning of the test to the asymptotic minimal value of the stress, is obtained.

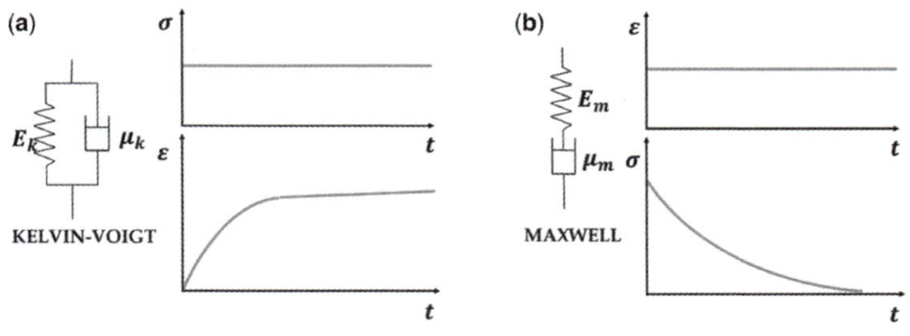

Figure 8.4 a) Schematic representation of Kelvin-Voigt model and the relative creep curve. When a constant stress is applied (upper graph), the dashpot μ_k in parallel with the spring E_k dampens the elastic response causing a retarded viscoelastic strain after (Reddy, 2013). **b)** Schematic representation of the Maxwell model and the relative stress-relaxation curve. When a constant strain is applied (upper graph), there is an immediate elastic response (represented by the spring E_m) followed by a retarded relaxation in the stress (caused by the dashpot μ_m) until it reaches zero after (Reddy, 2013).

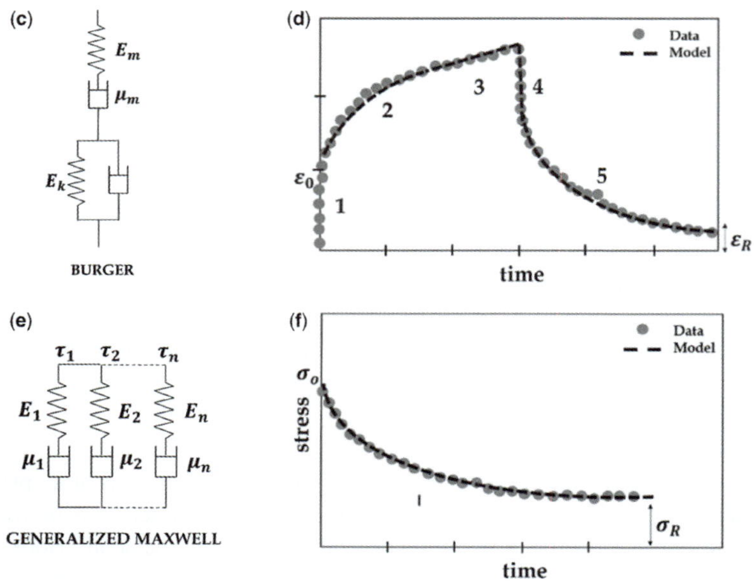

Figure 8.4 (*Continued*) c) Graphic representation for Burger's model. This model is generated by putting a Maxwell model in series with a Kelvin-Voigt model. E_m and E_k are the elastic elements while μ_m and μ_k are the viscous ones. Spring respond instantaneously to an applied stress (or strain), instead dashpot reduces the reaction speed. **d)** Illustration of a creep and recovery curve of a mixed-culture biofilm after rheometer analysis fitted with the Burger model after (Towler *et al.* 2003). The instantaneous elastic strain and recovery (region 1 and 4) corresponds to E_m and E_k. The retarded viscoelastic strain and recovery (region 2 and 5) are represented by μ_1 and the Kelvin-Voigt model. The purely viscous response (region 3) is represented by μ_m. **e)** Schematics of a generalized *n*-element Maxwell model. E_n represent the spring constants and τ_n the relaxation time constants, which are equal to μ_n/E_n. **f)** Illustration of a measured stress relaxation of a *P. aeruginosa* biofilm fitted with five Maxwell elements after (Peterson *et al.* 2013).

8.3 BIOFILM MECHANICS

Mechanical properties have been widely studied in order to understand the physical behaviour of biofilms when subjected to physical forces. This could help not only to control and predict biofilm attachment and growth in either mechanical or medical environments, but also to understand biofilm structure and functions. Due to the high complexity and variability of the biofilm structure it is difficult to find a single well-defined mechanical behaviour and a narrow range of values for the materials parameters. Instead, it is more likely that different behaviours can be defined by changing the conditions to which the biofilm is subjected (such as mechanical instruments, type of loading, temperature or biofilm growing conditions). We divide the findings into two main groups: macrorheological and microrheological studies. Macrorheological studies give great insights on the biofilm mechanical properties at a macroscopic scale through the measurement of bulk material parameters such as biofilm elastic moduli, viscosity and relaxation times. These measurements commonly involve the application of a controlled external force through mechanical tests (for example indenters or rheometers) or hydrodynamic means (such as flow cells). But, because of the microscale heterogeneity in the biofilm structure macroscale properties can only provide averages of the bulk material. Therefore, new technologies have been used and improved in order to reveal spatial distribution of biofilm viscoelastic parameters at the microscale. These includes atomic force microscopy (AFM); micromanipulators, microindenters, micro-cantilever, microfluidics chambers and particle-tracking microrheology (PTM). In addition, microrheological studies allow measurements of biofilm adhesive and cohesive strengths which are important parameters for assessing biofilm failure. Cohesive failure happens when the applied deformation disrupt only biofilm top layers leaving the underlying biofilm attached at the surface, whereas adhesive failure results in a completely detachment of the biofilm (Flemming *et al.* 2011). For a deeper description of the mechanical methods and instruments we suggest the following excellent reviews (Guélon *et al.* 2011; Bol *et al.* 2012).

8.3.1 Macrorheological studies

One of the first reports on biofilms viscoelasticity was in 1999 when Stoodley *et al.* performed rheological tests on mixed culture biofilms grown in a microfluids biofilm reactor (Stoodley *et al.* 1999). They measured the elongation of biofilm clusters caused by elevations in liquid shear stress. Creep and stress-strain curves demonstrated biofilms behave as viscoelastic solids when the applied fluid shear stress is below or near the stresses at which the biofilm was grown ($\tau = 5.09$ N/m^2), while at higher values of shear stress the biofilm can yield and behave as a viscoelastic fluid. This trend is similar to that of a non-Newtonian fluid. They also estimated G_{app} and E_{app} which were 27 Pa and in the range of 17 to 40 N/m^2 respectively. Biofilm viscoelastic fluid-like behaviour was also assessed on various *P. aeruginosa* biofilms grown in a flow cell subjected to laminar and turbulent flows (Stoodley *et al.* 2002; Klapper *et al.* 2002). Stress-strain curves showed the biofilms had an elastic-solid behaviour for short time scales but a linear viscous fluid like behaviour for long time scales. In addition, *P. aeruginosa* FRD1 strain biofilm grown under laminar flow displayed viscoelastic behaviour showing non-linear S-shaped curved with hysteresis and residual strain for low-shear stresses (Figure 8.1e), and a non-linear J-shaped curve for high-shear stresses (Figure 8.1f). Instead, *P. aeruginosa* PAO1 strain grown under turbulent flow showed a linear elastic response for low-shear stresses, while a non-linear viscoelastic response at high-shear stresses. The measured apparent shear modulus varied from 1 to 280 N/m^2 and increased with the applied shear. These finding demonstrated the high dependence of the biofilm viscoelastic properties on the biofilm growing conditions, applied force and bacterial strain. Uniaxial compression experiments on *P. aeruginosa* biofilms (Körstgens *et al.* 2001), performed with a rheometer, showed elastic-Hookean behaviour ($E_{app} = 6500 \pm 500$ Pa) for low loadings until a critical stress ($\sigma^* = 980 \pm 50$ Pa), then fluid-like behaviour was observed afterwards. In 2003; Towler *et al.* performed rheometers shear creep tests on mixed culture biofilms (Towler *et al.* 2003). The creep curves obtained from the experiments presented a residual strain when the load was removed, indicating viscoelastic fluid behaviour (Figure 8.4d). The measured biofilm G_{app} and μ_{app} were in the range of 0.2–24 Pa and 10–3000 Pa respectively. In a study of Vinogradov *et al.* (Vinogradov *et al.* 2004) the static and dynamic (sinusoidal oscillations) responses of *S. mutans* biofilms were studied performing rheological creep tests. Static creep curves showed a viscoelastic fluid behaviour with biofilm apparent shear modulus and viscosity values of $1.9 \pm 3.8 \times 10^2$ Pa and $2.8 \pm 6.4 \times 10^2$ Pa respectively. In addition, the decrease of the complex viscosity (η^*) with the frequency measured from dynamic creep tests suggested a shear-thinning viscoelastic fluid-like behaviour, similar to a polymeric solution (Menard, 1999). In 2005, a glass capillary system was used to assess the viscoelastic properties of *S. aureus* biofilms (Rupp *et al.* 2005) by measuring the deformation of individual biofilm clusters at different wall shear stresses values over a short time period. The resultant stress-strain and creep curves demonstrated biofilm viscoelastic behaviour in order to adapt the structure to different shear stresses. Stress-strain curves yielded a G_{app} of 4.9 ± 3.7 Pa and a μ_{app} of $3{,}500 \pm 2{,}900$ Pa/s. An interesting study of Lahaye and co-authors consisted in comparing the viscoelastic properties of extracted EPS solutions from *P. mirabilis* biofilms (Lahaye *et al.* 2007). They discovered that the water content inside the biofilm could vary the EPS viscoelastic behaviour by changing the types of interaction between the polymer chains. In 2009, Paramonova and co-authors performed low-load compression tests (with a LLCT instrument) on a wide range of biofilm species (Paramonova *et al.* 2009; Paramonova *et al.* 2009). In particular, stress-strain curves obtained from single or double species oral biofilms (Paramonova *et al.* 2009) showed both a non-linear behaviour with E_{app} ranging from 17 to 62 Pa and 29 to 182 Pa respectively. In the same year, Koza *et al.* measured the viscoelastic properties of a *P. fluorescens* biofilm air–liquid interface through amplitude sweep tests with an oscillatory rheometer (Koza *et al.* 2009). A loss factor (which is the ratio between G' and G'') less than 1 determined biofilms viscoelastic solid-like behaviour until a threshold stress value of 0.028 Pa. In addition, an apparent storage modulus of 0.75 ± 0.16 Pa and an apparent viscosity of 0.24 ± 0.05 Pa were obtained from the linear viscoelastic region. Afterwards, the G' decreased with increasing the shear stress indicating a shear-thinning behaviour. Similar results were obtained from the group of Ribbeck on *P. aeruginosa* biofilms (Lieleg *et al.* 2011). They assessed biofilm viscoelastic behaviour at large deformations discovering that biofilms are able to fully regain their initial stiffness after a period of recovery. This observation can explain the biofilm capability of dispersing in clusters rather than totally detaching from surfaces when subjected to high shear forces. In addition, they assessed the influence of chemical perturbation on the properties of *P. aeruginosa* biofilms. They found that biofilm elasticity was increased when adding multivalent cations (such as Fe^{3+} or Al^{3+}), decreased when adding citric acid or unchanged when adding citric acid and Fe^{3+} together. An increase or decrease in the cross linking interactions between trivalent ions and negatively charged groups in the EPS could possibly strength or weaken biofilm structure. These results are crucial for developing new strategies which consider the use of chemicals for biofilms removal. In 2012; Rmaile *et al.* (Rmaile *et al.* 2012) tested *S. mutans* biofilm viscoelasticity performing uniaxial indentation test with a Bose machine. They also assessed the effect of sugar concentration, osmotic-inducing agent (PEG-8000) and chelating agent (EDTA) on biofilm elasticity. Non-linear force-displacement curves and stress-relaxation measurements confirmed the viscoelastic behaviour of the biofilm samples. The biofilm elastic modulus and relaxation time were approximately 280 Pa and 11 sec. Osmotic and higher sugar concentration in the growth medium caused an increase in biofilm stiffness, while chelating effects caused a decrease. Since biofilms metabolize sugars to produce EPS glucans, a higher sugar supply will lead to a higher EPS production and therefore to a stiffer biofilm. Increasing stiffness due to the osmotic effects can be explained by "dewatering" of the biofilm. Similar results were obtained from Waters and co-authors who studied the mechanical properties through dynamic rheological tests of *in situ S. mutans* biofilms grown in the presence or absence of sucrose (Waters *et al.* 2013). Complex moduli measurements showed biofilms grown in the presence of sucrose behaved more like soft elastic solid materials than viscoelastic fluids enhancing the importance of the EPS matrix composition on the biofilm mechanical properties. More recently, rheological characterisation of biofouling layers under steady shear flow demonstrated biofilms thixotropic, shear-thinning viscoelastic behaviour for which solid-like behaviour appeared below a yield stress ($\tau^* = 10$ Pa) and fluidity at elevated values of τ^* (Patsios *et al.* 2015). The G' values (40–200 Pa) were bigger than those of G'' (20–30 Pa) indicating the marked elastic behaviour of the biofouling layer. Hysteresis in the stress-strain curves of heterotopic biofilms performing time-lapsed deformation experiments at different shear stresses using optical coherence tomography (OCT) confirmed a viscoelastic trend (Blauert *et al.* 2015). E_{app} (36.0 ± 2.6 Pa) was estimated from the linear part of the stress-strain curve while G_{app} (29.7 ± 1.7 Pa) from the optical coherence tomography (OCT) B-scans. *In situ* rheological characterisation of *S. epidermidis* biofilms probed through small-amplitude oscillatory deformation and non-linear creep tests showed that osmotic pressure (in terms of NaCl concentration) and temperature altered biofilm mechanical properties (Pavlovsky *et al.* 2013). In particular biofilms both showed a non-monotonic relationship between stiffness and the NaCl concentration and a hysteresis of the viscoelastic moduli with an increase in temperature. These finding advocates that biofilms viscoelastic properties are strongly determined by local environmental conditions. Researchers also showed that the storage modulus ($G' \sim 10$ Pa), unlike the loss modulus ($G'' \sim 1$ Pa), exhibited a power-law increase at increasing angular frequency. Similar results were obtained from the same group which showed a significant decrease in the yield stress and *S. epidermidis* biofilm elastic modulus under increasing temperatures (Pavlovsky *et al.* 2015). In addition they showed a correlation between the effect of temperature on cell viability (higher temperature caused a drastic decrease in cell viability) and the weakening of the biofilm. This could open new doors for controlling biofilm such as the development of non-invasive temperature-based strategies for the treatment of biofilm infections in the medical field.

8.3.2 Microrehological studies

In 1998, a micromanipulator with a T-shaped probe was developed by Chen and co-authors to measure the adhesive strength (σ_A), defined as the work required to detach the biofilm from the surface, of 29-days olds *P. fluorescens* biofilms grown under different fluid velocities (Chen *et al.* 1998). Investigators reported that the biofilm adhesive strength ($\sigma_A = 0.05$–0.2 J/m^2) increased as the fluid velocity increased ($u = 0.6$–1.6 m/s). In a later study using the same technique, Chen and co-authors reported that the adhesive strength of *P. fluorescens* biofilms increased as the biofilm age, fluid velocity, pH and roughness increased (Chen *et al.* 2005). Interestingly, the biofilm adhesive strength also increased with an increase in the glucose concentration, but when it exceeded 30 mg/l the adhesive strength decreased again. The glucose-adhesive strength non-linear relationship was explained suggesting that a too high glucose concentration might inhibit the irreversible adhesion and lead to a lower adhesive strength. In 2006, Cense and co-authors, using a microindenter apparatus to measure *S. mutans* biofilms failure confirmed the extracellular polysaccharide matrix pivotal role in the biofilm tensile strength (Cense *et al.* 2006). In addition, biofilms exhibited a viscoelastic solid-like behaviour during stress-relaxation tests. This conclusion differed from that of main macrorheological studies previously reported (where *S. mutans* biofilm was characterized as a viscoelastic liquid) enlightening the heterogeneity of the biofilm at different scales. They also discovered that the biofilm cells remained undamaged after testing enlightening EPS matrix role on the biofilm strength. One year later, PTM on *S. aureus* biofilms exhibited different mechanical environments between individual bacteria within the same biofilm and a dependence of the biofilm stiffness on the growth conditions – i.e., hardening during growth, and softening during starvation (Rogers *et al.* 2008). In 2009; Hohne *et al.* developed a flexible microfluidic rheometer which enabled measurements of *S. epidermidis* and *K. pneumonia* biofilms mechanical properties at the microscale (Hohne *et al.* 2009). Both biofilms exhibited typical non-linear exponential stress–strain trends but the *S. epidermidis* biofilm was much stiffer ($E_{app} = 3.2$ kPa) than *K. pneumonia* biofilm ($E_{app} = 1.1$ kPa) the variability in the mechanical properties in changing biofilm strain. Video holographic PTM enabled the measurements of the mechanical properties of a water-soluble and -insoluble polysaccharides of *S. mutans* biofilms (Cheong *et al.* 2009). The biofilm's water-soluble polysaccharides formed a shear-thinning fluid roughly ten times more viscous than water. By contrast, the N- type fraction formed an elastic gel with a storage modulus of 10 Pa but strongly shear-thinning as indicated by its dynamic viscosity. These studies show how the EPS functional roles are strictly connected with the macroscale biofilm physical and mechanical properties. Aggarwal *et al.* (Aggarwal *et al.* 2010) measured cohesive strength (σ_C) of *P. aeruginosa* and *S. epidermidis* performing tensile tests modifying the common microcantilever method of Poppele and Hozalski (Poppele & Hozalski 2003). The high variability of the measured cohesive strength values (59–18,900 Pa and 61–5,840 Pa for the *P. aeruginosa* and the *S. epidermidis* biofilms respectively) enlightened biofilm structural heterogeneity at the microscale. In another study, they measured the ultimate tensile strength and the elastic modulus of *S. epidermidis* biofilms using the same method (Aggarwal *et al.* 2010; Aggarwal & Hozalski, 2012).

Results showed biofilm viscoelastic behaviour characterized by a non-constant young modulus which increases with the strain rate applied. In 2012, using magnetic tweezers and micron-sized magnetic particles Galy *et al.* obtained a three-dimensional mapping of *E. coli* biofilm viscoelastic properties (Galy *et al.* 2012). A significant variation in the local mechanical properties within the structure revealed the three-dimensional mechanical heterogeneity of a biofilm. More recently, Chew *et al.* tracking the Brownian motion of beads within the biofilm (particle-tracking microrheology) in discovered that increasing or decreasing the amount of specific exopolysaccharides inside *P. aeruginosa* biofilms can change biofilm behaviour from more elastic to more viscous (Chew *et al.* 2014). One of the first AFM measurement on biofilms was performed by Fang and co-authors by measuring AFM tip-cell interaction forces of sulphate-reducing bacteria of an early stage 6h-old biofilm (Fang *et al.* 2000). Adhesion forces were found to be 25% higher at cell–cell and cell–substratum interface than on the middle of cells. Similar results were obtained in the study of Oh *et al.* on *E. coli* biofilms grown on glass slides (Oh *et al.* 2007). Force-distance curves showed a higher adhesion at the top of the bacterium surface (σ_A = 122.65 pN) than at the glass surface (σ_A = 51.79 pN). In addition, investigators also showed biofilm elasticity on the cell surface increased as biofilm became mature, suggesting that EPS could considerably enhancing the bacteria binding to surface. The elastic properties of three gram-negative (*E. coli* ML35, *E. coli* ZK1056 and *P. putida*) and two gram-positive (*B. subtilis* and *M. luteus*) bacterial strains were assessed calculating cellular spring constants (k_{cell}) from AFM force-distance curves (Volle *et al.* 2008). All the bacteria strains showed high k_{cell} values (0.16 to 0.41 N/m), demonstrating bacteria rigidity. In addition, extracellular polymeric substances were observed covering the biofilm cells outer layer, facilitating bacterial adhesion. In 2009, Oh end co-authors used AFM operating in a force mode to assess the influence of different substrates (aluminium, steel, rubber, and polypropylene) on 3 days-old *P. aeruginosa* biofilms attachment (Oh *et al.* 2009). The results demonstrated a higher bacteria adherence to polypropylene than to the other substrates (because of higher surface roughness). The researches also subjected biofilms to hot water treatment to measure changes in biofilm adhesion force which significantly decreased. Most of extra-cellular matrix was removed after water exposure, enlightening again matrix pivotal role in the biofilm attachment process. Lau *et al.* combined microbead force spectroscopy with atomic force microscopy showing significant changes in adhesion and viscoelastic properties of a *P. aeruginosa* PAO1 strain compared to a derived mutant *wapR*, which has defective expression of lipopolysaccharides (Lau *et al.* 2009b). Biofilm biophysical properties, genetic, growth and environmental factors are important aspect which need to be considered in order to understand bacterial biofilms persistence.

8.4 BIOFILM DETACHMENT

Environmental changes such as fluid velocity or shear stresses and chemical cues cause microbes in the biofilm to detach and disperse in the bulk fluid. Biofilm dispersal is of fundamental importance in the proliferation of biofilms, because detached cells could potentially transmit pathogens into the blood stream and cause bacteraemia and endocarditis (Lu, 2009) or release contaminating bacteria into the environment as seen in showers and cooling towers with legionella. Viscoelastic features such as elongation and recoil allows biofilms to tolerate even high-velocity fluid flows without detaching from the surface (Figure 8.5). A wide range of mechanical parameters such as cohesive or adhesion strength have been quantified in order to understand how different hydrodynamic conditions affect biofilm detachment. Parallel plate flow chamber reactor systems represent the traditional technique to measure such values (Ohashi & Harada, 1994; Stoodley *et al.* 2001; Stoodley *et al.* 2002). Flow devices such as jet impingement (Deshpande & Vaishnav, 1983; Vaishnav *et al.* 1983) and fluid dynamic gauging (Tuladhar *et al.* 2000; Chew *et al.* 2004) were developed to assess biofilm adhesion under well-defined hydrodynamic conditions. Micro jet impingement (MJI) on *S. mutans* biofilms grown in different sucrose concentrations (0.1%, 0.5% and 1%) demonstrated that sucrose affects biofilm adhesion strength (Kreth *et al.* 2004). In particular, the shear stress (measured from the zone of clearance) was about 20-fold higher between 0.1% and 0.5% sucrose and 1.6-fold between 0.5% and 1% sucrose. In 2005, Bayoudh and co-authors investigated the adhesion strength of *P. stutzeri* onto hydrophilic and hydrophobic glass surfaces using laminar microjet impingement at various Re numbers and exposure time combinations (Bayoudh *et al.* 2005). Researchers found that bacterial detachment strength decreased significantly with increasing exposure time and flow velocity. In addition, adhesion strength values measured for the biofilm grown onto the hydrophobic glass surface (343.9 Pa and 7,084 kPa at Re = 500 and Re = 1,500 respectively) were higher than those measured for the biofilm grown on hydrophilic ones (167.5 Pa and 3,347 kPa at Re = 500 and Re = 1,500 respectively). Similar experiments were performed on *P. Synechococcus* biofilms grown on three different substrata over a 4-week period but exposed to a pressure mode FDG in order to measure monitor biofilm strength (Salley *et al.* 2012). Interestingly, biofilm formed on stainless steel (SS) and on an indium tinoxide (ITO) on a polyethylene terephthalate (PET) substratum detached by cohesive mechanisms, whereas those on glass were removed by adhesive failure. Investigators related this difference to a change in substratum surface energies. Möhle *et al.* used fluid dynamic gauging to determine the cohesive strength of a heterotrophic mixed culture biofilms (Möhle *et al.* 2007). They found that biofilm thickness depended on biofilm hydrodynamic-growth conditions (shear stress) while biofilm cohesive strength (6.0 and 7.7 N m^{-2}) was independent. Interestingly, CLSM imaging after gauging showed a remaining layer of EPS glycoconjugates at the surface which could play an important role in biofilm mechanical strength. Paul and co-authors exposed biofilms grown under different hydrodynamic conditions to various shear stresses from a Couette Taylor Reactor (Paul *et al.* 2012). After a biofilm upper part detachment, which depended on the shear stress applied during biofilm grow, researchers discovered the presence of a basal layer that withstand shear stresses up to 13 Pa by compressing itself rather than detaching. Similar results were found by Hwang and co-authors on 64 and 116 h old *S. mutans* biofilms (Hwang *et al.* 2014). Researcher measured biofilm mechanical strength using a shear-induced biofilm mechanical strength testing combined with confocal laser scanning microscopy (CLSM) and biomass measurements. After an initial 50% biofilm removal the remaining biofilm resistance increased considerably even with a 10-fold increase in the shear stress. Confocal images showing EPS spatial distribution particularly near the surface and mechanical strength testing on biofilms threated with dextranase demonstrated EPS matrix role in modulation of biofilm mechanical stability. In 2012, Abe and co-authors imaged young drinking water biofilms in order to quantify the mechanical detachment shear stress exerted by an AFM tip (Abe *et al.* 2012). They developed a new technique consisting in a consecutive AFM scan were the biofilm itself was removed "layer by layer" by increasing the applied force. Therefore the cohesiveness of the biofilm removed layer was lower than the tip applied stress while the remaining biofilm was stronger. Interesting, high shear stress values (τ = 100 kPa) were measured for removing biofilm clusters greater than 200 μm^3 which means that biofilms cleaning in industrial sectors (pipes, hulls) might be not always efficient. The influence of surface roughness and temperature on *E. coli* biofilm detachment using a flow chamber was investigated by Fink *et al.* (Fink *et al.* 2015). Researchers found that more bacterial cells adhered on electro polished samples than on brushed samples and that the increasing liquid temperature and flow velocity increased biofilm detachment. Other mechanical methods used to remove biofilms in industry and medical fields consist in applying low volume, high-velocity water droplets (Cense *et al.* 2006) or entrained air bubbles (Parini *et al.* 2005; Parini & Pitt, 2006). In particular, exposing *S. mutans* dental biofilms to high-velocity (ms) microbursts investigators demonstrated not only biofilm viscoelastic behaviour (Rmaile *et al.* 2014; Rmaile, 2015), but also fluid-like behaviour on the order of milliseconds in response (Fabbri *et al.*). These techniques have shown positive results due to the droplets' impact pressure, hydrodynamic shear stresses and the surface tension effects of the passage of an air-water interface over a solid surface (Busscher *et al.* 2010).

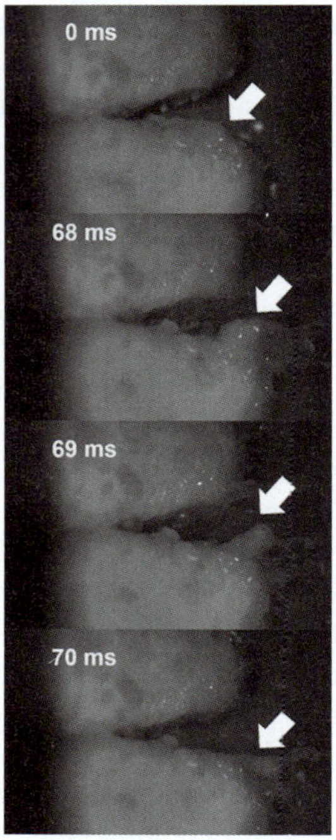

Figure 8.5 Four frames from a high-speed camera movie showing the viscoelastic elongation of a 3 days-old *S. mutans* dental biofilm, grown between two central incisors from a typodont model, during high-velocity ($u > 60$ m/s) microspray exposure. A biofilm cluster (white arrow) elongated under shear-flow without detaching from the tooth surface. Original data shown.

8.5 MATERIAL MODELLING OF BIOFILM MECHANICS

From the previous paragraphs is evident that despite biofilm large physiological heterogeneity, which precluded the use of a single standard experimental procedure, researches produced a vast amount of useful data regarding biofilm materials properties. This stimulated modellers to mathematical model biofilms mechanics in order to predict and control biofilms behaviour which in turn can help the development of new experimental strategies.

The earliest modelling approach of biofilm mechanical properties was presented by Klapper *et al.* who modelled *P. aeruginosa* biofilm bulk structure as cross-linked polymer viscoelastic gel using the linear Jeffrey's stress-strain constitutive law (Klapper *et al.* 2002). The authors considered EPS responsible for the viscoelastic behaviour because of the internal and physicochemical bonds. In absence of an external stress the matrix remained in an isotropic configuration in which the attractive and repulsive forces between the chains are in balance. When a loading force is applied on the biofilm matrix, alignment and stretching of both polymer chains and intermolecular bonds took place until the yield limit is reached. After this point, intermolecular bonds broke and matrix polymers started flowing. Finally, when the shear is removed the intermolecular bonds reformed and the biofilm simulated a gel-like behaviour.

Biofilm time-dependent strain and creep compliance responses at different scales of testing were interpreted using the Maxwell (Wloka *et al.* 2004) or Burger (Vinogradov *et al.* 2004; Towler *et al.* 2003) linear viscoelastic models or by combining multiple elements of Kelvin-Voigt and Maxwell (Figure 8.4) (Cense *et al.* 2006; Lau *et al.* 2009a; Ehret & Böl, 2013; Pavlovsky *et al.* 2013). In particular, Lau *et al.* modelling the creep curves using a combined viscoelasticity model showed that the elastic parameters in *P. aeruginosa* biofilms were drastically reduced by a lipopolysaccharide deficiency (ref). The good agreement between experimental data and the material models suggested biofilm viscoelastic behaviour similar to synthetic polymers (Janmey *et al.* 1990) and other biological fluids such as saliva and mucus (Fung, 2013). However, due to the vast amount of biofilm materials models available in the literature, stress-relaxation constants covered a wide range of values (5 to 100s) (Peterson *et al.* 2014). Peterson and co-authors performed principal component analysis (PCA) on a four-element Maxwell model showing that only three principal components are enough to model biofilms time-dependent behaviour: the fastest one (<3 s) was linked to the water due to its smallest viscosity; the intermediate one (3 to 70 s) was associated with the matrix polysaccharides; and the slowest component (10 to 25 s) incorporated eDNA structures (Peterson *et al.* 2013). Materials model were also used to investigate biofilm mechanical behaviour after the exposure to chemicals/antimicrobials (Jones *et al.* 2011; He *et al.* 2014). In particular, He and co-authors (He *et al.* 2014) assessed dental biofilm expansion caused after brushing and relative increased antimicrobials penetration using a three-element Maxwell model. Results showed an increase of the slow relaxation element which is related to a more open biofilm structure (after brushing). These results demonstrated that by changing oral biofilm mechanical properties we can enhance antimicrobial penetration inside the biofilm and have a better therapeutical effect.

A weakness of the dashpot and spring models is that it lacks capturing biofilm structural heterogeneity (such as density, porosity, bacterial growth and nutrient transfer). The advances in computing science have allowed the development of new complex and multidisciplinary models which combine microbiology, biochemistry and fluid mechanics. We can divide these models in three main categories: 1) one-dimensional continuum (1D) biofilm models, 2) advanced multi-dimensional biofilm models (which include Cellular Automata models (CA) and Individual-based models (IbM)) and 3) multi-physics and mechanical biofilm models. For a deeper mathematical description of these models we recommend the following reviews (Horn & Lackner, 2014; Klapper & Dockery, 2010; Wang & Zhang, 2010). The first two categories are mainly centred to model biofilm growth dynamics, substrate concentration, bacterial reproduction and the cell-to-cell communication. Only a few studies have included simplified mechanical analysis in 1D models finding that shear-stress influences not only biofilm detachment but also microbial growth rate (Stewart, 1993; Peyton & Characklis, 1992; Horn *et al.* 2003). One particularly elegant two-dimensional CA model solved two different detachment processes (erosion and sloughing) at the biofilm surface caused by fluid-flow-induced stresses (Picioreanu *et al.* 2001). By treating the biofilm as elastic solid researchers showed that sloughing significantly increased biofilm-surface roughness compared to the erosion process. In addition, simulations showed that high-growth rates can cause a faster biofilm detachment compares to slow-growth ones. Three hypothetical mechanisms of detachment were incorporated into a 3D CA framework (Chambless & Stewart, 2007). Results showed that detachment was highly related to the biofilm structure. In addition, nutrient depletion in the biofilm generated a higher amount of sloughing events.

Multi-physic and mechanical models are a new generation of models, where the physics of the biofilm system interacts with the adjacent fluid in order to model biofilm viscoelasticity and biofilm/fluid interactions. In 2007, Alpkvist and Klapper modelled biofilm bulk fluid-driven mechanical response and detachment by developing a ball-spring viscoelastic model (Figure 8.6a) (Alpkvist & Klapper, 2007). Subsequently, researchers used a two-phase flow model (Cogan & Guy, 2010) to understand the combined motion of the fluid and the biofilm (Cogan & Keener, 2004; Zhang *et al.* 2008b; Zhang *et al.* 2008a; Cogan, 2008). In particular, Zhang *et al.* were able to simulate deformations, detachment phenomena, rippling and streaming events (Zhang *et al.* 2008b). In another study, Laspidou and colleagues developed a unified multi-component cellular automata model (UMCCA) using finite element analysis (FEA) to test biofilm mechanical properties under compression, tension and shear tests (Laspidou *et al.* 2014). Towler *et al.* studied biofilm response to a turbulent flow field using FEA (Towler *et al.* 2007). Navier-Stokes equation were used to describe flow hydrodynamic while a single biofilm cluster was modelled using the linear viscoelastic Burge model. Results showed that biofilms with a lower Young modulus were more predisposed to lift forces and that biofilm induced-detachment was most effective during the initial relaxation period. A similar finite element model was created in order to assess the influence of the Young modulus, Re number and biofilm structure on biofilm detachment during fluid flow (Böl *et al.* 2009). Researches used stacks from CLSM images to reconstruct the biofilm structure and St. Venant-Kirchhoff and Arruda-Boyce models to represent the biofilm mechanics (Figure 8.6b). Interestingly, the biofilm detachment rate resulted to be independent from a change of Young modulus but was highly dependent on the biofilm structure (such as growth) and Re numbers.

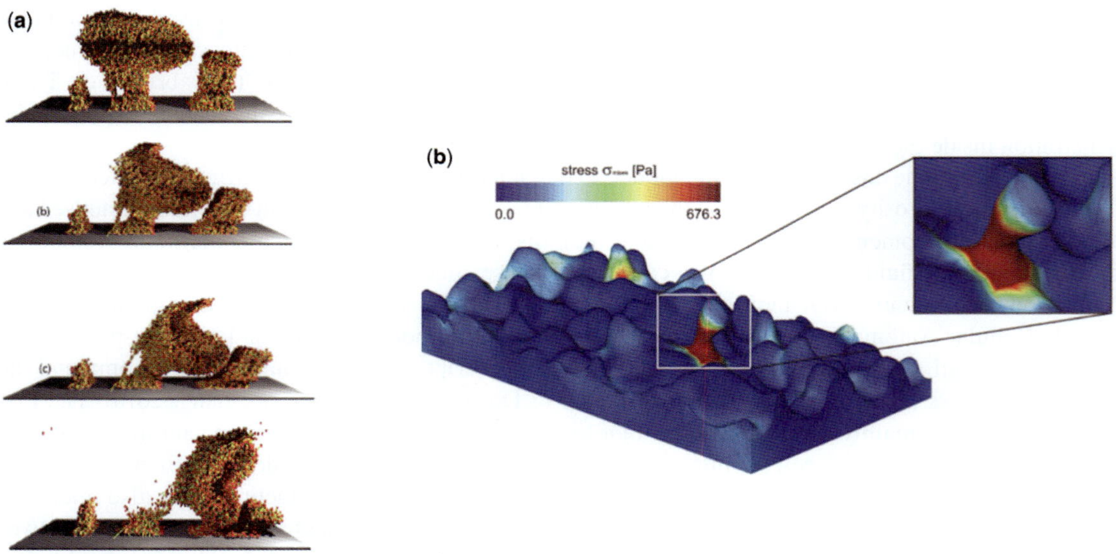

Figure 8.6 a) 3D simulation generated using the immersed boundary method showing biofilm detachment under impact of flow (used with authors permission (Alpkvist & Klapper, 2007)). **b)** 3D-finite element model of a biofilm showing shear-stresses distribution at the surface *y* (used with authors permission (Böl *et al.* 2009)).

8.6 ARE BIOFILM MECHANICS AND FUNCTION CORRELATED?

We have seen that there is compelling evidence which shows biofilms are complex biological materials composed by bacteria and hydrated polymeric substances in water. The matrix is itself a multi-component element which gives flexibility to the biofilm structure by changing shape in response to an applied force. As a result, biofilms behave as non-linear viscoelastic materials whose viscous or elastic responses dominates depending on the environmental and growth conditions.

Reviews of the literature on biofilm mechanics showed that changes in the EPS composition caused by various external factors (such as change in the nutrient intake, temperature shift, enzyme exposure or a specific gene knock-out) considerably change biofilm mechanical behaviour (sometimes from viscoelastic fluids they became solid elastic) or increase/decrease biofilm detachment. In addition, in the natural, industrial and medical environment biofilms are often exposed to a wide range of hydrodynamic forces (compressive, tensile or shear) which significantly alter biofilm physical morphology. Wave-like pattern biofilms have been found inside endotracheal tubes (Inglis, 1993) and venous catheter (Rusconi *et al.* 2010). Flow-induced slow-moving (Stoodley *et al.* 1999) and fast-moving ripples-like structures (Figure 8.7) have been discovered at the surface of in vitro biofilms as well as similar ripples marks have been observed in modern streams (Battin *et al.* 2003) and in ancient fossils (Noffke, 2011; Thomas *et al.* 2013). It has been hypothesized that these structures are related to biofilm viscoelasticity as an adapted trait that allows bacteria to flows and remain attached to surfaces when subjected to steady or transient overlying shear forces.

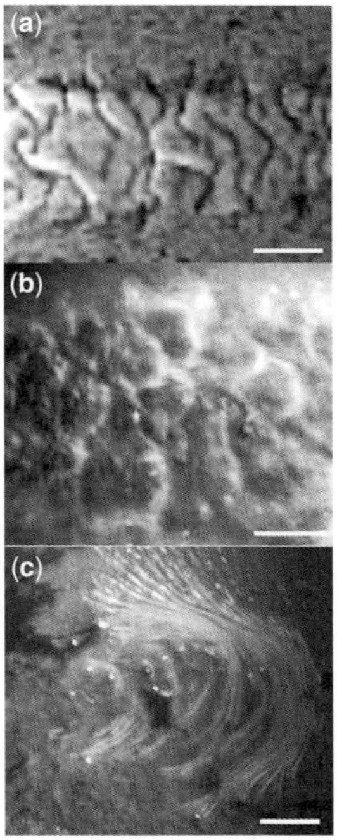

Figure 8.7 High magnification pictures of different ripples morphologies generated at the surface of 3-days old *S. mutans* **(a)**, 3-days old *S. epidermidis* **(b)** and 7-days old *P. aeruginosa* **(c)** biofilms during the exposure to high-velocity microsprays. Flow was left to right. Scale bar = 2 mm. Showing that high-velocity biofilm-fluid interactions can cause the generation of high-velocity ripples provides further insight into the mechanical behaviour of biofilms as viscoelastic complex liquids. Original data shown.

Herewith, mechanical properties are determined by an interaction between a structure's composition and external environmental changes which may have a possible survival meaning. Viscoelasticity may be considered an emergent property of populations of prokaryotes when in biofilms and its commonality to a diverse range of bacterial species with different EPS polymers suggests a highly conserved trait and thus an integral fitness component for survival on surfaces. Biofilm adaptation to persist on surfaces involves structural and physiological changes which are represented in the viscoelasticity of a biofilm. This has important consequences for mechanical removal control strategies in the medical and industrial fields.

Moreover, recent studies on infectious and oral biofilms also related viscoelastic behaviour with antimicrobial/chemicals penetration in biofilms (Peterson *et al.* 2015). This means that the mechanical behaviour allows biofilm not only to withstand physical stresses but also chemical ones, possibly increasing their chance of survival. This hypothesis opens up new research paths on biofilm mechanical properties which might help the designing of new strategies for controlling and preventing biofilms, as well as revealing new insights into the role of biofilm formation in prokaryotic biology.

ACKNOWLEDGMENTS

This work was financially supported in part by EPSRC DTP EP/K503130/1 award and in part by Philips Oral Healthcare, Bothell, WA. USA.

REFERENCES

Abe Y., Skali-Lami S., Block J.-C. and Francius G. (2012). Cohesiveness and hydrodynamic properties of young drinking water biofilms. *Water Research*, **46**, 1155–1166.

Aggarwal S., Poppele E. H. and Hozalski R. M. (2010). Development and testing of a novel microcantilever technique for measuring the cohesive strength of intact biofilms. *Biotechnol Bioeng*, **105**, 924–934.

Aggarwal S. and Hozalski R. M. (2012). Effect of strain rate on the mechanical properties of Staphylococcus epidermidis biofilms. *Langmuir*, **28**, 2812–2816.

Aggarwal S. and Hozalski R. M. (2010). Determination of biofilm mechanical properties from tensile tests performed using a micro-cantilever method. *Biofouling*, **26**, 479–486.

Alpkvist E. and Klapper I. (2007). Description of mechanical response including detachment using a novel particle model of biofilm/flow interaction. *Water Sci. Technol*, **55**, 265–273.

Banks H. T., Hu S. and Kenz Z. R. (2010). A brief review of elasticity and viscoelasticity. In: DTIC Document.

Banyavichyus R. B., Barauskas A., Marma A. and Askadskii A. A. (1981). Features of viscoelastic behavior of materials based on heat-resistant polymers. *Mechanics of Composite Materials*, **16**, 644–649.

Barnes H. A., Hutton J. F. and Walters K. (1989). An Introduction to Rheology. Elsevier.

Batchelor G. K. (2000). An Introduction to Fluid Dynamics. Cambridge University Press.

Battin T. J., Kaplan L. A., Newbold J. D., Cheng X. and Hansen C. (2003). Effects of current velocity on the nascent architecture of stream microbial biofilms. *Applied and Environmental Microbiology*, **69**, 5443–5452.

Bayoudh S., Ponsonnet L., Ouada H. B., Bakhrouf A. and Othmane A. (2005). Bacterial detachment from hydrophilic and hydrophobic surfaces using a microjet impingement. *Colloids and Surfaces A: Physicochemical and Engineering Aspects*, **266**, 160–167.

Blauert F., Horn H. and Wagner M. (2015). Time-resolved biofilm deformation measurements using optical coherence tomography. *Biotechnology and Bioengineering.*

Bol M., Ehret A. E., Albero A. B., Hellriegel J. and Krull R. (2012). Recent advances in mechanical characterisation of biofilm and their significance for material modelling. *Crit. Rev. Biotechnol.*, **33**(2), 145–171.

Böl M., Möhle R. B., Haesner M., Neu T. R., Horn H. and Krull R. (2009). 3D finite element model of biofilm detachment using real biofilm structures from CLSM data. *Biotechnology and Bioengineering*, **103**, 177–186.

University of Cambridge. Elasticity in Biological Materials. Accessed Jan 2016. http://www.doitpoms.ac.uk/tlplib/bioelasticity/index.php.

Cense A. W., Peeters E. A. G., Gottenbos B., Baaijens F. P. T., Nuijs A. M. and van Dongen M. E. H. (2006). Mechanical properties and failure of streptococcus mutans biofilms, studied using a microindentation device. *Journal of Microbiological Methods*, **67**, 463–472.

Cense A. W., Van Dongen M. E. H., Gottenbos B., Nuijs A. M. and Shulepov S. Y. (2006). Removal of biofilms by impinging water droplets. *Journal of Applied Physics*, **100**, 124701–124708.

Chambless J. D. and Stewart P. S. (2007). A three-dimensional computer model analysis of three hypothetical biofilm detachment mechanisms. *Biotechnology and Bioengineering*, **97**, 1573–1584.

Characklis W. G. (1980). Biofilm development and destruction. Final Report, EPRI CS-1554, Project RP902-1. Electric Power Research Institute, Palo Alto, CA.

Chen M. J., Zhang Z. and Bott T. R. (2005). Effects of operating conditions on the adhesive strength of pseudomonas fluorescens biofilms in tubes. *Colloids and Surfaces B: Biointerfaces*, **43**, 61–71.

Chen M. J., Zhang Z. and Bott T. R. (1998). Direct measurement of the adhesive strength of biofilms in pipes by micromanipulation. *Biotechnology Techniques*, **12**, 875–880.

Cheong F. C., Duarte S., Lee S.-H. and Grier D. G. (2009). Holographic microrheology of polysaccharides from streptococcus mutans biofilms. *Rheologica Acta*, **48**, 109–115.

Chew J. Y. M., Paterson W. R. and Wilson D. I. (2004). Fluid dynamic gauging for measuring the strength of soft deposits. *Journal of Food Engineering*, **65**, 175–187.

Chew S. C., Kundukad B., Seviour T., van der Maarel J. R. C., Yang L., Rice S. A., Doyle P. and Kjelleberg S. (2014). Dynamic remodeling of microbial biofilms by functionally distinct exopolysaccharides. *mBio*, **5**, p. e01536–14.

Chhabra R. P. (2010). Non-newtonian fluids: an introduction. In: Rheology of Complex Fluids. Springer, New York.

Cogan N. G. (2008). Two-fluid model of biofilm disinfection. *Bulletin of Mathematical Biology*, **70**, 800–819.

Cogan N. G. and Robert D. G. (2010). Multiphase flow models of biogels from crawling cells to bacterial biofilms. *HFSP Journal*, **4**, 11–25.

Cogan N. G. and Keener J. P. (2004). The role of the biofilm matrix in structural development. *Mathematical Medicine and Biology*, **21**, 147–166.

Costerton J. W., Stewart P. S. and Greenberg E. P. (1999). Bacterial biofilms: a common cause of persistent infections. *Science*, **284**, 1318–1322.

Cotterell B. and Williams G. (2010). Fracture and life. London: Imperial College Press.

Cuadrado D. G., Perillo G. M. E. and Vitale A. J. (2014). Modern microbial mats in siliciclastic tidal flats: Evolution, structure and the role of hydrodynamics. *Marine Geology*, **352**, 367–380.

de Beer D., Stoodley P., Roe F. and Lewandowski Z. (1994). Effects of biofilm structures on oxygen distribution and mass transport. *Biotechnology and Bioengineering*, **43**, 1131–1138.

De Leenheer P. and Cogan N. G. (2009). Failure of antibiotic treatment in microbial populations. *Journal of Mathematical Biology*, **59**, 563–579.

Deshpande M. D. and Vaishnav R. N. (1983). Wall stress distribution due to jet impingement. *Journal of Engineering Mechanics*, **109**, 479–493.

Ehret A. E. and Böl M. (2013). Modelling mechanical characteristics of microbial biofilms by network theory. *Journal of the Royal Society Interface*, **10**, 20120676.

Fabbri S., Johnston D. A., Rmaile A., Gottenbos B., De Jager M., Aspiras M., Starke E. M., Ward M. and Stoodley P. (2015). Streptococcus mutans biofilm transient viscoelastic fluid behaviour during high-velocity microsprays. *Journal of the Mechanical Behavior of Biomedical Materials*, **59**, 197–206.

Fang H. H. P., Chan K.-Y. and Xu L.-C. (2000). Quantification of bacterial adhesion forces using atomic force microscopy (AFM). *Journal of Microbiological Methods*, **40**, 89–97.

Fink R., Oder M., Rangus D., Raspor P. and Bohinc K. (2015). Microbial adhesion capacity. Influence of shear and temperature stress. *International Journal of Environmental Health Research*, 1–14.

Flemming H.-C. and Wingender J. (2010). The biofilm matrix. *Nature Reviews Microbiology*, **8**, 623–633.

Flemming H.-C., Wingender J. and Szewzyk U. (2011). Biofilm Highlights. Springer-Verlag Berlin Heidelberg.

Fung Y.-C. (2013). Biomechanics: Mechanical Properties of Living Tissues. Springer Science & Business Media.

Fux C. A., Costerton J. W., Stewart P. S. and Stoodley P. (2005). Survival strategies of infectious biofilms. *Trends Microbiol*, **13**, 34–40.

Galy O., Latour-Lambert P., Zrelli K., Ghigo J.-M., Beloin C. and Henry N. (2012). Mapping of bacterial biofilm local mechanics by magnetic microparticle actuation. *Biophys J*, **103**, 1400–1408.

Goehring L. (2013). Pattern formation in the geosciences. *Philosophical Transactions of the Royal Society A: Mathematical, Physical and Engineering Sciences*, 371.

Gresham B. (2008). Viscosity: A fluid's resistance to flow. *Tribology & Lubrication Technology*, **64**, 55–57.

Guélon T., Mathias J.-D. and Stoodley P. (2011). Advances in biofilm mechanics. In: H.-C. Flemming, J. Wingender and U. Szewzyk (eds.), Biofilm Highlights. Springer Berlin Heidelberg.

Hagadorn J. W. and Bottjer D. J. (1997). Wrinkle structures: Microbially mediated sedimentary structures common in subtidal siliciclastic settings at the proterozoic-phanerozoic transition. *Geology*, **25**, 1047–1050.

Hall-Stoodley L., Costerton J. W. and Stoodley P. (2004). Bacterial biofilms: From the natural environment to infectious diseases. *Nat Rev Micro*, **2**, 95–108.

He Y., Peterson B. W., Ren Y., van der Mei H. C. and Busscher H. J. (2014). Antimicrobial penetration in a dual-species oral biofilm after noncontact brushing: an in vitro study. *Clinical Oral Investigations*, **18**, 1103–1109.

Hohne D. N., Younger J. G. and Solomon M. J. (2009). Flexible microfluidic device for mechanical property characterization of soft viscoelastic solids such as bacterial biofilms. *Langmuir*, **25**, 7743–7751.

Horn H. and Lackner S. (2014). Modeling of biofilm systems: a review. *Adv. Biochem. Eng. Biotechnol.*, **146**, 53–76.

Horn H., Reiff H. and Morgenroth E. (2003). Simulation of growth and detachment in biofilm systems under defined hydrodynamic conditions. *Biotechnology and Bioengineering*, **81**, 607–617.

Hosford W. F. (2010). Mechanical Behavior of Materials. Cambridge University Press.

Hwang G., Klein M. I. and Koo H. (2014). Analysis of the mechanical stability and surface detachment of mature streptococcus mutans biofilms by applying a range of external shear forces. *Biofouling*, **30**, 1079–1091.

Inglis T. J. J. (1993). Evidence for dynamic phenomena in residual tracheal tube biofilm. *British Journal of Anaesthesia*, **70**, 22–24.

Irgens F. (2008). Continuum Mechanics. Springer Science & Business Media.

Janmey P. A., Hvidt S., Lamb J. and Stossel T. P. (1990). Resemblance of actin-binding protein/actin gels to covalently crosslinked networks. *Nature*, **345**(6270), 89–92.

Jones W. L., Sutton M. P., McKittrick L. and Stewart P. S. (2011). Chemical and antimicrobial treatments change the viscoelastic properties of bacterial biofilms. *Biofouling*, **27**, 207–215.

Klapper I., Rupp C. J., Cargo R., Purvedorj B. and Stoodley P. (2002). Viscoelastic fluid description of bacterial biofilm material properties. *Biotechnology and Bioengineering*, **80**, 289–296.

Klapper I. and Dockery J. (2010). Mathematical description of microbial biofilms. *SIAM review*, **52**, 221–265.

Konovalova A. and Sogaard-Andersen L. (2011). Close encounters: Contact-dependent interactions in bacteria. *Mol. Microbiol.*, **81**, 297–301.

Korstgens V. (2001). Influence of calcium ions on the mechanical properties of a model biofilm of mucoid Pseudomonas aeruginosa. *Water Sci. Technol.*, **43**, 49–57.

Koza A., Hallett P. D., Moon C. D. and Spiers A. J. (2009). Characterization of a novel air–liquid interface biofilm of pseudomonas fluorescens SBW25. *Microbiology*, **155**, 1397–1406.

Kreth J., Hagerman E., Tam K., Merritt J., Wong D. T. W., Wu B. M., Myung N. V., Shi W. and Qi F. (2004). Quantitative analyses of streptococcus mutans biofilms with quartz crystal microbalance, microjet impingement and confocal microscopy. *Biofilms*, **1**, 277–284.

Körstgens V., Flemming H. C., Wingender J. and Borchard W. (2001). Uniaxial compression measurement device for investigation of the mechanical stability of biofilms. *Journal of Microbiological Methods*, **46**, 9–17.

Lahaye E., Aubry T., Kervarec N., Douzenel P. and Sire O. (2007). Does water activity rule P. mirabilis periodic swarming? I. biochemical and functional properties of the extracellular matrix. *Biomacromolecules*, **8**, 1218–1227.

Lakes R. S. (1998). Viscoelastic Solids. CRC press.

Laspidou C. S., Spyrou L. A., Aravas N. and Rittmann B. E. (2014). Material modeling of biofilm mechanical properties. *Mathematical Biosciences*, **251**, 11–15.

Lau P. C., Dutcher J. R., Beveridge T. J. and Lam J. S. (2009a). Absolute quantitation of bacterial biofilm adhesion and viscoelasticity by microbead force spectroscopy. *Biophys J.*, **96**, 2935–2948.

Lau P. C. Y., Dutcher J. R., Beveridge T. J. and Lam J. S. (2009b). Absolute quantitation of bacterial biofilm adhesion and viscoelasticity by microbead force spectroscopy. *Biophys J.*, **96**, 2935–2948.

Lieleg O., Caldara M., Baumgärtel R. and Ribbeck K. (2011). Mechanical robustness of pseudomonas aeruginosa biofilms. *Soft matter*, **7**, 3307–3314.

Liu I. S. (2013). Continuum Mechanics. Springer Science & Business Media.

Lu J. (2009). Defect of cell wall construction may shield oral bacteria's survival in bloodstream and cause infective endocarditis. *Med Hypotheses*, **73**, 1055–1057.

Meyers, Marc A., and Krishan Kumar Chawla. 2009. *Mechanical behavior of materials* (Cambridge university press, Cambridge).

Mezger T. G. (2006). The Rheology Handbook: For Users of Rotational and Oscillatory Rheometers. Vincentz Network GmbH & Co KG.

Möhle R. B., Langemann T., Haesner M., Augustin W., Scholl S., Neu T. R., Hempel D. C. and Horn H. (2007). Structure and shear strength of microbial biofilms as determined with confocal laser scanning microscopy and fluid dynamic gauging using a novel rotating disc biofilm reactor. *Biotechnology and Bioengineering*, **98**, 747–755.

Noffke N. (2011). Microbially induced sedimentary structures. *Encyclopedia of Astrobiology*, 1045–1048.

Noffke N., Decho A. W., and Stoodley P. (2013). Slime through time: The fossil record of prokaryote evolution. *PALAIOS*, **28**, 1–5.

Noffke N., Gerdes G., Klenke T. and Krumbein W. E. (2001). Microbially induced sedimentary structures: A new category within the classification of primary sedimentary structures. *Journal of Sedimentary Research*, **71**, 649–656.

Oh Y. J., Jo W., Yang Y. and Park S. (2007). Influence of culture conditions on escherichia coli O157: H7 biofilm formation by atomic force microscopy. *Ultramicroscopy*, **107**, 869–874.

Oh Y. J., Lee N. R., Jo W., Jung W. K. and Lim J. S. (2009). Effects of substrates on biofilm formation observed by atomic force microscopy. *Ultramicroscopy*, **109**, 874–880.

Ohashi A. and Harada H. (1994). Adhesion strength of biofilm developed in an attached-growth reactor. *Water Science and Technology*, **29**, 281–288.

Ohashi A. and Harada H. (1996). A novel concept for evaluation of biofilm adhesion strength by applying tensile force and shear force. *Water Science and Technology*, **34**, 201–211.

Paramonova E., Kalmykowa O. J., van der Mei H. C., Busscher H. J. and Sharma P. K. (2009). Impact of hydrodynamics on oral biofilm strength. *J. Dent. Res.*, **88**, 922–926.

Paramonova E., Krom B. P., van der Mei H. C., Busscher H. J. and Sharma P. K. (2009). Hyphal content determines the compression strength of candida albicans biofilms. *Microbiology*, **155**, 1997–2003.

Parini M. R., Eggett D. L. and Pitt W. G. (2005). Removal of streptococcus mutans biofilm by bubbles. *Journal of Clinical Periodontology*, **32**, 1151–1156.

Parini M. R. and Pitt W. G. (2006). Dynamic removal of oral biofilms by bubbles. *Colloids and Surfaces B: Biointerfaces*, **52**, 39–46.

Patsios S. I., Goudoulas T. B., Kastrinakis E. G., Nychas S. G. and Karabelas A. J. (2015). A novel method for rheological characterization of biofouling layers developing in membrane bioreactors (MBR). *Journal of Membrane Science*, **482**, 13–24.

Paul E., Ochoa J. C., Pechaud Y., Liu Y. and Liné A. (2012). Effect of shear stress and growth conditions on detachment and physical properties of biofilms. *Water Research*, **46**, 5499–5508.

Pavlovsky L., Sturtevant R. A., Younger J. G., and Solomon M. J. (2015). Effects of temperature on the morphological, polymeric, and mechanical properties of staphylococcus epidermidis bacterial biofilms. *Langmuir*, **31**, 2036–2042.

Pavlovsky L., Younger J. G. and Solomon M. J. (2013). In situ rheology of staphylococcus epidermidis bacterial biofilms. *Soft Matter*, **9**, 122–131.

Peterson B. W., He Y., Ren Y., Zerdoum A., Libera M. R., Sharma P. K., van Winkelhoff A.-J., Neut D., Stoodley P. and van der Mei H. C. (2015). Viscoelasticity of biofilms and their recalcitrance to mechanical and chemical challenges. *FEMS Microbiology Reviews*, **39**, 234–245.

Peterson B. W., Busscher H. J., Sharma P. K. and Van Der Mei H. C. (2014). Visualization of microbiological processes underlying stress relaxation in pseudomonas aeruginosa biofilms. *Microscopy and Microanalysis*, **20**, 912–915.

Peterson B. W., van der Mei H. C., Sjollema J., Busscher H. J. and Sharma P. K. (2013). A distinguishable role of eDNA in the viscoelastic relaxation of biofilms. *MBio*, **4**, e00497–13.

Peyton B. M. and Characklis W. G. (1992). Kinetics of biofilm detachment. *Water Science & Technology*, **26**, 1995–1998.

Picioreanu C., Van Loosdrecht M. C. M. and Heijnen J. J. (2001). Two-dimensional model of biofilm detachment caused by internal stress from liquid flow. *Biotechnology & Bioengineering*, **72**, 205–218.

Ponnamma D. and Thomas S. (2014). Non-linear Viscoelasticity of Rubber Composites and Nanocomposites: Influence of Filler Geometry and Size in Different Length Scales. Springer International Publishing Switzerland.

Poppele E. H. and Hozalski R. M. (2003). Micro-cantilever method for measuring the tensile strength of biofilms and microbial flocs. *Journal of Microbiological Methods*, **55**, 607–615.

Reddy J. N. (2013). An Introduction to Continuum Mechanics. Cambridge University Press.

Rmaile A., Carugo D., Capretto L., Aspiras M., De Jager M., Ward M. and Stoodley P. (2014). Removal of interproximal dental biofilms by high-velocity water microdrops. *J Dent Res*, **93**, 68–73.

Rmaile A., Carugo D., Capretto L., Zhang X., Wharton J. A., Thurner P. J., Aspiras M., Ward M. and Stoodley P. (2012). Microbial tribology and disruption of dental plaque bacterial biofilms. *Wear.*

Rmaile A., Carugo D., Capretto L., Wharton J. A., Thurner P. J., Aspiras M., Ward M., De Jager M. and Stoodley. (2015). An experimental and computational study of the hydrodynamics of high-velocity water microdrops for interproximal tooth cleaning. *Journal of the Mechanical Behavior of Biomedical Materials.*

Rogers S. S., Van Der Walle C. and Waigh T. A. (2008). Microrheology of bacterial biofilms in vitro: Staphylococcus aureus and pseudomonas aeruginosa. *Langmuir*, **24**, 13549–13555.

Rupp C. J., Fux C. A. and Stoodley P. (2005). Viscoelasticity of staphylococcus aureus biofilms in response to fluid shear allows resistance to detachment and facilitates rolling migration. *Applied and Environmental Microbiology*, **71**, 2175–2178.

Rusconi R., Lecuyer S., Guglielmini L. and Stone H. A. (2010). Laminar flow around corners triggers the formation of biofilm streamers. *Journal of the Royal Society Interface*, **7**, 1293–1299.

Salley B., Gordon P. W., McCormick A. J., Fisher A. C., and Wilson D. I. (2012). Characterising the structure of photosynthetic biofilms using fluid dynamic gauging. *Biofouling*, **28**, 159–173.

Salta M., Wharton J. A., Blache Y., Stokes K. R. and Briand J.-F. (2013). Marine biofilms on artificial surfaces: structure and dynamics. *Environmental Microbiology*, **15**, 2879–2893.

Schramm G. (1994). A Practical Approach to Rheology and Rheometry. Haake Karlsruhe.

Schultz M. P. and Swain G. W. (2000). The influence of biofilms on skin friction drag. *Biofouling*, **15**, 129–139.

Stewart P. S. (1993). A model of biofilm detachment. *Biotechnology and Bioengineering*, **41**, 111–117.

Stoodley P., Cargo R., Rupp C. J., Wilson S. and Klapper I. (2002). Biofilm material properties as related to shear-induced deformation and detachment phenomena. *J Ind Microbiol Biotechnol*, **29**, 361–367.

Stoodley P., Lewandowski Z., Boyle J. D. and Lappin-Scott H. M. (1999). Structural deformation of bacterial biofilms caused by short-term fluctuations in fluid shear: An in situ investigation of biofilm rheology. *Biotechnol Bioeng*, **65**, 83–92.

Stoodley P., Wilson S., Hall-Stoodley L., Boyle J. D., Lappin-Scott H. M. and Costerton J. W. (2001). Growth and detachment of cell clusters from mature mixed-species biofilms. *Appl Environ Microbiol*, **67**, 5608–5613.

Stoodley P., Lewandowski Z., Boyle J. D. and Lappin-Scott H. M. (1998). Oscillation characteristics of biofilm streamers in turbulent flowing water as related to drag and pressure drop. *Biotechnology and Bioengineering*, **57**, 536–544.

Stoodley P., Lewandowski Z., Boyle J. D., and Lappin-Scott H. M. (1999). The formation of migratory ripples in a mixed species bacterial biofilm growing in turbulent flow. *Environmental Microbiology*, **1**, 447–55.

Thomas K., Herminghaus S., Porada H. and Goehring L. (2013). Formation of kinneyia via shear-induced instabilities in microbial mats. *Philosophical Transactions of the Royal Society A: Mathematical, Physical and Engineering Sciences*, **371**, 20120362.

Towler B. W., Rupp C. J., Cunningham A. B. and Stoodley P. (2003). Viscoelastic properties of a mixed culture biofilm from rheometer creep analysis. *Biofouling*, **19**, 279–285.

Towler B. W., Cunningham A., Stoodley P. and McKittrick L. (2007). A model of fluid–biofilm interaction using a Burger material law. *Biotechnology and Bioengineering*, **96**, 259–271.

Tuladhar T. R., Paterson W. R., Macleod N. and Wilson D. I. (2000). Development of a novel non-contact proximity gauge for thickness measurement of soft deposits and its application in fouling studies. *The Canadian Journal of Chemical Engineering*, **78**, 935–947.

Vaishnav R. N., Patel D. J., Atabek H. B., Deshpande M. D., Plowman F. and Vossoughi J. (1983). Determination of the local erosion stress of the canine endothelium using a jet impingement method. *Journal of Biomechanical Engineering*, **105**, 77–83.

Vinogradov A. M., Winston M., Rupp C. J. and Stoodley P. (2004). Rheology of biofilms formed from the dental plaque pathogen streptococcus mutans. *Biofilms*, **1**, 49–56.

Volle C. B., Ferguson M. A., Aidala K. E., Spain E. M. and Nunez M. E. (2008). Spring constants and adhesive properties of native bacterial biofilm cells measured by atomic force microscopy. *Colloids and Surfaces B: BioInterfaces*, **67**, 32–40.

Wang Q. and Zhang T. (2010). Review of mathematical models for biofilms. *Solid State Communications*, **150**, 1009–1022.

Waters M. S., Kundu S., Lin N. J. and Lin-Gibson S. (2013). Microstructure and mechanical properties of in situ streptococcus mutans biofilms. *ACS Applied Materials & Interfaces*.

Wloka M., Rehage H., Flemming H. C. and Wingender J. (2004). Rheological properties of viscoelastic biofilm extracellular polymeric substances and comparison to the behavior of calcium alginate gels. *Colloid and Polymer Science*, **282**, 1067–1076.

Zhang T., Cogan N. G. and Wang Q. (2008a). Phase field models for biofilms. I. Theory and one-dimensional simulations. *SIAM Journal on Applied Mathematics*, **69**, 641–669.

Zhang T., Cogan N. and Wang Q. (2008b). Phase field models for biofilms. II. 2-D numerical simulations of biofilm-flow interaction. *Commun. Comput. Phys.* **4**(1), 72–101.

Chapter 9

Travelling through slime – bacterial movements in the EPS matrix

J.-C. Piard[1], S.-Y. Kim[1], J. Deschamps[1], Y. Li[2], C. Dorel[3], A. Gruss[1], A. Trubuil[2] and R. Briandet[1]

[1]Micalis Institute, INRA, AgroParisTech, Université Paris-Saclay, 78350 Jouy-en-Josas, France

[2]INRA, UR1404, MalAGE, Jouy-en-Josas, France

[3]INSA-Lyon, LGCIE, Villeurbanne F-69621, France

ABSTRACT

Recently, it was shown that sessile biofilm cells could share the EPS matrix with motile planktonic populations. The presence of motile cells creates a transient vascularisation network in the matrix so that cells in the depth of the biofilm can be exposed to the exterior conditions. This can have profound implications in term of natural biofilm traits and biofilm population dynamics. EPS-swimmers can also be exploited in "green" anti-biofilm strategies *e.g.*, to increase biocide action and to deliver antibiofilm agents *in situ*. This chapter reviews the available description of EPS-swimmers (including experimental considerations), the proof of concept of their efficacy in biofilms control strategies and related future lines of investigations.

Keywords: biofilm control, EPS vascularisation, biocide action, real-time confocal imaging

9.1 INTRODUCTION: BACTERIAL MOVEMENTS INVOLVED IN THE BIOFILM LIFE-CYCLE

Our representation of natural biofilms essentially comes from microscopic images that do not allow visualisation of the dynamic dimensions of these microbial landscapes. Biofilms are vibrant ecosystems composed of various organisms that are continuously in motion. Bacteria can move in a wide range of spatial and temporal scales, ranging from sub-microns to millimetres and from microseconds to hours (Molaei *et al.* 2014). Some planktonic cells swim in quest of better environments, such as a new interface to conquer. Once associated to a surface, bacteria move to spread and compete for space and nutrients with other biofilm inhabitants. Associated in the biofilm EPS, they move passively during biofilm expansion and under shear stress. Bacteria have developed a wealth of strategies to keep moving in these different environments; they can swim, glide, swarm, oscillate, vibe (Kearns, 2010). Some of these movements involve specific extracellular molecular effectors including flagella, pili, polysaccharides, surfactant or enzymes. Of particular interest for this review are flagella which are rigid extracellular helical filaments that propel motile cells by rotation in a liquid volume (swimming) or across a solid surface (swarming). In liquid, motile cell velocity can reach up to 160 μm/s. Individual cell trajectory is not continuous, and involves a succession of cell runs and tumbles: a cell swims in one direction for a short period (≈1 s), moves erratically for a small fraction of second, and then swims steadily again in a different direction (Berg, 1973). The direction is not random, it results from the integration of signals generated by extracellular sensors that reckon the amount of molecules of interest in the surrounding microenvironment and their gradients (chemotaxis); cells go where grass is greener (Sourjik *et al.* 2004).

Flagella appear at distinct location and in numbers that are characteristic of each bacterial species and of its corresponding environment: polar flagella are required for fast swimming in liquid (*Pseudomonas aeruginosa, Helicobacter pylori, Campylobacter jejuni*), and peritrichous flagellation (Figure 9.1) for moving through more viscous environments or over surfaces (*Escherichia coli, Bacillus subtilis)* (Martinez *et al.* 2014; Schuhmacher *et al.* 2015). The lyme disease *Borrelia burgdorferi* cells are even able to swim in viscous medium with internal flagella (Vig & Wolgemuth, 2012)! Their intracellular rotations trigger undulations of the cell body in the form of a traveling flat wave (Charon *et al.* 2012). Swarming on surface generally requires both movements of multicellular groups and production of surfactant to reduce the friction with the surface (Kearns, 2010). The number of flagella per swarming cell is generally higher than per swimming cell and can reach several hundreds (Shinoda *et al.* 1977). Consistent with this, in a recent evolution experimental approach in *P. aeruginosa* swarming cells, van Ditmarsch *et al.* demonstrated the emergence of 2 or more polar flagella in hyperswarming subpopulations, in contrast to single polar flagella in WT strain (van Ditmarsch *et al.* 2013). Flagella synthesis and functioning is energetically costly for the cells (≈2% of a cell metabolic resource are required for flagella synthesis) and hence it is a highly regulated process in response to environmental signals (including nutrients, viscosity, temperature). During biofilm formation, it is assumed that there is a motility-to-biofilm transition for swimming and swarming cells. In the short term, flagella can be functionally regulated (e.g., inhibition of flagella rotation) and over the long term, flagella gene transcription is inhibited triggering flagella dilution through biofilm cell growth (Guttenplan & Kearns, 2013).

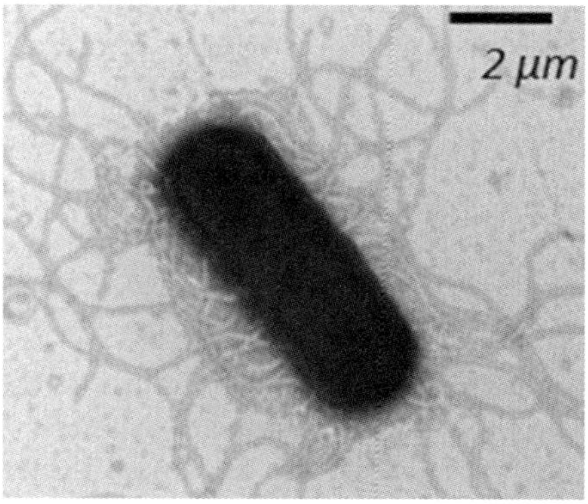

Figure 9.1 Peritrichous flagella of *Bacillus licheniformis* observed using transmission electronic microscopy after negative staining of bacteria (courtesy of A. Canette, Inra Mima2).

Even without flagella, some bacteria are able to migrate on surfaces. Type 4 pili are long proteinaceous filaments with lengths that can exceed 10 μm and that often extend from the cell poles. They are involved in twitching motility with a velocity that can reach 1 μm/s by a sequence of pilus extension, attachment to a surface and retraction (Jarrell & McBride, 2008). Even in absence of flagella or pili, some bacteria are able to glide actively on surfaces. Velocity of gliding cells on surface is lower, typically 3 μm/min. In *Myxococcus xanthus*, the mechanisms behind those movements are still controversial and 3 motion models are still proposed: slime secretion, focal adhesion and helical rotor (Nan & Zusman, 2011).

Once trapped in the biofilm EPS matrix, bacteria are still prone to movements. Every single cell in the biofilm matrix is exposed to thermal oscillations (Rogers *et al.* 2008). The volume of Brownian motion for sessile biofilm cells is generally limited to the volume of a cell size and depends on the local matrix visco-elastic properties (Houry *et al.* 2012). Each time a cell divides, it generates a sliding force on the neighbouring cells that are pushed to the edge of the biofilm. Applying confocal single cell observation to *P. aeruginosa* PAO1 for several hours, Rice *et al.* estimated that cells exhibited a typical velocity of 1 μm/h (24 μm/day) in the biofilm (Rice *et al.* 2003). A subpopulation of emigrating cells (cells that exit the field of view during the observation period) reached 6 μm/h in the same experiment, suggesting the existence of different populations of cells with various migration velocities. Cells trapped in the EPS can be subjected to a collective motion when the biofilm is exposed to hydrodynamic variations and viscoelastic deformations (Galy *et al.* 2012). When the flow is sustained, filamentous streamers can form, tethered at one end to the surface while the rest is floating and oscillating in the flow (Karimi *et al.* 2015). In different species, swimming cells trapped in hollow voids in the center of mature biofilms have been observed (Webb *et al.* 2003; Marchal *et al.* 2011). This is part of active dispersal processes starting by localized cell death in the biofilm, typically mediated by the expression of prophages and/or of autotoxic-proteins (Rice *et al.* 2009; Mai-Prochnow *et al.* 2004; Mai-Prochnow *et al.* 2006). Cells lysis provides resources and nutrients for a resistant subpopulation that will disperse after evacuation from breaking up point (McDougald *et al.* 2011).

9.2 EXISTENCE OF FLAGELLA PROPELLED MOTILE BACTERIAL SUBPOPULATIONS IN THE EPS MATRIX

As mentioned previously, bacterial motility has been identified as an important factor for the planktonic to biofilm transition (and reverse), and biofilm cells are described as non-motile in the EPS matrix (Guttenplan *et al.* 2013). But are the highly hydrated gelatinous biofilm EPS always acting as an impermeable shield to the invasion of flagella propelled free cells? In other words, can sessile-biofilm cells coexist with motile-planktonic populations in the biofilm matrix? Taking advantage of real-time confocal imaging at the single cell scale, recent papers described the unexpected existence of motile bacterial cells in the EPS slime. These populations have been observed in single strain biofilms both with Gram-positive (*Bacillus thuringiensis* Bt 407) and Gram-negative bacteria (*Yersinia enterolitica* CIP 106676, *Campylobacter jejuni* 81–176) (Houry *et al.* 2012; Turonova *et al.* 2015). In those axenic biofilms, a minor subpopulation of motile planktonic cells (typically <1%) shares the matrix with a vast majority of sessile biofilm cells. Planktonic *Bt* 407 cells genetically tagged with a fluorescent reporter and deposited on the top of an isogenic biofilm were able to infiltrate the matrix within minutes, showing that the motile subpopulation is not derived from a hyper-swimming mutant population. This biophysical process is dependent on the balance between the velocity of the swimmers (flagella length, number and speed of rotation) and the visco-elastic properties of the host matrix. It is hence difficult to predict the ability of motile bacteria to swim in a specific biofilm EPS only from their planktonic swimming performance. In the case of *Staphylococcus aureus* as a host biofilm, we observed that bacilli that were unable to form a 30 mm disk in a 0.25% motility agar assay

(24h of incubation at 37°C) were generally not able to swim in the biofilm matrix. In its own isogenic biofilm, *Bt* 407 cell motility is dependant of the matrix density and cohesion that evolve with biofilm age: the velocity of the class of highest frequency decreases from 8 μm/s in 24h-biofilm to 2 μm/s in 72h-biofilm, along with a 3 fold increase in exopolysaccharide density (Houry *et al.* 2012). The invasion of the matrix by planktonic cells is limited in *Vibrio cholera* biofilms by a secreted protein (RbmA) which binds mother-daughter cell pairs to each other and to polysaccharide components of the matrix, creating a robust protecting shield (Nadell *et al.* 2015).

Motile bacteria were also observed in heterologous systems with different combinations of bacilli swimmers and host biofilm matrices: *Bt* 407 was able to swim in the EPS biofilm matrix of *S. aureus* (Figure 9.2), *Enterococcus faecalis, Y. enterolitica, P. aeruginosa, Listeria monocytogenes* (Houry *et al.* 2012). Isolates of eight *Bacillus* species (*B. thuringiensis, B. licheniformis, B. megaterium, B. polymyxa, B. pumilus, B. spaericus B. cereus and B. subtilis*) were able to infiltrate and swim in *S. aureus* matrix with a speed up to 30 μm/s (Figure 9.3). Swimmers trajectories in this matrix varies according to the bacilli strain as estimated by the large variation of the mean persistency values that ranged from 0.2 to 0.6 (Li *et al.* 2015).

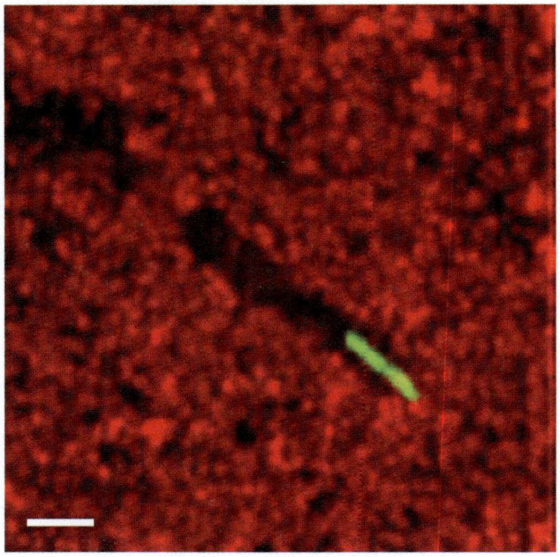

Figure 9.2 A green GFP expressing bacilli cell added to a *Staphylococcus aureus* biofilm (red cells) swim and creates transient pores throughout the matrix (scale bar = 5 μm).

The swimmers movements create transient pores in the EPS matrix. *Bt* 407 and some other chain forming bacilli were able to displace locally sessile cells of *S. aureus* and form macroscopic hollow voids of up to 10 μm diameter within a few minutes. These movements weaken the protective EPS shield and open the door to biofilm invasion by organisms larger than bacteria such as amoeba in the environment or leucocytes in infected host tissues. Smaller pores of around the size of a bacterium were also detected for ~1 second after the swimmer pass. With the big cells of *B. sphaericus*, swimmers preferentially reuse visited space in contrast to small cells of *B. licheniformis* that visit new space within biofilms (Li *et al.* 2015). Those local tunnelling events create a vascularization network able to increase the molecular transfer at the population level, exposing biofilm cells to ambient concentrations of substrates. This molecular mixing within the biofilm likely limits spatially driven heterogeneities in the biostructure. This may have a profound ecological impact on biofilm traits in natural environments. As described in the next paragraph, these properties can also be exploited to increase the delivery and efficacy of biocides in biofilm control strategies.

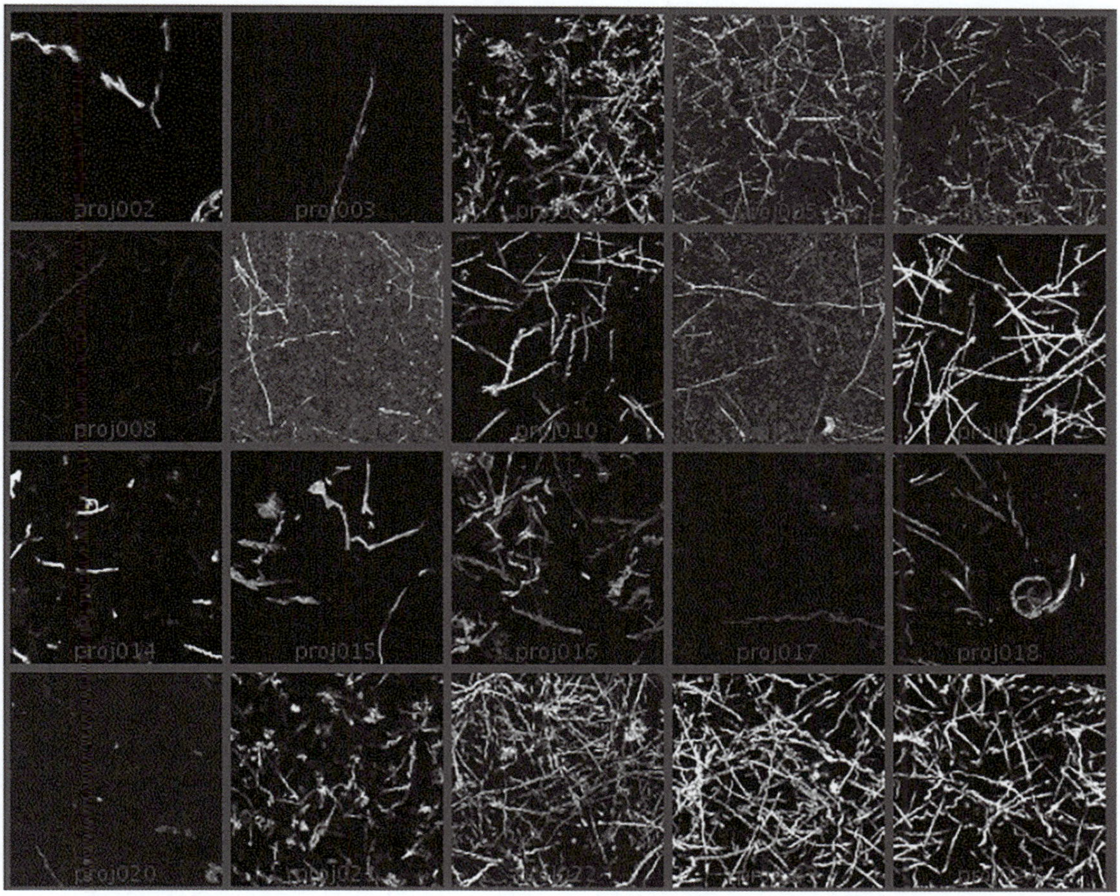

Figure 9.3 Thirty second time projections of 20 different bacilli swimmers in *S. aureus* biofilm matrix. Light regions are associated with the visited spaces of the swimming cells. The diversity of behaviors within the *Bacillus* species is illustrated. Each image size is 184 x 184 μm.

BOX 1: HOW TO DETECT AND EXPERIMENTALLY TRACK BACTERIAL SWIMMERS IN THE SLIME?

Analysis of motile bacteria in the biofilm EPS necessitates a high-speed visualization technique with single cell scale resolution. Real-time laser based derived microscopic approaches are today the best options for tracking swimmers in biofilms. Commercial compatible technologies include conventional confocal laser scanning microscopy (CLSM), spinning disk and light sheet fluorescence microscopy.

(*Continued*)

BOX 1: HOW TO DETECT AND EXPERIMENTALLY TRACK BACTERIAL SWIMMERS IN THE SLIME? (*Continued*)

The full pipeline of analysis integrates 4 steps:

(1) *Growth of the host-biofilm.* The biofilm in which the swimmers will be added need to be grown in a device compatible with *in situ* microscopic images e.g., flowcells (under a constant growth medium flow) or microscopic grade microtiterplate devices (static conditions).
(2) *Fluorescent labelling.* To visualise the swimmers in an exogenic biofilm, the ideal situation is a genetic labelling with a fluorescent protein reporter (e.g., green GFP, red mCherry) that is highly specific and exhibits low interference with cell movements. Alternatively, planktonic swimmers can be labelled with low dose of chemical dye (e.g., green syto 9 or red syto 61) just before their introduction in the biofilm. Note that the dye can rapidly interfere with the swimming performance of the cells and diffuse non-specifically in the biofilm, lowering the labelling specificity. Robust validation protocols should be applied with such chemical strategy. Counter labelling of the sessile biofilm cells can be performed similarly with genetic or chemical dye with a fluorescence signal distinguishable from the swimmers (red if the swimmers are green and vice versa). In addition specific dyes can be applied to contrast eDNA (non permeant nucleic acid dye) or polysaccharides (fluorescent lectins) in the matrix.
(3) *Real-time laser based microscopic acquisitions.* A suspension with the fluorescent swimmers is introduced in the supernatant of the host biofilm at a minimum concentration of 10^6 CFU/ml. Time-lapse microscopic acquisitions can be launched 1 min after addition of the swimmers. For 2D acquisition, a plane a few micrometers above the solid substrate allow to select swimming populations parallel to the solid surface and to increase their persistency in the focal plane. Typical CLSM parameters of acquisition for 2D movies are images of 512×512 pixels at 800 Hz for 3 minutes. 3D tracking necessitates a high-speed microscope (8000 Hz and more).
(4) *Estimating the swimmers trajectories by image analysis.*

From a dual colours denoised image sequence (e.g., host-biofilm cells labelled in green with GFP, bacilli swimmers labelled in red with syto 61), it is possible to extract single cells trajectories and derived quantitative descriptors such as swimmers speed and persistence with available routines. The difficulty here is the density of moving elements and their high speed relatively to the microscope acquisition frame rate. This generates important confusion in bacterial trajectories. To overcome this limitation, a bar coding procedure using a normalized intensity histogram measure for each swimmer was used to feed a linking cost function. In addition, better tracking performances were obtained when correlating swimmers trajectories with the transient pores they form in the host-biofilm (used as "track of the past") (Li *et al.* 2015).

9.3 EXPLOITATION OF BACTERIAL MOTILITY IN BIOFILM CONTROL

The high level of tolerance of biofilms to biocides requires the use of high amounts of biocides in cleaning and disinfection. Although inefficient in complete biofilm control this leads to the pollution of the environment, represents a high economic cost (US$ 11 billion of worldwide formulated biocides as estimated for 2016 by *Biocide Information Limited*), and is likely to lead to the emergence and spreading of antibiotic/biocide resistant organisms in the environment. The major mechanism involved in biofilm resistance to biocides is due to the so called diffusion/reaction limitation mediated by the matrix resulting in the death of bacteria located at the surface of biofilms and the survival of bacteria located in the inner part of biofilms (Bridier *et al.* 2011). The presence of a dense bacterial community associated to a compact structure of EPS-containing matrix hampers the diffusion of biocides within the internal layers of the biofilm (Flemming, 2011). Also, biocides are often highly reactive chemicals likely to interact with the EPS components of the matrix thus hampering their activity towards bacterial cells (Davison *et al.* 2010). Bacterial swimmers have great potential in attempts to circumvent diffusion/reaction phenomena through the creation of channels within the matrix allowing a better irrigation of the deepest layers of biofilms by biocides. Likewise swimmers are potent candidates for the *in situ* delivery of antimicrobials in the depth of biofilms and/or of enzymes able to degrade components of the matrix thus allowing higher diffusion of biocides and decreased interference of biocides with some of these components. Recent research results down those lines are summarized below.

9.3.1 Enhanced biocide action

The ability of swimmers to create channels allowing solute access to the deeper layers of biofilms was first demonstrated using a high-molecular-weight fluorescent macromolecular marker as solute (FITC-dextran, 250 kDa) in biofilms of motile or not motile *Bacillus thuringiensis* strains. The entry of FITC-dextran in biofilms monitored through time-lapse confocal imaging showed a 1-log higher rate of incorporation in motile than in non-motile biofilms suggesting that swimmers create a certain level of fluidity within biofilms allowing molecular exchanges between the external medium and the biofilm itself (Houry *et al.* 2012). In the rationale to assess whether this phenomenon would improve biocide spreading and action within biofilms, *Staphylococcus aureus* biofilms were submitted to the swimming of 2 bacilli either separately or in combination, and to subsequent exposure to a biocide (benzalkonium chloride, BAC) (Figure 9.4) (Houry *et al.* 2012). While the biocide treatment in the absence of bacilli decreased by 8-fold only the *S. aureus* population, biofilms pretreated with 1 swimmer strain decreased by 100-fold and those treated with both swimmers by 300-fold. This demonstrates that swimmers can improve biocide efficacy in biofilms and opens doors to novel strategies aiming to reduce the amount of biocide used to control detrimental biofilms.

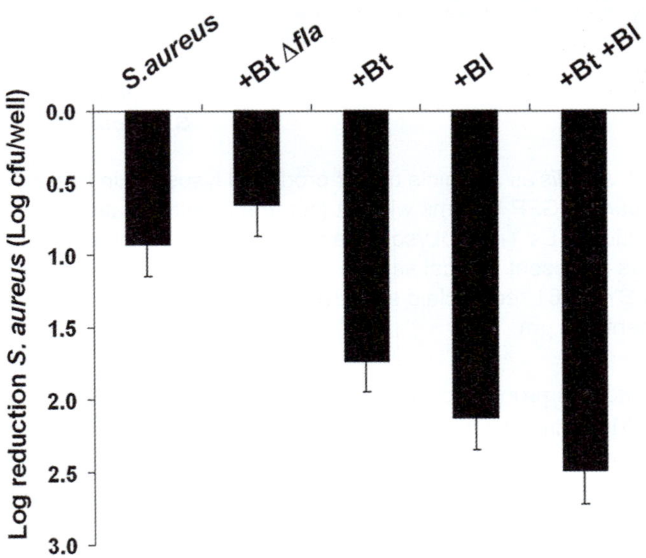

Figure 9.4 Enhancement of biocide action in *S. aureus* biofilms pretreated with swimmers. *S. aureus* biofilms were untreated (*S. aureus*) or treated with swimmers *B. thuringiensis* (Bt) or *B. licheniformis* (Bl) or a mixture of both (Bt + Bl) or with the non-motile Bt*Δfla* prior to BAC (750 ppm) treatment for 5 min and subsequent colony-forming unit determination. The indicated Log reduction in *S. aureus* population refers to the *S. aureus* biofilm that was not treated with BAC (adapted from Houry *et al.* 2012).

9.3.2 Delivery of antibacterials

9.3.2.1 Targeting cell viability within biofilms

A further exploitation of swimmers in biofilm control would be to replace exogeneous biocides by swimmer-produced bactericidal peptides/proteins that would be delivered by the swimmer throughout the biofilm. A proof of principle down this line has been shown using lysostaphin, a bacteriolysin that specifically hydrolyzes the interpeptide bridges of the peptidoglycan of *S. aureus* (Houry *et al.* 2012). Treatment of *S. aureus* biofilms with wild type *B. subtilis* did not impair the 3D structure of *S. aureus* biofilms while treatment with a *B. subtilis* derivative that produced lysostaphin eradicated, and even replaced, those biofilms within 24 h (Figure 9.5). This was linked to the swimming ability of *B. subtilis* since a swimming deficient derivative of this strain did not affect the biofilm integrity of *S. aureus*.

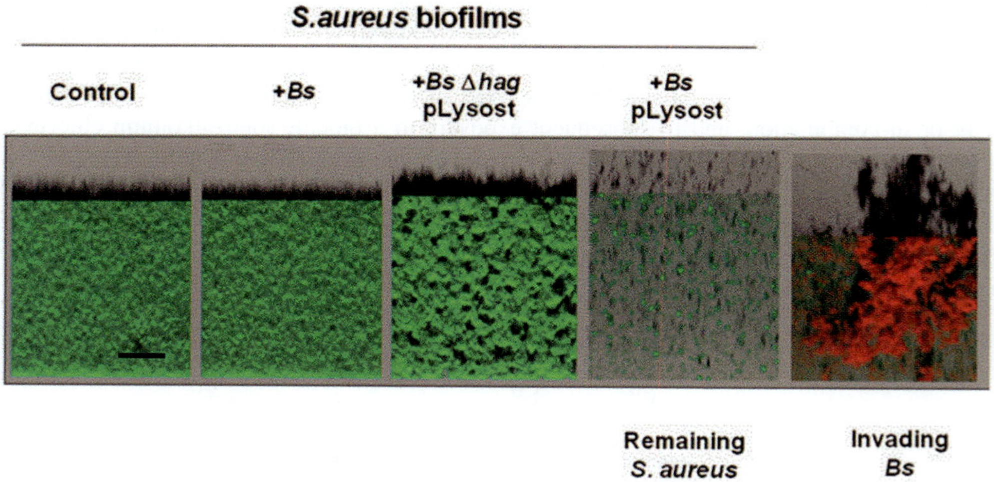

Figure 9.5 Swimming *B. subtilis* as a vehicle of self-produced lysostaphin to eradicate *S. aureus* biofilms. Confocal images of S. *aureus* GFP biofilms without (control) or with treatment by motile *B. subtilis* (+Bs), motile lysostaphin producing Bs (+Bs pLysost), and nonmotile lysostaphin producing Bs (+Bs Δ*hag* pLysost). Upper shadows represent vertical section projections of the biofilms. Scale bar 30 μm. Invading Bs were labelled using SYTO 61 red nucleic acid dye (adapted supporting information from Houry *et al.* 2012). Scale bar represents 30 μm.

In an attempt to extend this proof of concept to natural bacterial isolates, a collection of bacilli were screened for motility (Mot$^+$) and bacteriocin production (Bac$^+$). Bacteriocins form a broad range of proteinaceous peptides/proteins that typically exhibit bactericidal activity through pore formation in target cell membranes (Cotter *et al.* 2013). In contrast with antibiotics, bacteriocins are only active against bacteria that are closely related to the producing strain. This allows to target detrimental or spoilage bacteria using a specific bacteriocin thus avoiding drawbacks otherwise encountered with the use of antibiotics (e.g., emergence and dissemination of antibiotic resistance genes, and collateral killing of non-pathogen species such as commensals) (Riley & Wertz, 2002). The first screening test of bacilli for swimming motility performed using an agar medium swimming test yielded 125 swimming candidate strains whose swimming phenotype was characterized further by CSLM biofilm matrix swimming assay using *S. aureus* biofilms (Li *et al.* 2015). This provided a series of descriptors (swimming speed, persistency, trajectories, size of the created pores, etc.) for each swimmer strain. These swimming candidates were further screened for bacteriocin production using an *S. aureus* strain as indicator. About a third of the tested strains appeared to produce a bacteriocin of different activity level towards *S. aureus*. Selected Mot^+ Bac^+ strains were further shown to efficiently challenge an *S. aureus* biofilm (Figure 9.6) suggesting that relevant natural isolates may be used as a reservoir to pick up swimming and bacteriocin producing strains that have the properties required in dedicated applications.

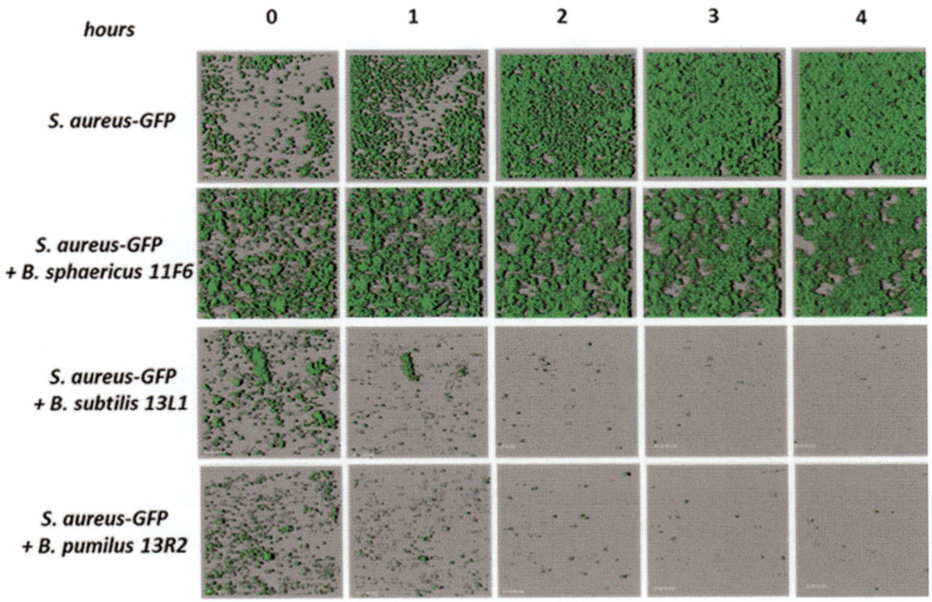

Figure 9.6 Eradication of a *S. aureus* biofilm using motile (Mot^+) bacteriocin producing (Bac^+) bacilli. Confocal images of *S. aureus* GFP biofilms without (*S. aureus*-GFP) or with treatment by Mot^+ Bac^- *B. sphaericus* 11F6 strain, Mot^+ Bac^+ *B. subtilis* 13L1 strain or Mot^+ Bac^+ *B. pumilus* 13R2 strain. Each image size is 184×184 µm.

Although promising, this strategy of exploiting Mot^+ and Bac^+ natural isolates may be unsuccessful if the bacteriocins produced by the swimmers do not exhibit the activity spectrum required in the targeted application. This bottleneck could be circumvented by derived strategies consisting in using one motile strain for pore formation and another non-motile bacteriocin producing strain or a bacteriocin *per se* as an additive. Trials down those lines have been conducted using lysostaphin, and although this is a non-bacteriocin antimicrobial, promising results were obtained using those strategies. As observed above with *Bacillus subtilis*, a *B. thuringiensis* Bt Δ*fla* pLysost up to 7 log/ml was unable to displace a preformed *S. aureus* biofilm (Figure 9.7) (Briandet *et al.* 2014). In contrast, when the *S. aureus* biofilm was pretreated with a Mot^+ Lys^- *B. thuringiensis* strain during 4 h before treatment with the Bt Δ*fla* pLysost strain, the pores created by the swimming strain potentialized the activity of lysostaphin released by the non-swimming strain and the bactericidal effect towards the *S. aureus* biofilm was positively correlated to the number of cells of the lysostaphin producer used (Figure 9.7) (Briandet *et al.* 2014). This suggests that bacterial strains producing potent antimicrobials and deficient in swimming ability can be exploited along with a swimmer strain assuring the spreading of antimicrobials throughout biofilms.

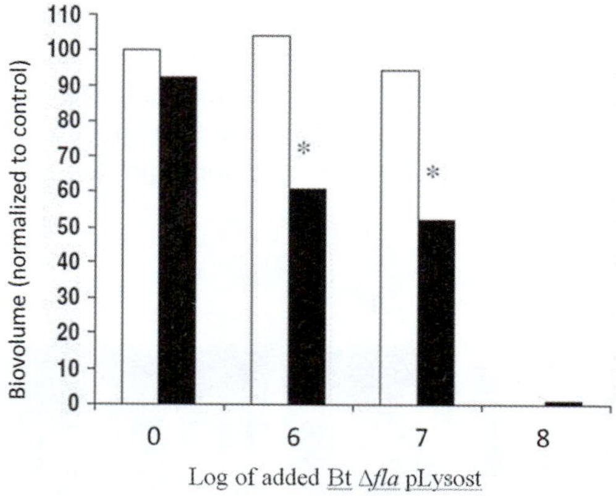

Figure 9.7 *S. aureus* biofilms were exposed to Bt Δ*fla* pLysost strain (a non-motile lysostaphin producing strain) after a 4 h pretreatment with non-motile (white bars) or motile (black bars) isogenic bacilli swimmers. Stars indicated a significant decrease in *S. aureus* biovolume associated with the EPS-swimmers pretreatment (adapted from Briandet *et al.* 2014).

To further widen the field of application of swimmers, attempts have been made to assess whether the anti-biofilm activity of an antibacterial compound could be enhanced using swimmers. Again, this has been performed using pure lysostaphin as a model antimicrobial compound. While increasing amounts of lysostaphin applied straight on *S. aureus* biofilms showed a limited biocidal activity on biofilm cells, pretreatment of those biofilms to swimming Bt prior to the lysostaphin treatment markedly increased the observed biocidal effect (Briandet *et al.* 2014).

9.3.2.2 Targeting matrix integrity within biofilms

Instead of targeting bacterial viability, swimmers have also been assessed for their potential in the delivery of an enzyme that targets the ECM. Dispersin, a glycoside hydrolase that specifically cleaves β-1,6-glycosidic linkages of acetylglucosamine polymers found in biofilm matrices (Itoh *et al.* 2005) was chosen and a swimming *B. subtilis* strain expressing a dispersin gene from *Aggregatibacter actinomycetemcomitans* was constructed (http://2012.igem.org/Team:Lyon-INSA). Treatment of *S. aureus* biofilms with this strain followed by washing allowed a 30-fold reduction of *S. aureus* biofilms while untreated biofilms were unaffected by the washing step (Figure 9.8).

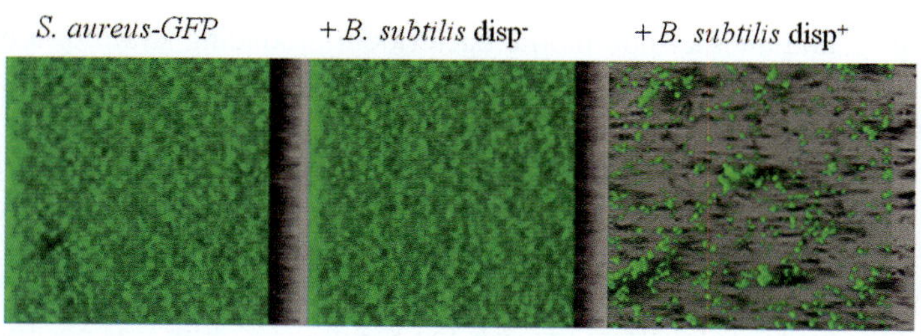

Figure 9.8 Treatment of *S. aureus*-GFP biofilms with *Bacillus subtilis* swimmers expressing dispersin (disp+) followed by washing allowed a 30-fold reduction of *S. aureus* biofilms biovolume, while untreated biofilms, or biofilm treated with the isogenic dispersin non-producing strain (disp-) were unaffected by the washing step.

9.4 FUTURE LINES OF INVESTIGATIONS

While the presented data were obtained using simplified mono- or bi-species biofilm models, natural biofilms are composed of much wider plurimicrobial communities likely yielding more complex and robust EPS matrix (Flemming, 2011). This raises the question of the ability of swimmers to infiltrate such biostructures. A first step to answer this question will be to assess the performance of the above-characterized swimmers in model multispecies biofilms as well in natural complex biofilms. This requires overcoming technical bottlenecks such as the transfer of native biofilms and the maintenance of their microbiological balance under laboratory conditions. Current work to this end is being performed and preliminary results indicate that swimming can occur in some plurimicrobial biofilms (unpublished data).

As illustrated through this chapter, swimmers have potential in the control of surface-associated pathogens and examples with swimmers facilitating the irrigation of the matrix with antibiofilm agents targeting either biofilm cells or the EPS matrix have been provided. In these fields, swimmers have also potential in the delivery of specific bacteriophages (for medical phage therapy), or matrix-destabilizing compounds (polysaccharide hydrolases, proteases, DNase) (Hogan *et al.* 2015). They could also be used to interfere with the social organisation within biofilms through i) the delivery of Quorum-sensing inhibitors (Aggarwal *et al.* 2015) and ii) the local dilution of Quorum-sensing mediators through the swimmers-created pores within the EPS matrix.

In the field of biotechnology, the use of biofilms for protein/peptide production has been proposed. Each image size is 184 × 184 μm (Absalon *et al.* 2012; Pongtharangkul & Demirci, 2006). Swimmers have potent benefit in these issues through i) superior nutrient acquisition and higher detoxification for the embedded biofilm populations contributing to the global bacterial fitness, and ii) increased release of the proteins from the biofilm network. Finally, applications making use of swimmers are also foreseen in probiotic issues as their ability to swim throughout mucosa-associated gels such as mucus would provide them with a competitive advantage to reach epithelial cells. This perspective is reinforced by the recent finding of flagellum-mediated motility in lactic acid bacteria (Cousin *et al.* 2015).

ACKNOWLEDGEMENTS

Financial support was provided by the French National Research Agency ANR-12-ALID-0006 programme ANR GreenSwimmers. Alexis Canette from the INRA MIMA2 Imaging Center is acknowledged for electron microscopy as well as all the bright IGEM team of the “Biofilm killers” project from INSA Lyon.

REFERENCES

Aggarwal C., Jimenez J. C., Lee H., Chlipala G. E., Ratia K. and Federle M. J. (2015). Identification of Quorum-sensing inhibitors disrupting signaling between Rgg and short hydrophobic peptides in streptococci. *MBio.*, **6**(3), e00393–e00415.

Berg H. C. (1973). Flagellar motion. In: *E. coli* in Motion, Series: Biological and Medical Physics, Biomedical Engineering, AIP Press, Springer, USA, pp. 39–47.

Briandet R., Gruss A., Aymerich S., Gohar M., Deschamps J. and Houry A. (2014). Method for removing a microorganism biofilm. Patent. 03/06/14 US 20140065122 A1.

Bridier A., Briandet R., Thomas V. and Dubois-Brissonnet F. (2011). Resistance of bacterial biofilms to disinfectants: a review. *Biofouling*, **27**(9), 1017–1032.

Charon N., Cockburn A., Li C., Liu J., Miller K., Miller M., Motaled M. and Wolgemuth C. (2012). The unique paradigm of spirochete motility and chemotaxis. *Annu. Rev. Microbiol.*, **66**, 349–370.

Cotter P. D., Ross R. P. and Hill C. (2013). Bacteriocins - a viable alternative to antibiotics? *Nat. Rev. Microbiol.*, **11**(2), 95–105.

Cousin F. J., Lynch S. M., Harris H. M., McCann A., Lynch D. B., Neville B. A., Irisawa T., Okada S., Endo A. and O'Toole P. W. (2015). Detection and genomic characterization of motility in *Lactobacillus curvatus*: confirmation of motility in a species outside the *Lactobacillus salivarius* clade. *Appl. Environ. Microbiol.*, **81**(4), 1297–1308.

Davison W. M., Pitts B. and Stewart P. S. (2010). Spatial and temporal patterns of biocide action against *Staphylococcus epidermidis* biofilms. *Antimicrob. Agents Chemother.*, **54**(7), 2920–2927.

Flemming H.-C. (2011). The perfect slime. *Colloids Surf B Biointerfaces*, **86**(2), 251–259.

Galy O., Latour-Lambert P., Zrelli K., Ghigo J.-M., Beloin C., Henry N. (2012). Mapping of bacterial biofilm local mechanics by magnetic microparticle actuation. *Biophys. J.* **103**(6), 1400–1408.

Guttenplan S. B. and Kearns D. B. (2013). Regulation of flagellar motility during biofilm formation. FEMS *Microbiol. Rev.*, **37**(6), 849–71.

Hogan S., Stevens N. T., Humphreys H., O'Gara J. P. and O'Neill E. (2015). Current and future approaches to the prevention and treatment of staphylococcal medical device-related infections. *Curr. Pharm. Des.*, **21**(1), 100–113.

Houry A., Gohar M., Deschamps J., Tischenko E. Aymerich S., Gruss A. and Briandet R. (2012). Bacterial swimmers that infiltrate and take over the biofilm matrix. *Proc. Natl. Acad. Sci. USA*, **109**(32), 13088–13093.

Itoh Y., Wang X., Hinnebusch B. J., Preston J. F. 3rd and Romeo T. (2005). Depolymerization of beta-1,6-N-acetyl-D-glucosamine disrupts the integrity of diverse bacterial biofilms. *J. Bacteriol.*, **187**(1), 382–387.

Jarrell K. F. and McBride M. J. (2008). The surprisingly diverse ways that prokaryotes move. *Nat. Rev. Microbiol.*, **6**(6), 466–476.

Karimi A., Karig D., Kumar A. and Ardekani A. M. (2015). Interplay of physical mechanisms and biofilm processes: review of microfluidic methods. *Lab Chip.*, **15**(1), 23–42.

Kearns D. B. (2010). A field guide to bacterial swarming motility. *Nat. Rev. Microbiol.*, **8**(9), 634–44.

Li Y., Kim S. Y., Deschamps J., Briandet R. and Trubuil A. (2015). Spatio-temporal Interaction of Bacteria Mixture within Biofilms. *Procedia Environmental Sciences*, **26**, 11–18.

Mai-Prochnow A., Evans F., Dalisay-Saludes D., Stelzer S., Egan S., James S., Webb J. S. and Kjelleberg S. (2004). Biofilm development and cell death in the marine bacterium *Pseudoalteromonas tunicata. Appl. Environ. Microbiol.*, **70**(6), 3232–3238.

Mai-Prochnow A., Webb J. S., Ferrari B. C. and Kjelleberg S. (2006). Ecological advantages of autolysis during the development and dispersal of *Pseudoalteromonas tunicata* biofilms. *Appl. Environ. Microbiol.*, **72**(8), 5414–5420.

Marchal M., Briandet R., Halter D., Koechler S., DuBow M. S., Lett M.-C. and Bertin P. N. (2011). Subinhibitory arsenite concentrations lead to population dispersal in Thiomonas sp. *PLoS One*, **6**(8), e23181.

Martinez V. A., Schwarz-Linek J., Reufer M., Wilson L. G., Morozov A. N. and Poon W. C. (2014). Flagellated bacterial motility in polymer solutions. *Proc. Natl. Acad. Sci. USA*, **111**(50), 17771–17776.

McDougald D., Rice S. A., Barraud N., Steinberg P. D. and Kjelleberg S. (2011). Should we stay or should we go: mechanisms and ecological consequences for biofilm dispersal. *Nat. Rev. Microbiol.*, **10**(1), 39–50.

Molaei M. and Sheng J. (2014). Imaging bacterial 3D motion using digital in-line holographic microscopy and correlation-based de-noising algorithm. *Opt. Express*, **22**(26), 32119–32137.

Nadell C. D., Drescher K., Wingreen N. S. and Bassler B. L. (2015). Extracellular matrix structure governs invasion resistance in bacterial biofilms. *ISME J.*, **9**(8), 1700–1709.

Nan B. and Zusman D. R. (2011). Uncovering the mystery of gliding motility in the Myxobacteria. *Annu. Rev. Genet.*, **45**, 21–39.

Pongtharangkul T. and Demirci A. (2006). Effects of pH profiles on nisin production in biofilm reactor. *Appl. Microbiol. Biotechnol.*, **71**(6), 804–811.

Rice A. R., Hamilton M. A. and Camper A. K. (2003). Movement, replication, and emigration rates of individual bacteria in a biofilm. *Microb. Ecol.*, **45**(2), 163–172.

Rice S. A., Tan C. H., Mikkelsen P. J., Kung V., Woo J., Tay M., Hauser A., McDougald D., Webb J. S. and Kjelleberg S. (2009). The biofilm life cycle and virulence of Pseudomonas aeruginosa are dependent on a filamentous prophage. *ISME J.*, **3**(3), 271–282.

Riley M. A. and Wertz J. E. (2002). Bacteriocins: evolution, ecology, and application. *Ann. Rev. Microbiol.*, **56**, 117–137.

Rogers S. S., van der Walle C. and Waigh T. A. (2008). Microrheology of bacterial biofilms in vitro: *Staphylococcus aureus* and *Pseudomonas aeruginosa*. *Langmuir*, **24**(23), 13549–13555.

Schuhmacher J. S., Thormann K. M. and Bange G. (2015). How bacteria maintain location and number of flagella? *FEMS Microbiol. Rev.*, **39**(6), 812–822.

Shinoda S. and Okamoto K. (1977). Formation and function of *Vibrio parahaemolyticus* lateral flagella. *J. Bacteriol.*, **129**(3), 1266–1271.

Sourjik V. and Berg H.-C. (2004). Functional interactions between receptors in bacterial chemotaxis. *Nature*, **428**(6981), 437–441.

Turonova H., Briandet R., Rodrigues R., Hernould M., Hayek N., Stintzi A., Pazlarova J. and Tresse O. (2015). Biofilm spatial organization by the emerging pathogen *Campylobacter jejuni*: comparison between NCTC 11168 and 81–176 strains under microaerobic and oxygen-enriched conditions. *Front Microbiol.*, **6**, 709.

Van Ditmarsch D., Boyle K. E., Sakhtah H., Oyler J. E., Nadell C. D., Déziel É., Dietrich L. E. and Xavier J. B. (2013). Convergent evolution of hyperswarming leads to impaired biofilm formation in pathogenic bacteria. *Cell Rep.*, **4**(4), 697–708.

Vig D. K. and Wolgemuth C. W. (2012). Swimming dynamics of the lyme disease spirochete. *Phys. Rev. Lett.*, **109**(21), 218104.

Webb J. S., Thompson L. S., James S., Charlton T., Tolker-Nielsen T. Koch B., Givskov M. and Kjelleberg S. (2003). Cell death in *Pseudomonas aeruginosa* biofilm development. *J. Bacteriol.*, **185**(15), 4585–4592.

Chapter 10

Why and how biofilms cause biofouling – the "hair-in-sink"-effect

J. S. Vrouwenvelder[1,2,3], C. Dreszer[3,4], R. Valladares Linares[2], J. C. Kruithof[3], C. Mayer[5] and H.-C. Flemming[4,6,7]

[1]Department of Biotechnology, Faculty of Applied Sciences, Delft University of Technology, Van der Maasweg 9, 2629 HZ Delft, The Netherlands, j.s.vrouwenvelder@tudelft.nl

[2]King Abdullah University of Science and Technology (KAUST), Water Desalination and Reuse Center (WDRC), Division of Biological and Environmental Science and Engineering (BESE), Thuwal 23955–6900, Saudi Arabia, johannes.vrouwenvelder@kaust.edu.sa

[3]Wetsus, European Centre of Excellence for Sustainable Water Technology, Oostergoweg 9, 8911 MA Leeuwarden, The Netherlands, joop.kruithof@wetsus.nl

[4]University of Duisburg-Essen, Faculty of Chemistry, Biofilm Centre, Universitaetsstrasse 5, 45141 Essen, Germany, hc.flemming@uni-due.de

[5]University of Duisburg-Essen, Faculty of Chemistry, CeNIDE, Physical Chemistry, Universitaetsstrasse 5, 45141 Essen, Germany, christian.mayer@uni-due.de

[6]Singapore Centre for Environmental Life Sciences Engineering (SCELSE), and School of Biological Sciences, Nanyang Technological University, 60 Nanyang Drive, SBS-01N-27, Singapore 637551
[7]IWW Water Centre, Moritzstrasse 26, 45467 Mülheim, Germany

ABSTRACT

Biofilms can interfere strongly with membrane filtration performance both by increased transmembrane pressure drop and increase of feed-concentrate pressure drop. Intuitively, the transmembrane pressure drop effect seems proportional to biofilm thickness. However, only few data about the intrinsic hydraulic resistance of biofilms are available, with the surprising result that the contribution of the biofilm to the transmembrane pressure drop seems of relevance only for ultrafiltration, but much less for nanofiltration and negligible for reverse osmosis membranes. Using optical coherence tomography, it could be shown that variations in permeate flux lead to lower biofilm thickness at higher permeate flux, indicating that biofilms are compressible. As a consequence, it was observed that the thicker biofilm at low permeate flux (20 L m^{-2} h^{-1}) exerted a lower hydraulic resistance than the thinner biofilm at high permeate flux (60 L m^{-2} h^{-1}) – both at the same bacterial cell density. This was investigated in a transparent membrane biofouling monitor. It could be demonstrated that exclusively the biofilm extracellular polymeric substances (EPS) and not the biofilm bacterial cells cause the hydraulic resistance of biofilms. The thinner biofilm had the same EPS content as the thicker biofilm, but a higher EPS concentration due to biofilm compression. Therefore, not biofilm thickness but EPS concentration dominates the hydraulic biofilm resistance. The mechanism on which this effect is based can be explained by a model developed from the Hagen-Poiseuille law in which the boundary effects of EPS polymers during forced water passage are included. This phenomenon, which we graphically term the "hair-in-sink-effect", depends not only on the concentration of the EPS but also on the nature of the EPS molecules which can vary greatly among biofilms from different feed waters and strongly influence interactions between these molecules. The biofilm compression may be caused by transport of water from the biofilm gel to the bulk water and a physical rearrangement of the biofilm strand structure. Membrane operational aspects can increase the impact of biofouling on membrane performance. To the author's best knowledge, for the first time, a theoretical understanding of the hydraulic resistance of biofilms has been presented.

Keywords: Biofouling, membranes, EPS, hydraulic resistance, OCT, Hagen-Poiseulle

10.1 INTRODUCTION

High quality water can be produced with membrane filtration systems. One of the most serious problems in nanofiltration (NF) and reverse osmosis (RO) applications is biofouling – biofilm formation causing unacceptable operational problems (Ridgway *et al.* 1983; Tasaka *et al.* 1994; Vrouwenvelder *et al.* 1998; Schneider *et al.* 2005; Shannon *et al.* 2008). Biofilms may interfere with membrane performance in three ways: (i) increase of transmembrane pressure drop (TMP), (ii) increase of feed-channel (feed concentrate) pressure drop (FCP), and (iii) concentration polarization, causing a decrease of salt rejection. An increase in TMP indicates a decline in permeate flux (water production). Biofouling, excessive growth of biomass, is an operationally defined problem affecting the performance negatively of these membrane systems, influencing the amount and quality of the produced fresh water and costs, e.g. by shortening lifetime of membranes (Melo & Flemming, 2010). Principally, all membrane systems which are not run completely sterile will have biofilms on their surfaces, but not all have biofouling (Vrouwenvelder *et al.* 2008). Only, if the effect of biofilms exceed an individually set operational effect on membrane performance, there is biofouling (Flemming, 2011). Therefore, the effect of biofilms on performance parameters is crucial for the occurrence of biofouling. Biofilms, representing a layer on top of the separation membrane, add to the hydraulic resistance in the membrane system. On that background, it is worth considering the hydraulic resistance contributed by the biofilm. Intuitively, the biofouling effect is caused by the additional hydraulic transmembrane resistance of a biofilm which acts as a secondary membrane on top of the separation membrane, exerted by extracellular polymer substances (EPS, Flemming & Wingender, 2010) and biofilm cells.

Figure 10.1 shows schematically the way of water transport through the biofilm, crossing the EPS matrix and the bacterial cells imbedded. Plausibly, both will contribute to hydraulic resistance.

As the major component of biofilms is water (up to 95%) and biofilms can be considered a hydrogel (Flemming & Wingender, 2010), it is an interesting question how the hydrodynamic resistance is caused. In view of the technical and economical relevance of biofouling, it is surprising how few data exist about the actual hydraulic resistance of biofilms that may negatively affect the TMP and membrane performance. If the details of water transport through biofilms are understood, it might be possible to engineer biofilm properties and possibly increase their permeability chemically, as suggested by McDonogh *et al.* (1994).

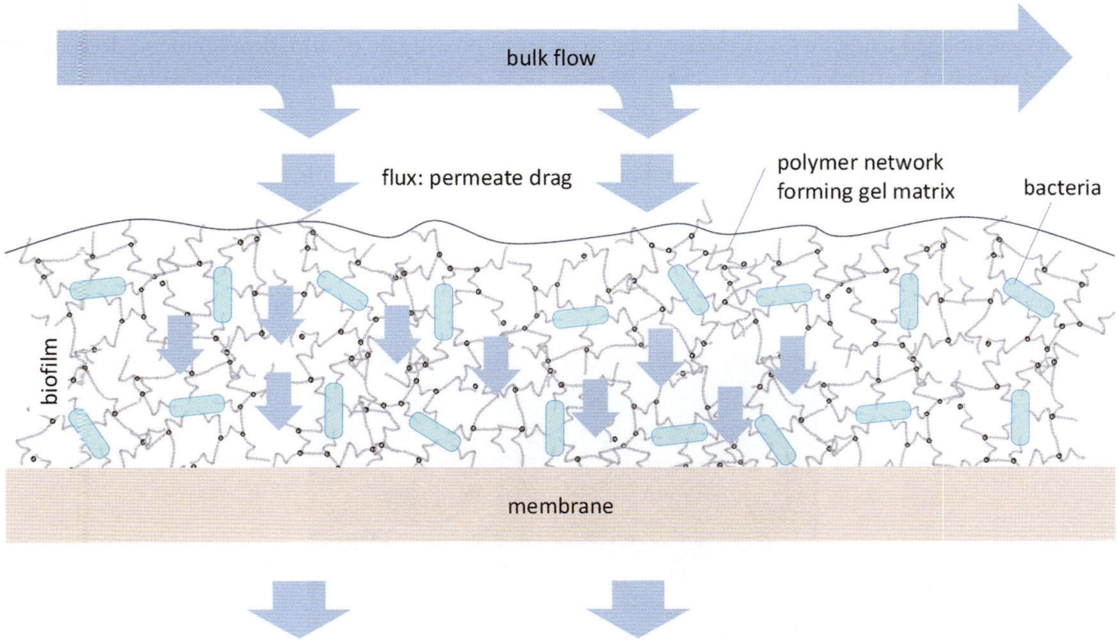

Figure 10.1 Schematic depiction of the way of water transport through a membrane with a biofouling layer consisting of extracellular polymeric substances (EPS) and bacterial cells. The arrows indicate water transport.

The objectives of this study were to (i) quantify the contribution of biofilm accumulation on hydraulic biofilm resistance as a function of e.g. time, water flux (rate of water production) and water flux variations, to (ii) develop a model for specifically singling out the contribution of both EPS and bacterial cells in a biofilm, and (iii) develop a model which allows to understand the mechanism of water passage through biofilms. For these investigations, membrane fouling simulators provided testing systems representative for practice (Vrouwenvelder *et al.* 2006, 2007; Dreszer *et al.* 2014a). Optical coherence tomography, a non-destructive technique to visualize even relative thick biofilms three dimensionally, was employed to investigate the response of biofilms on membranes to hydraulic water pressure (Dreszer *et al.* 2014b).

10.2 MATERIALS AND METHODS

10.2.1 Experimental set-up

Membrane fouling simulators (MFSs) are used to monitor and study in membrane systems biofilm/biofouling development and control strategies (Creber *et al.* 2010a; Staal *et al.* 2011; Prest *et al.* 2012; Miller *et al.* 2012; Araújo *et al.* 2012; Dreszer *et al.* 2014c; Bucs *et al.* 2014; Farhat *et al.* 2015). In this study, the membrane fouling simulator (MFS) was combined with optical coherence tomography (Figure 10.2) to assess *in-situ* the (i) hydraulic biofilm resistance, (ii) biofilm morphology and (iii) performance parameters feed-channel pressure drop and transmembrane pressure drop (Dreszer *et al.* 2013, 2014a). At the end of the study, the accumulated biofilm was analyzed: thickness, total bacterial cell number, total organic carbon (TOC) and extracellular polymeric substances (EPS). A biofouling monitor was imaged using an OCT microscope in order to investigate biofilm development, detachment and the impact of permeate flux variation *in-situ*. OCT enables non-destructive imaging of biofilm structures (Wagner *et al.* 2010; Derlon *et al.* 2012; Valladares Linares *et al.* 2015; West *et al.* 2016). The MFS system was operated at constant crossflow velocity, and a permeate flux of 20 and 60 L m^{-2} h^{-1}. To enhance biofilm accumulation the feed water was supplemented with a carbon concentration of 1 mg L^{-1}. The development of TMP was monitored using a sensitive differential pressure transmitter (Vrouwenvelder *et al.* 2009a). An ultrafiltration membrane was chosen because of its low inherent hydraulic resistance which allows for separating the inherent hydraulic resistance of biofilms on top of this membrane.

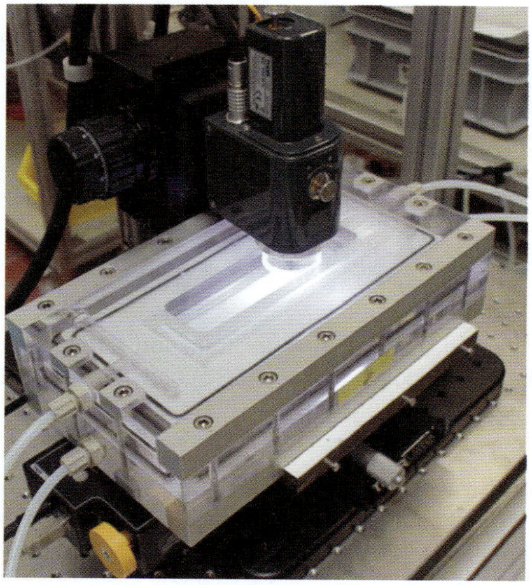

Figure 10.2 Monitor used for *in-situ* biofilm thickness measurements with optical coherence tomography (OCT). OCT measurements were performed during monitor operation at constant crossflow velocity (0.1 m s^{-1}) and permeate production (20 or 60 L m^{-2} h^{-1}), (Dreszer *et al.* 2014b).

10.2.2 Feed water

Drinking water prepared from anaerobic groundwater, subsequently treated by aeration, rapid sand filtration, deacidification, softening and rapid sand filtration at treatment plant Spannenburg in The Netherlands, is distributed without primary chemical disinfection and without a disinfectant residual. This drinking water was used as feed water source for the crossflow filtration system experiments. The feed water had a total bacterial cell number of 3×10^5 cells mL^{-1} as determined by microscopical quantification of DAPI-stained fluorescent particles. The number of colony forming units on R2A media was 2×10^3 CFU mL^{-1} at 25°C after 10 days incubation. A concentrated nutrient solution was dosed into the feed water prior to the filtration cell at a flow rate of 0.12 L h^{-1}, using a peristaltic pump (Masterflex, Cole Palmer, Vernon Hills, USA). To suppress bacterial growth in the nutrient dosage solution, the pH-value of the nutrient solution was set at 11 by sodium hydroxide addition. The dosing flow rate of the nutrient solution

(0.12 L h^{-1}) was low compared to the feed water flow rate (28.3 L h^{-1}). Therefore, the high pH-value of the nutrient solution had no effect on the feed water pH of 7.8, and also no effect on the permeability. As nutrients, a solution of sodium acetate, sodium nitrate and sodium dihydrogen orthophosphate in the mass ratio for C:N:P of 100:20:10, respectively, were employed (Dreszer *et al.* 2013, 2014b). The concentration of sodium acetate added to the feed water was 1 mg L^{-1} of carbon, the amount of nitrogen and phosphorus was derived from the carbon concentration. Acetate is commonly used as biodegrable carbon source in microbial and biofilm growth studies (Van der Kooij *et al.* 1995; Vrouwenvelder *et al.* 2009b, c). Phosphate was dosed to avoid phosphate limiting conditions (Vrouwenvelder *et al.* 2010). Fresh nutrient solutions were prepared every 2 days. All chemicals were purchased in analytical grade from Boom B.V. (Meppel, The Netherlands) and were dissolved in deionized water.

10.2.3 Optical coherence tomography (OCT)

In-situ imaging of the feed channel surface of the membrane was performed using a spectral domain optical coherence tomograph (Thorlabs Ganymede OCT System). The OCT was fitted with a 5× telecentric scan lens (Thorlabs LSM03BB) which provides a maximum scan area of 100 mm^2. The OCT engine was configured to provide high resolution images with a sensitivity of 106 dB at 1.25 kHz A-scan rate. Volumetric images were created using the maximum intensity profile algorithm included in the instrument software (Thorlabs SD-OCT system software version 3.2.1) for a rectangular area of 2 mm × 5 mm using 200 B-scans and 500 A-scans of 619 pixels corresponding to a physical depth of 1.1 mm. The OCT engine was configured to provide high resolution images. A similar procedure for the generation of biofilm thickness was already applied by Derlon *et al.* (2012). A total OCT imaging area of 10,000 µm × 8 µm distributed over 10 locations of the monitor were imaged.

10.2.4 Biofilm thickness

Biofilm thickness was calculated based on the number of pixels found between the top edge of the biofilm and the upper membrane surface of each OCT image, using a pixel scaling factor to obtain the biofilm thickness in microns. The biofilm thickness range was calculated for the total OCT scanned areas, limited in value by the thickest and thinnest sections of the biofilm in the scan.

10.3 RESULTS

10.3.1 Biofouling layer formation: filtration or biofilm growth?

A first question to be answered was, if the biofouling layer is only the result of planktonic cells accumulated on the membrane surface by vertical transport or actually a grown biofilm. The impact of microbial cells without an EPS matrix was investigated by comparison of a grown biofilm with a filter cake of similar cell number per cm^2 (Dreszer *et al.* 2013). In this experiment, the filter cake was generated by accumulation of cells added to the feed water onto the membrane surface (Figure 10.2, Dreszer *et al.* 2013).The filtration time was 2 h, so significant microbial growth was avoided. At a similar cell number, the resistance of the cell filter cake was five times less than that of the grown biofilm. The amount of EPS in the bacterial filter cake was small compared to the biofilm (Dreszer *et al.* 2013).

Figure 10.3 clearly shows that the thickness of the fouling layer was caused by growth, not by a filter cake of microorganisms on the membrane. Furthermore, very interesting and important result is that it is equally clear that the hydraulic resistance was not caused by biofilm cells but by the EPS matrix (Dreszer *et al.* 2013). Biofilm bacterial cells play no role in the hydraulic resistance of biofilms.

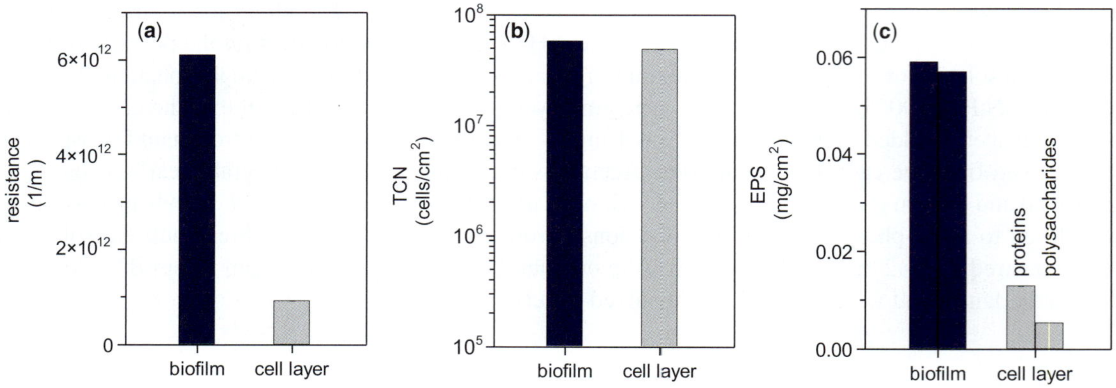

Figure 10.3 Comparison of effects of grown biofilm vs. filter cake of bacterial cells. (a) Resistance, (b) total bacterial cell numbers and (c) EPS of biofilm and cell layer with the same cell density as the biofilm (Dreszer *et al.* 2013).

10.3.2 Impact of permeate flux change on biofilm hydraulic resistance and thickness?

Using optical coherence tomography, it could be shown that variations in permeate flux lead to lower biofilm thickness at higher permeate flux (Figure 10.4), indicating that biofilms are compressible. As a consequence, it was observed that the thinner biofilm at higher permeate flux (60 L m^{-2} h^{-1}) exerted a higher hydraulic resistance than the thicker biofilm at lower permeate flux (20 L m^{-2} h^{-1}) – both with the same concentration of bacterial cells (Dreszer *et al.* 2014b).

A permeate flux increase caused an increased hydraulic resistance (Figure 10.4a) and thinner biofilm thickness (Figure 10.4b). Interestingly, subsequent flux restoration did not fully restore the biofilm resistance and thickness (Figure 10.4c, Valladares Linares *et al.* 2015).

Comparison of the biofilm resistance at 20 L m^{-2} h^{-1} before and after the temporarily permeate flux elevation showed a 16% higher hydraulic biofilm resistance after the flux changes (Figure 10.5a), (Dreszer *et al.* 2014b). The biofilm thickness was highest at a permeate flux of 20 L m^{-2} h^{-1} prior to flux increase (Figure 10.5b). After increasing the flux to 60 L m^{-2} h^{-1} the biofilm thickness decreased, while returning to the original flux caused an increase of the biofilm thickness again. Comparison of the biofilm thickness at 20 L m^{-2} h^{-1} before and after flux elevation showed a 25% lower biofilm thickness after the temporary flux elevation (Figure 10.5b). Interestingly, reverse to the lower flux did not completely restore the original hydraulic biofilm resistance and thickness. The results of the experimental studies indicate that permeate flux changes may lead to more permanent compression of biofilms and, thus, to an increase of their hydraulic resistance, impacting negatively the membrane performance (Dreszer *et al.* 2014b; Valladares Linares *et al.* 2015).

Measurements of the average biofilm thickness were made after time intervals of 100 milliseconds (ms). Figure 10.6 shows the normalized values for the biofilm thickness in time for the 2 flow regimes: (i) increasing the permeate flux from 20 to 60 L m^{-2} h^{-1} at $t = 0$ s and restoring the original flux from 60 to 20 L m^{-2} h^{-1} after 46 s until the end of the experiment after 145 s (Valladares Linares *et al.* 2015). Replicate experiments confirmed the results of the study. The biofilm thickness change rate for compaction and relaxation differed, while restoring the original flux did not completely restore the original biofilm thickness. Biofilm compaction showed to be faster than biofilm relaxation.

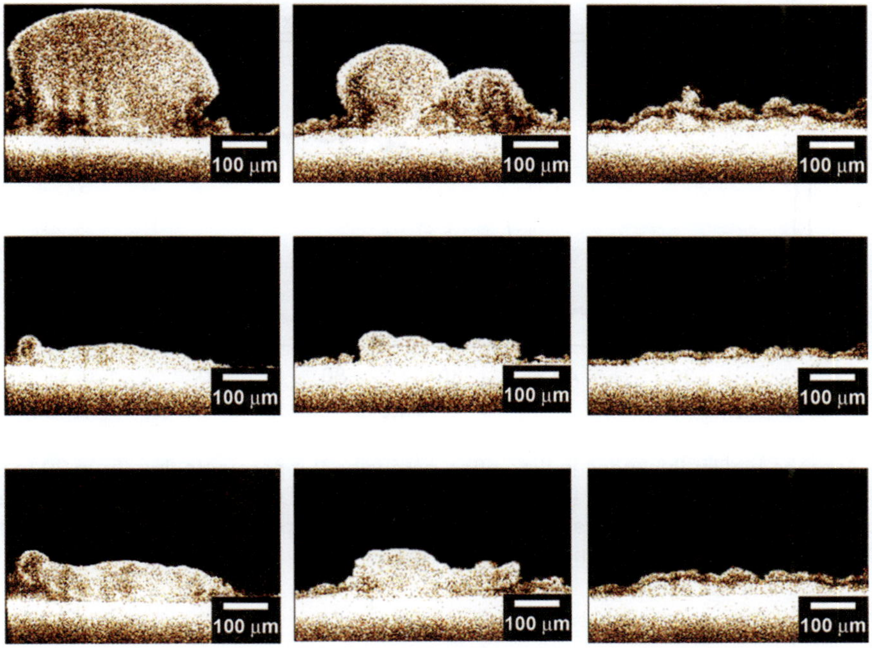

Figure 10.4 Impact of permeate flux increase on biofilm thickness. OCT observations of the biofilm at three different locations at permeate flux 20 L m^{-2} h^{-1} (top), and after permeate flux increase to 60 L m^{-2} h^{-1} (center) and after restoring the original flux (bottom). The lateral scan length is 8 mm. The images were made at a stable flux and transmembrane pressure and constant crossflow velocity (0.1 m s^{-1}), (Valladares Linares *et al.* 2015). The biofilm thickness could be influenced by compaction and detachment.

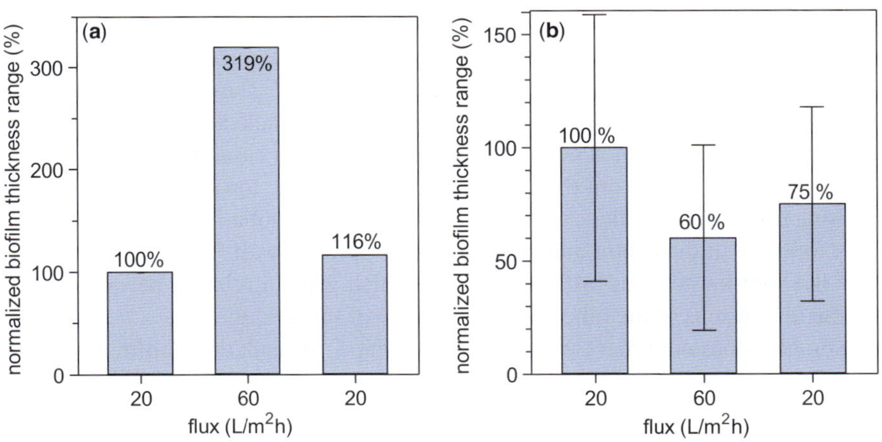

Figure 10.5 Normalized hydraulic biofilm resistance (a) and biofilm thickness range (b), at a permeate flux of 20 L m^{-2} h^{-1}, after flux increase to 60 L m^{-2} h^{-1} and, after restoring the original flux of 20 L m^{-2} h^{-1}. The error bars show the total range of biofilm thickness in the total OCT scan area, indicating the membrane was fully covered with a grown biofilm (Dreszer *et al.* 2014b).

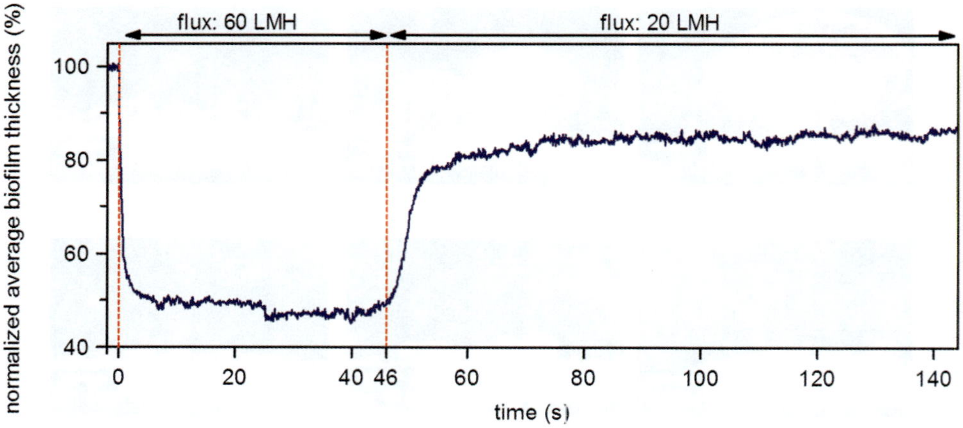

Figure 10.6 Average biofilm thickness in time after increasing the permeate flux (from 20 to 60 L m^{-2} h^{-1}) and restoring the original flux (from 60 to 20 L m^{-2} h^{-1}) determined with OCT. Individual frames (B-scans) from the sequence were taken at intervals of 100 ms, for each 255 pixels (A-scans). The crossflow velocity was constant: 0.10 m s^{-1} (Valladares Linares *et al.* 2015).

10.3.3 Understanding the reason of hydraulic resistance of biofilms: Model for the mechanism of water permeation

The question is now: how does the EPS matrix as a highly hydrated gel cause hydraulic resistance? Measurements of the self-diffusion rate of water inside biofilms have shown that by far the largest fraction of the bulk water exhibits free diffusion with rates similar to those of pure water (Vogt *et al.* 2000). This given, the physical reason for the hydraulic resistance of the biofilm cannot be assigned to a reduced mobility of individual water molecules. Instead, the actual drag which develops for flowing bulk water must derive from the hindrance of a collective motion of a large water volume. Basically, the molecular network which presents the mechanical frame of the biofilm provides within the time-scale studied (minutes) a set of fixed points in space remaining immobile during the water flow process. These fixed points cause a constant loss of momentum from the flowing bulk water towards the macromolecular EPS frame. The situation is somewhat comparable to bulk water flowing through a pipe: in this case, the water is slowed down by a constant transport of momentum from the centre of the flow profile towards the walls of the pipe. As a result, the resulting distribution of flow velocities resembles a parabolic function of rotational symmetry with the highest velocity in the centre and the lowest one near the wall. The overall resistance of a pipe is described by the well-known Hagen-Poiseuille law which gives a relation between the flow rate, the pressure drop, the dimensions of the tube and the viscosity of the medium.

Of course, a biofilm does not reflect the geometry of a pipe. Instead, the biofilm molecular framework may be seen as a three-dimensional (3D) network of molecular chains equally distributed over the total volume. In a first approximation, the 3D network may be idealized by a rectangular lattice with an equal spacing in all three dimensions in space (Figure 10.7). The biofilm molecular network would be defined by a elementary cubicle at a molecular scale with a side length a, the edges of the cube being formed by molecular chains of a diameter l. If we assume a water flow along one of the principal axes (vertical in Figure 10.7), the water molecules would be forced to pass through the meshes formed by the horizontal faces of the cubes. Within each of these meshes, a flow velocity profile would form caused by the momentum transport to the horizontal chains (Figure 10.7, right). The effect of the vertical chains (grey colour in Figure 10.7) on the water flow profile could be neglected.

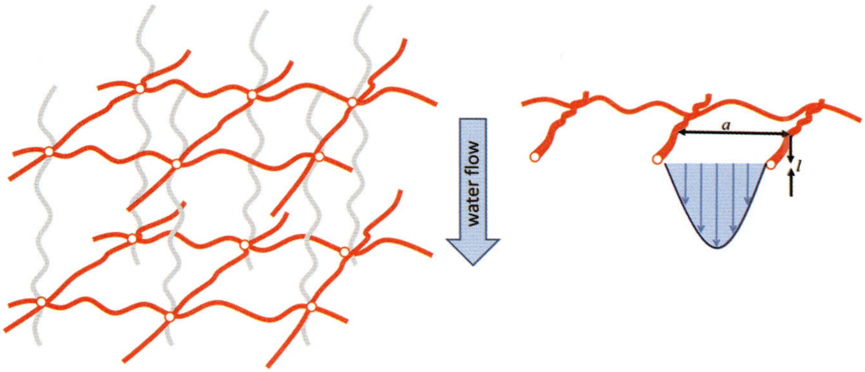

Figure 10.7 Schematic representation of a biofilm gel structure approximated by a rectangular network of EPS macromolecules in an aqueous environment. The sketch on the right indicates the water flow profile which develops in each single EPS mesh with perpendicular orientation (shown in red). In first approximation, the EPS strands oriented parallel to the flow direction (gray color) are assumed to not contribute to the drag.

In a rough approximation, the situation in each of these meshes would resemble a water flow in a very short rectangular "pipe" with a square cross-section of the size "a by a" and a length of l. For such a geometry, with a given pressure drop of ΔP and a fluid viscosity of η, the law of Hagen-Poiseuille predicts a volume flow velocity v of

$$v = \frac{Ka^4 \Delta P}{12\eta l} \tag{1}$$

Here, the constant K refers to the cross-sectional geometry and amounts to $K \approx 0.4218$ in case of a square (Bruus, 2008).

If we consider a biofilm section with a total area of A and a thickness of d, this volume flow is scaled proportionally to the number of parallel meshes A/a^2 and inversely proportional to the number of serial meshes d/a, giving

$$v = \frac{Ka^3 \Delta PA}{12\eta ld} \tag{2}$$

Assuming, for simplicity, a square cross section of the EPS polymer chains and accounting for the fact that each of the 12 strands forming the sides of a cubic elementary cell contributes with one fourth to its volume, the mesh size a can be derived from the volume fraction V_{rel} of the polymer in the gel according to

$$a = l\,(3/V_{rel})^{1/2} \tag{3}$$

With Eq. (2), this results in

$$v = \frac{Kl^2 \left(\frac{3}{V_{rel}}\right)^{3/2} \Delta PA}{12\eta d} \tag{4}$$

This means that the water volume flow at a given pressure drop and a given gel polymer type would scale with $V_{rel}^{-3/2}$. So, for example, an increase of the polymer content by a factor of four would lead to a decrease of the volume flow by a factor of eight.

With a definition of the hydraulic resistance R according to

$$R = \frac{\Delta PA}{\eta v} \tag{5}$$

Eq. (2) turns into

$$R = \frac{12ld}{Ka^3} \tag{6}$$

Finally, with Eq. (3), we get

$$R = \frac{12d}{Kl^2\left(\frac{3}{V_{rel}}\right)^{3/2}} \tag{7}$$

This given, the experimental values for the hydraulic resistance shown in Figure 10.2 can be used to estimate the volume fraction of polymer in the given biofilm. The strand thickness of the polymer of course varies with the type of EPS. However, since this is meant to be a rough estimation, it is legitimate to insert a typical value at this point. Atomic force microscopy (AFM) measurements on single dextran molecules show typical strand thicknesses around 1 nm (Zellerman *et al.* 2013). Most of the biofilm is water. Using the 1 nm thickness strand together with the hydraulic resistance of $6 \cdot 10^{12}\ m^{-1}$ (as given in Figure 10.3) and a biofilm thickness of 200 μm (see Figure 10.4), Eq. 7 shows that the EPS polymer volume fraction is 3% of the biofilm matrix. This 3% value is in the typical range for biofilms and therefore quite plausible.

Eq. (7) also allows to predict the effect of biofilm compression on the hydraulic resistance. Scaling the biofilm thickness d by a factor of x leads to a corresponding change of the polymer volume fraction by a factor of $1/x$. If both variations enter Eq. (7), this results in a total variation of the hydraulic resistance by a factor of $x^{-1/2}$. In other words: if the biofilm is compressed to one fourth of its original thickness, we must expect an increase of the resistance by a factor of two. This relation may be used to interpret the data shown in Figure 10.5: after the high flux period, the biofilm thickness had been reduced to ¾ of its original value. According to Eq. (7) and the relation given above, this should lead to an increase of resistance by a factor of $(3/4)^{-1/2} \approx 1.15$. This coincides well with the observed relative resistance of 116% at this point (Figure 10.5). However, this does not explain for the very high relative resistance during the high permeate flux period itself. A possible interpretation of this observation may be a strongly inhomogeneous compression. Presumably, the flux of 60 $L\ m^{-2}\ h^{-1}$ leads to a very high compression of the lower biofilm layer which is in direct contact to the membrane, while the layer closer to the bulk liquid is only slightly compressed. The resulting compression gradient could lead to very high local concentrations of the polymer network near the membrane surface which in turn cause a dramatic increase of the local resistance. This interpretation coincides with the partial reversibility of the biofilm compression: after the reduction of the flux, the gel matrix relaxes and its resistance again follows the assumptions given above. The biofilm compression may be caused by transport of water from the biofilm gel to the bulk water and a physical rearrangement of the biofilm strand structure.

10.4 DISCUSSION

From the results, it is clear that the hydraulic resistance of biofilms in membrane processes is caused by the EPS matrix and not by the bacterial cells embedded in that matrix. Thus, counter-intuitively, they can be neglected in modelling considerations.

Very interestingly, flux increase leads to a thinner biofilm which in turn exerts a higher hydraulic resistance. This can be graphically described as "hair-in-sink-effect": with pressure, the concentration of EPS components increased (like hair in the sink) and, thus, a higher hydraulic resistance results. Therefore, contrary to intuition, biofilm thickness is not a suitable parameter to assess biofilm resistance – thinner biofilms can have a higher hydraulic resistance than thicker ones. Particularly interesting is the observation, that – after increasing the permeate flux and returning to the original flux after time – the biofilm does not return to its original thickness but remains slightly thinner. This is an important observation: the biofilm behaves not perfectly as a spring, which is consistent with many studies that have shown that biofilms are viscoelastic rather than elastic solids and could explain some of the biofilm thickness lack of recovery addressed here. For membrane management this means that flux variations can lead to a compaction of the biofilm and, on a long term, to increased hydraulic resistance as a biofouling effect.

To the author´s best knowledge, for the first time, a theoretical approach to understand of the hydraulic resistance of biofilms has been presented by application of the Hagen-Poiseulle-law. This model for the mechanism of water transport through biofilms gives a physical explanation for the drag of bulk water flow and allows for an estimation of the polymer content and leads to a very good prediction of the consequence of irreversible biofilm compaction.

These results can explain why biofilms may contribute to increased transmembrane resistance at increased flux and suggest that fluctuations in the flow regime may lead to increasing effects of biofilms, thus, to more biofouling.

Membrane operational aspects can increase the impact of biofouling on membrane performance. Our study showed that feed pressure variations affect both the permeate flux and the hydraulic biofilm resistance (Figure 10.5). In practice, in case of membrane fouling the feed pressure needs to increase to maintain constant permeate production, causing the biofilm to compact thereby increasing the biofilm hydraulic resistance. In addition to biofilm compaction itself, there may be open channels through the biofilm to the substratum that might be closed off as the biofilm is compressed (Vinogradov *et al.* 2015), contributing to the higher hydraulic resistance of a compacted biofilm. In other words, the hydraulic biofilm resistance could increase due to the effect of compression on the molecular and the microscale: (i) decreased porosity in the EPS matrix and (ii) closing of larger biofilm channels. It has been shown that the crossflow velocity of water in membrane systems influences the impact of accumulated biomass on the feed-concentrate pressure drop: the impact of accumulated biofilm on the feed-concentrate pressure drop is higher at a high crossflow velocity, increasing operational costs (Vrouwenvelder *et al.* 2009d). A high crossflow velocity enables also early warning of biofouling since the impact of accumulated biofilm on membrane performance is higher (Vrouwenvelder *et al.* 2011a). Clearly, current membrane operational aspects can increase the negative impact of accumulated biofilm on membrane performance. Low and stable permeate flux operated membrane systems may (i) require a larger footprint than high flux operated membrane systems but may (ii) reduce the impact of fouling on performance providing a more stable and robust membrane performance with less maintenance requirements to control fouling. Future membrane operation and design should aim to minimize the impact of accumulated biofilm on membrane performance and employ advanced cleaning strategies (e.g. Cornelissen *et al.* 2007; Creber *et al.* 2010b; Vrouwenvelder *et al.* 2011b; Van Loosdrecht *et al.* 2012).

ACKNOWLEDGEMENTS

The research reported in this paper was supported by King Abdullah University of Science and Technology (KAUST), Wetsus and Evides Industriewater.

REFERENCES

Araújo P. A., Miller D. J., Correia P. B., van Loosdrecht M. C. M., Kruithof J. C., Freeman B. D., Paul D. R. and Vrouwenvelder J. S. (2012). Impact of feed spacer and membrane modification by hydrophilic, bactericidal and biocidal coating on biofouling control. *Desalination*, **295**, 1–10.

Bruus H. (2008). Theoretical Microfluidics. Oxford University Press, Oxford.

Bucs Sz. S., Valladares Linares R., Van Loosdrecht M. C. M., Kruithof J. C. and Vrouwenvelder J. S. (2014). Impact of organic nutrient load on biomass accumulation and pressure drop increase in membrane systems. *Water Research*, **67**, 227–242.

Cornelissen E. R., Vrouwenvelder J. S., Heijman S. G. J., Viallefont X. D., Van Der Kooij D. and Wessels L. P. (2007). Periodic air/water cleaning for control of biofouling in spiral wound membrane elements. *Journal of Membrane Science*, **287**(1), 94–101.

Creber S. A., Vrouwenvelder J. S., Van Loosdrecht M. C. M. and Johns M. L. (2010a). Chemical cleaning of biofouling in reverse osmosis membranes evaluated using magnetic resonance imaging. *Journal of Membrane Science*, **362**(1–2), 202–210.

Creber S. A., Pintelon T. R. R., Schulenburg D. A. W. G., Vrouwenvelder J. S., van Loosdrecht M. C. M. and Johns M. L. (2010b). Magnetic resonance imaging and 3D simulation studies of biofilm accumulation and cleaning on reverse osmosis membranes. *Food and Bioproducts Processing*, **88**(4), 401–408.

Derlon N., Peter-Varbanets M., Scheidegger A., Pronk W. and Morgenroth E. (2012). Predation influences the structure of biofilm developed on ultrafiltration membranes. *Water Res.*, **46**, 3323–3333.

Dreszer C., Vrouwenvelder J. S., Paulitsch-Fuchs A. H., Zwijnenburg A., Kruithof J. C. and Flemming H.-C. (2013). Hydraulic Resistance of Biofilms. *Journal of Membrane Science*, **429**, 436–447.

Dreszer C., Flemming H. C., Wexler A. D., Zwijnenburg A., Kruithof J. C. and Vrouwenvelder J. S. (2014a). Development and testing of a transparent membrane biofouling monitor. *Desalination Water Treat.*, **52**, 1807–1819.

Dreszer C., Wexler A. D., Drusová S., Overdijk S. T., Zwijnenburg A., Flemming H.-C., Kruithof J. C. and Vrouwenvelder J. S. (2014b). *In-situ* biofilm characterization in membrane systems using Optical Coherence Tomography: formation, structure, detachment and impact of flux change. *Water Research*, **67**, 243–254.

Dreszer C., Flemming H.-C., Wexler A. D., Zwijnenburg A., Kruithof J. C. and Vrouwenvelder J. S. (2014c). Impact of biofilm accumulation on transmembrane and feed channel pressure drop: effects of crossflow velocity, feed spacer and biodegradable substrate. *Water Research*, **50**, 200–211.

Farhat N. M., Staal M., Siddiqui A., Borisov S., Bucs Sz. S. and Vrouwenvelder J. S. (2015). Early non-destructive biofouling detection and spatial distribution: application of oxygen sensing optodes. *Water Research*, **83**, 10–20.

Flemming H.-C. (2011). Microbial biofouling – unsolved problems, insufficient approaches and possible solutions. In: Biofilm Highlights, H.-C. Flemming, J. Wingender and U. Szewzyk (eds.), Springer Int., Heidelberg, New York, pp. 81–109.

Flemming H.-C. and Wingender J. (2010). The Biofilm Matrix. *Nat. Rev. Microbiol.*, **8**, 623–633.

McDonogh R., Schaule G. and Flemming H.-C. (1994). The permeability of biofouling layers on membranes. *J. Membr. Sci.*, **87**, 199–217.

Melo L. and Flemming H.-C. (2010). Mechanistic aspects of heat exchanger and membrane biofouling and prevention. In: The Science and Technology of Industrial Water Treatment, Z. Amjad (ed.), Taylor and Francis, London, pp. 365–380.

Miller D. J., Araújo P. A., Correia P. B., Ramsey M. M., Kruithof J. C., van Loosdrecht M. C. M., Freeman B. D., Paul D. R., Whiteley M. and Vrouwenvelder J. S. (2012). Short-term adhesion and long-term biofouling testing of polydopamine and poly(ethylene glycol) surface modifications of membranes and feed spacers. *Water Research*, **46**, 3737–3753.

Prest E. I., Staal M., Kühl M., van Loosdrecht M. C. M. and Vrouwenvelder J. S. (2012). Quantitative measurement and visualization of biofilm O_2 consumption rates in membrane filtration systems. *Journal of Membrane Science*, **392–393**, 66–75.

Ridgway H. F., Kelly A., Justice C. and Olson B. H. (1983). Microbial fouling of reverse-osmosis membranes used in advanced wastewater-treatment technology, chemical, bacteriological, and ultrastructural analyses. *Appl. Environ. Microbiol.*, **45**(3), 1066–1084.

Schneider R. P., Ferreira L. M., Binder P., Bejarano E. M., Goes K. P., Slongo E., Machado C. R. and Rosa G. M. Z. (2005). Dynamics of organic carbon and of bacterial populations in a conventional pretreatment train of a reverse osmosis unit experiencing severe biofouling. *J. Membr. Sci.*, **266**(1–2), 18–29.

Shannon M. A., Bohn P. W., Elimelech M., Georgiadis J. G., Marinas B. J. and Mayes A. M. (2008). Science and technology for water purification in the coming decades. *Nature*, **452**, 301–310.

Staal M., Prest E. I., Vrouwenvelder J. S., Rickelt L. F. and Kuhl M. (2011). A simple optode based method for imaging O2 distribution and dynamics in tap water biofilms. *Water Research*, **45**, 5027–5037.

Tasaka K., Katsura T., Iwahori H. and Kaiyama Y. (1994). Analysis of RO elements operated at more than 80 plants in Japan. *Desalination*, **96**(1–3), 259–272.

Valladares Linares R., Wexler A. D., Bucs Sz. S., Dreszer C., Zwijnenburg A., Flemming H.-C., Kruithof J. C. and Vrouwenvelder J. S. (2015). Compaction and relaxation of biofilms. *Desalination and Water Treatment*, 1–13.

Van der Kooij D., Vrouwenvelder H. S. and Veenendaal H. R. (1995). Kinetic aspects of biofilm formation on surfaces exposed to drinking water. *Water Science & Technology*, **32**(8), 61–65.

Van Loosdrecht M. C. M., Bereschenko L. A., Radu A. I., Kruithof J. C., Picioreanu C., Johns M. L. and Vrouwenvelder J. S. (2012). New approaches to characterizing and understanding biofouling of spiral wound membrane systems. *Water Science & Technology*, **66**(1), 88–94.

Vinogradov A. M., Winston M., Rupp C. J. and Stoodley P. (2015). Rheology of biofilms formed from the dental plaque pathogen *Streptococcus mutans*. *Biofilms*, **1**(01), 49–56.

Vogt M., Flemming H.-C. and Veeman W. (2000). Diffusion in Pseudomonas aerugiosa biofilms: a pulsed field gradient NMR study. *J. Biotechnol.*, **77**(1), 137–146.

Vrouwenvelder J. S. and Van Der Kooij D. (2001). Diagnosis, prediction and prevention of biofouling of NF and RO membranes. *Desalination*, **139**(1–3), 65–71.

Vrouwenvelder H. S., Van Paassen J. A. M., Folmer H. C., Hofman J. A. M. H., Nederlof M. M. and Van der Kooij D. (1998). Biofouling of membranes for drinking water production. *Desalination*, **118**(1–3), 157–166.

Vrouwenvelder J. S., van Paassen J. A. M., Wessels L. P., van Dam A. F. and Bakker S. M. (2006). The membrane fouling simulator: a suitable tool for fouling prediction and control. *J. Mem. Sci.*, **281**, 316–324.

Vrouwenvelder J. S., Bakker S. M., Cauchard M., Le Grand R., Apacandie´ M., Idrissi M., Lagrave S., Wessels L. P., van Paassen J. A. M., Kruithof J. C. and van Loosdrecht M. C. M. (2007). The membrane fouling simulator: a suitable tool for prediction and characterisation of membrane fouling. *Water Science & Technology*, **55**(8–9), 197–205.

Vrouwenvelder J. S., Manolarakis S. A., van der Hoek J. P., van Paassen J. A. M., van der Meer W. G. J., van Agtmaal J. M. C., Prummel H. D. M., Kruithof J. C. and van Loosdrecht M. C. M. (2008). Quantitative biofouling diagnosis in full scale nanofiltration and reverse osmosis installations. *Water Research*, **42**(19), 4856–4868.

Vrouwenvelder J. S., Van Paassen J. A. M., Kruithof J. C. and Van Loosdrecht M. C. M. (2009a). Sensitive pressure drop measurements of individual lead membrane elements for accurate early biofouling detection. *Journal of Membrane Science*, **338**(1–2), 92–99.

Vrouwenvelder J. S., Hinrichs C., Van der Meer W. G. J., Van Loosdrecht M. C. M. and Kruithof J. C. (2009b). Pressure drop increase by biofilm accumulation in spiral wound RO and NF membrane systems: role of substrate concentration, flow velocity, substrate load and flow direction. *Biofouling*, **25**(6), 543–555.

Vrouwenvelder J. S., Graf von der Schulenburg D. A., Kruithof J. C., Johns M. L. and van Loosdrecht M. C. M. (2009c). Biofouling of spiral wound nanofiltration and reverse osmosis membranes: a feed spacer problem. *Water Research*, **43**(3), 583–594.

Vrouwenvelder J. S., Hinrichs C., Van der Meer W. G. J., Van Loosdrecht M. C. M. and Kruithof J. C. (2009d). Pressure drop increase by biofilm accumulation in spiral wound RO and NF membrane systems: role of substrate concentration, flow velocity, substrate load and flow direction. *Biofouling*, **25**(6), 543–555.

Vrouwenvelder J. S., Beyer F., Dahmani K., Hasan N., Galjaard G., Kruithof J. C. and Van Loosdrecht M. C. M. (2010). Phosphate limitation to control biofouling. *Water Research*, **44**(11), 3454–3466.

Vrouwenvelder J. S., Van Loosdrecht M. C. M. and Kruithof J. C. (2011a). Early warning of biofouling in spiral wound nanofiltration and reverse osmosis membranes. *Desalination*, **265**(1–3), 206–212.

Vrouwenvelder J. S., Van Loosdrecht M. C. M. and Kruithof J. C. (2011b). A novel scenario for biofouling control of spiral wound membrane systems. *Water research*, **45**, 3890–3898.

Wagner M., Taherzadeh D., Haisch C. and Horn H. (2010). Investigation of the mesoscale structure and volumetric features of biofilms using optical coherence tomography. *Biotechnol. Bioeng.*, **107**, 844–853.

West S., Wagner M., Engelke C. and Horn H. (2015). Optical coherence tomography for the in situ three-dimensional visualization and quantification of feed spacer channel fouling in reverse osmosis membrane modules. *Journal of Membrane Science*, **498**, 345–352.

Zellermann A.-M., Bergmann D. and Mayer C. (2013). Cation induced conformation changes in hyaluronate solution. *Eur. Polym. J.*, **49**, 70–79.

Chapter 11

Unique and baffling aspects of the matrix: EPS syneresis and glass formation during desiccation

Decho, Alan W.

Department of Environmental Health Sciences, Arnold School of Public Health, University of South Carolina, Columbia, SC. 29208. USA. Email: awdecho@mailbox.sc.edu

ABSTRACT

The challenge of surviving near-complete desiccation, then resuming cellular- and extracellular-activities upon rehydration, is encountered by biofilms across many environments. Since many activities occur outside of cells, the molecules involved must be preserved in some manner during desiccation within the matrix of extracellular polymeric substances (EPS). These include extracellular-enzymes (*e*-enzymes), -DNA (*e*-DNA), -vesicles, -signals, and -antibiotics. How this occurs is not yet understood. Unique changes in EPS matrix architecture occur during the progression from full-hydration to one of near-complete desiccation; changes that may offer insight into biofilm survival. Anhydrophilic microbial mats in hypersaline ponds, which undergo periodic desiccation and salinity fluctuations, offer a natural laboratory for observing matrix changes. Several discernable steps are becoming apparent during matrix desiccation. First, during lower salinities (30–100 mg/L), which are optimal for growth, EPS is highly-hydrated and allows ample diffusive-exchange (for microbial cells) with overlying water. As salinities in overlying-waters increase due to evaporation, EPS on the mat surface condenses, exchange with overlying-water is progressively-reduced, and a *'hydrophobic-skin'* begins to form. This may occur due to an increase in hydrogen-bonding among EPS-polymers, perhaps through spontaneous *syneresis.* This slows intrusion of hypersaline waters into the mat. Finally, as salinities of overlying-waters exceed 200 mg/L, evaporite deposits of gypsum and halite, respectively, form on the mat surface. The underlying EPS attain a hard *'plastic-like'* condition, called a *glass state*, within which cells and extracellular molecules are embedded, and remain so until rehydration. We propose that the formation of an organic glass by components of EPS facilitate the preservation of important biomolecules and vesicles during desiccation so that upon rehydration, microbes can quickly resume activities.

Keywords: EPS, microbial mats, quorum sensing, desiccation, syneresis, glass

11.1 INTRODUCTION

The matrix of extracellular polymeric substances (EPS) surrounding cells is the defining emergent property of the biofilm (Flemming & Wingender, 2010). Its presence facilitates most of the activities that distinguish biofilm cells from their free-living counterparts (Flemming *et al.* 2007). Yet, the matrix itself is arguably the least understood component of biofilms. The physical structure of EPS ranges from a dissolved-state (i.e. colloidal) to a loose-flowing slime to a solid hydrated gel-state. The different physical states occur due to interactions with divalent cations (Ca^{2+}, Mg^{2+}), anions (PO_4^{2-}, SO_4^{2-}), weaker-force interactions (e.g., Van der Waals), and cross-linking among the matrix components such as proteins, nucleic acids, lipids and carbohydrates (Sutherland, 1990; Mayer *et al.* 1999). There are also more-specialized molecules such as extracellular-DNA (*e*DNA), amyloid proteins, proteoglycans, and glycoproteins, whose functions are also not yet fully understood. This has been reviewed elsewhere in this issue (*see* Neu and Lawrence chapter). It is realized that the matrix contributes important stabilizing properties to the microenvironments in proximity to cells. These stabilizing attributes, in part, have expanded the domain of microbial life to more-extreme environments, where the persistence of other life forms becomes limited. In this Chapter, the challenge of environmental desiccation to biofilm bacteria is explored. Specifically, the extracellular matrix of the biofilm is examined as a means to enhance the survival of cells and key extracellular molecules during desiccation, and facilitate subsequent recovery and function upon rehydration. This will be relevant to understanding life under extreme conditions, the search for life beyond Earth, and of more-immediate practical importance in preserving molecules and cells in the dry state.

11.2 THE CHALLENGE OF DESICCATION

11.2.1 Desiccation at the level of cells and individual molecules

Hydration is essential to all life processes. Desiccation, defined here as the partial or total loss of water, represents a physiological challenge to all life systems. This occurs when water is progressively-reduced in a biological system (e.g. a cell), and ultimately when the hydration shells surrounding individual biological molecules are removed (Alpert, 2005). The loss of water from cells or just outside of cells, incurs lethality when biological molecules making up the cells are either damaged upon dehydration or are unable to function properly upon rehydration. For example, cell membranes, proteins, lipids become denatured because of an accumulation of hydrophobic bonding that occurs during dehydration, and cannot properly re-hydrate to their native (active) conformations once water is again present. In other words, they remain in a denatured state, and are unable to function as needed in the biological system.

Most biological molecules require 'bound-water' as part of their active conformational structure. Bound-water consists of water molecules bonded to hydroxyl groups and other functional groups on the molecule. Proteins, for example, contain significant amounts of 'bound water' (*ca.*, approx. 0.25 g of H_2O per g protein) (Jaenicke, 1991). When water is bound to specific moieties of biological molecules, it contributes to their proper structure, orientation, stability and activity. That is why removal of water results in significant conformational changes to a protein. Interestingly however, NMR studies indicate that hydrogen-bonds between adjacent water molecules (in solution) are transient and water molecules, whether bound- or free are shifting positions and locations on a very fast timescale (*ca.* one picosecond). For example, the residence time of a 'bound-water' molecule when linked to a hydroxyl group on a sugar is approximately one nanosecond (ns) (Slade & Levine, 1991). Therefore, bound-water, while necessary to the proper steric structure of a molecule, is highly-transient under hydrated conditions, and continually shifting positions through electrostatic and H-bond interactions (Richards *et al.* 1989).

Molecules typically maintain a hydration sphere in immediate proximity to the molecule, as measured by their hydrated radius, which ensures an ample supply of water molecules. Two general hypotheses have been put forth to explain how molecules may retain activity after desiccation and rehydration. The first, called the anchor hypothesis, states that when desiccation occurs, bound-water remains 'bound' to the molecule. Therefore, a given molecule is not fully dehydrated. A second hypothesis, called the water-replacement hypothesis, suggests that other molecules (e.g. certain disaccharides) take the place of water in order to maintain the proper steric conformation of the molecule. These will be discussed in the context of EPS further below.

The periodic absence of water is a stressor that influences the activities and persistence of biofilms. Such conditions are considered to be a form of environmental-extreme. Therefore, it is surprising that certain microorganisms, and even plants/animals, are quite well adapted to periodic desiccation, and appear to be not adversely-affected by it. The ability to reversibly undergo desiccation and rehydration is called *anhydrobiosis* (Potts, 2001; Alpert, 2005). Across many environments, however, biofilms may persist and even flourish despite periodic desiccation. Certain bacteria and other microorganisms can survive desiccation for varying periods of time; ranging from just a few minutes to over several centuries. Some everyday examples of anhydrophilic biofilms include those present on the surfaces of mineral statues, building roofing-tiles and sidewalks (Keshari & Adhikary, 2013). Common occurrences in natural systems include the upper reaches of the ocean's rocky intertidal zones (Smith *et al.* 2004), where rock surfaces are exposed to air for hours each day. The biofilms of terrestrial and desert biocrusts are epi- or endo-lithic communities, which together span over 28% percent of the Earth's terrestrial surface, yet receive water as infrequent rainfall, or condensation (Garcia & Pringault, 2001; Belnap & Lange, 2003; Smith *et al.* 2004; Nagy *et al.* 2005; Pointing & Belnap 2012; Rajeev *et al.* 2013) and hypersaline ponds (Paerl *et al.* 2000).

11.2.2 EPS and hydration-maintenance

Much has been learned regarding the dehydration and rehydration of cells, however, comparatively little is known regarding the effects of these same processes on molecules just outside the cell. The contributing role of EPS in the persistence of bacteria and other microorganisms under water stress has been recognized for some time, and nicely summarized in the seminal reviews of Malcom Potts and colleagues (Potts, 1994, 1999, 2001, 2004). It is considered that the presence of EPS slows the rate of water loss and maintains a hydrated environment near cells for longer periods of time than in its absence. This appears crucial because the rate of desiccation directly affects desiccation tolerance and survival. Slow rates of dehydration allow cells an increased time to adapt for desiccation and enhances their survival upon rehydration. A large number of laboratory studies have documented increased EPS production by bacteria in response to water-limitation (Roberson & Firestone, 1992; Hill *et al.* 1994; Ophir & Gutnick, 1994; Shaw *et al.* 2003; Tamaru *et al.* 2005; Chang *et al.* 2007).

The EPS matrix of most biofilms provides a highly-hydrated milieu for its component cells. On average over 95–98% of a hydrated EPS matrix, by weight, is water. Such estimates consist of simple measurements of EPS weight before and after removal of water by evaporation. The EPS consist of a molecular network of polymers suspended with water-filled spaces in between them (Decho, 1999). When examined at a much closer level, using fourier-transform infared spectroscopy (FT-IR) and later proton-nuclear magnetic resonance (^{1}H-NMR), it becomes more-obvious that only a small portion of the water in a biofilm is actually bound to EPS; the majority being 'free-water' that is contained within the spaces of the polymer network (Schmitt *et al.* 1995; Schmitt & Flemming, 1998, 1999; Vogt *et al.* 2000). This observation has led to the assertion that the biofilm matrix is essentially 'immobilized water' (Sutherland, 1990), and is thought to be a strategy for retaining water in proximity to cells through a hydraulic decoupling to slow wetting or drying and protect cells from rapid fluctuations (Or *et al.* 2007).

Interestingly, however, free-water (e.g. distilled water in a beaker) does not actually consist of 'free' (i.e. individual) water molecules. Currently, water is thought to occur in clusters, having 6–8 or more water molecules bound to each other by transient hydrogen-bonds at any given time (Smith *et al.* 2005; Tokmachev *et al.* 2010). This would imply that diffusion of a molecule or ion through the water-filled spaces of the EPS matrix may actually be diffusing in between these clusters (Decho, 2015). A much closer look at a *Pseudomonas aeruginosa* biofilm by Flemming and colleagues (Vogt *et al.* 2000) was able to decipher three major water fractions having different diffusion coefficients ranging from 2×10^{-9} to 1×10^{-13} $m^2 s^{-1}$: a highly-mobile bulk water, an EPS-bound water, and an intracellular water fraction. This study indicated that the diffusivities of water (or molecules or ions) cannot be simply reduced to a single value. The bound-water fraction in EPS is of significance as dehydration occurs. Free (i.e. bulk)-water is removed first, then 'bound water' is removed. The sugar monomers of EPS typically contain an abundance of water-binding hydroxyl groups. Upon further dehydration, removal of bound-water allows more binding group (in proximity) to interact.

However, during desiccation complete removal of water from a molecule does not always occur. When certain polymers are extensively dried, they still may retain some water. Hyaluronic acid is an acidic polysaccharide polymer that is found in certain streptococcal bacterial capsules, (and also a major component of the human extracellular matrix) (Wessels *et al.* 1991). In its fully-hydrated form, N-H and −O-C=O pairs are surrounded by groups of 25 H_2O molecules (Maréchal *et al.* 2003). However, even when extensively dried at 104°C, the dried polymer is found to still retain water, as 'water-wires' consisting of 4–5 H_2O molecules that are anchored to carboxyl groups. This illustrates some of the unique, moiety-specific aspects of bound-water on EPS polymers, even under dehydration conditions. Other bacterial EPS likely operate similarly. Further examinations of polymers at this level of resolution are strongly warranted to understand the dynamic and adaptable nature of EPS.

Specific polysaccharides and proteins have been studied to understand their roles with respect to desiccation (Hill *et al.* 1997; Helm *et al.* 2000). The cyanobacterium *Nostoc commune* is tolerant of repeated desiccation and hydration. Potts and colleagues (Shaw *et al.* 2003) studied the kinetics of water uptake and loss by the extracellular glycans of this strain. They found that after seven-years of desiccation, dried samples took over 30 h to rehydrate fully. When rehydrated, the matrix showed a 20- to 40-fold increase in weight. However, after several repeated cycles of drying and rehydration, the time required to fully rehydrate was reduced to only 3 h.

During the drying process, it was found that the 'rate' of drying is important in tolerance success because it will affect the ability of bacteria to prepare for desiccation. When dried rather slowly, a unique 39 kDa water-stress protein (Wsp) and a superoxide dismutase (sodF) were produced in *N. commune* and secreted in high relative amounts as cells approach desiccation (Scherer & Potts, 1989; Wright *et al.* 2005). During desiccation, oxidative damage to DNA, proteins and other molecules is a primary driver in loss of functionality, even if the molecules are preserved in an intact conformational (i.e. natured) state. The composition of the Wsp protein is low in levels of tyrosine, methionine and lysine; amino acids that are sensitive to oxidation (Scherer & Potts, 1989), which occurs during desiccation. This affords Wsp with greater resistance to oxidative damage. These proteins are produced in response to UV A/B wavelengths, which generate oxidants during desiccation through photon interactions with EPS (Hill *et al.* 1994; Wright *et al.* 2005). In ephemeral lakes that are periodically exposed to desiccation, microbes biosynthesize wax esters. Upon rehydration, the wax esters are utilized by surviving bacteria to synthesize membrane lipids (Finkelstein *et al.* 2010). We still remain in the infancy of understanding how molecules within EPS influence desiccation tolerance, especially in natural systems.

11.3 MICROBIAL MATS: AN EPS-MACROSTRUCTURE HAVING MICROSCALE ARCHITECTURE

11.3.1 Mat systems

11.3.1.1 An EPS-based analog of Earth's earliest life

The ability of biofilms to survive cyclical periods of desiccation is an adaptation that likely developed early, persisted throughout the evolution of microbial life, and has contributed to the architecture of present-day biofilms. Mats represent one of the Earth's earliest ecosystems, and one that has been refined through geologic time. Evidence now suggests that mats, their EPS signatures called *MISS* (i.e. microbially-induced sedimentary structures), and marine stromatolites have existed since early in the Precambrian era (3.4 Gyrs), and are considered to be among one of the earliest detectable fossil forms of life on Earth (Noffke *et al.* 2013). Microbial mats therefore, represent a complex biofilm system with a long evolutionary history of exposure to desiccation and rehydration. Therefore, mats represent a persistent natural system from which to study EPS interactions during the desiccation process. The present-day mat analogs of these structures (Reid *et al.* 2000) and terrestrial roll-up structures (Beraldi-Campesi & Garcia-Pichel, 2011) have been studied in great detail in an effort to understand the early evolution of bacterial communities and to interpret macrofossil evidences.

11.3.2 Present-day mats

In order to begin understanding *in situ* changes in EPS, it is necessary to briefly cover certain cyclical changes that occur in the natural environments of microbial mats. Photosynthetic microbial mats are diverse, horizontally-layered structures and consist of cyanobacteria, whose carbon-fixation forms the driver for the mat community production and recycling. Other major functional groups of aerobic- and anaerobic heterotrophs such as fermenters, sulfate –reducers, sulfide-oxidizers, purple-sulfurs, and archaea utilize the photosynthetic organic-carbon production and couple this to the efficient recycling of elements, such as H, O, P, S, and N, by the highly-diverse microbial community of the mat (Des Marais, 2003; Fourcans *et al.* 2004; Goh *et al.* 2009). Photosynthetic bacteria typically optimize their vertical positions with respect to light gradients; some even utilize infared (>900 nm) wavelengths (Jørgensen & Des Marais, 1988). Together, these groups secrete an abundant network of EPS that offer coherency, optical enhancement (i.e. forward scattering of light), and protection to cells (Decho, 2000a; Decho *et al.* 2003).

11.3.2.1 Anhydrophilic hypersaline environments

The microbial mats located within hypersaline ponds, such as those of San Salvador Island, Bahamas (24°05′N, 74°30′W), consist of well-defined horizontal layers of different microbial functional groups that have been previously described (Paerl *et al.* 2003; Pinckney & Paerl 1997; Pinckney *et al.* 1995). The surface mat layer is composed of densely-packed arrays of microbial cells embedded within an EPS gel matrix that coats the mat surface. Immediately below this is a dense layer of cyanobacteria containing *Lyngbia* sp., *Oscillatoria* sp., *Microcoleus* sp.), green cocoid cyanobacteria *Pleuorcapsa* sp., and purple sulfur bacteria (e.g. *Chromatium* sp.). The well-defined layers of the mat system typically vary in relative thickness across transects and by season (Paerl *et al.* 2000). The mat community here faces a multitude of stresses, including salinity fluctuations ranging from less than 20 mg/L (NaCl) to as high as 340 mg/L in the overlying-water (i.e. recall seawater is approx. 32 mg/L), intense irradiance sometimes exceeding 2200 μE m^{-2} s^{-1}) and chronic nutrient depletion (Pinckney *et al.* 1995). During summer months, temperatures and salinities reach their highest annual levels, and are followed by a slow (e.g. weeks/months) evaporation of water in the ponds and underlying mats. Sudden heavy rainfalls during late summer begin the 'wet season' and contribute sudden large pulses of freshwater to the ponds. The microbial mat communities of these anhydrophilic ponds appear well-adapted to the seasonally-driven wet and dry conditions.

Active-growth and permeable EPS

During the times of relatively lower-salinities (30–90 mg/L NaCl), the mat community is most-active (Pinckney *et al.* 1995). It is at these times, the community forms 'polymeric towers' (PT) on the mat surface (Figure 11.1A). The process begins during the rainy season with moderate nutrient influx (Paerl *et al.* 2000) and results in the rapid grown of cyanobacterial-rich surface mats. Cyanobacteria produce an abundant and coherent EPS gel, which forms the structural architecture for the upward growth of the PTs. During high-daytime photosynthesis (Pinckney *et al.* 1995), bubbles are observed on mat surfaces. The orogeny of such structures is not entirely certain but likely result from upward migration of O_2 bubbles produced by high-photosynthesis and trapped with EPS. Others have shown that cyanobacteria produce similar structures under laboratory conditions in order to reduce O_2 toxicity to cells (Grötzschel & de Beer, 2002). Therefore, the PTs may be a consequence of rapid growth at more ideal (for cyanobacteria) salinities, combined with an increased access to nutrients.

11.3.2.2 Higher-salinity, syneresis and development of EPS hydrophobic-skin

Enhanced EPS Production during Water-Limitation

As desiccations approaches, prolonging the retention of water in a biofilm will likely enhance the survival of the microbial cells, especially as salinity increases. In general, slower rates of water-loss improve survival because they provide more time for a cell to physiologically adapt to the impending water limitation (Potts, 1994). Examining single-species cultures of *Bacillus subtilis*, a Gram-positive soil isolate, it was found that they produce a hydrophobic cap on their colonies. It is composed of an exopolysaccharide and an amyloid fiber-forming protein (TasA), and is assembled using a third smaller protein (BslA). Together, this forms an elastic, protective cap on the biofilm, perhaps to aid in water retention (Hobley *et al.* 2013). Similarly, the formation of a hydrophobic surface shield on a mat or biofilm may be helpful to cells in slowing water-loss, called *matric water-stress*, and even in slowing the intrusion of hypersaline (ions) water, called *osmotic water-stress*.

In a hypersaline pond, one might assume that the salinity of pond water overlying the mat, and mat-porewater should be similar. Recent observations, however, have shown that porewater-salinities within mats often do not correspond to those in the overlying-water (Decho, in prep). Rather, porewater-salinities remain generally lower than those in the overlying water, with the differences being most-pronounced at higher-salinities. During low-salinity conditions, salinity differences in the overlying water and porewater are negligible (i.e. only a few mg/ L higher than those in the mat-pore water). However, as the salinity in the overlying-water gradually increases (to 200–300 mg/L) due to evaporation, the EPS on the mat surface condense to form a more-coherent EPS surface coating (see below). This coincides with the precipitation of abundant evaporite deposits of gypsum and halite on the mat surface. At this point, salinities in the overlying water are now approximately 50–100 mg/L higher than in mat-porewater. During this period, occasional brief rains and wind slowly remove much of the evaporites from the dry, exposed mat-surface, leaving a coherent, leathery-EPS surface. Taken together, these observations suggest that the mat surface, and more specifically the EPS, is forming an ionic barrier between the overlying-water and the underlying mat; one that becomes increasingly pronounced during hypersaline conditions. This is demonstrated when intrusion of fluorescent lanthanide ions into the mat becomes negligible (Decho in prep). How this barrier works is an interesting question. The physical compaction of biofilms has been shown to occur on water-filtration membranes, as determined by optical coherence tomography. The biofilm matrix exhibits a flexible compaction and relaxation over varying time scales (Valladares-Linares *et al.* 2015) and has been referred to a the "hair-in-sink-effect (see Vrouwenvelder *et al.*, Chapter). This further illustrates the pliant nature of the EPS.

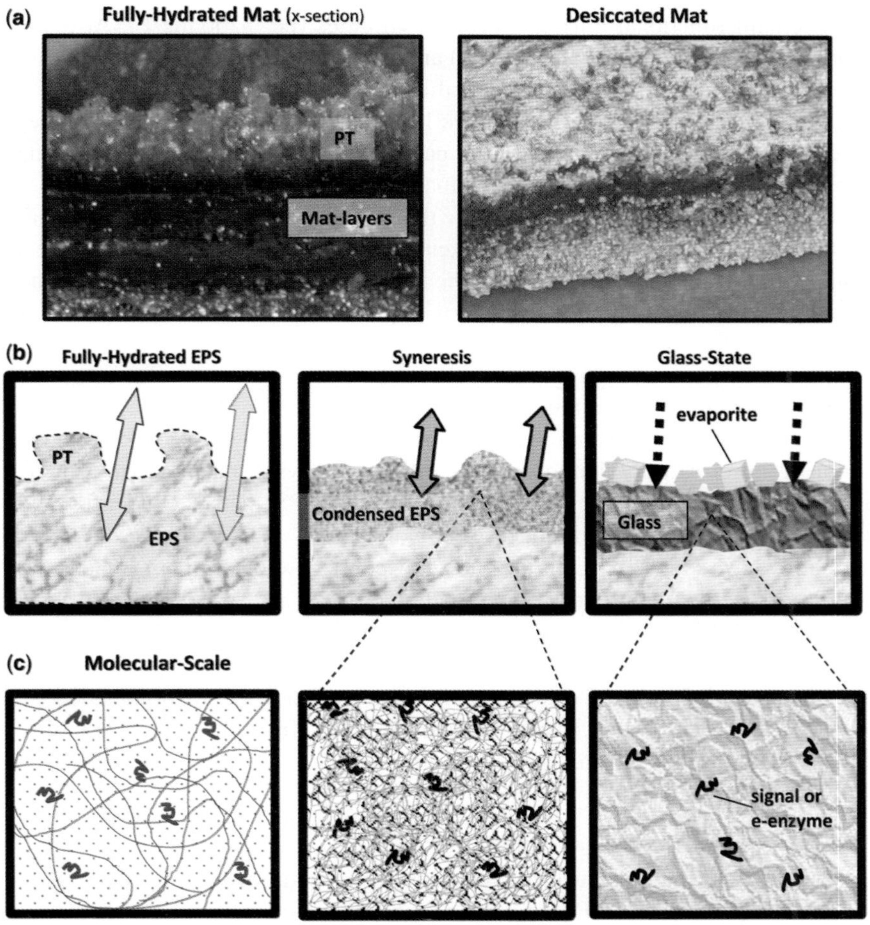

Figure 11.1 (**a**) Cross-sectional images showing fully-hydrated and desiccated mats from a hypersaline pond (San Salvador Is, Bahamas). In fully-hydrated state, polymeric towers (PT) are readily visible on surface. Desiccated mat is covered with evaporites of halite and gypsum. (**b**) A schematic illustrating the proposed preservation of microbial activities via glass-formation during desiccation. When EPS is fully-hydrated there is much diffusive exchange with overlying water. As salinity increases the process of syneresis likely ensues, and condenses EPS through increasing H-bond formation, and reduces diffusive-transfer with overlying water. In formation of a glass-state, most water is removed from surface EPS. (**c**) At a molecular scale, important extracellular molecules, such as signals, and e-enzymes, are "frozen" within glass to protect their intact steric conformation.

Potential mechanisms for ion-exclusion during high salinities

As evaporation occurs throughout the dry season in a hypersaline system there are increases in salinity, and the selective precipitation of salts (i.e. evaporate deposits) on the mat surface. Much of the Ca^{2+} is typically removed as gypsum ($CaSO_4 \cdot 2H_2O$) or smaller amounts of calcite ($CaCO_3$). Gypsum begins to precipitate when salinity concentrations reach about 160 mg/L (NaCl). At very high salinities (>300 mg/L) halite (NaCl) begins to precipitate. This allows less soluble minerals to be removed from solution, as a function of ionic concentration. Extreme Archaea halophiles additionally have specific polar phospholipids (e.g. PGP-Me), which are not present in halo-tolerant forms (Tenchov *et al.* 2006), and permit their cell membranes that remain functional during very-high salinities (i.e. >200 mg/L). At the immediate mat surface, this may allow different groups of bacteria and Archaea to become functionally active during the changing ionic conditions in anhydrophilic hypersaline systems.

Just as salinities in the overlying-water rise above 120 mg/L, the underlying mat itself begins to change, and begins to exclude (or significantly reduce the passive transport of) ions from entering into the mat. This implies that NaCl is being excluded from the mat. The slowing of ion intrusion likely occurs through changes in the EPS matrix, and would at least slow the rate of osmotic stress to cells, while concurrently conserving water within the mat. How this ion-exclusion mechanism might occur is a central question. One mechanism, currently under investigation, is the formation of a hydrophobic skin on the surface of the mat. Here, the polymers of an amphiphilic EPS on the uppermost surface of the mat, re-orient themselves during high-salinities to form the hydrophobic skin.

These initial investigations now suggest that the EPS of the surface mat layer may be a reversible ionic-gateway that buffers the mat communities (Decho, unpublished). The pliant EPS gateway differently responds to low- and high-salinity conditions. It allows the mat to be open for exchange of ions such as O_2, CO_2 and during lower salinity conditions (i.e. <100 mg/L) when much active growth and oxygenic photosynthesis are occurring. Then at higher-salinities, the EPS matrix condenses and in doing so, conserves water within the biofilm, and reduces influx of 'salts'.

Laboratory studies have shown that cells migrate in response to water-limitation and/or salinity changes. *Escherichia coli* cells actively-migrate to more-hydrated areas and self-organize prior to desiccation, which enhances survival (Gomez-Gomez *et al.* 2014). Cyanobacteria in natural desert crusts are observed to follow a vertical migration towards more-hydrated areas during repeated hydration/dehydration cycles (Pringault & Garcia-Pichel, 2004) and cyanobacteria have been shown to exhibit halotaxis in hypersaline mats (Kohls *et al.* 2010). These migrations may remove cells from the uppermost matrix during the formation of the EPS hydrophobic skin.

Syneresis in EPS gel matrix?

The extracellular matrix on the mat surface condenses to form an impervious, leathery-like layer. How does this process occur? How does EPS go from a highly-hydrated state to one of less-water and more-dense molecular packing? One likely possibility under current investigation, is the process of syneresis. By definition, *syneresis* is the expulsion of a liquid from a gel. In hypersaline ponds and just prior to desiccation, free-water becomes more limiting and ionic concentrations continue to increase. Syneresis is a process that can quickly (or slowly) condense EPS. The squeezing-out of water occurs due to the formation of hydrogen-bonds (and van der Waals interactions) among adjacent polymers (Saartratra *et al.* 2005; Amatayakul *et al.* 2006). Using the cheese-making process as an example, the formation of 'curd' (i.e. the solid phase) occurs because of the removal of immobilized water and hydrophilic molecules. This is thought to be initially triggered by an imbalance in intermolecular forces, and results in bonds being formed between hydrophobic sites, which then condenses and aggregates polymeric proteins and polysaccharides to form a relatively-dehydrated 'curd', with the concurrent loss (i.e. squeezing out) of water and hydrophilic molecules (i.e. the 'whey') into a surrounding solution phase (Pereira *et al.* 2009). The trigger(s) for syneresis is not fully understood, even though it has been a carefully-manipulated process that is used in the food industry. The extent and rates of syneresis by carrageenan gels have been precisely measured using ^{113}Cs-NMR (Ramakrishnan *et al.* 2004). Here, several factors contribute to syneresis. The presence of salts in the gel enhances the syneresis process, when compared with similar gels lacking salts at the same pH, and results in much of the salt being pushed out of the gel. In other artificial hydrogels, the swelling and contraction in response to both ions and pH, drives a reversible-transition of the gel (Zhang *et al.* 2007). In this study, changes in free-water content were monitored by differential scanning calorimetry (DSC), while FT-IR was used to show the process was linked to formation/dissociation of carboxyl groups. While laboratory-generated hydrogels do not necessarily reflect EPS gels under natural conditions, the results provide one such mechanism as to how such a reversible compaction and swelling might work. Other studies examining polysaccharides and proteins involved in yogurt consistency behavior have shown that acetylation strengthens the stability, elastic modulus and viscous-modulous of the colloids (Cui *et al.* 2014). The degree of acetylation was measured by FT-IR, examining the ratio of amides: C-O ether stretches (Smith, 1996). While these results are from yogurt-derived EPS, such changes leading to dehydrated zones may contribute to the microscale variability in natural EPS that is observed in many types of biofilms. These localized zones in EPS, called microdomains, have been noted by several authors (Decho, 2000b; Lawrence *et al.* 2007). The mat surface EPS is enriched in hydrophobic moieties, as determined by hydrophobic probes, and that may contribute to a hydrophobic skin (Decho, in prep). The relative enrichment of deoxy-sugars has been shown to increase the hydrophobic nature of some EPS (Underwood *et al.* 2004). However, in order to more-fully understand the compaction of mat EPS at high salinities, the use of non-invasive approaches such as portable Raman and FT-IR spectroscopy, and selective lectin probes coupled with confocal microscopy (Smith, 1996; Neu & Lawrence, 2009; Neu *et al.* 2010) will be required for examinations of natural EPS.

11.3.2.3 Desiccation protection by EPS glass formation

Compatible-solutes in EPS

As desiccation progresses, external ionic concentrations increase, and microbial cells experience osmotic stresses from a net loss of water (from cells). Most microorganisms continuously sense their external osmolarity conditions, often using proteins that are hydrophilic extensions of the cell wall or membrane (Heermann & Jung, 2004). This allows cells to respond accordingly in the perception of osmotic changes. In response, cells produce and/or accumulate internal compatible solute-molecules or ions in order to maintain an osmotic balance with the external environment. Halophilic/halotolerant microorganisms, which are common inhabitants in mats, have evolved two basic intracellular pathways in order to compensate for increases in salt stress: 1) a KCl-type and 2) a compatible-solute type of osmoadaptation (Galinski & Trüper, 1994; Roeβler & Müller, 2001). Both of these types accumulate within cells to osmotically-balance external ion concentrations. Many archaea in hypersaline ponds have internal K^+ ion concentrations, which balance the high external salinities (da Costa *et al.* 1998; Roeβler & Müller, 2001), with some cells containing up to 5 M KCl (Potts, 1994). Bacteria, in contrast, utilize a range of compatible solutes such as glycerol, glycine-betaine, trehalose, glycine, B-glutamate and proline to compensate for osmotic challenges (Welsh & Herbert, 1999; Potts, 1994; Welsh, 2000; Sleator & Hill, 2001; Tanne *et al.* 2014). Compatible solutes are most-often highly-soluble and of low molecular-weight. In addition, enzymes in halophilic organisms tend have structural adaptations that afford them stability under the comparatively-high ionic concentrations (Madern *et al.* 2000). Osmoprotection in such extreme environments such as the high-salinity systems of hypersaline ponds and bays is crucial to the survival of microorganisms (Goh *et al.* 2010). Some compatible solutes also offer protective roles to cells during desiccation, and therefore, have been called *xeroprotectants* (Julca *et al.* 2012). For example, some xeroprotectants can function as antioxidants. It is significant, however, that certain of these solutes, while of obvious importance to cells, also accumulate outside of cells in the EPS. This may occur either by direct cellular secretion or through leakage from dead cells. As explained below, the extracellular accumulation in EPS of the osmolyte trehalose can exert protective effects on the preservation of important extracellular molecules during desiccation.

Maintaining steric conformation of extracellular biomolecules during desiccation

Why do bacteria (and Archaea) preserve extracellular molecules? There are many important extracellular biomolecules such as *e*-DNA, *e*-enzymes, antibiotics, signals, sensory molecules and vesicles, that exist in the extracellular mileau (Böeckelmann *et al.* 2006; Decho *et al.* 2009; Mashburn & Whiteley, 2005; Schooling & Beveridge, 2006) and therefore, during desiccation must remain in an intact conformation that approximates their hydrated state. This is necessary for the biomolecules in order to regain their functionality when rehydration occurs. It is realized that during desiccation either a) a portion of the water remains 'anchored' on biomolecules, or b) water is entirely removed and 'replaced' by some other moiety. The loss of water is disruptive to the steric conformation of most molecules and hence affects their ability to function. Removal of water induces the formation of hydrogen-bonds, where water was once bound, and results in the molecule being permanently denatured and rendered non-functional by water-loss. Therefore, a central theme in desiccation has been to understand how biological molecules persist during loss of water, and can again be functional once water is restored? Potts and colleagues (Hill *et al.* 1997) demonstrated that an extracellular polysaccharide inhibited fusion of membrane vesicles upon drying. In the context of a biofilm, the preservation extracellular molecules, such as *e*-enzymes, antibiotics, signaling and sensory molecules, that are localized within the EPS, in addition to the cells themselves is of obvious importance (Decho *et al.* 2010; Montgomery *et al.* 2013).

Two major hypotheses

Two general hypotheses have been developed to explain how biological molecules are protected in their proper structural conformations during desiccation, so they can regain functionality upon rehydration. As we shall see the hypotheses are not mutually-exclusive, and all can involve small molecules called *xeroprotectants* (Julca *et al.* 2012). They are the: 1) water-replacement hypothesis; 2) anchor or water-entrapment hypothesis. The mechanisms appear to be similar, but they are fundamentally different in 'how' they operate, and both can be facilitates by the formation of an organic glass state in proximity to the molecules.

The *water-replacement hypothesis* states that if/when bound-water is removed by desiccation, certain molecules (i.e. xeroprotectants) take their place, and in doing so, retain the proper 3-dimensional configuration of the molecule (Clegg *et al.* 1982; Oliver *et al.* 2001). These are poly-hydroxylated compounds such as glycerol, trehalose and sucrose. The hydroxyl groups on xeroprotectants form hydrogen-bonds with the biomolecules, where water was once bound. During desiccation and removal of water, polar groups on the phospholipids of cell-membranes become fused together and cannot rehydrate to their proper steric forms. Membrane proteins also become denatured. Binding of xeroprotectants avoids the merging of the lipids, and denaturation of proteins, respectively, during dehydration. This theory involves the removal of shell water from the biomolecule and being replaced by the binding of the poly-hydroxylated compound (e.g. trehalose). This requires H-bonding between the hydroxyl on the sugar and biomolecules in question, where water once occupied, and provides the dry biomolecule with a physical conformation resembling the hydrated-state. The *water-entrapment hypothesis* does not involve inhibition of molecular motion. Rather, according to this hypothesis, *sugars concentrate water close to the surfaces of biomolecules*, and in doing so preserve their intact properties. Thus the molecules remain intact because they remain hydrated (Belton & Gil, 1994). In nature, it is likely that a combination of these mechanisms, and perhaps others, contribute to the preservation of extracellular biological molecules exposed to desiccation within the EPS matrix.

Glass-formation or vitrification

From a structure/function standpoint, the formation of a glass state (see below) ensures reduced chemical reactivity (of biomolecules) and its associated deterioration (Aksan & Toner, 2004). A glass involves the formation of an amorphous solid in which the hydrated conformation of a biomolecule is essentially frozen (Angell, 1995). Vitrification, therefore, prevents chemical reactions from occurring that are destructive to the biomolecules trapped within the glass. A glass corresponds to a physical solid, but is a thermodynamic liquid having a very high-viscosity and flows very, very slowly (e.g. 30 μm per century) (Buitink & Leprince, 2004). (Note: old glass windows in buildings also flow slowly due to gravity, and for this reason will be thicker at the bottom, compared with the top after being in-place for >100 years). More importantly, and from a biological standpoint, a glass is an amorphous solid (i.e. having no defined structure), so molecules localized within a glass will have little/no mobility, not even molecular motion (Crowe *et al.* 1998; Crowe, 2002). This conformational rigidity allows a molecule's steric configuration to be preserved. If a molecule, such as an enzyme, is localized in a glass while in a natured-state, it can potentially remain so until rehydration ensues. While glass formation seems a necessary attribute of anhydrophilic systems, it is not solely sufficient to maintain the natured-state of a molecule(s). When polysaccharide-rich polymers are exposed to heating and cooling, high ionic concentrations or loss of water, they may begin to enter a glass-state. Glass formation occurs as a result of loss of free-water and a concurrent increase in H-bonding (Buitink & Leprince, 2004). It was first suggested long ago that EPS may contribute to the stabilization of cells during storage in the dry state (Potts, 2001), and certain EPS have properties similar to those of amorphous glasses (Billi & Potts, 2002; Narvaez-Reinaldo *et al.* 2010; Potts, 1994). Interestingly some of the water within a bacterial glass could remain localize within the hydrated radius of a molecule, until rehydration.

The presence of a glass-state is detected by the *glass-transition temperature* (T_g) using differential scanning calorimetry (DSC). In the glass-state, molecular motion (i.e. translational energy) is essentially frozen (Angell, 2008). The T_g, measures where the heat-capacity (i.e. ability of sample to absorb heat) abruptly changes during heating. This occurs when a glass reaches a high-enough temperature to lose its molecule rigidity, and molecules are once again moving.

The relative size of a molecule influences its glass transition temperature. High molecular weight polymers often have an elevated T_g. However, they are not effective in preserving other molecules owing to their size. Larger polymers also form glasses, but owing to their larger size, the glass matrix will have a higher molecular-free volume (Buitink & Leprince, 2004). There is a likelihood that small disaccharides, such as trehalose, may interact with larger polymers in glass formation. For example, vitrification can be accelerated by incorporating high-mw dextrans with trehalose (Aksan & Toner, 2004). Fourier-transform infared (FT-IR) spectroscopy has verified the increasing formation of hydrogen bonds during glass formation (Carpenter & Crowe, 1989; Wolkers *et al.* 2004). By examining small changes in the locations of OH spectral bands, the studies showed that the relative length and strength of hydrogen bonds increases with increasing molecular sizes of sugars. Since such a matrix will consist of a random network of hydrogen bonds, the longer and weaker bonds would be expected to exhibit higher rates of molecular mobility within the matrix – and be less stable in preserving the structure of biomolecules enclosed within the matrix.

Small molecules such as peptides, disaccharides and oligosaccharides typically form the most-effective glasses for preserving other molecules (Allison *et al.* 1999; Shirke & Ludescher, 2005). A number of disaccharides such as sucrose, maltose, lactose and trehalose readily form glasses (Haque *et al.* 2006; Crowe, 2009). Studies of glasses made of the sugar maltose show that at stable network of many hydrogen bonds is formed at the T_g, which immobilizes hydroxyl groups (van den Dries *et al.* 1998). Glasses can be formed by di-saccharide sugars such as trehalose (glu-glu), and lactose (glu-gal), and are extensively used in dried foods to preserve (the structures of proteins that affect) taste, flavor and shelf-life. The glass-state preserves important molecules by preventing their denaturation in the absence of water (Crowe *et al.* 1996, 1998). The organic glass retards damaging chemical reactions and deleterious physical events, such as crystal formation (Crowe, 2002), and may stop most chemical reactions that require molecular diffusion (Burke, 1986). Previous studies have shown that trehalose is commonly-utilized by soil bacteria during periods of drought (Zhang & Yan, 2012) in order to maintain the membrane stability of cells (Crowe *et al.* 1988) and preserve the structure of proteins (Leslie *et al.* 1995).

Formation of an organic glass and trehalose?

The study of trehalose in glass formation provides a useful starting point for understanding microbial survival in natural EPS and biofilms. However, it will require much further study to decipher the mechanisms of protection provided during desiccation. It has been observed that when many microbial cells are dried slowly, they synthesize large quantities of the disaccharide trehalose (Hershkovitz *et al.* 1991; Avonce *et al.* 2006; Crowe, 2014). Trehalose is especially efficient for microbiological systems and can account for up to 20% of the dry weight of a cell during desiccation (Clegg, 2001). Trehalose is a multifunctional molecule for microbial cells. It can serve as an energy and carbon reserve, a stabilizer and protectant for proteins and cell membranes, protects against heat and damage by oxygen radicals, and is even used as a sensing compound and structural component of certain bacterial cell walls such as *Mycobacterium tuberculosis* (Potts, 1994; Elbein *et al.* 2003).The importance of trehalose in stabilizing cell membranes cannot be over emphasized (Crowe & Tablin, 2004) because it has strong implications to the EPS matrix. For example, extracellular vesicles are often abundant, especially in microbial mat EPS (Mashburn & Whiteley, 2005; Schooling & Beveridge, 2006). Vesicles often contain important molecules such as antibiotics, signaling molecules, enzymes and nucleic acids. The vesicles may additionally be packed with trehalose and form 'protective glass packets' for enclosed biomolecules.

While there are three major forms of trehalose, only the α,α-trehalose is a naturally-occurring form. Its T_g is quite high (T_g = 85–120°C), depending on its internal water content (Albertorio *et al.* 2007), and therefore forms a stable amorphous solid, even at relatively high temperatures. The unique properties of trehalose make it extremely useful to biological systems (Feofilova *et al.* 2014; Leprince & Buitink, 2010; Schebor *et al.* 2010). It forms a glass while in a dried state, and the temperature at which it undergoes a glass/liquid transition is very-high compared with other sugars.

The type of linkage between two sugars affects the T_g. Trehalose is especially efficient because the α, α-1,1 glycosidic bond linking the two glucose monomers is highly stable and forms a V-shaped cleft between the two adjacent glucose monomers. The dynamic conformation of trehalose is less flexible than any other disaccharides due to anisotropic hydration (Choi *et al.* 2006). The very-low hydration adjacent to the glycosidic oxygen imparts the conformational rigidity, which makes trehalose an excellent sugar-template for stable interactions with hydrogen-bonded water molecules (Liu *et al.* 1997; Choi *et al.* 2006). Early examinations of trehalose indicated that it has very high ability to bind water, as determined by NMR (Kawai *et al.* 1992). A single trehalose molecule is known to bind up to 10 water molecules (Magazu *et al.* 1997). Since it is a non-reducing sugar, it does not undergo reactions with proteins and other molecules. It is non-hygroscopic over a wide humidity range, yet it is freely-soluble in water. Trehalose readily forms hydrogen bonds with polar residues in proteins and phospholipids of membranes, and confers a physical state to a dried molecule, that resembles the presence of water. In short, it acts as a water analog while the molecule (e.g. protein) is dry (Jain & Roy, 2009). Therefore, a dynamic reduction of water occurs when another molecule binds with trehalose.

The formation of organic glasses within the extracellular matrix is currently being explored and verified, and offers one potential mechanism for the preservation of microbial cells and important extracellular molecules during the often-frequent desiccation and rehydration that are experienced by biofilms.

11.3.2.4 Rehydration and rapid resumption of activities

Preservation of cellular and extracellular activities

Some desiccated microbial communities are able to rapidly regain metabolic activity when rehydrated, even after prolonged periods (i.e. months to years) desiccation. There are many anecdotal reports of microbes surviving for centuries, however, remain unconfirmed. Recent confirmations show that cyanobacteria were able remain desiccated up to 13 yrs without significant DNA damage to cells, and were successfully rehydrated (Shirkey *et al.* 2003).

How rapidly can a desiccated system recover? Interestingly, some dried mat communities, upon the addition of water, exhibit a relatively rapid (e.g., minutes to hours) recovery from anhydrobiosis. Laboratory studies have shown that certain processes, such as lipid biosynthesis, are very quickly restored from desiccated cells upon rehydration (Taranto *et al.* 1993), some as rapid as 2 minutes (Harel *et al.* 2004). In other natural saline systems, rehydration was followed by bacterial respiration in just 10–30 min, and then followed by photosynthesis in just 2 h (Anati 1999; Fleming *et al.* 2007). In other desiccated marine microbial mats, consisting largely of cyanobacteria, photosynthesis was restored in just 15 min, with increased chlorophyll detected, and followed by a cyanobacterial migration towards the surface after 48 h (Chennu *et al.* 2015). Similar relatively rapid recoveries were observed in desert crust biofilms (Abed *et al.* 2014). In laboratory studies, preservation of extracellular enzyme activities were facilitated by the presence of trehalose during heating and other stressors (Skillman *et al.* 1999; Melo *et al.* 2001). The important area of extracellular enzyme activities within EPS requires further investigation.

Potts (1994) indicated that when cells are dehydrated more slowly, they have greater chances of recovery after rehydration. Slower dehydration provides increased time for the cell to synthesize trehalose and other protective measures. For example, Potts and colleagues (Billi *et al.* 2000; Shirkey *et al.* 2003) showed that EPS slowed the evaporative loss of water; a process that may afford extra time for cells and extracellular enzymes, such as superoxide dismutase to prepare for impending desiccation. This reduced the production of oxidant radicals. During desiccation, there exist the potential for UV-induced oxidative damage to nucleic acids, but also proteins. As mentioned earlier, polysaccharide exposure to UV photons often results in release of oxidative radicals, whose damaging effects will occur during desiccation but also upon rehydration (Shirkey *et al.* 2003). They also must address how molecules are stabilized and protected against oxidation (during desiccation) (Crowe, 2014).

11.4 CONCLUSIONS

The ability of biofilms to undergo dehydration and desiccation, then resume metabolic activities upon rehydration is an extra-ordinary capability restricted to only a few animals and plants, but is present in a multitude of microorganisms (Crowe, 2014). In biofilms, this ability is likely enhanced by the properties of its EPS matrix. However, there remains considerable exploration to formulate hypotheses, and to guide our understanding. The present review has provided a framework, and indeed some speculative inference to begin understanding how the EPS matrix of a biofilm contributes to the survival and revival of dried microbial communities; a process that is abundantly common across many environments. From an ecological perspective, desiccation represents a major disturbance for most microbial organisms; one that requires a degree of adaptation prior to dehydration (Barthes *et al.* 2015). Microorganisms, especially those within biofilms, have evolved to survive in extreme environments through a number of adaptations. Microbial mats consist of a structured, laminated community that allows microorganisms to survive and thrive in extreme environments. It is proposed that the formation of an organic glass by components of EPS facilitate the preservation of important biomolecules and vesicles during desiccation. The remarkable pliancy of the EPS responds differently to changing ionic conditions and affords greater stability to the mat community when challenged by osmotic, irradiation, and temperature changes. Finally, future investigations must address the large suite of genes, proteins and metabolites that are involved in desiccation protection; which has been collectively termed 'the desiccome' (Leprince & Buitink, 2010).

ACKNOWLEDGEMENTS

This work was supported by several grants from the US National Science Foundation (BME-1032579, DMR-1206072, EF-0723707, MCB-0132528, EF-0723707). We thank the staff of the Gerace Research Centre (San Salvador, Bahamas) for hospitality and efficiency during the course of field research, and for helpful discussions by many EPS-based researchers, and those in other fields as well.

REFERENCES

Abed R. M. M., Polerecky L., Al-Habsi A., Oeetjec J., Strous M. and de Beer D. (2014). Rapid recovery of cyanobacterial pigments in desiccated biological soil crusts following addition of water. *PLoS ONE*, **9**, 2112372. Doi: 10.1371/journal.pone.0112372.

Aksan A. and Toner M. (2004). Isothermal desiccation and vitrification kinetics of trehalose-dextran solutions. *Langmuir*, **20**, 5521–5529.

Albertorio F., Chapa V. A., Chen X., Diaz A. J. and Cremer P. S. (2007). The α,α-($1 \rightarrow 1$) linkage of trehalose is key to anydrobiotic preservation. *Journal of American Chemical Society*, **129**, 10567–10574.

Allison S. D., Chang B., Randolph T. W. and Carpenter J. F. (1999). Hydrogen bonding between sugar and protein is responsible for inhibition of dehydration-induced protein unfolding. *Archives Biochemistry Biophysics*, **365**, 289–298.

Alpert P. (2005). The limits on frontiers of desiccation-tolerant life. *Integrative Comparative Biology*, **45**, 685–695.

Amatayakul T., Sherkat F. and Shah N. P. (2006). Syneresis in set yogurt as affected by EPS starter cultures and levels of solids. *International Journal of Dairy Technology*, **59**(3), 216–221.

Anati D. A. (1999). The salinity of hypersaline brines: concepts and misconceptions. *International Journal of Salt Lake Research*, **8**, 55–70.

Angell C. A. (1995). Formation of glasses from liquids and biopolymers. *Science*, **267**, 1924–1935.

Angell C. A. (2008). Insights into phases of liquid water from study of its unusual glass-forming properties. *Science*, **319**, 582–587.

Avonce N., Mendoza-Vargas A., Morett E. and Iturriaga G. (2006). Insights on the evolution of trehalose biosynthesis. *BMC Evolutionary Biology*, **6**, 109. doi: 10.1186/1471-2148-6-109.

Barthes A., Ten-Hage L., Lamy A., Rols J.-L. and Leflaive J. (2015). Resilience of aggregated microbial communities subjected to drought-small-scale studies. *Microbial Ecology*, **70**, 9–20.

Belnap J. and Lange O. L. (2003). Biological Soil Crusts: Structure, Function, and Management, Ecological Studies Series 150, I. T. Baldwin et al. (eds), Springer, Berlin.

Belton P. S. and Gil A. M. (1994). IR and Raman-spectroscopic studies of the interaction of trehalose with hen egg-white lysozyme. *Biopolymers*, **34**(7), 957–961.

Beraldi-Campesi H. and Garcia-Pichel F. (2011). The biogenicity of modern terrestrial roll-up structures and its significance for ancient life on land. *Geobiology*, **9**, 10–23.

Billi D. and Potts M. (2002). Life and death of dried prokaryotes. *Research in Microbiology*, **153**, 7–12.

Billi D., Wright D. J., Helm R. F., Prickett T., Potts M. and Crowe J. H. (2000). Engineering desiccation tolerance in *Escherichia coli. Applied Environmental Microbiology*, **66**, 1680–1684.

Böeckelmann, U., Janke A.; Kuhn R., Neu T. R., Wecke J., Lawrence J. R. and Szewzyk U. (2006). Bacterial extracellular DNA forming a defined network-like structure. *FEMS Microbiology Letters*, **262**, 31–38.

Buitink J. and Leprince O. (2004). Glass formation in plant anhydrobiotes: survival in the dry state. *Cryobiology*, **48**, 215–228.

Burke M. J. (1986). The glassy state and survival of anhydrous biological systems. In: Membranes, Metabolism and Dry Organisms, A.C. Leopold (ed.), Cornell University Press, Ithaca, N.Y. pp. 358–363.

Carpenter J. F. and Crowe J. H. (1989). Infared spectroscopic studies on the interaction of carbohydrates with dried proteins. *Biochemistry*, **28**, 3916–3922.

Chang W.-S., van de Mortel M., Nielsen L., de Guzman G. N., Li X. and Halverson L. J. (2007). Alginate production by *Psuedomonas putida* creates a hydrated microenvironment and contributes to biofilm architecture and stress tolerance under water-limiting conditions. *Journal of Bacteriology*, **189**(22), 8290–8299.

Chennu A., Grinham A., Polerecky L., de Beer D. and Al-Najjar A. A. (2015). Rapid reactivation of cyanobacterial photosynthesis and migration upon rehydration of desiccated marine microbial mats. *Frontiers in Microbiology*, **6**, 1472. doi: 10.3389/fmicb.2015.01472.

Choi Y., Cho K. W., Jeong K. and Jung S. (2006). Molecular dynamics simulations of trehalose as a dynamic reducer' for solvent water molecules in the hydration shell. *Carbohydrate Research*, **341**, 1020–1028.

Clegg J. S (2001). Cryptobiosis- a peculiar state of biological organization. *Comparative Biochemistry Physiology B Biochemistry Molecular Biology*, **128**, 613–624.

Clegg J., Seitz P., Seitz W. and Hazlewood C. F. (1982). Cellular responses to extreme water loss: the water-replacement hypothesis. *Cryobiology*, **19**, 306–316.

Crowe L. M. (2002). Lessons from nature: the role of sugars in anhydrobiosis. *Comparative Biochemistry Physiology Part A: Molecular and Integrative Physiology*, **131**(3), 505–513.

Crowe J. H. (2009). Preservation of biomaterials in the dry state: lessons from nature. *In Vitro Cell Developmental Biology-Animals*, **45**, S18–S18.

Crowe J. H. (2014). Anhydrobiosis: an unsolved problem. *Plant, Cell and Environment*, **37**, 1491–1493.

Crowe J. H. and Tablin F. (2004). Stabilization of cells in the dry state by trehalose. *In Vitro Cell Developmental Biology-Animals*, **40**, 11A–11A.

Crowe J. H., Crowe L. M., Carpenter J. F., Rudolph A., Wistrom C. A., Spargo B. and Anchordoguy T. (1988). Interactions of sugars with membranes. *Biochimica et Biophysica Acta (BBA)-Reviews on Biomembranes*, **947**(2), 367–384.

Crowe L. M., Reid D. S. and Crowe J. H. (1996). Is trehalose special for preserving dry biomaterials. *Biophysical Journal*, **71**, 2087–2093.

Crowe J. H., Carpenter J. F. and Crowe L. M. (1998). The role of vitrification in anhydrobiosis. *Annual Review of Physiology*, **60**, 73–103.

Cui M., Fang L., Zhou H. and Yang H. (2014). Effects of amino acids on the physiochemical properties of potato starch. *Food Chemistry*, **151**, 162–167.

da Costa M. S., Santos H. and Galinski E. A. (1998). An overview of the role and diversity of compatible solutes in Bacteria and Archaea. *Advances in Biochemical Engineering and Biotechnology*, **61**, 117–153.

Decho A. W. (1999). Imaging an alginate polymer gel matrix using atomic force microscopy. *Carbohydrate Research*, **315**(3), 330–333.

Decho A. W. (2000a). Microbial biofilms in intertidal systems: an overview. *Continental Shelf Research*, **20**(10), 1257–1273.

Decho A. W. (2000b). Exopolymer microdomains as a structuring agent for heterogeneity within microbial biofilms. In: Microbial Sediments, R.E. Riding and S.M. Awramik (eds), Springer-Verlag, Berlin, pp. 9–15.

Decho A. W. (2015). Localization of quorum sensing by extracellular polymeric substances (EPS): considerations of *in situ* signaling. In: The Physical Basis of Bacterial Quorum Communication, S.J. Hagen (ed.), Biological and Medical Physics, Biomedical Engineering, doi: 10.1007/978-1-4939-1402-9_6.

Decho A. W., Kawaguchi T., Allison M. A., Louchard E. M., Reid R. P., Stephens F. C., Voss K. J., Wheatcroft R. A., Taylor B. B. (2003). Sediment properties influencing upwelling spectral reflectance signatures: the "Biofilm gel effect". *Limnology Oceanography*, **48**, 431–443.

Decho A. W., Visscher P. T., Ferry J., Kawaguchi T., He L., Przekop K. M., Norman R. S. and Reid R. P. (2009). Autoinducers extracted from microbial mats reveal a surprising diversity of *N*-acylhomoserine lactones (AHLs) and abundance changes that may relate to diel pH. *Environmental Microbiology*, **11**(2), 409–420.

Decho A. W., Norman R. S. and Visscher P. T. (2010). Quorum sensing in natural environments: emerging views from microbial mats. *Trends in Microbiology*, **18**(2), 73–80.

Des Marais D. J. (1995). The biogeochemistry of hypersaline microbial mats. *Advances in Microbial Ecology*, **14**, 251–274.

Des Marais D. J. (2003). Biogeochemistry of hypersaline microbial mats illustrates the dynamics of modern microbial ecosystems and the early evolution of the biosphere. *Biological Bulletin*, **204**, 160–167.

Elbein A. D., Pan Y. T., Pastuszak I. and Carroll D. (2003). New Insights on trehalose: a multifunctional molecule. *Glycobiology*, **13**(4), 17R–27R.

Feofilova E. P., Usov A. I., Mysyakina I. S. and Kochkina G. A. (2014). Trehalose: chemical structure, biological functions and practical application. *Microbiology*, **83**(3), 184–194.

Finkelstein D. B., Brassell S. C. and Pratt L. M. (2010). Microbial biosynthesis of wax esters during desiccation: adaptation for colonization of the earliest environments? *Geology*, **38**(3), 247–250.

Fleming E. D. and Castenholz R. W. (2007). Effects of periodic desiccation on the synthesis of the UV-screening compound, scytonemin, in cyanobacteria. *Environmental Microbiology*, **9**(6), 1448–1455.

Fleming E. D., Bebout B. M. and Castenholz R. W. (2007a). Effects of salinity and light intensity on the resumption of photosynthesis in rehydrated cyanobacterial mats from Baja California Sur, Mexico. *Journal of Phycology*, **43**, 15–24.

Flemming H. C. and Wingender J. (2010). The biofilm matrix. *Nature Reviews in Microbiology*, **8**(9), 623–633.

Flemming H-C., Neu T. R. and Wozniak D. J. (2007b). The EPS matrix: the "House of Biofilm Cells". *Journal of Bacteriology*, **189**(22), 7945–7947. doi: 10.1128/JB.00858-07.

Fourcans A., Garcia et Oteyza T., Wieland A., Solé A., Diestra E., van Bleijswijk J., Grimalt J. O., Kühl M., Esteve I., Muyzer G., Caumette P. and Duran R. (2004). Characterization of functional bacterial groups in a hypersaline microbial mat community (Salins-de-Giraud, Camargue, France). *FEMS Microbiology Ecology*, **51**, 55–70.

Galinski E. A. and Trüper H. G. (1994). Microbial behaviour in salt-stressed ecosystems. *FEMS Microbiology Reviews*, **15**(2–3), 95–108.

Garcia-Pichel F. and Pringault O. (2001). Cyanobacteria track the water in desert soils. *Nature*, **413**, 380–381.

Goh F., Allen M. A., Leuko S., Kawaguchi T., Decho A. W., Burns B. P. and Neilan B. A. (2009). Determining the specific microbial populations and their spatial distribution within the stromatolite ecosystem of Shark Bay. *ISME Journal*, **3**(4), 383–396.

Goh F., Barrow K. D., Burns B. P. and Neilan B. A. (2010). Identification and regulation of novel compatible solutes from hypersaline stromatolite-associated cyanobacteria. *Archives Microbiology*, **192**, 1031–1038.

Gomez-Gomez J. M., Medina J., Hochberg D., Mateo-Marti E., Martinez-Frias J. and Ruli F. (2014). Drying bacterial biosaline patterns capable of vital reanimation upon rehydration: novel hibernating biomineralogical life formations. *Astrobiology*, **14**(7), 589–602.

Grötzschel S. and de Beer D. (2002). Effect of oxygen concentration on photosynthesis and respiration in two hypersaline microbial mats. *Microbial Ecology*, **44**, 208–216.

Haque M. K., Kawai K. and Suzuki T. (2006). Glass transition and enthalpy relaxation of amorphous lactose glass. *Carbohydrate Research*, **341**, 1884–1889.

Harel Y., Ohad I. and Kaplan A. (2004). Activation of photosynthesis and resistance to photoinhibition in cyanobacteria within biological desert crust. *Plant Physiology*, **136**, 3070–3079.

Heermann R. and Jung K. (2004). Structural features and mechanisms for sensing high osmolarity in microorganisms. *Current Opinions in Microbiology*, **7**, 168–174.

Helm R. F., Huang Z., Edwards D., Leeson H., Peery W. and Potts M. (2000). Structural characterization of the released polysaccharide of the desiccation-tolerant *Nostoc commune* DHR-1. *Journal of Bacteriology*, **182**, 974–982.

Hershkovitz N., Oren A., Cohen Y. (1991). Accumulation of trehalose and sucrose in cyanobacteria exposed to matric water stress. *Applied Environmental Microbiology*, **57**, 645–648.

Hill D. R., Hladum S. L., Scherer S. and Potts M. (1994). Water stress proteins of *Nostoc commune* (cyanobacteria) are secreted by UV-A/B absorbing pigments and associate with 1,4-β-D-xylanohydrolase activity. *Journal of Biological Chemistry*, **269**, 7726–7734.

Hill D. R., Keenan T. W., Helm R. F., Potts M., Crowe L. M. and Crowe J. H. (1997). Extracellular polysaccharide of *Nostoc commune* (Cyanobacteria) inhibits fusion of membrane vesicles during desiccation. *Journal of Applied Phycology*, **9**, 237–248.

Hobley L., Ostrowski A., Rao F. V., Bromley K. M., Porter M., Prescott A. R., MacPhee C. A., van Aalten D. M. F. and Stanley-Wall N. R. (2013). BslA is a self-assembling bacterial hydrophobin that coats the *Bacillus subtilus* biofilm. *Proceedings of the National Academy of Sciences USA*, **110**(33), 13600–13605.

Jaenicke R. (1991). Protein stability and molecular adaptation to extreme conditions. *European Journal of Biochemistry*, **202**, 715–729.

Jain N. K. and Roy I. (2009). Effect of trehalose on protein structure. *Protein Science*, **18**(1), 24–36.

Jorgensen B. B. and Des Marais D. J. (1988). Optical properties of benthic photosynthetic communities: fiber optic studies of cyanobacterial mats. *Limnology Oceanography*, **33**, 99–113.

Julca I., Alaminos M., Gonzalez-Lopez J. and Manzanera M. (2012). Xeroprotectants for the stabilization of biomaterials. *Biotechnology Advances*, **30**, 1641–1654.

Kawai H., Sakurai M., Inoue Y., Chujo R. and Kobayashi S. (1992). Hydration of oligosaccharides: anomalous hydration ability of trehalose. *Cryobiology*, **29**(5), 599–606.

Keshari N. and Adhikary S. P. (2013). Characterization of cyanobacteria isolated from biofilms on stone monuments at Santiniketan, India. *Biofouling*, **29**(5), 525–536.

Kohls K., Abed R. M. M., Polerecky L., Weber M. and de Beer D. (2010). Halotaxis of cyanobacterial in an intertidal hypersaline microbial mat. *Environmental Microbiology*, **12**, 567–575.

Lawrence J. R., Swerhone G. D. W., Kuhlicke U. and Neu T. R. (2007). *In situ* evidence for microdomains in the polymer matrix of bacterial microcolonies. *Canadian Journal of Microbiology*, **53**, 450–458.

Leprince O. and Buitink J. (2010). Desiccation tolerance: from genomics to the field. *Plant Sciences*, **179**, 554–564.

Leslie S. B., Israeli E., Lighthart B., Crowe J. H. and Crowe L. M. (1995). Trehalose and sucrose protect both membranes and proteins in intact bacterial during drying. *Applied Environmental Microbiology*, **61**(10), 3592–3597.

Liu Q., Schmidt R. K., Teo B., Karplus P. A. and Brady J. W. (1997). Molecular dynamics studies of the hydration of α,α-Trehalose. *Journal of the American Chemical Society*, **119**(4), 7851–7862.

Madern D., Ebel C. and Zaccai G. (2000). Halophilic adaptation of enzymes. *Extremophiles*, **4**, 91–98.

Magazu S., Migliardo P., Musolino A. M. and Sciortino M. T. (1997). Alpha, alpha-trehalose-water solutions. Hydration phenomena and anomalies in the acoustic properties. *Journal of Physical Chemistry B*, **101**(13), 2348–2351.

Maréchal Y., Milas M. and Rinaudo M. (2003). Hydration of hyaluron polysaccharide observed by IR spectrometry. III. Structure and mechanism of hydration. *Biopolymers*, **72**, 162–173.

Mashburn, L. M. and Whiteley, M. (2005). Membrane vesicles traffic signals and facilitate group activities in a prokaryote. *Nature*, **437**, 422–425.

Mayer C., Moritz R., Kirschner C., Borchard W., Maibaum R., Wingender J. and Flemming H-C. (1999). The role of intermolecular interactions: studies on model systems for bacterial biofilms. *International Journal of Biological Macromolecules*, **26**, 3–16.

Melo E. P., Faria T. Q., Martins L. O., Goncalves A. M. and Cabral J. M. S. (2001). Cutinase unfolding and stabilization by trehalose and mannosylglycerate. *Proteins-Structure Function and Genetics*, **42**(4), 542–552.

Montgomery K., Charlesworth J. C., LeBard R., Visscher P. T. and Burns B. P. (2013). Quorum sensing in extreme environments. *Life*, **3**(1), 131–148.

Nagy M., Perez A. and Garcia-Pichel F. (2005). The prokaryotic diversity of biological soil crusts in the Sonoran Desert of Arizona. *FEMS Microbiology Ecology*, **54**, 233–245.

Narvaez-Reinaldo J., Barba I., Gonzalez-Lopez J., Tunnacliffe A. and Manzanera M. (2010). Rapid methods for isolation of desiccation-tolerant strains and xeroprotectants. *Applied Environmental Microbiology*, **76**, 5254–5262.

Neu T. R. and Lawrence J. R. (2009). Extracellular polymeric substances in microbial biofilms. In: Microbial Glycobiology: Structures, Relevance and Applications, A.P. Moran, O. Holst and P.J. Brennan (eds), pp. 735–758.

Neu T. R. and Lawrence J. R. (2015). Innovative techniques, sensors and approaches for imaging biofilms at different scales. *Trends in Microbiology*, **23**(4), 233–242.

Neu T. R., Manz B., Volke F., Dynes J. J., Hitchcock A. P. and Lawrence J. R. (2010). Advanced imaging techniques for assessment of structure, composition and function in biofilm systems. *FEMS Microbiology Ecology*, **72**, 1–21.

Noffke N., Decho A. W. and Stoodley P. (2013). Slime through time- the fossil record of prokaryote evolution. *Palaios*, **28**(1), 1–5.

Oliver A. E., Leprince O., Wolkers W. F., Hincha D. K., Heyer A. G. and Crowe J. H. (2001). Non-disaccharide-based mechanisms of protection during drying. *Cryobiology*, **43**, 151–167.

Ophir T. and Gutnick D. L. (1994). A role for exopolysaccharides in the protection of microorganisms from desiccation. *Applied Environmental Microbiology*, **60**(2), 740–745.

Or D., Phutane S. and Dechesne A. (2007). Extracellular polymeric substances affecting pore-scale hydrologic conditions from bacterial activity in unsaturated soils. *Vadose Zone Journal*, **6**(2), 298–305.

Paerl H. W., Pinckney J. L. and Steppe T. F. (2000). Cyanobacterial-bacterial mat consortia: examining the functional unit of microbial survival and growth in extreme environments. *Environmental Microbiology*, **2**(1), 11–26.

Paerl H. W., Steppe T. F., Buchan K. C. and Potts M. (2003). Hypersaline cyanobacterial mats as indicators of elevated tropical hurricane activity and associated climate change. *AMBIO: A Journal of Human Environment*, **32**(2), 87–90.

Pereira C. I., Gomes A. M. P. and Malcata F. X. (2009). Microstructure of cheese: processing, technological and microbiological considerations. *Trends in Food Sciences and Technology*, **20**, 213–219.

Pinckney J. L. and Paerl H. W. (1997). Anoxygenic photosynthesis and nitrogen fixation by a microbial mat community in a bahamian hypersaline lagoon. *Applied Environmental Microbiology*, **63**(2), 420–426.

Pinckney J. L., Paerl H. W. and Bebout B. M. (1995). Salinity control of benthic microbial mat production in a Bahamian hypersaline lagoon. *Journal of Experimental Marine Biology Ecology*, **187**(2), 59–66.

Pointing S. B. and Belnap J. (2012). Microbial colonization and controls in dryland systems. *Nature Reviews in Microbiology*, **10**, 551–562.

Potts M. (1994). Desiccation tolerance of prokaryotes. *Microbiology Reviews*, **58**(4), 755–805.

Potts, M. (1999). Mechanisms of desiccation tolerance in cyanobacteria. *European Journal of Phycology*, **34**(4), 319–326.

Potts M. (2001). Desiccation tolerance: a simple process? *Trends in Microbiology*, **9**, 553–559.

Potts M. (2004). Nudist colonies: a revealing glimpse of cyanobacterial extracellular polysaccharide. *Journal of Phycology*, **40**, 1–3.

Pringault O. and Garcia-Pichel F. (2004). Hydrotaxis of cyanobacteria in desert crusts. *Microbial Ecology*, **47**, 363–373.

Rajeev L., da Rocha U. N., Klitgord N., Luning E. G., Fortney J., Axen S. D., Shih P. M., Bouskill N. J., Bowen B. P., Kergeld C. A., Garcia-Pichel F., Brodie E. L., Northen T. R. and Mukhopadhyay A. (2013). Dynamic cyanobacterial response to hydration and dehydration in a desert biological soil crust. *ISME Journal*, **7**, 2178–2191.

Ramakrishnan S., Gerardi C., Prud'homme R. K. (2004). Syneresis of carrageenan gels: NMR and Rheology. *Soft Materials,* **2**(2/3), 145–153.

Reid R. P., Visscher P. T., Decho A. W., Stolz J. F., Bebout B. M., Dupraz C., Macintyre I. G., Paerl H. W., Pinckney J. L., Prufert-Bebout L., Steppe T. F. and DesMarais D. J. (2000). The role of microbes in accretion, lamination and early lithification of modern marine stromatolites. *Nature*, **406**, 989–992.

Richards W. G., King P. M. and Reynolds C. A. (1989). Solvation effects. *Protein Engineering*, **2**, 319–327.

Roberson E. B. and Firestone M. K. (1992). Relationship between desiccation and exopolysaccharide production in a soil *Pseudomonas* sp. *Applied Environmental Microbiology*, **58**(4), 1284–1291.

Roeβler M. and Müller V. (2001). Osmoadaptation in bacteria and archaea: common principles and differences. *Environmental Microbiology*, **3**(12), 743–754.

Saartratra S., Puttanlekb C., Rungsardthong V. and Uttapap D. (2005). Paste and gel properties of low-substitution acetylated canna starches. *Carbohydrate Polymers*, **61**, 211–221.

Schebor C., Mazzobre M. F. and del Pilar Buera M. (2010). Glass transition and time-dependent crystallization behavior of dehydration bioprotectant sugars. *Carbohydrate Research*, **345**, 303–308.

Scherer S. and Potts M. (1989). Novel water stress protein from a desiccation-tolerant cyanobacterium. *Journal of Biological Chemistry*, **264**, 12546–12553.

Schmitt J. and Flemming H-C. (1998). FTIR-spectroscopy in microbial and material analysis. *International Biodeterioration and Biodegradation*, **41**(1), 1–11.

Schmitt J. and Flemming H.-C. (1999). Water binding in biofilms. *Water Science and Technology.*, **39**(7), 77–82.

Schmitt J., Nivens D., White D. C. and Flemming H-C. (1995). Changes in biofilm properties in response to sorbed substances – an FTIR-ATR study. *Water Sciences and Technology*, **32**(8), 149–155.

Schooling, S. R. and Beveridge T. J. (2006). Membrane vesicles: an overlooked component of the matrices of biofilms. *Journal of Bacteriology*, **188**, 5945–5957.

Shaw E., Hill D. R., Brittain B., Wright D. J., Tauber U., Marand H., Helm R. F. and Potts M. (2003). Unusual water flux in the extracellular polysaccharide of the cyanobacterium *Nostoc commune. Applied Environmental Microbiology*, **69**, 5679–5684.

Shirke S. and Ludescher R. D. (2005). Molecular mobility and the glass transition in amorphous glucose, maltose and maltotriose. *Carbohydrate Research*, **340**, 2654–2660.

Shirkey B., Kovarcik D. P., Wright D. J., Wilmoth G., Prickett T. F., Helm R. F., Gregory E. M. and Potts M. (2000). Active Fe-containing superoxide dismutase and abundant *sodF* mRNA in *Nostoc commune* (Cyanobacteria) after years of desiccation. *Journal of Bacteriology*, **182**, 189–197.

Shirkey B., McMaster N. J., Smith S. C., Wright D. J., Rodriguez H. and Jaruga P. (2003). Genomic DNA of *Nostoc commune* (Cyanobacteria) becomes covalently modified during long-term (decades) desiccation but is protected from oxidative damage and degradation. *Nucleic Acids Research*, **31**, 2995–3005.

Skillman L., Sutherland I. W. and Jonse M. V. (1999). The role of exopolysaccharides in dual species biofilm development. *Journal of Applied Microbiology*, **85**, S13–S18.

Slade L. and Levine H. (1991). Beyond water activity: recent advances based on an alternative approach to the assessment of food quality and safety. *Critical Reviews in Food Science and Nutrition*, **30**, 115–360.

Sleator R. D. and Hill C. (2002). Bacterial osmoadaptation: the role of osmolytes in bacterial stress and virulence. *FEMS Microbiology Reviews*, **26**(1), 49–71.

Smith, B. C. (1996). Fundamental of Fourier transform infared spectroscopy. CRC Press, N.Y., 202p.

Smith S. M., Abed R. M. M. and Garcia-Pichel F. (2004). Biological soil crusts of sand dunes in Cape Cod National Seashore, Massachusetts, USA. *Microbial Ecology*, **28**, 200–208.

Smith J. D., Cappa C. D., Wilson K. R., Cohen R. C., Geissler P. L. and Saykally R. J. (2005). Unified description of temperature-dependent hydrogen bond rearrangements in liquid water. *Proceedings of the National Academy of Sciences USA*, **102**, 14171–14174.

Sutherland I. W. (1990). Biotechnology of microbial exopolysaccharides. Cambridge University Press, NY. 163p.

Tamaru Y., Takani Y., Yoshida T. and Sakamoto T. (2005). Crucial role of extracellular polysaccharides in desiccation and freezing tolerance in the terrestrial cyanobacterium *Nostoc commune. Applied Environmental Microbiology*, **71**, 7327–7333.

Tanne C. Golovina E. A., Hoekstra F. A., Meffert A. and Galinski E. A. (2014). Glass-forming property of hydroxyectoine is the cause of its superior function as a desiccation protectant. *Frontiers in Microbiology*, **5**, Article 150, doi: 10.3389/fmicb.2014.00150.

Taranto P. A., Keenan T. W. and Potts M. (1993). Rehydration induces rapid onset of lipid biosynthesis in desiccated *Nostoc commune* (Cyanobacteria). *Biochimica et Biophysica Acta*, **1168**, 228–237.

Tenchov B., Vescio E. M., Sprott G. D., Zeidel M. L. and Mathai J. C. (2006). Salt tolerance of Archaeal extremely halophilic lipid membranes. *Journal of Biological Chemistry*, **281**, 10016–10023.

Tokmachev A. M., Tchougreeff A. L. and Dronskowski R. R. (2010). Hydrogen-bond networks in water clusters (H_2O) (20): an exhaustive quantum-chemical analysis. *Chemistry Physical Chemistry*, **11**(2), 384–388.

Underwood G. J. C., Boulcott M., Raines C. A. and Waldron K. (2004). Environmental effects of exopolymer production by marine benthic diatoms: dynamics, changes in composition, and pathways of production. *Journal of Phycology*, **40**(2), 293–304.

Valladares-Linares R., Wexler A. D., Bucs Sz. S., Dreszer C., Zwijnenburg A., Flemming H.-C., Kruithof J. C. and Vrouwenvelder J. S. (2015). Compaction and relaxation of biofilms. *Desalination and Water Treatment*. 1–13. doi: 10.1080/19443994.2015.1057036.

van den Dries I. J., van Dusschoten D. and Hemminga M. A. (1998). Mobility in maltose-water glasses studied with H-1 NMR. *Journal of Physical Chemistry B*, **102**(51), 10483–10489.

Vogt M., Flemming H-.C. and Veeman W. S. (2000). Diffusion in *Pseudomonas aeruginosa* biofilms: a pulsed field gradient NMR study. *Journal of Biotechnology*, **77**(1), 137–146.

Welsh, D. T. (2000). Ecological significance of compatible solute accumulation by micro-organisms: from single cells to global climate. FEMS Microbiology Reviews, **24**(3), 263–290.

Welsh D. T. and Herbert R. A. (1999). Osmotically induced intracellular trehalose, but not glycine betaine accumulation promotes desiccation tolerance in *Escherichia coli. FEMS Microbiology Letters*, **174**, 57–63.

Wessels M. R., Moses A. E., Goldberg J. B. and DiCesare T. J. (1991). Hyaluronic acid capsule is a virulence factor for mucoid group A streptococci. *Proceedings of the National Academy of Sciences. USA*, **88**(19), 8317–8321.

Wolkers W. F., Oliver A. E., Tablin F. and Crowe J. H. (2004). A fourier-transform infrared spectroscopy study of sugar glasses. *Carbohydrate Research*, **339**, 1077–1085.

Wright D. J., Smith S. C., Jordar V., Scherer S., Jervis J., Warren A., Helm R. F. and Potts M. (2005). UV irradiation and desiccation modulate the three-dimensional extracellular matrix of *Nostoc commune* (Cyanobacteria). *Journal of Biological Chemistry*, **280**(48), 40271–40281.

Zhang K., Luo Y. and Li Z. (2007). Synthesis and characterization of a pH- and ionic strength-responsive hydrogel. *Soft Materials*, **5**(4), 183–195.

Chapter 12

Extracellular factors involved in biofilm matrix formation by Rhizobia

Patricia L. Abdian[1] and Angeles Zorreguieta

Laboratorio de Microbiología Molecular y Celular, Fundación Instituto Leloir, IIBBA-CONICET. Av. Patricias Argentinas 435, C1405BWE, Buenos Aires, Argentina

[1]Corresponding author: pabdian@leloir.org.ar

ABSTRACT

Rhizobia are soil bacteria that play an important role in biological nitrogen fixation through their symbiotic interaction with legumes. Rhizobia grow in varied environmental conditions during their life cycle and are able to produce different types of biofilms. Rhizobial biofilm formation may represent an advantageous way to survive long periods of nutrient starvation or desiccation when bacteria encounter the bulk soil, but it may also aid in the interaction with the host plant, increasing the inoculum of bacteria on plant tissues susceptible to infection.

Many of the extracellular polymeric substances (EPS) produced by rhizobia have been previously studied in the context of the symbiotic interaction with the legume host, but only in some cases their importance in biofilm matrix formation have been determined. The best characterized polymers are exopolysaccharides, and although several secreted proteins were identified only a few were linked to matrix formation. This chapter summarizes most of the cell associated or extracellular components (exopolysaccharides, capsular polysaccharides, lipopolysaccharide, proteins) produced by several rhizobial species that have been involved in biofilm matrix development. Other components such as glucomannan, flagella and Nod factors are also discussed, as recent evidence points them as polymers/compounds that participate in biofilm matrix formation or attachment to different surfaces.

Keywords: rhizobia, legume, rhizosphere, biofilm, matrix, exopolysaccharide, adhesin, glycanase

12.1 OVERVIEW OF RHIZOBIA AND THEIR SYMBIOTIC RELATION WITH LEGUMES

The term "rhizobia" (*rhiza* = root and *bios* = life) has been long used to describe a non-taxonomic group of bacteria with the ability to nodulate leguminous plants. Since the first isolation of root-nodule bacteria

(Beijerinck, 1888) microorganisms from various legume species were classified within the genus *Rhizobium*. As new rhizobial isolates emerged, nodulation tests on diverse legume hosts led to the concept of cross-inoculation groups at the dawn of rhizobial taxonomy. But this classification based on host-specificity was replaced in the mid-20th century by the evaluation of morphological, nutritional and metabolic traits that proved to be more reliable. Actual taxonomic classification of rhizobia also includes phylogeny, biochemical, genetic as well as molecular techniques that help in the differentiation between species. Over the years the number of genera has increased gradually reaching six defined genera (*Rhizobium, Bradyrhizobium, Sinorhizobium, Mesorhizobium, Azorhizobium* and *Allorhizobium*) (van Berkum & Eardly, 1998). Rhizobia are Gram-negative aerobic bacilli that belong to the Alphaproteobacteria. The discovery of members of the Betaproteobacteria capable of establishing symbiosis with legumes and inducing nitrogen-fixing root nodules suggested that horizontal gene transfer occurred (Moulin *et al.* 2001). This added to the fact that the study of legumes other than those of agricultural importance is expanding, will probably result in new species and genera of rhizobia to arise.

The ability of rhizobia to convert atmospheric nitrogen into ammonia in association with legumes scores them as a group of ecological and agronomical importance. It has been estimated that the rhizobial symbiosis with the agriculturally important legumes contribute nearly half of the nitrogen fixed annually by soil ecosystems (Tate, 1995). Biological nitrogen fixation (BNF) represents a renewable source of nitrogen for agriculture and has a potential role in land remediation. However, BNF importance has decreased in the last decades due to the massive use of nitrogen fertilizers despite the negative effects of these compounds on aquatic systems (water pollution, eutrophication) (Sprent & Sprent, 1990).

The symbiotic relationship legume-rhizobium is established in nitrogen-starved environments. It begins with a complex dialog between both partners, in which the host plant roots secrete flavonoids/isoflavonoids and related compounds into the soil that elicit the production of lipo-chitooligosaccharide Nod factors by compatible rhizobia. The association is mainly controlled by the host plant and requires a coordinated exchange of signals (Janczarek *et al.* 2015). The infection process is initiated by bacterial adhesion to root hairs, followed by curling and degradation of epidermal cell wall. The bacteria, trapped in the curl, induce the formation of an infection thread that extends, ramifies and leads bacteria to cortical root cells. These host cells divide to give rise to the root nodule primordium. The bacteria inside cortical cells differentiate into bacteroids, within a low oxygen environment, in which nitrogen fixation can take place.

12.2 RHIZOBIAL BIOFILM FORMATION

It is now well established that in natural environments bacteria exist mainly as biofilm communities (Costerton *et al.* 1987) and rhizobia do not escape the rule. The formation of biofilms by rhizobia was approached by Fujishige *et al.* (2006) recognizing that rhizobia are able to form biofilms on different surfaces (biotic and abiotic). This is consistent with the complex life cycle of rhizobia, including survive the harsh conditions of the bulk soil for extended periods, and also grow rapidly on host plant tissues to increase the chances of host infection and nodulation.

The rhizosphere, defined as the region of soil immediately adjacent to the root surface and extending few millimeters into the soil (Bringhurst *et al.* 2001), is a rich environment containing root exudates (mainly organic acids and sugars) that support a variety of microbial life forms. Root surface characteristics vary along its length, being the actively growing root tissues responsible for higher exudation rates into the soil (Ramey *et al.* 2004). This contrasts bulk soil conditions, in which microbial communities are thought to be carbon limited (Cheng *et al.* 1996) and exposed to desiccation. Rhizobial communities inhabit all these microenvironments, forming biofilm structures that vary from extended and dense biofilms to small bacterial aggregates. Biofilms or bacterial aggregates of different rhizobia have been observed on the root epidermis, but they were usually associated with root hairs (Rinaudi & Gonzalez, 2009; Santaella *et al.* 2008; Williams *et al.* 2008). Confocal microscopy of curled root hairs revealed the presence of rhizobial microcolonies within the curl and also rhizobial cells migrating down infection threads as biofilm-like filaments (Gage, 2002; Ramey *et al.* 2004). Rhizobia can also form biofilms on sand and soil particles as shown for *Sinorhizobium meliloti* using a buried slide assay (Fujishige *et al.* 2006). Therefore, it seems evident that the surface properties and the secreted polymers will vary, having a strong influence on the resulting biofilm structure.

The switch from planktonic to biofilm formation under unfavorable conditions in rhizobia was assessed; nutrient limitation has been established as a key factor in the transition to a sessile mode of life both in *S. meliloti* and *Rhizobium leguminosarum* growing on an inert surface (Fujishige *et al.* 2006; Rinaudi *et al.* 2006; Russo *et al.* 2006). But the relation between rhizobial biofilm formation and the establishment of a productive rhizobium-legume symbiosis remains elusive. Root tissues are nutrient-rich environments and rhizobia are still able to develop biofilms on them. It was proposed that plant factors may promote biofilm formation on root surfaces and overcome the inhibitory effect of nutrients availability (Rinaudi & Giordano, 2010).

An interesting study on various mutants defective in exopolysaccharides biosynthesis (Williams *et al.* 2008) revealed that the molecular determinants of biofilm formation in vitro are different from those involved in biofilms growing on plant roots. Bacterial cellulose and glucomannan (Laus *et al.* 2006) are important for root hair colonization, but not for in vitro biofilm formation (Williams *et al.* 2008). On the other hand, plant lectins enhance bacterial attachment to root hairs (Laus *et al.* 2006; van Rhijn *et al.* 2001) and enzymes in root exudates promote changes in bacterial surface polysaccharides, modifying bacterial adhesion properties (Dazzo *et al.* 1982).

The examination of mixed cultures and soil conditions such as pH, water potential, concentrations of Ca^{2+} and Mg^{2+}, would add another grade of complexity to the study of biofilms on roots. It is therefore necessary to develop appropriate techniques to study rhizobial biofilm production under more natural conditions. Anyhow, it seems plausible that biofilms on root surfaces would position rhizobia near the susceptible tissues of the host, increasing the bacterial inoculum for plant infection, and an effective legume nodulation (Downie, 2010).

12.3 RHIZOBIAL COMPONENTS OF THE BIOFILM MATRIX

The mechanisms that lead to the assembly of the biofilm matrix constitute a rather unexplored subject of biofilm research. Furthermore, despite the many functions that were associated with the biofilm matrix (Flemming & Wingender, 2010), little is known about the interactions between the constituent macromolecules. The study of matrix components and their interactions at the molecular level would form the basis to understand the dynamics of the biofilm matrix, its remodeling in response to environmental or community changes and finally the biofilm disassembly mechanisms that allow the dispersion of resident cells and the colonization of new habitats.

This chapter will focus on both the extracellular polymeric substances (EPS) produced by rhizobia that have been identified as components of the biofilm matrix and the surface associated polysaccharides that are involved in adhesion and cohesion between cells (Table 12.1). Most of the studies have been performed under laboratory conditions and are restricted to some genera. In addition, Nod factors, recently proposed as structural components of the matrix (Fujishige *et al.* 2008) and flagella, as surface appendages that may contribute to surface adhesion (Fujishige *et al.* 2006; Xie *et al.* 2012) are presented in sections 12.3.3 and 12.3.4 respectively.

Table 12.1 Rhizobial extracellular polymeric substances.

EPS Component	Organism	Reference
Succinoglycan	*S. meliloti*	Fujishige *et al.* (2006) Rinaudi *et al.* (2009)
Galactoglucan		Rinaudi *et al.* (2009) Sorroche *et al.* (2010)
ML β-glucan		Pérez-Mendoza *et al.* (2015)
Acidic exopolysaccharide	*R. leguminosarum*	Fujishige *et al.* (2006) Russo *et al.* (2006) Williams *et al.* (2008) Mongiardini (2008)
	Rhizobium sp.	Santaella *et al.* (2008)
	Mesorhizobium tianshanense	Wang *et al.* (2008)
	Bradyrhizobium japonicum	Pérez-Giménez *et al.* (2009)
Lipopolysaccharide	*R. leguminosarum*	Vanderlinde *et al.* (2009) Russo *et al.* (2015)
	B. japonicum	Lee *et al.* (2010)
Glucomannan	*R. leguminosarum*	Williams *et al.* (2008)
Cellulose	*R. leguminosarum*	Laus *et al.* (2005) Williams *et al.* (2008)
Lectin	*R. leguminosarum*	Russo *et al.* (2006) Mongiardini *et al.* (2008) Abdian *et al.* (2013)
Glycanase	*R. leguminosarum*	Russo *et al.* (2006) Robledo *et al.* (2012)

The information, although still fragmentary, has been useful to state that the biofilms produced by rhizobia, will be different depending on the environment and the nature of the surface. These variations can only be attributed to a different matrix composition. An update on available information and a model of matrix assembly in *R. leguminosarum* (Figure 12.2) are presented.

12.3.1 Rhizobial extracellular polysaccharides

12.3.1.1 Exopolysaccharides

Rhizobial exopolysaccharides are secreted to the milieu in copious amounts and accomplish several functions to the cells that produce them. They have been extensively studied due to their importance in the establishment of a successful symbiosis with legumes. Aspects of their chemical structure, their biosynthesis and its regulation have been reviewed by Skorupska *et al.* (2006) and Janczarek (2011).

These polymers are heteropolysaccharides (composed of different types of monosaccharides) assembled by the polymerization of repeating units in a complex process that requires the activity of enzymes in different compartments of the cell. Some rhizobial exopolysaccharides are composed of octasaccharide repeating units (Figure 12.1) modified at certain residues by the addition of non-carbohydrate substituents or decorations (acetate, pyruvate or succinate).

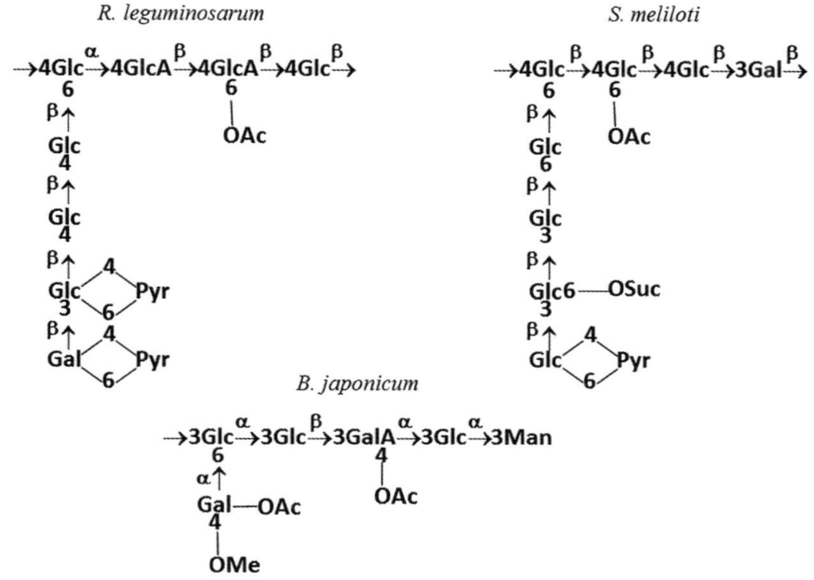

Figure 12.1 Examples of rhizobial exopolysaccharide repeating units. *Abbreviations*: Glc, glucose; Gal, galactose; Man, mannose; GlcA, glucuronic acid; GalA, galacturonic acid; Pyr, pyruvate; Suc, succinate, Me, methyl and Ac, acetyl.

The structures depicted in Figure 12.1 represent just a few examples, but rhizobial exopolysaccharides are diverse, showing variations in sugar composition, side-chain length, number and position of decorations and degree of polymerization. All these structural variations have a profound impact in the dynamics of the polymers. The physicochemical behavior and conformational changes of the polymer chains in aqueous solution of the exopolysaccharides produced by different strains of *R. leguminosarum* were determined using optical techniques (Coviello *et al.* 1995; Dentini *et al.* 1988). Addition of salt induced exopolysaccharide chains to undergo disorder-order transitions, with a highly cooperative character. The molecules adopted a stiff double helical conformation. However, longer side chains in the polymer, impeded the coil to helix transitions, and thereby destabilized polysaccharide ordered structures. Moreover, modification of minor substituents or decorations plays an important role, affecting the net charge of the polymers, with consequences in the solubility and supermolecular structure of the polysaccharide (Coviello *et al.* 1995).

The gelation property of the exopolysaccharide produced by *R. leguminosarum* bv. *phaseoli* strain 8002 was investigated by Coviello *et al.* (1997). They argued that the thermoreversible gelation of the polysaccharide can be induced by ions depending on the concentration of the polymer. It seems obviously important that the ability to form gels as well as the correlation between polymer structure and its physicochemical properties will influence the physical integrity of the biofilm matrix as well as the processes taking place in it (Sutherland, 2001).

The genome of *R. leguminosarum* bv. *viciae* contains at least seven different clusters of genes involved in polysaccharide biosynthesis. Among them, the acidic exopolysaccharide has been shown to be important for biofilms formed on abiotic surfaces. Three-dimensional biofilm structures formed on polystyrene were observed, and staining revealed the matrix of exopolysaccharide (Fujishige *et al.* 2006). Moreover, examination of GFP-tagged cells by confocal laser scanning microscopy (CLSM) over a 4-day experiment allowed visualization of the progression of biofilm formation in *R. leguminosarum* (Russo *et al.* 2006). It was shown that initial attachment to the glass surface occurred via one cell pole. Then microcolonies formed by cell division and also by recruitment of cells. Cellular aggregates grew and progressed to complex structures, made up of layers of cells interacting mainly through lateral contacts interlaced with water channels. These complex arrangements were called "honeycomb" structures and represent a mature *R. leguminosarum* biofilm. Mutation of *pssA*, a gene coding for the glycosyltransferase that catalyzes the first step in the biosynthesis of the acidic exopolysaccharide, resulted in a reduced adhesion to abiotic surfaces. The *pssA* mutant was unable to develop microcolonies and after four days, only a flat lawn of bacteria was observed on the base of the chamber, devoid of any structure (Russo *et al.* 2006). Therefore, it was concluded that the biosynthesis of the acidic exopolysaccharide by *R. leguminosarum* is essential for development of a structured biofilm on an inert surface.

The role of the acidic exopolysaccharide produced by *R. leguminosarum* on biofilm formation on roots was also investigated (Williams *et al.* 2008). Vetch (*Vicia hirsuta*) seedlings inoculated with GFP-labelled bacteria showed the attachment of individual bacterial cells to root hairs, followed by formation of a cap structure (by recruitment of several bacteria) on the root tips. It is interesting to note that this cap of bacteria on root tips developed in a 90-minute incubation time, and was detected in more than 70% of the root hairs in the growing root hair zone. Evaluation of a *pssA* mutant revealed no attachment or cap formation. This suggests that the acidic exopolysaccharide also plays a role in root hair attachment and biofilm formation on root tips.

Rhizobium sp.YAS34 produces an exopolysaccharide composed of a heptasaccharide repeating unit (Glc, Gal and GlcA in relation 3:3:1) whose structure and properties have been established (Villain-Simonnet *et al.* 2000a, b). The synthesis of this polymer was not essential for biofilm formation on roots of non-legume hosts, but was critical for colonization of the basal part of the root. Indeed, a mutant lacking this exopolysaccharide was still able to develop a biofilm on plant surfaces, though at a lower rate and with an altered pattern of colonization (Santaella *et al.* 2008). Interestingly, using a fluorescent lectin and CLSM, the wild type cells were observed embedded in a three-dimensional exopolysaccharide network that also enclosed soil organic matter and mineral particles. It is worth to mention that in this work, formation of biofilm-like structures was observed several days (even 15–30 days) after root or seed inoculation while in *R. leguminosarum*, only the initial stages were monitored (see above). Synthesis of the acidic exopolysaccharides (and its impact on biofilm formation and root colonization) would depend on nutrient fluctuations and differential production of root exudates along the root tissues (Santaella *et al.* 2008). It was also shown that exopolysaccharide production modifies soil structure by cementing soil particles around the root systems. In this way, rhizobial populations could persist in the absence of a legume host, waiting in a protected environment the opportunity to establish an effective symbiosis.

Exopolysaccharide biosynthesis by the *Mesorhizobium* genus has been scarcely explored. Genes involved in exopolysaccharide production were identified in *Mesorhizobium tianshanense* using transposon mutagenesis (Wang *et al.* 2008). This bacterium nodulates at least eight species including *Glycyrrhiza uralensis* (Asian licorice) and is widely distributed in dry soils. As in many other bacterial species it was found that mutants in exopolysaccharide biosynthesis were also impaired in biofilm formation. More recently Kelly *et al.* (2013) isolated and characterized several exopolysaccharide mutants of *Mesorhizobium loti* and evaluated their symbiotic phenotypes on *Lotus corniculatus* and *Lotus japonicus*. They found that mutants in early or late exopolysaccharide biosynthetic genes were disrupted at the stage of infection thread development, suggesting that the exopolysaccharide was not essential for attachment to roots. Recently, co-inoculation experiments showed that *M. loti*, can selectively guide (other) endophytic bacteria towards nodule primordia and analysis of exopolysaccharide mutants showed that these polysaccharides are key factors for nodule infection (Zgadzaj *et al.* 2015). However, biofilm-related phenotypes of these mutants were not assessed.

Seneviratne and Jayasinghearachchi (2003) first described biofilm formation by *Bradyrhizobium* sp. on diverse biotic and abiotic surfaces. *B. japonicum*, that nodulates *Glycine max* (soybean), produces different exopolysaccharides depending on the genotype and the carbon source used for growth (Karr *et al.* 2000), but the involvement of these exopolysaccharides on biofilm matrix formation has not been addressed in detail. There is only one report on a mutant that produces low levels of a shorter exopolysaccharide lacking galactose (Parniske *et al.* 1993) and also showed a reduced ability to form biofilms (Pérez-Giménez *et al.* 2009).

Succinoglycan produced by *S. meliloti* (Figure 12.1) is one of the best studied rhizobial exopolysaccharides. It is secreted in two major fractions (HMW, high molecular weight and LMW, low molecular weight) that differ in their degree of polymerization. The conditions for establishing biofilms by *S. meliloti* on different surfaces were investigated (Fujishige *et al.* 2006). The wild type strain Rm1021 developed microcolonies on plastic that evolved to what the authors described as "compact towers of cells" at ten days post-inoculation. Examination of an *exoY* mutant derived from Rm1021 revealed that biofilm formation was 50–60% reduced. This mutant is affected in galactosyl-1-P-transferase activity, responsible for the first step in succinoglycan subunit biosynthesis. The absence of succinoglycan in the mutant strain had a striking impact on the biofilm three-dimensional architecture as examined by microscopy, only small and compact cell aggregates that were easily detached from the surface were observed. On the other hand, strain *exoS96*, a succinoglycan overproducer (Cheng & Walker, 1998), was found to have increased ability to form biofilms. Despite the secondary defects displayed by *exoS96* mutant (Yao *et al.* 2004) the increase in biofilm formation was attributed to the augmented succinoglycan production.

But *S. meliloti* makes a second exopolysaccharide called galactoglucan, a polymer composed of a disaccharide repeating unit of acetylated glucose and pyruvylated galactose (Glazebrook & Walker, 1989; Her *et al.* 1990; Zhan *et al.* 1989). This exopolysaccharide is produced by most strains of *S. meliloti* under conditions of phosphate starvation (Zhan *et al.* 1991) as would prevail in the soil. It was observed that spontaneous mutants in *expR* (like strain Rm1021) do not produce galactoglucan. Therefore the use of Rm1021 as a wild type strain was questioned (Rinaudi & Gonzalez, 2009). ExpR, a member of the LuxR family of proteins, is a transcriptional regulator involved in quorum-sensing-dependent gene expression (Hoang *et al.* 2004). Examination of the status of the *expR* locus showed the disruption of *expR* by an insertion sequence (IS*Rm*2011–1) in several *S. meliloti* strains used in genetic laboratory studies (Pellock *et al.* 2002). It was proposed that in the natural environment, IS elements may modulate polysaccharide-dependent phase variation in *S. meliloti*. However, under laboratory conditions, IS genomic fingerprint patterns are highly stable, leading to a non-mucoid phenotype and the absence of galactoglucan production. A second spontaneous mutation, *expR101*, is required to turn on production of galactoglucan. This mutation results from the precise excision of the IS from *expR* restoring the coding sequence. The resulting strain, Rm8530 (*expR*$^+$) was studied and its ability to form biofilms was evaluated (Rinaudi & Gonzalez, 2009). The Rm8530 strain now can regulate several bacterial phenotypes based on population density and is able to produce both HMW and LMW galactoglucan. It was shown that the symbiotically active LMW fraction of galactoglucan is essential for the proper and abundant formation of biofilms by *S. meliloti* in vitro and in vivo. In fact, the authors argued that that it is galactoglucan (and not succinoglycan) that provides the matrix for biofilm. This apparent discrepancy on the role of succinoglycan to that reported by Fujishige *et al.* (2006) might arise from the different genetic backgrounds evaluated, strain Rm1021 is indeed a mutant strain, with a defective ExpR/Sin quorum-sensing system.

Bacterial autoaggregation and the consequent sedimentation are processes that depend on the physical interaction between bacterial cells. Self-aggregation can be modulated by exopolysaccharides. Besides, cell-cell interactions are important in the development of microcolonies and a mature biofilm attached to different surfaces. In line with the findings of Rinaudi and Gonzalez (2009), it was reported that galactoglucan is directly involved in autoaggregation of planktonic *S. meliloti* cells, while succinoglycan plays no obvious role in this process (Sorroche *et al.* 2010). Interestingly, it was further shown that galactoglucan plays a crucial role in both autoaggregation (and therefore cell-cell interaction) and biofilm formation in several undomesticated *S. meliloti* strains. Furthermore, a positive correlation between both processes was observed, suggesting that cell-cell interactions in planktonic aggregates and within the biofilm depend, under assayed conditions, on the same physical adhesive forces (Sorroche *et al.* 2012)

Very recently, a novel polysaccharide was discovered in *S. meliloti* strain 8530 (Pérez-Mendoza *et al.* 2015). This compound was identified as a linear mixed-linkage $(1 \rightarrow 3)(1 \rightarrow 4)$-$\beta$-D-glucan (ML β-glucan) not previously described in bacteria but resembling ML β-glucans found in plants and lichens. Production of this β-glucan was only observed when intracellular cyclic diguanylate (c-di-GMP) levels were artificially increased. Biosynthesis of this polymer was found to be regulated not only by c-di-GMP at post-translational level, but also by the exopolysaccharide/quorum sensing ExpR/SinI regulatory system. Induction of the β-glucan synthesis generated wrinkled colonies, bacterial aggregation and dense biofilms and provided a competitive advantage for attachment and colonization of the alfalfa root surface. Phylogenetic analyses suggest that other *Rhizobiales* might be potentially able to produce this polysaccharide. It would be interesting to determine the environmental conditions that activate production of this glucan and therefore establish its physiological relevance in biofilm formation both in vitro and in vivo.

An intriguing theory about the phenomenon which leads to cell aggregation was proposed by Dorken *et al.* (2012). Examination of a *S. meliloti* succinoglycan overproducer strain that displayed enhanced aggregation showed that the cell clusters were not covered by this polysaccharide. Instead, it was proposed that a crowding mechanism was taking place and that aggregation was driven by depletion attraction. Supporting this hypothesis, changes in succinoglycan structure and molecular weight, or even the addition of a different bacterial exopolysaccharide (xanthan) did not alter the sedimentation rate of *S. meliloti* cells. Therefore depletion attraction was proposed as the driving force promoting the self-organization of cells into clusters and subsequent formation of biofilm-like structures, in which laterally aligned cells prevail. In fact, depletion attraction has been previously described as a force operating in the organization of organelles and cellular structures in eukaryotes (Marenduzzo *et al.* 2006).

12.3.1.2 Capsular polysaccharides

The capsular polysaccharide forms a hydrated matrix around the bacterial cell that fulfills many important functions, including protection from predation, desiccation and other environmental stresses, suppression of plant-defense response and adhesion to solid surfaces.

Among rhizobia, different types of capsule have been described: the tightly cell-associated KPSs and the loosely-bound CPSs. To date, all known KPSs have been described in *Sinorhizobium* species. These polysaccharides contain a high proportion of 3-deoxy-D-manno-2-octulosonic acid (Kdo) and are structurally analogous to group II K antigens found in *Escherichia coli* (Reuhs *et al.* 1993). K-antigens are strain-specific antigens and their production could be dependent on bacterial growth conditions The role of KPS has been extensively studied in the *S. meliloti*-alfalfa symbiosis, but its involvement in biofilm formation remained unexplored. A recent report showed that mutants of *Sinorhizobium fredii* deficient in the synthesis of KPS are not affected in attachment or biofilm formation on a plastic surface (Rodríguez-Navarro *et al.* 2014).

Other types of capsular polysaccharides (non-K-antigenic) are produced by *R. leguminosarum* (Sherwood *et al.* 1984) and *Bradyrhizobium* strains (Mort & Bauer, 1982). These CPSs are structurally identical to the exopolysaccharides produced by these genera and only minor differences in the degree of non-carbohydrate substitutions were reported (Laus *et al.* 2004). Mutations affecting the biosynthesis of exopolysaccharide will also affect the production of CPS, thus, it is difficult to unequivocally ascertain the role of these polymers in biofilm formation, especially when genetic approaches are undertaken. However, it is reasonable to predict that the production of an acidic polysaccharide surrounding the bacterial cell will inevitably influence cell-cell interaction and microcolony development.

In *R. leguminosarum* bv. *trifolii* cultivated on agar plates the development of the capsule was followed by the ability of cells to bind the white clover lectin Trifoliin A (Sherwood *et al.* 1984). Using immunofluorescence and electron microscopy on bacteria grown on agar plates, it was shown that CPS begins to develop at one cell pole by day 3 and completely surrounds the cell by day 5. At day 7 the CPS begins to decrease showing that the capsule is a dynamic structure rather than a static protective sheath. The capsule produced by *B. japonicum* was also studied by immunoelectron microscopy (Tsien & Schmidt, 1981), but in this case the capsule was produced during growth in liquid cultures and was observed most frequently attached to one pole of the cell.

Rhizobium leguminosarum produces a neutral gel-forming polysaccharide (GPS) that began to accumulate towards the end of the exponential phase of growth (Laus *et al.* 2006; Zevenhuizen & van Neerven, 1983). This cell-bound polymer is composed of Gal, Glc and Man in a 4:1:1 ratio and it seems to replace the acidic CPS capsule in the late stationary phase of growth. A *gelA* mutant affected in GPS biosynthesis was tested for attachment and biofilm formation in vitro and on root hairs (Williams *et al.* 2008). The mutant formed biofilms indistinguishable from those of the wild-type strain, showing that GPS is not an essential component of the biofilm matrix under the assay conditions.

12.3.1.3 Lipopolysaccharide

In Gram-negative bacteria, the lipopolysaccharide (LPS) is an integral part of the outer membrane. Many aspects of its structure and biosynthesis, as well as its role in the symbiotic interaction, have been described in various species of *Rhizobium* and *Sinorhizobium* (Carlson *et al.* 2010). LPS consists of three distinct structural components: the lipid A, an acylated saccharide anchored to the lipid membrane; the core oligosaccharide; and the variable O-chain polysaccharide, which protrudes from the cell surface. Rhizobial LPS retains the same general architecture, although some structural traits are unique: the very long chain fatty acid (VLCFA) 27-hydroxyoctacosanoic acid in lipid A, shared with other members of the order Rhizobiales (*Brucella* spp., *Bartonella* spp. and *Legionella* spp.) that cause chronic intracellular infections; the absence of a heptosyl (Hep) residue attached to the 3-deoxy-Dmanno-2-octulosonic acid (Kdo) residue that links the core to lipid A; the O-polysaccharide structures enriched in deoxyhexoses and methyl-deoxyhexoses and its significant structural variability even among strains of the same species.

The changing conditions that rhizobia encounter in the infection of the host plant, as well as during life in the bulk soil, require the ability to change their cellular surface, including the LPS structure. LPS structural modifications were involved in the endocytic infection process, the differentiation of bacteria into bacteroids (Kannenberg & Carlson, 2001) and the inhibition of the host defense reaction (Perotto *et al.* 1994). However, LPS also accomplishes important roles in biofilm development. The study of a mutant of *R. leguminosarum* bv. *viciae* strain 3841, deficient in VLCFA-modified lipopolysaccharide, showed pleiotropic effects, producing an altered biofilm on a plastic surface (Vanderlinde *et al.* 2009). The mutant and wild type strains formed biofilms with similar mean cell densities, but examination of cellular interactions with the polystyrene surface showed that unlike the typical polar attachment of wild type cells, the mutant cells were attached to the plastic by their sides. Staining of glutaraldehyde-fixed biofilms with a fluorescent lectin (TRITC-ConA), showed a lectin distribution across the cell surface in the wild type cells. Remarkably, the lectin was preferentially located at one cell pole in the VLCFA-deficient mutant cells. As asymmetrical distribution of cell-surface polysaccharides seems important for normal polar attachment of rhizobia to surfaces (Laus *et al.* 2006), the authors suggest that the VLCFA modification of LPS is directly or indirectly involved in this process and therefore affects the normal organization of *R. leguminosarum* onto surfaces.

A detailed observation of cell-surface and cell-cell interactions in several *R. leguminosarum* LPS mutants affected in biofilm formation was recently reported by Russo *et al.* (2015). From structural data, it could be predicted that the absence of the outermost part of LPS, would render a more hydrophilic structure (Carlson *et al.* 2010) thus making *R. leguminosarum* more proficient in attaching to hydrophilic surfaces. LPS mutants showed a 40–50% decrease in the biofilms attached to polystyrene, while attachment to glass was enhanced. It was also shown, that the O-polysaccharide core moiety of LPS of *R. leguminosarum* is essential for the establishment of lateral and intimate cell-cell interactions. Observation of mutant GFP-labelled bacteria by CLSM revealed premature formation of microcolonies on a glass surface, with abnormal interactions. The typical lateral cell-to-cell contacts that prevail in the wild type biofilm were replaced by abundant chains of cells interacting through their poles. Microcolonies of the O-chain core-defective mutants evolved to loose and ramified structures instead of the compact "honeycomb" structures developed by wild type cells in mature biofilms (Russo *et al.* 2006). A 30–40% decrease in acidic CPS was measured in the LPS mutants that could partially account for the altered biofilm phenotypes. Quantitative analysis of CLSM images of mature wild type and mutant biofilms, showed a 70% reduction in surface colonization by LPS mutants. A cell coverage reduction of 3–5% observed in the outermost layers of the biofilm corresponded to layer 50 in the wild type but layer 150 in the LPS mutants, accounting for the thicker (but much less compact) biofilms formed by the mutant strains. The O-polysaccharide and core structures of LPS also influenced adhesion properties to root surfaces. Although comparable amounts of wild type and LPS mutant cells were able to colonize the root epidermis, the proportion of bacteria attached directly to the epidermis was lower in the LPS mutants, forming aberrant star-like microcolonies.

By contrast, a report on an O-antigen-deficient mutant of *Bradyrhizobium japonicum* showed increased biofilm forming ability on an hydrophobic surface as polyvinyl chloride (PVC) (Lee *et al.* 2010). This phenotype was attributed to a more hydrophobic surface in the mutant strain due to the lack of the O-polysaccharide portion of the LPS (Park & So, 2000). This phenotype would be different to that of *R. leguminosarum* O-antigen core-defective mutant, which is predicted to exhibit an external reduced hydrophobicity (Russo *et al.* 2015)

It seems likely that the interplay between LPS components and other surface components are involved in the overall surface physicochemical properties. Indeed, the differential biofilm phenotypes of LPS mutants from different species suggest that the resulting adhesive abilities will depend on the interaction between the hydrophobic nature of the surface, the exposed LPS structure and the production of other extracellular factors involved in biofilm formation. Another process that could be relevant for the biofilm matrix is that LPS may be released from the cell in the form of vesicles (Beveridge, 1999; Schooling & Beveridge, 2006). Therefore, LPS would be crucial not only for the intimate contact between cells or with the adhering surface, but at distance as well. The formation of membrane vesicles in rhizobial biofilms remains an unexplored area of research.

12.3.1.4 Glucomannan

A neutral and high molecular weight polysaccharide isolated from the surface of *R. leguminosarum* strain RBL5523 was exclusively localized on one pole of the bacterial cell, more precisely, on the pole involved in attachment to the root surface (Laus *et al.* 2006). Although its chemical structure is still lacking, carbohydrate composition analysis showed that the polysaccharide is mainly composed of mannose and glucose, with minor amounts of galactose and rhamnose (Laus *et al.* 2006). The high conservation of the cluster of genes involved in glucomannan biosynthesis in rhizobia and agrobacteria and its absence in the closely related *Brucella* spp. that are pathogens of animals and humans (Williams *et al.* 2008), suggests a widespread role for glucomannan in the interaction with the host plant.

Plant-made lectins are well-characterized proteins, their binding affinity (Gebauer *et al.* 1979; Gebauer *et al.* 1981) as well as their sugar binding site, have been defined by structural and mutational analysis (Rini *et al.* 1993; van Eijsden *et al.* 1992). Pea (*P. sativum*) lectin and vetch (*V. sativa*) lectin are almost identical proteins, with high affinity for glucose/mannose-containing saccharides. These proteins were used in lectin interaction experiments with all known rhizobial surface polysaccharides, showing that glucomannan is the only lectin-binding polysaccharide produced by *R. leguminosarum* RBL5523 (Laus *et al.* 2006). Furthermore, a mutant strain deficient in the production of glucomannan was impaired in attachment and root hair infection. Analysis of attachment at varying pH led to the idea that rhizobia use different mechanisms for docking at the root surface that ultimately would be determined by the acidic or alkaline condition of the environment. The glucomannan-lectin mechanism operates under slightly acidic conditions that assure retention of plant lectins on the root-hair tip. At an alkaline pH a mechanism involving a bacterial protein was identified (see section 12.3.2.1).

As mentioned in previous sections, Williams *et al.* (2008) used the genome sequence of *R. leguminosarum* bv. *viciae* strain 3841 to identify genes predicted to be involved in polysaccharide biosynthesis and isolated mutants in those genes. Polysaccharide analyses indicated that the novel *gmsA* locus was specifically involved in glucomannan synthesis. The insertional mutant in this locus was strongly affected in polar attachment and cap formation on root hairs at acidic pH values, but the biofilm phenotype on a glass surface was unchanged. Interestingly, the glucomannan deficient-mutant was outcompeted by the wild type strain in co-inoculation experiments to host plants, indicating that this polysaccharide is important for competitive nodulation. These observations are in line with those of Laus *et al.* (2006) and reinforce the idea that increased adhesion and biofilm formation on transgenic legumes expressing specific root lectins (van Rhijn *et al.* 2001) could enhance nodulation capacity with obvious effects on plant growth and productivity.

12.3.1.5 Cellulose

Cellulose is composed of β-(1$\rightarrow$ 4)-linked D-glucose residues, forming long unbranched chains (Bielecki *et al.* 2005). In the early 70s, floc formation was recognized in several gram-negative bacteria grown in broth that correlated with the production of cellulose microfibrils (Deinema & Zevenhuizen, 1971). Later on, the production of extracellular cellulose microfibrils was also examined in several rhizobia, and its relation to flocculation was established (Napoli *et al.* 1975). A role for these microfribils in the infection process preceding the establishment of symbiosis was proposed; rhizobial cellulose would enhance infection by causing aggregation of infective cells and mediating adsorption to the root hair surface.

Over the next decade, a sequence of steps in the root attachment process was described in *R. leguminosarum* bvs. *trifolii* (Dazzo *et al.* 1984) and *viciae* (Smit *et al.* 1987). The first step was described as a specific reversible interaction between bacterial and root surfaces, while during the second step, production of cellulose microfibrils tether bacteria firmly to the root hair. Cellulose-deficient mutants were impaired in the formation of cap-like aggregates on root hair tips, but they were still able to nodulate the host plant (Ausmees *et al.* 1999; Smit *et al.* 1987). Laus *et al.* (2005) argued that a cellulose-defective mutant was less able to infect older root hairs, implying that the developmental stage of root hairs is another factor to take into account. However, it seems that cellulose production and/or cellulose-mediated cap formation facilitate the successful infection of legumes, increasing the inoculum for infection of older root hairs (Laus *et al.* 2005).

More recently in the comprehensive mutational study on *R. leguminosarum* bv. *viciae* 3841 by Williams *et al.* (2008) it was shown that cellulose synthesis is not required for biofilm formation in static cultures but, as expected, was responsible for bacterial cap formation on pea root hairs. Besides, cellulose synthesis was not important for competitive nodule infection. One possibility is that bacterial cellulose plays a more significant role in the onset of infection in the natural and competitive rhizosphere environment.

12.3.2 Extracellular proteins

The formation of biofilms on either, roots, root hairs, or abiotic surfaces (including soil particles) also involve proteins secreted outside the bacterial cell. To date, two types of functions for rhizobial extracellular proteins in biofilm formation were reported. Some of them act as adhesins in the early attachment events. Other secreted proteins that are completely released from the cell surface or remain cell surface-associated seem to participate in the remodeling of the biofilm matrix.

12.3.2.1 Adhesins

Rhicadhesin is an acidic calcium-binding protein (14 kDa) that was first isolated from the surface of *R. leguminosarum* bv. *viciae* (Smit *et al.* 1989b). This protein was then found to be widely distributed in the *Rhizobiaceae* (Dardanelli *et al.* 2003; Swart *et al.* 1993), but the coding gene still remains unknown, precluding full characterization. Rhicadhesin mediates attachment to root hairs (Smit *et al.* 1992) nonspecifically at alkaline pH (Laus *et al.* 2006). At pH below 6.5, rhicadhesin is released from the bacterial cell surface due to calcium loss, and thus the lectin-glucomannan mechanism of attachment described above is operative.

A galactose-specific lectin BJ38 was isolated from *B. japonicum* that mediates specific binding to soybean roots (Ho *et al.* 1990). BJ38 was immunolocalized to one pole of the cell surface of wild type *B. japonicum*. This protein binds selectively to soybean emergent root hairs, probably because only these regions expressed the carbohydrate structures recognized by BJ38 (Ho *et al.* 1994). It would be interesting to address the role of BJ38 in biofilm formation, but again lack of sequence information hampers additional studies.

12.3.2.2 Lectins

In *R. leguminosarum* proteins exported via the PrsDE type I secretion system were found to be important for the development of a mature biofilm (Russo *et al.* 2006). PrsDE is responsible for the secretion of a family of proteins named as "Rap" for Rhizobium adhering proteins (Ausmees *et al.* 2001), which contain one or two *Rhizobium*-adhering or "Ra" domains. Biochemical and biophysical characterization of RapA2 (24 kDa), the smallest member of the family, led to the definition of the protein as a calcium-binding lectin, of extreme structural flexibility that specifically recognizes the acidic exopolysaccharides produced by *R. leguminosarum* biovars (Abdian *et al.* 2013). As RapA2 is solely composed of two Ra domains, it was proposed that these domains would confer carbohydrate-binding ability to the proteins that bear them. The Raps include the RapA1 and RapA2 paralogues, the glycanases PlyA, PlyB, and PlyC (described in the section below); the RapB protease (P.L. Abdian *et al.*, unpublished) and others of unknown function such as RapC, and RapD (Ausmees *et al.* 2001; Russo *et al.* 2006). RapA has been involved in autoaggregation of *Rhizobium* cells (Ausmees *et al.* 2001) and adhesion to plant roots (Frederix *et al.* 2014; Mongiardini *et al.* 2008). It was proposed that RapA may contribute to the assembly of the polysaccharide in the biofilm matrix and that its activity may be coordinated with the Ply glycanases, which cleave the acidic polysaccharides on the bacterial surface (Abdian *et al.* 2013).

12.3.2.3 Glycanases

In *R. leguminosarum* the lengths of the acidic exopolysaccharide molecules are limited by the PlyA and PlyB glycanases secreted by PrsDE (Finnie *et al.* 1997; Krehenbrink & Downie, 2008). As a consequence, both the *plyB* mutant and more severely, the *plyA plyB* double mutant produce longer chains of exopolysaccharide (Finnie *et al.* 1998) and were impaired in the formation of a typical biofilm on a glass surface, generating small and dense clusters of bacteria instead of the typical "honeycomb" structures (Russo *et al.* 2006). Thus the activity of the Ply glycanases on the exopolysaccharide strongly influences the assembly of the biofilm matrix. An interesting observation is that the Ply glycanases are activated on the cell surface (Zorreguieta *et al.* 2000), probably acting on nascent polysaccharide chains and thus restricting the cleavage to the region adjacent to the bacterial surface. A similar observation was made for the ExoK and ExsH glycanases from *S. meliloti* (York & Walker, 1997); in this case, mature accumulated succinoglycan was refractory to depolymerization by these glycanases. The authors proposed that the glycanases might be active only during a limited period, in which succinoglycan has not changed to its final conformation or aggregation state. Although the activity of ExoK and ExsH glycanases, has not been studied in the context of biofilm formation, it is expected that an effect on the regulation of the molecular weight distribution of succinoglycan would impact matrix development.

The surface-associated CelC2 cellulase of has a key role in cellulose biosynthesis by modulating the length of the cellulose fibrils that mediate firm adhesion to root tips and lead to cap formation. Strains that lacked or overproduced CelC2 showed increased or reduced amounts of external cellulose microfibrils, respectively. It was proposed that a defect in cellulose processing would collapse the biofilm structure (Robledo *et al.* 2012).

12.3.3 Nod factors

Nod factors are molecular signals secreted by rhizobia and play a key role for the establishment of symbiosis (Geurts & Bisseling, 2002). They are a family of lipo-chitooligosaccharides composed by a tetrameric or pentameric N-acetylglucosamine backbone, always carrying a fatty acid at the non-reducing end and substituted with acetyl-, fucosyl-, and sometimes sulfate residues (D'Haeze & Holsters, 2002). Numerous genes participate in the biosynthesis of Nod factors; such as the common *nod* genes (*nodABC*) required for the production of the core structure, the host-specificity *nod* genes (*nodEF, nodG, nodH, nodPQ, nodL*) that add the modifications to the core, and regulatory *nod* genes (*nodD, nolR, nodVW*), transcriptional regulators of the above genes (Debelle *et al.* 2001; Schlaman *et al.* 2006; Spaink, 2000). The biosynthesis of Nod factors is induced by plant flavonoids. But the classic picture of Nod factors acting only as signaling molecules changed, they were recently recognized as critical extracellular components for establishing a mature rhizobial biofilm (Fujishige *et al.* 2008). *S. meliloti* mutants deleted of any one of the common *nod* genes exhibited reduced (50–70%) biofilm formation both in vivo and in vitro. Confocal imaging of wild type GFP-labelled rhizobia showed reduced cell-cell contacts and cell-root interactions in the *nod* mutants. Overexpression of common *nod* genes in *S. meliloti* enhanced biofilm formation (25%), so did the same series of *nod* plasmids when introduced into *Agrobacterium tumefaciens*, another member of the Rhizobiaceae. These findings show that the core structure of Nod factors contributes to biofilm formation, perhaps by associating with the bacterial cell surface and contributing to cell-to-cell interactions (Fujishige *et al.* 2008). Interestingly, the authors proposed that the biofilm function of Nod factors may reflect an ancestral trait, as this property is encoded by common *nod* genes found in all nodulating rhizobia, including the *β*-rhizobia. Moreover, although a plant-derived inducer of *nod* genes as luteolin stimulates biofilm formation, the presence of this flavonoid was not required to observe the *nod* dependent-biofilm phenotypes, implying a different control mechanism (flavonoid-independent) to regulate Nod factor production and giving further support to this hypothesis. In line with these observations, it was recently shown that, in *S. fredii*, flavonoids induce transition from monolayer to microcolony, in a NodD1 dependent manner (Perez-Montaño *et al.* 2014). Several lines of evidence led these authors to propose a model in which, exudation of flavonoids in the rhizosphere, besides inducing the synthesis of Nod factors, which are necessary for the microcolony-type biofilm, potentiate via NodD1 a quorum sensing system, leading to full development of the microcolony-type biofilm.

12.3.4 Flagella

Several common soil bacteria, including *S. meliloti*, *Rhizobium lupini* and *B. japonicum* possess "complex" flagellar filaments that differ from the common "plain" type in their fine structure, amino acid composition and rigidity. Complex flagella are considerably fragile, but at the same time, suitable for motility in viscous media (Maruyama *et al.* 1978). In rhizobia these complex flagella are often peritrichously inserted, except for *B. japonicum* that possesses two different flagellar systems, one subpolar and the other lateral, which are formed by different sets of flagellins (Althabegoiti *et al.* 2008). On the other hand, *R. leguminosarum* strains have plain flagella that can exhibit subpolar or peritrichous location (Tambalo *et al.* 2010).

The roles of flagella in rhizobial biofilm formation are not completely understood, but in all cases studied to date, there is a parallel between the presence of functional flagella and the ability to form biofilms. It has been proposed that flagella might act in biofilm formation both as surface adhesins (Giron *et al.* 2002) and as providers of force-generating motility (Lemon *et al.* 2007).

However, pioneering works of the research group headed by Kijne, Lugtenberg and co-workers did not find any evidence of a possible role of flagella in attachment of *R. leguminosarum* bv. *viciae* to pea root hair tips (Smit *et al.* 1989a). On the other hand, Xie *et al.* (2012) reported that root exudates induced *R. leguminosarum* bv. *viciae* to attach polarly to a glass surface, and identified an arabinogalactan protein (AGP) in pea exudate as the plant partner of the interaction. The root-exudate-induced polar attachment was visualized by CLSM and was used to screen several *R. leguminosarum* mutants defective in the production of surface components. Mutations affecting flagella delayed but did not block polar attachment. Therefore the authors proposed that flagella can promote polar attachment, but they would not be essential for it to occur.

Fujishige *et al.* (2006) used the microtiter plate assay method to examine biofilms of *S. meliloti* flagellar mutants, and observed a 50% reduction compared to the wild type strain Rm1021. When these flagella-deficient mutants were inoculated on alfalfa and white sweetclover roots, a delay in nodule development was observed. The altered symbiotic phenotypes of flagellar mutants suggest a defective colonization of root surfaces.

12.4 CONCLUDING REMARKS

Several reports described bacterial aggregation on roots and cap formation long before rhizobial biofilm formation was recognized. At that time, a great deal of work focused in the structural characterization of rhizobial surface components, driven by the interest in the determinants of host-specificity. All that knowledge gathered is now extremely useful in the study of EPS and the assembly of the biofilm matrix. However, information about the relevance of certain EPS in biofilm formation is still scarce and sometimes puzzling. We are just beginning to explore the matrix at the molecular level, being aware of the oversimplification of the system, but aiming, in the near future, to identify its components, their interactions and hopefully their temporal and spatial dynamics. A model of *R. leguminosarum* matrix assembly, showing the possible interactions between acidic exopolysaccharides with secreted proteins is depicted in Figure 12.2. It will be interesting to gain insight into the interplay between different biofilm matrix components and "modulating" protein factors in other bacterial species. Besides, further studies on the regulation of both EPS production and matrix building will be also important to understand the underlying molecular mechanisms of biofilm matrix development.

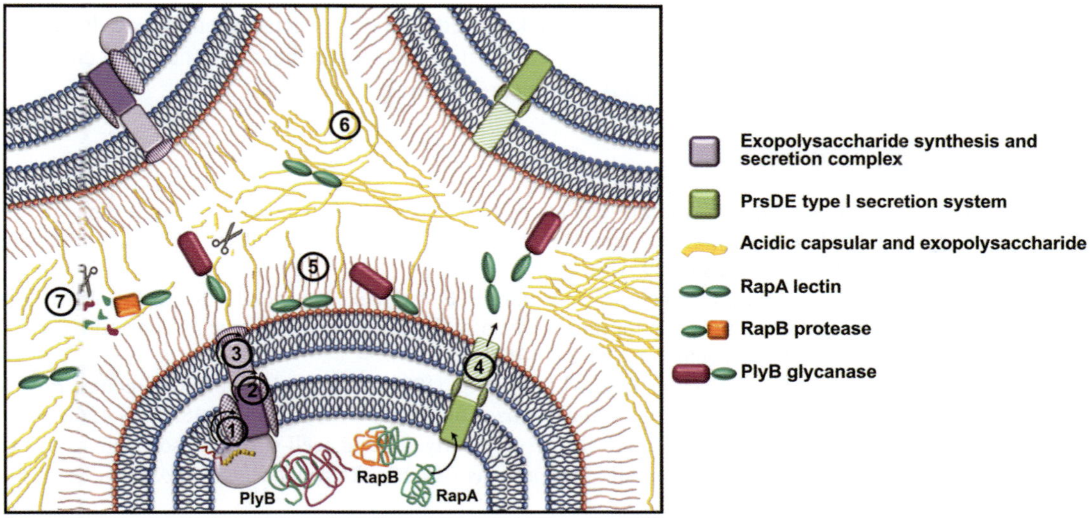

Figure 12.2 A hypothetical model of the interaction among EPS components in *R. leguminosarum.* **1, 2, 3)** represent acidic exopolysaccharides biosynthesis and secretion. **4)** Secretion of Rap proteins via the PrsDE typeI secretion system and protein folding in the extracellular milieu, promoted by the higher calcium concentration. The Ra domains are shown in green. **5)** Capsule formation at the cell pole. **6)** Matrix assembly by spontaneous interaction of exopolysaccharide molecules and formation of a network. The RapA lectin and the Ply glycanases may control acidic exopolysaccharides processing. **7)** Extracellular proteolysis may be a control mechanism of the activity of secreted proteins.

Since the first report on rhizobial biofilm formation (Seneviratne & Jayasinghearachchi, 2003) increasing efforts were made to understand the adoption of a biofilm lifestyle by several rhizobia. The complex life cycle of rhizobia that survive in soil for extended periods of time, proliferate in complex communities in the rhizosphere or grow in intimate association with plant roots, make one suspect the difficult task underneath the comprehension of rhizobial biofilm formation. Some insights into the environmental signals, the mechanisms and the molecules involved have been addressed, but much work is still to be done. For instance, calcium concentration in the rhizosphere affects not only bacterial aggregation, but also regulates anchoring of rhicadhesin on the cell surface (Smit *et al.* 1991), probably the activity of the RapA lectin and other Rap(s) (Abdian *et al.* 2013) and other calcium-dependent polymers. Moreover, the pH in the rhizosphere has a strong influence in the initial attachment to roots, determining the mechanism that will operate and the molecules involved. Of course plant factors, that were not subject of much discussion here, will affect rhizobial attachment and biofilm formation (Xie *et al.* 2012). In fact the molecular determinants of biofilm formation in vivo and vitro are different, maybe due the influence of plant-derived substances and root-surface components. Moreover, plant secreted molecules may contribute to the biofilm matrix, even in the absence of plants (Pérez-Giménez *et al.* 2009). Undoubtedly, the challenge for future studies will be to develop new methodologies to explore rhizobial biofilm formation in their natural habitats. The area that aims to elucidate the molecular basis of biofilm matrix formation might as well benefit from these approaches.

ACKNOWLEDGEMENTS

We are grateful to our colleagues at the Laboratorio de Microbiología Molecular y Celular and the Fundación Instituto Leloir community. We also acknowledge the Agencia Nacional de Promoción Científica y Tecnológica (ANPCyT) and the Consejo Nacional de Investigaciones Científicas y Tecnológicas (CONICET) for financial support. P.L.A. and A.Z. are career members of CONICET.

REFERENCES

Abdian P. L., Caramelo J. J., Ausmees N. and Zorreguieta A. (2013). RapA2 is a calcium-binding lectin composed of two highly conserved cadherin-like domains that specifically recognize *Rhizobium leguminosarum* acidic exopolysaccharides. *J Biol Chem*, **288**, 2893–2904.

Althabegoiti M. J., Lopez-Garcia S. L., Piccinetti C., Mongiardini E. J., Perez-Gimenez J., Quelas J. I., Perticari A. and Lodeiro A. R. (2008). Strain selection for improvement of *Bradyrhizobium japonicum* competitiveness for nodulation of soybean. *FEMS Microbiol Lett*, **282**, 115–123.

Ausmees N., Jonsson H., Hoglund S., Ljunggren H. and Lindberg M. (1999). Structural and putative regulatory genes involved in cellulose synthesis in *Rhizobium leguminosarum* bv. *trifolii*. *Microbiology*, **145**, 1253–1262.

Ausmees N., Jacobsson K. and Lindberg M. (2001). A unipolarly located, cell-surface-associated agglutinin, RapA, belongs to a family of *Rhizobium*-adhering proteins (Rap) in *Rhizobium leguminosarum* bv. *trifolii*. *Microbiology*, **147**, 549–559.

Beijerinck M. W. (1888). Die Bacterien der Papilionaceenknölchen. *Botanische Zeitung*, **46**, 797–804.

Beveridge T. J. (1999). Structures of gram-negative cell walls and their derived membrane vesicles. *J Bacteriol*, **181**, 4725–4733.

Bielecki S., Krystynowicz A., Turkiewicz M. and Kalinowska H. (2005). Bacterial cellulose. In: Polysaccharides and Polyamides in the Food Industry, A. Steinbuchel and S. K. Rhee (eds), Wiley-VCH Verlag, Weinhein, Germany, pp. 31–85.

Bringhurst R. M., Cardon Z. G. and Gage D. J. (2001). Galactosides in the rhizosphere: utilization by *Sinorhizobium meliloti* and development of a biosensor. *Proc Natl Acad Sci USA*, **98**, 4540–4545.

Carlson R. W., Forsberg L. S. and Kannenberg E. L. (2010). Lipopolysaccharides in *Rhizobium*-legume symbioses. *Subcell Biochem*, **53**, 339–386.

Costerton J. W., Cheng K. J., Geesey G. G., Ladd T. I., Nickel J. C., Dasgupta M. and Marrie T. J. (1987). Bacterial biofilms in nature and disease. *Annu Rev Microbiol*, **41**, 435–464.

Coviello T., Burchard W., Dentini M., Crescenzi V. and Morris V. J. (1995). Solution properties of exocellular polysaccharides of *Rhizobium leguminosarum*. Static and dynamic light-scattering: characterization in dilute solution. *J Polym Sci B Polym Phys*, **33**, 1833–1846.

Coviello T., Burchard W., Geissler E. and Maier D. (1997). Static and dynamic light scattering by a thermoreversible gel from *Rhizobium leguminosarum* 8002 exopolysaccharide. *Macromolecules*, **30**, 2008–2015.

Cheng H. P. and Walker G. C. (1998). Succinoglycan is required for initiation and elongation of infection threads during nodulation of alfalfa by *Rhizobium meliloti*. *J Bacteriol*, **180**, 5183–5191.

Cheng W., Zhang Q., Coleman D. C., Ronald Carroll C. and Hoffman C. A. (1996). Is available carbon limiting microbial respiration in the rhizosphere? *Soil Biol Biochem*, **28**, 1283–1288.

Dardanelli M., Angelini J. and Fabra A. (2003). A calcium-dependent bacterial surface protein is involved in the attachment of rhizobia to peanut roots. *Can J Microbiol*, **49**, 399–405.

Dazzo F. B., Truchet G. L., Sherwood J. E., Hrabak E. M., Abe M. and Pankratz S. H. (1984). Specific phases of root hair attachment in the *Rhizobium trifolii*-clover symbiosis. *Appl Environ Microbiol*, **48**, 1140–1150.

Dazzo F. B., Truchet G. L., Sherwood J. E., Hrabak E. M. and Gardiol A. E. (1982). Alteration of the trifoliin A-binding capsule of *Rhizobium trifolii* 0403 by enzymes released from clover roots. *Appl Environ Microbiol*, **44**, 478–490.

Debelle F., Moulin L., Mangin B., Denarie J. and Boivin C. (2001). *nod* genes and Nod signals and the evolution of the *Rhizobium* legume symbiosis. *Acta Biochim Pol*, **48**, 359–365.

Deinema M. H. and Zevenhuizen L. P. (1971). Formation of cellulose fibrils by gram-negative bacteria and their role in bacterial flocculation. *Arch Mikrobiol*, **78**, 42–51.

Dentini M., Coviello T., Burchard W. and Crescenzi V. (1988). Solution properties of exocellular microbial polysaccharides. 3. Light scattering from gellan and from the exocellular polysaccharide of *Rhizobium trifolii* (strain TA-1) in the ordered state. *Macromolecules*, **21**, 3312–3320.

D'Haeze W. and Holsters M. (2002). Nod factor structures, responses, and perception during initiation of nodule development. *Glycobiology*, **12**, 79–105.

Dorken G., Ferguson G. P., French C. E. and Poon W. C. (2012). Aggregation by depletion attraction in cultures of bacteria producing exopolysaccharide. *J R Soc Interface*, **9**, 3490–3502.

Downie J. A. (2010). The roles of extracellular proteins, polysaccharides and signals in the interactions of rhizobia with legume roots. *FEMS Microbiol Rev*, **34**, 150–170.

Finnie C., Hartley N. M., Findlay K. C. and Downie J. A. (1997). The *Rhizobium leguminosarum prsDE* genes are required for secretion of several proteins, some of which influence nodulation, symbiotic nitrogen fixation and exopolysaccharide modification. *Mol Microbiol*, **25**, 135–146.

Finnie C. Zorreguieta A., Hartley N. M. and Downie J. A. (1998). Characterization of *Rhizobium leguminosarum* exopolysaccharide glycanases that are secreted via a type I exporter and have a novel heptapeptide repeat motif. *J Bacteriol*, **180**, 1691–1699.

Flemming H. C. and Wingender J. (2010). The biofilm matrix. *Nat Rev Microbiol*, **8**, 623–633.

Frederix M., Edwards A., Swiderska A., Stanger A., Karunakaran R., Williams A., Abbruscato P., Sanchez-Contreras M., Poole P. S. and Downie J. A. (2014). Mutation of *praR* in *Rhizobium leguminosarum* enhances root biofilms, improving nodulation competitiveness by increased expression of attachment proteins. *Mol Microbiol*, **93**, 464–478.

Fujishige N. A., Kapadia N. N., De Hoff P. L. and Hirsch A. M. (2006). Investigations of *Rhizobium* biofilm formation. *FEMS Microbiol Ecol*, **56**, 195–206.

Fujishige N. A., Lum M. R., De Hoff P. L., Whitelegge J. P., Faull K. F. and Hirsch A. M. (2008). *Rhizobium* common *nod* genes are required for biofilm formation. *Mol Microbiol*, **67**, 504–515.

Gage D. J. (2002). Analysis of infection thread development using Gfp- and DsRed-expressing *Sinorhizobium meliloti*. *J Bacteriol*, **184**, 7042–7046.

Gebauer B., Schiltz E., Schimpl A. and Rudiger H. (1979). Purification and characterization of a mitogenic lectin and a lectin-binding protein from *Vicia sativa*. *Hoppe Seylers Z Physiol Chem*, **360**, 1727–1735.

Gebauer G., Schiltz E. and Rudiger H. (1981). The amino-acid sequence of the alpha subunit of the mitogenic lectin from *Vicia sativa*. *Eur J Biochem*, **113**, 319–325.

Geurts R. and Bisseling T. (2002). *Rhizobium* nod factor perception and signalling. *Plant Cell*, **14**, S239–S249.

Giron J. A., Torres A. G., Freer E. and Kaper J. B. (2002). The flagella of enteropathogenic *Escherichia coli* mediate adherence to epithelial cells. *Mol Microbiol*, **44**, 361–379.

Glazebrook J. and Walker G. C. (1989). A novel exopolysaccharide can function in place of the calcofluor-binding exopolysaccharide in nodulation of alfalfa by *Rhizobium meliloti*. *Cell*, **56**, 661–672.

Her G. R., Glazebrook J., Walker G. C. and Reinhold V. N. (1990). Structural studies of a novel exopolysaccharide produced by a mutant of *Rhizobium meliloti* strain Rm1021. *Carbohydr Res*, **198**, 305–312.

Ho S. C., Schindler M. and Wang J. L. (1990). Carbohydrate binding activities of *Bradyrhizobium japonicum*. II. Isolation and characterization of a galactose-specific lectin. *J Cell Biol*, **111**, 1639–1643.

Ho S. C., Wang J. L., Schindler M. and Loh J. T. (1994). Carbohydrate binding activities of *Bradyrhizobium japonicum*. III. Lectin expression, bacterial binding, and nodulation efficiency. *Plant J*, **5**, 873–884.

Hoang H. H., Becker A. and Gonzalez J. E. (2004). The LuxR homolog ExpR, in combination with the Sin quorum sensing system, plays a central role in *Sinorhizobium meliloti* gene expression. *J Bacteriol*, **186**, 5460–5472.

Janczarek M. (2011). Environmental signals and regulatory pathways that influence exopolysaccharide production in rhizobia. *Int J Mol Sci*, **12**, 7898–7933.

Janczarek M., Rachwał K., Marzec A., Grządziel J. and Palusińska-Szysz M. (2015). Signal molecules and cell-surface components involved in early stages of the legume–*rhizobium* interactions. *Appl Soil Ecol*, **85**, 94–113.

Kannenberg E. L. and Carlson R. W. (2001). Lipid A and O-chain modifications cause *Rhizobium* lipopolysaccharides to become hydrophobic during bacteroid development. *Mol Microbiol*, **39**, 379–391.

Karr D. B., Liang R. T., Reuhs B. L. and Emerich D. W. (2000). Altered exopolysaccharides of *Bradyrhizobium japonicum* mutants correlate with impaired soybean lectin binding, but not with effective nodule formation. *Planta*, **211**, 218–226.

Kelly S. J., Muszynski A., Kawaharada Y., Hubber A. M., Sullivan J. T., Sandal N., Carlson R. W., Stougaard J. and Ronson C. W. (2013). Conditional requirement for exopolysaccharide in the *Mesorhizobium-Lotus* symbiosis. *Mol Plant Microbe Interact*, **26**, 319–329.

Krehenbrink M. and Downie J. A. (2008). Identification of protein secretion systems and novel secreted proteins in *Rhizobium leguminosarum* bv. *viciae*. *BMC Genomics*, **9**, 55.

Laus M. C., Logman T. J., Van Brussel A. A., Carlson R. W., Azadi P., Gao M. Y. and Kijne J. W. (2004). Involvement of *exo5* in production of surface polysaccharides in *Rhizobium leguminosarum* and its role in nodulation of *Vicia sativa* subsp. *nigra*. *J Bacteriol*, **186**, 6617–6625.

Laus M. C., van Brussel A. A. and Kijne J. W. (2005). Role of cellulose fibrils and exopolysaccharides of *Rhizobium leguminosarum* in attachment to and infection of *Vicia sativa* root hairs. *Mol Plant Microbe Interact*, **18**, 533–538.

Laus M. C., Logman T. J., Lamers G. E., Van Brussel A. A., Carlson R. W. and Kijne J. W. (2006). A novel polar surface polysaccharide from *Rhizobium leguminosarum* binds host plant lectin. *Mol Microbiol*, **59**, 1704–1713.

Lee Y. W., Jeong S. Y., In Y. H., Kim K. Y., So J. S. and Chang W. S. (2010). Lack of O-polysaccharide enhances biofilm formation by *Bradyrhizobium japonicum*. *Lett Appl Microbiol*, **50**, 452–456.

Lemon K. P., Higgins D. E. and Kolter R. (2007). Flagellar motility is critical for *Listeria monocytogenes* biofilm formation. *J Bacteriol*, **189**, 4418–4424.

Marenduzzo D., Finan K. and Cook P. R. (2006). The depletion attraction: an underappreciated force driving cellular organization. *J Cell Biol*, **175**, 681–686.

Maruyama M., Lodderstaedt G. and Schmitt R. (1978). Purification and biochemical properties of complex flagella isolated from *Rhizobium lupini* H13-3. *Biochim Biophys Acta*, **535**, 110–124.

Mongiardini E. J., Ausmees N., Perez-Gimenez J., Julia Althabegoiti M., Ignacio Quelas J., Lopez-Garcia S. L. and Lodeiro A. R. (2008). The rhizobial adhesion protein RapA1 is involved in adsorption of rhizobia to plant roots but not in nodulation. *FEMS Microbiol Ecol*, **65**, 279–288.

Mort A. J. and Bauer W. D. (1982). Application of two new methods for cleavage of polysaccharides into specific oligosaccharide fragments. Structure of the capsular and extracellular polysaccharides of *Rhizobium japonicum* that bind soybean lectin. *J Biol Chem*, **257**, 1870–1875.

Moulin L., Munive A., Dreyfus B. and Boivin-Masson C. (2001). Nodulation of legumes by members of the beta-subclass of Proteobacteria. *Nature*, **411**, 948–950.

Napoli C., Dazzo F. and Hubbell D. (1975). Production of cellulose microfibrils by *Rhizobium*. *Appl Microbiol*, **30**, 123–131.

Park K. M. and So J. S. (2000). Altered cell surface hydrophobicity of lipopolysaccharide-deficient mutant of *Bradyrhizobium japonicum*. *J Microbiol Methods*, **41**, 219–226.

Parniske M., Kosch K., Werner D. and Müller P. (1993). *ExoB* mutants of *Bradyrhizobium japonicum* with reduced competitiveness for nodulation of *Glycine max*. *Molecular Plant Microbe Interactions*, **6**, 99–106.

Pellock B. J., Teplitski M., Boinay R. P., Bauer W. D. and Walker G. C. (2002). A LuxR homolog controls production of symbiotically active extracellular polysaccharide II by *Sinorhizobium meliloti*. *J Bacteriol*, **184**, 5067–5076.

Pérez-Giménez J., Mongiardini E. J., Althabegoiti M. J., Covelli J., Quelas J. I., Lopez-Garcia S. L. and Lodeiro A. R. (2009). Soybean lectin enhances biofilm formation by *Bradyrhizobium japonicum* in the absence of plants. *Int J Microbiol, 2009*, **719367**.

Pérez-Mendoza D., Rodriguez-Carvajal M. A., Romero-Jimenez L., Farias Gde A., Lloret J., Gallegos M. T. and Sanjuan J. (2015). Novel mixed-linkage beta-glucan activated by c-di-GMP in *Sinorhizobium meliloti*. *Proc Natl Acad Sci USA*, **112**, E757–765.

Perez-Montaño F., Jimenez-Guerrero I., Del Cerro P., Baena-Ropero I., Lopez-Baena F. J., Ollero F. J., Bellogin R., Lloret J. and Espuny R. (2014). The symbiotic biofilm of *Sinorhizobium fredii* SMH12, necessary for successful colonization and symbiosis of *Glycine max* cv Osumi, is regulated by Quorum Sensing systems and inducing flavonoids via NodD1. *PLoS One*, **9**, e105901.

Perotto S., Brewin N. J. and Kannenberg E. L. (1994). Cytological evidence for a host defense response that reduces cell and tissue invasion in pea nodules by lipopolysaccharide-defective mutants of *Rhizobium leguminosarum* strain 3841. *Mol Plant Microbe Interact*, **7**, 99–112.

Ramey B. E., Koutsoudis M., von Bodman S. B. and Fuqua C. (2004). Biofilm formation in plant-microbe associations. *Curr Opin Microbiol*, **7**, 602–609.

Reuhs B. L., Carlson R. W. and Kim J. S. (1993). *Rhizobium fredii* and *Rhizobium meliloti* produce 3-deoxy-D-manno-2-octulosonic acid-containing polysaccharides that are structurally analogous to group II K antigens (capsular polysaccharides) found in *Escherichia coli*. *J Bacteriol*, **175**, 3570–3580.

Rinaudi L. V. and Giordano W. (2010). An integrated view of biofilm formation in rhizobia. *FEMS Microbiol Lett*, **304**, 1–11.

Rinaudi L. V. and Gonzalez J. E. (2009). The low-molecular-weight fraction of exopolysaccharide II from *Sinorhizobium meliloti* is a crucial determinant of biofilm formation. *J Bacteriol*, **191**, 7216–7224.

Rinaudi L., Fujishige N. A., Hirsch A. M., Banchio E., Zorreguieta A. and Giordano W. (2006). Effects of nutritional and environmental conditions on *Sinorhizobium meliloti* biofilm formation. *Res Microbiol*, **157**, 867–875.

Rini J. M., Hardman K. D., Einspahr H., Suddath F. L. and Carver J. P. (1993). X-ray crystal structure of a pea lectin-trimannoside complex at 2.6 A resolution. *J Biol Chem*, **268**, 10126–10132.

Robledo M., Rivera L., Jimenez-Zurdo J. I., Rivas R., Dazzo F., Velazquez E., Martinez-Molina E., Hirsch A. M. and Mateos P. F. (2012). Role of *Rhizobium* endoglucanase CelC2 in cellulose biosynthesis and biofilm formation on plant roots and abiotic surfaces. *Microb Cell Fact*, **11**, 125.

Rodríguez-Navarro D. N., Rodríguez-Carvajal M. A., Acosta-Jurado S., Soto M. J., Margaret I., Crespo-Rivas J. C., Sanjuan J., Temprano F., Gil-Serrano A., Ruiz-Sainz J. E., *et al.* (2014). Structure and biological roles of *Sinorhizobium fredii* HH103 exopolysaccharide. *PLoS One*, **9**, e115391.

Russo D. M., Williams A., Edwards A., Posadas D. M., Finnie C., Dankert M., Downie J. A. and Zorreguieta A. (2006). Proteins exported via the PrsD-PrsE type I secretion system and the acidic exopolysaccharide are involved in biofilm formation by *Rhizobium leguminosarum*. *J Bacteriol*, **188**, 4474–4486.

Russo D. M., Abdian P. L., Posadas D. M., Williams A., Vozza N., Giordano W., Kannenberg E., Downie J. A. and Zorreguieta A. (2015). Lipopolysaccharide O-chain core region required for cellular cohesion and compaction of in vitro and root biofilms developed by *Rhizobium leguminosarum*. *Appl Environ Microbiol*, **81**, 1013–1023.

Santaella C., Schue M., Berge O., Heulin T. and Achouak W. (2008). The exopolysaccharide of *Rhizobium* sp. YAS34 is not necessary for biofilm formation on *Arabidopsis thaliana* and *Brassica napus* roots but contributes to root colonization. *Environ Microbiol*, **10**, 2150–2163.

Schlaman H. R., Olsthoorn M. M., Harteveld M., Dorner L., Djordjevic M. A., Thomas-Oates J. E. and Spaink H. P. (2006). The production of species-specific highly unsaturated fatty acyl-containing LCOs from *Rhizobium leguminosarum* bv. *trifolii* is stringently regulated by *nodD* and involves the *nodRL* genes. *Mol Plant Microbe Interact*, **19**, 215–226.

Schooling S. R. and Beveridge T. J. (2006). Membrane vesicles: an overlooked component of the matrices of biofilms. *J Bacteriol*, **188**, 5945–5957.

Seneviratne G. and Jayasinghearachchi H. S. (2003). Mycelial colonization by Bradyrhizobia and Azorhizobia. *J Biosci*, **28**, 243–247.

Sherwood J. E., Vasse J. M., Dazzo F. B. and Truchet G. L. (1984). Development and trifoliin A-binding ability of the capsule of *Rhizobium trifolii*. *J Bacteriol*, **159**, 145–152.

Skorupska A., Janczarek M., Marczak M., Mazur A. and Krol J. (2006). Rhizobial exopolysaccharides: genetic control and symbiotic functions. *Microb Cell Fact*, **5**, 7.

Smit G., Kijne J. W. and Lugtenberg B. J. (1987). Involvement of both cellulose fibrils and a Ca^{2+} -dependent adhesin in the attachment of *Rhizobium leguminosarum* to pea root hair tips. *J Bacteriol*, **169**, 4294–4301.

Smit G., Kijne J. W. and Lugtenberg B. J. (1989a). Roles of flagella, lipopolysaccharide, and a Ca^{2+} -dependent cell surface protein in attachment of *Rhizobium leguminosarum* biovar *viciae* to pea root hair tips. *J Bacteriol*, **171**, 569–572.

Smit G., Logman T. J., Boerrigter M. E., Kijne J. W. and Lugtenberg B. J. (1989b). Purification and partial characterization of the *Rhizobium leguminosarum* biovar *viciae* Ca^{2+} -dependent adhesin, which mediates the first step in attachment of cells of the family *Rhizobiaceae* to plant root hair tips. *J Bacteriol*, **171**, 4054–4062.

Smit G., Tubbing D. J., Kijne J. and Lugtenberg B. J. (1991). Role of Ca^{2+} in the activity of rhicadhesin from *Rhizobium leguminosarum* biovar *viciae*, which mediates the first step in attachment of *Rhizobiaceae* cells to plant root hair tips. *Arch Microbiol*, **155**, 278–283.

Smit G., Swart S., Lugtenberg B. J. and Kijne, J. W. (1992). Molecular mechanisms of attachment of *Rhizobium* bacteria to plant roots. *Mol Microbiol*, **6**, 2897–2903.

Sorroche F. G., Rinaudi L. V., Zorreguieta A. and Giordano W. (2010). EPS II-dependent autoaggregation of *Sinorhizobium meliloti* planktonic cells. *Curr Microbiol*, **61**, 465–470.

Sorroche F. G., Spesia M. B., Zorreguieta A. and Giordano W. (2012). A positive correlation between bacterial autoaggregation and biofilm formation in native *Sinorhizobium meliloti* isolates from Argentina. *Appl Environ Microbiol*, **78**, 4092–4101.

Spaink H. P. (2000). Root nodulation and infection factors produced by rhizobial bacteria. *Annu Rev Microbiol*, **54**, 257–288.

Sprent J. I. and Sprent P. (1990). Nitrogen fixing organisms, pure and applied aspects. Chapman and Hall Ltd., London, United Kingdom.

Sutherland I. W. (2001). Exopolysaccharides in biofilms, flocs and related structures. *Water Sci Technol*, **43**, 77–86.

Swart S., Smit G., Lugtenberg B. J. and Kijne J. W. (1993). Restoration of attachment, virulence and nodulation of *Agrobacterium tumefaciens chvB* mutants by rhicadhesin. *Mol Microbiol*, **10**, 597–605.

Tambalo D. D., Bustard D. E., Del Bel K. L., Koval S. F., Khan M. F. and Hynes M. F. (2010). Characterization and functional analysis of seven flagellin genes in *Rhizobium leguminosarum* bv. *viciae*. Characterization of *R. leguminosarum* flagellins. *BMC Microbiol*, **10**, 219.

Tate R. L. (1995). Symbiotic nitrogen fixation. In: Soil microbiology, John Wiley and Sons, Inc., New York, N.Y., pp. 307–333.

Tsien H. C. and Schmidt E. L. (1981). Localization and partial characterization of soybean lectin-binding polysaccharide of *Rhizobium japonicum*. *J Bacteriol*, **145**, 1063–1074.

van Berkum P. and Eardly B. (1998). Molecular Evolutionary Systematics of the *Rhizobiaceae*. In: The *Rhizobiaceae*, H. Spaink, A. Kondorosi and P. J. Hooykaas (eds), Springer, Netherlands, pp. 1–24.

van Eijsden R. R., Hoedemaeker F. J., Diaz C. L., Lugtenberg B. J., de Pater B. S. and Kijne J. W. (1992). Mutational analysis of pea lectin. Substitution of Asn125 for Asp in the monosaccharide-binding site eliminates mannose/glucose-binding activity. *Plant Mol Biol*, **20**, 1049–1058.

van Rhijn P., Fujishige N. A., Lim P. O. and Hirsch A. M. (2001). Sugar-binding activity of pea lectin enhances heterologous infection of transgenic alfalfa plants by *Rhizobium leguminosarum* biovar *viciae*. *Plant Physiol*, **126**, 133–144.

Vanderlinde E. M., Muszynski A., Harrison J. J., Koval S. F., Foreman D. L., Ceri H., Kannenberg E. L., Carlson R. W. and Yost C. K. (2009). *Rhizobium leguminosarum* biovar *viciae* 3841, deficient in 27-hydroxyoctacosanoate-modified lipopolysaccharide, is impaired in desiccation tolerance, biofilm formation and motility. *Microbiology*, **155**, 3055–3069.

Villain-Simonnet A., Milas M. and Rinaudo M. (2000a). A new bacterial exopolysaccharide (YAS34). II. Influence of thermal treatments on the conformation and structure. Relation with gelation ability. *Int J Biol Macromol*, **27**, 77–87.

Villain-Simonnet A., Milas M. and Rinaudo M. (2000b). A new bacterial polysaccharide (YAS34). I. Characterization of the conformations and conformational transition. *Int J Biol Macromol*, **27**, 65–75.

Wang P., Zhong Z., Zhou J., Cai T. and Zhu J. (2008). Exopolysaccharide biosynthesis is important for *Mesorhizobium tianshanense*: plant host interaction. *Arch Microbiol*, **189**, 525–530.

Williams A., Wilkinson A., Krehenbrink M., Russo D. M., Zorreguieta A. and Downie J. A. (2008). Glucomannan-mediated attachment of *Rhizobium leguminosarum* to pea root hairs is required for competitive nodule infection. *J Bacteriol*, **190**, 4706–4715.

Xie F., Williams A., Edwards A. and Downie J. A. (2012). A plant arabinogalactan-like glycoprotein promotes a novel type of polar surface attachment by *Rhizobium leguminosarum*. *Mol Plant Microbe Interact*, **25**, 250–258.

Yao S. Y., Luo L., Har K. J., Becker A., Ruberg S., Yu G. Q., Zhu J. B. and Cheng H. P. (2004). *Sinorhizobium meliloti* ExoR and ExoS proteins regulate both succinoglycan and flagellum production. *J Bacteriol*, **186**, 6042–6049.

York G. M. and Walker G. C. (1997). The *Rhizobium meliloti exoK* gene and *prsD/prsE/exsH* genes are components of independent degradative pathways which contribute to production of low-molecular-weight succinoglycan. *Mol Microbiol*, **25**, 117–134.

Zevenhuizen L. P. T. M. and van Neerven A. R. W. (1983). Gel-forming capsular polysaccharide of *Rhizobium leguminosarum* and *Rhizobium trifolii*. *Carbohydr Res*, **124**, 166–171.

Zgadzaj R., James E. K., Kelly S., Kawaharada Y., de Jonge N., Jensen D. B., Madsen L. H. and Radutoiu S. (2015). A legume genetic framework controls infection of nodules by symbiotic and endophytic bacteria. *PLoS Genet*, **11**, e1005280.

Zhan H. J., Levery S. B., Lee C. C. and Leigh J. A. (1989). A second exopolysaccharide of *Rhizobium meliloti* strain SU47 that can function in root nodule invasion. *Proc Natl Acad Sci USA*, **86**, 3055–3059.

Zhan H. J., Lee C. C. and Leigh J. A. (1991). Induction of the second exopolysaccharide (EPSb) in *Rhizobium meliloti* SU47 by low phosphate concentrations. *J Bacteriol*, **173**, 7391–7394.

Zorreguieta A., Finnie C. and Downie J. A. (2000). Extracellular glycanases of *Rhizobium leguminosarum* are activated on the cell surface by an exopolysaccharide-related component. *J Bacteriol*, **182**, 1304–1312.

Chapter 13

Transparent exopolymeric particles: an important EPS component in seawater

Oliver Wurl[1] and Michael Cunliffe[2]

[1]*Oliver Wurl, Carl-von-Ossietzky University Oldenburg, Institute of Chemistry and Biology of the Marine Environment, 26382 Wilhelmshaven, Germany*

[2]*Michael Cunliffe, Marine Biological Association of the United Kingdom, The Laboratory, Plymouth PL1 2PB, UK*

ABSTRACT

Transparent exopolymer particles (TEP), abundant gel-particles in the ocean, are of much scientific interests due to their active role in marine biogeochemical cycles and the microbial loop. We discuss the multiple sources of pre-cursor material, mainly dissolved polysaccharides, for the aggregation to TEP, and the distribution in the water column. As phytoplankton is the main producers of dissolved polysaccharides, TEPs are enriched in the upper water column, especially after blooms. TEPs control the distribution and availability of nutrients, but plays through their stickiness especially an active role in the formation sinking aggregates, called marine snow, and, therefore, facilitate the carbon export. However, if not ballasted, TEP is positive buoyant and rise up to the sea surface, where its enrichment can form biofilm-like habitats at the boundary layer between the ocean and atmosphere. TEP occurs also in sea-ice closely associated with phytoplankton embedded in the brine solution, and TEP acts as protective agent against cold temperature and high osmosis pressure. We complete the discussion of TEP with its role in microbial communities, including bacteria, phyto- and zooplankton.

Keywords: Transparent exopolymeric particles (TEP), marine waters, biofilm, microbial loop

13.1 INTRODUCTION

Extracellular polymeric substances (EPS) in the ocean have increasingly attracted the attention of the marine scientific community due to their active role in biogeochemical cycles and the microbial loop (Passow, 2002). Transparent exopolymer particles (TEP, Figure 13.1), the most abundant EPS in the ocean, are mainly formed by abiotic coagulation of dissolved polysaccharides excreted by phytoplankton communities in the sunlit surface layer of the ocean (Passow, 2002). However, only a smaller fraction of the marine dissolved organic carbon (DOC) become TEP (Engel & Passow, 2001), and the pre-cursor material requires distinct properties to polymerize and aggregate to TEP. This dissolved material further referred to pre-cursur material throughout this report. The dissolved pre-cursor material forms colloidal polymers by annealing and gelation, which then aggregates to TEP (Verdugo *et al.* 2004). Typical size range of TEP covers <1 mm to 100 s of micrometers with the larger fraction being in the lower range below 50 μm.

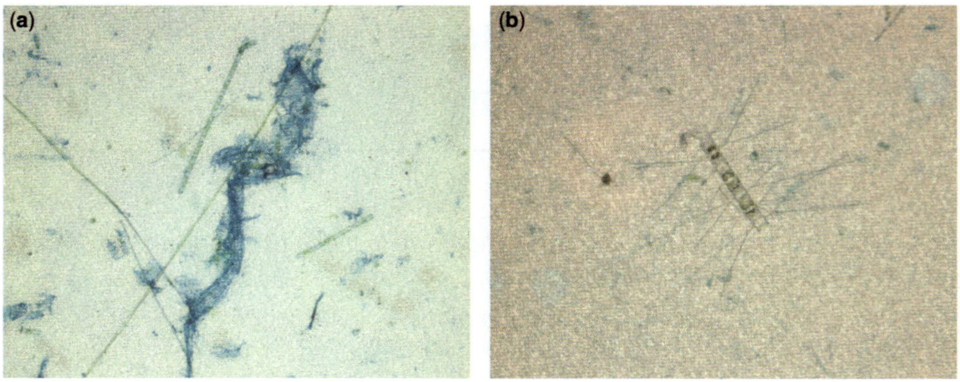

Figure 13.1 a) Stained TEP particle, b) smaller TEP particles associated with *Chaetoceros spp.* Both images taken from filtered seawater samples from the Greenland Sea (O. Wurl).

The spectrophotometric (Passow & Alldredge, 1995) and microscopic (Logan *et al.* 1994) technique allows semi-quantitative determination of TEP in seawater. Detailed procedures are described by Engel (2009). The stain Alcian Blue interferes with high salt concentration, and, therefore, seawater samples have to be filtered gently before staining TEP retained on the polycarbonate membranes. More recent developments seem to be promising for routine TEP measurements avoiding time-consuming filtration step (Claquin *et al.* 2008; Villacorte *et al.* 2015). With further methodological developments, routine measurement of TEP can be established, a crucial step to improve our understanding of the cycling of TEP and its ecological role.

TEP possesses strong surface-active properties and through their "stickiness," TEP acts as a glue matrix for other solid particles (i.e., detritus). This process forms larger aggregates, and, if sufficiently heavy, sinks to the bottom of the sea. During phytoplankton blooms this aggregation process occurs intensively (Thornton, 2002) and from the viewpoint of a diver the sinking aggregates looks like snow flakes, and so it is called marine snow in the field of oceanography. As marine snow plays an important role in the export of carbon and nutrients from the surface to the deep ocean (Passow *et al.* 2001), TEP, as glue matrix, is a key environmental driver for life in the deep sea (section 13.2.2). If unballasted, TEP are positively buoyant and ascend toward the surface to accumulate in the sea surface microlayer (SML), contributing a gelatinous composition to this interfacial film between the ocean and atmosphere (Wurl *et al.* 2011; Cunliffe *et al.* 2013; section 13.2.3). By bubble bursting at the sea surface, TEP may transfer to the atmosphere as part of aerosol particles, and, therefore, potentially be part of an influential factor in maritime cloud formation. Much effort are currently undertaken to understand the role of "atmospheric TEP", and we will provide an outline about current hypotheses in section 13.2.3. The multiple ecological functions of TEP in the marine environments includes also as cryo-protectant for sea-ice algae reviewed in section 13.2.4. TEP can represent up to 30% of the particulate organic carbon (POC) in the mixed layer (ML) of the water column (Harlay *et al.* 2010; Malpezzi *et al.* 2013). Considering that marine POC represent 50% of its global mass, TEP is a key component in the marine carbon cycle and oases for microbial life in the vast ocean (section 13.3).

13.2 SOURCES AND DISTRIBUTIONS OF TEP IN THE OCEAN

13.2.1 Sources

Multiple sources exist for precursor material and TEP, including bacterial production as attachment mechanism (Biddanda *et al.* 1986) or refractory dissolved organic matter (DOM) (Lechtenfeld *et al.* 2015), coral reefs forming mucus layers (Wild *et al.* 2004), benthic organisms (Ma *et al.* 2012) and kelp (Mazilla *et al.* 2013). However, the major source of TEP is polymerization of dissolved polysaccharides (Passow, 2000) excreted by phytoplankton communities (A in Figure 13.2). The excretion of polysaccharides is a normal process occurring during metabolism, but especially towards the end of blooms when the community becomes nutrient limited and get into a senescent stage (Corzo *et al.* 2000). However, the excretion of organic matter depends on phytoplankton species (Myklestad, 1997), growth stage (Engel *et al.* 2004) and environmental conditions (Corzo *et al.* 2000). High excretion rates and TEP production are often associated with diatoms (Thornton, 2001), coccolithophores (Engel *et al.* 2004), cyanobacteria (Berman-Frank *et al.* 2007) and prymnesiophyceae (Hong *et al.* 1997) leading to the formation of cell-TEP aggregates. Species of these groups are distributed throughout the world's oceans, and play a leading role in the global carbon cycle. For example, large colonies of the cyanobacterium Trichodesmium floating on the sea surface can be observed by satellites (Dupouy *et al.* 2000), and the floating colonies were associated with very high TEP concentrations (Wurl *et al.* 2011). Embedding cells in a low density TEP matrix probably supports their assimilation of atmospheric nitrogen. However, differences among single species also exist complicated to elucidate the ecological function of TEP. For example, Van Oostende *et al.* (2012) reported higher production of TEP during the diploid (means having two sets of chromosomes) stage compared to the haploid (means have a single set of chromosomes) stage of the life cycle of the coccolithophore of E. huxleyi.

Bacterial interaction with phytoplankton communities can stimulate the production of dissolved polysaccharides (Lechtenfeld *et al.* 2015), which may become readily available for TEP formation (Grossart *et al.* 1996; Gaerdes *et al.* 2011). Bacterial communities can also alter the composition of dissolved material (Oostende *et al.* 2012), and, therefore, have a direct influence of TEP formation and their physical properties (i.e., stickiness). Other sources include bacterial production of polysaccharide-rich and TEP-like mucus (biological glue), which is used for attachment to food-rich particles (Biddanda, 1986). As EPS is a key component to build up the network of phytoplankton colonies, disintegration of colonies is a major source of TEP. Grazing by zooplankton and protists on phytoplankton and bacteria can also release substantial amounts of dissolved pre-cursor material readily available for coagulation to TEP (E and G in Figure 13.2), a process called "sloppy feeding" (Møller, 2003). Zooplankton may also incidentally ingest TEP as part of their hunt on attached bacteria (Ling & Passow, 2003), but no evidence exists for direct uptake of TEP as food particle (Prieto *et al.* 2001). The larger fraction of the TEP pool is very small (Verdugo *et al.* 2004) but incidentally ingested by zooplankton (Decho, 1990). On the other hand, phytoplankton may reduce grazing pressure by releasing TEP to their colonies (Dutz *et al.* 2005). During grazing on colonies zooplankton ingest inevitably TEP replacing phytoplankton as food source (Passow & Alldredge, 1999). The same authors observed an increase in the abundance of larger TEP in the presence of copepods, probably copepods exudate pre-cursor material or TEP (Schuster & Herndl, 1995), or more likely enhance aggregation of existing TEP by forcing a current of water towards them to capture food particles. Copepods can reject particles, like TEP, during the feeding process (Young *et al.*

1997). Additional source of dissolved pre-cursor material includes viral lysis of host cells, an inherent part of the marine microbial loop, and thereby releasing organic matter to the dissolved phase (Middelboe *et al.* 1996). The released organic matter can also stimulate the growth of new bacterial communities (Noble *et al.*1999), potentially including colonizing aggregates. Benthic biological processes in sediments can be also a source of TEP (Ma *et al.* 2012).

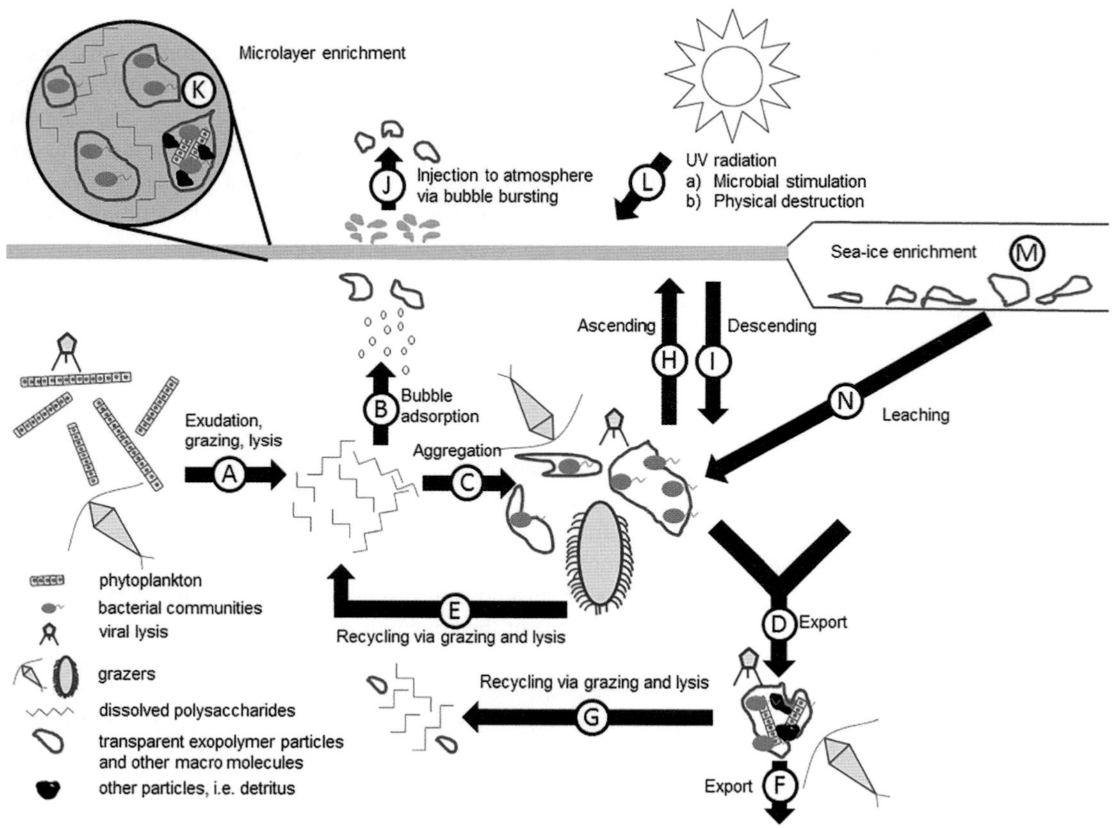

Figure 13.2 Conceptual model of TEP cycling in the ocean. Capital letters identify processes discussed in the text. Modified and reprinted from Wurl *et al.* 2011 with kind permission from John Wiley and Sons.

Conversion of dissolved pre-cursor material to polymer-gels occur spontaneously (Chin *et al.* 1998), but Schuster and Herndl (1995) pointed out that at a turbulence of 0.06 $cm^2\ s^{-3}$ an increase of TEP formation can be observed. This level of turbulence is comparable to those observed in the upper water layers of the ocean. In this context, it is widely acknowledged that turbulent flows are the key force in the formation of TEP. Aggregation requires that particles collide by Brownian motion, advection, differential settling and then stick together (Kepkay, 1994; Jackson, 1995; Jackson & Burd, 1998). For example, aggregation of dissolved polysaccharides to TEP can be described by the coagulation theory of Smoluchowski (1917), which has been used by Engel *et al.* (2004) to model TEP formation in mesocosms and later by Wurl *et al.* (2011) in the field. Beside the concentrations of polysaccharides and TEP, and their stickiness, the so-called carbon-specific collision kernels (collision frequency functions) are crucial factors describing a function of the size and motility of the particles and of the turbulent forces (Kiørboe, 1997). The formation rate of TEP increased during the induced growth of E. huxleyi (Engel *et al.* 2004) probably mainly due to the increase in the concentration of polysaccharides, major precursor material, and TEP by planktonic excretion. Wurl *et al.* (2011) did not observe significant differences in the formation rates of TEP in the surface (<50 m) and deep waters (>1000 m). TEPs are typically enriched in the upper water layers (see next section), but Wurl *et al.* (2011) observed a more constant distribution of polysaccharides in the upper 1000 m of the ocean. The derived rates are scaled by the square of their concentrations but only by the first order of the TEP concentration. Further research is required to obtain better estimates of stickiness factors and specific collision kernels at different depth to understand conversion of dissolved material into TEP in the ocean.

Bubbles, induced into the surface waters by breaking waves, are important for the formation of particulate material (Johnson & Cooke, 1980) including TEP (Zhou *et al.* 1998). Zhou *et al.* (1998) concluded that dissolved polysaccharides accumulate at the air-water interface of ascending bubbles and react to polymers and TEP. A fraction of the formed TEP may be transferred to the aerosol upon bubble bursting at the sea surface (J in Figure 13.2) supported by the presence of gel-like material in airborne aggregates (Leck & Bigg, 2005). However, production of TEP in the sea-surface microlayer (K in Figure 13.2) and by bubbles rising through the subsurface water (B in Figure 13.1) exceeds the losses via bubble bursting as supported by frequent enrichments of TEP in the sea-surface microlayer (see section 13.2.3).

13.2.2 Distribution in the water column

Highest concentrations of TEP have been associated with phytoplankton blooms (reviewed by Passow, 2002) due to the highest abundance of pre-cursor material. Updated data are presented in Table 13.1. As discussed it is well established that phytoplankton and their activity is related to the TEP concentration. Highest concentrations are found in eutrophic bays and estuaries (Ramaiah & Furuya, 2002; Prieto *et al.* 2006; Wurl & Holmes, 2008), or during spring blooms (Radic *et al.* 2005). High TEP concentrations were also detected during iron fertilization experiments, which trigger the growth of phytoplankton (Ramaiah *et al.* 2005). Relationships between TEP and chlorophyll-a (chl-a), as a proxy for phytoplankton biomass, have been described as $TEP = a\ (chl\text{-}a)^b$. Log transformation allows linear regression of the data with b as slope. The factor b is variable but usually below unity suggesting that either losses of TEP is counter-balanced by the production and accumulation of TEP or that production rates of TEP decreases with the progress of blooms. Declined TEP production with progress of bloom supports the hypothesis that TEP formation is a function of growth rather than standing stock ((Waite *et al.* 1995; Engel *et al.* 2002), but not the observation that excretion of polysaccharides increase towards the end of blooms when the community becomes nutrient limited and get into a senescent stage (Corzo *et al.* 2000). However, Wurl *et al.* (2011) reported that TEP concentrations were highest above the fluorescence maxima, another in-situ proxy for phytoplankton biomass. The reason for that is potential upward transport of unballasted and, therefore, positive buoyant TEP (Azetsu-Scott & Passow, 2004). TEP is ubiquitous in the ocean, even at times with low photosynthetic activities. Additional formation and recycling processes are due to abotic aggregation of aged dissolved organic (C, E, in Figure 13.1). Aged, or more recalcitrant, organic material are produced by the microbial communities, for example via remineralization processes of sinking material (Jiao *et al.* 2010). Grazing by zooplankton, for examples on senescent phytoplankton blooms, may also release substantial amounts of dissolved organic material, which can spontaneously aggregate (Chin *et al.* 1998) without observing high photosynthetic activities. However photosynthetic organisms remain to be a major source of TEP and its precursor material.

Table 13.1 Typical concentration range and occurence pattern of TEP in the ocean (nd = not detectable, ML = mixed layer). Data before 2002 reviewed by Passow (2002).

Location	TEP Concentration [μg Xeq L^{-1}]	Remark	Reference
Tokio Bay	14–1774	Eutrophic bay. Highest concentration in inner bay.	Ramaiah and Furuya (2002)
Baltic Sea	145–322		Engel *et al.* (2002)
North Atlantic	28.5 ± 10.2	Rather uniform distribution in the upper 50 m	Engel (2004)
California Currents	131–290	Rather uniform distribution in surface water	Azetsu-Scott and Passow (2004)
Bransfield Strait, Antarctica	nd - 346	Spatial distribution related to autotrophic biomass	Corzo *et al.* (2005)
Adriatic Sea	4–14800 (average 570)	Concentrations related to microphytoplankton. Highest conentrations during spring	Radic *et al.* (2005)
Western subarctic Pacific	40–189	Highest concentration during iron fertilization	Ramaiah *et al.* (2005)
Dona Paula bay, west coast of India	1.3–149.1	Concentrations higher during monsoon season	Bhaskar and Bhosle (2006)
Alboran Sea	<25–121 (offshore) 506–560 (inshore)	Spatial and temporal variability of TEP mediated by the hydrodynamics and primary production	Prieto *et al.* (2006)
Guaymas Basin, Gulf of California (deep-sea hydrothermal vents)	8–6451	Highest concentration associated with microbial mat-covered sediments.	Prieto and Cowen (2007)
Singapore and Johor Strait	196–8026	Highest concentration within mangrove catchments	Wurl and Holmes (2008)
Gulf of Aqaba (Israel)	23–228	TEP as link for carbon fluxes between the primary producers and heterotrophic bacteria	Bar-Zeev *et al.* (2009)
Southern Ocean	nd - 48	TEP in mixed layer related to chlorophyll-a, and below mixed layer to bacterial production	Ortega-Retuerta *et al.* (2009a)
California Currents	1975 ± 1056 (above ML) 138 ± 115 (below ML)	TEP concentration significant higher above mixed layer (ML)	Wurl *et al.* (2009)
Mediterrean Sea	4.5–94.3	Bacterial production as significant source of TEP	Ortega-Retuerta *et al.* (2010)
Bay of Biscay (France)	15.3–75.2	TEP represented up to 26% of the particulate organic carbon in the upper 40 m of the water column	Harlay *et al.* (2010)
Pearl River Estuary (China)	85.0–1234.9	TEP concentrations decreased gradually towards the open sea	Sun *et al.* (2010)

(*Continued*)

Table 13.1 Typical concentration range and occurence pattern of TEP in the ocean (nd = not detectable, ML = mixed layer). Data before 2002 reviewed by Passow (2002) (*Continued*).

Location	TEP Concentration [μg Xeq L^{-1}]	Remark	Reference
Subarctic North Pacific	18.1 ± 11.9 (above ML) 7.3 ± 5.1 (below ML)		
Subtropical North Pacific	14.2 ± 16.3 (above ML) 5.8 ± 4.6 (below ML)	TEP concentration significant higher above mixed layer (ML), but concentrations and formation rates did not vary significantly between the different study regions.	Wurl *et al.* (2011)
Northwest Passage (Arctic Sea)	10.5 ± 5.86 (above ML) 7.4 ± 4.0 (below ML)		
Under Arctic sea ice	33.8 ± 28.7 (above ML) 19.8 ± 16.5 (below ML)		
Shelf regions in Atlantic Ocean	11–1101	High TEP concentrations associated with stratified, coccolithophore-rich water masses.	Van Oostende *et al.* (2012)
Cheseapeake Bay	37 to 2820	TEP represented 32 ± 16% of particulate organic carbon	Malpezzi *et al.* (2013)
Western tropical North Pacific	18–69	Rather uniform distribution in the upper 300 m	Kodama *et al.* (2014)

TEP concentrations are typically high in the mixed layer (ML) (Ortega-Retuerta *et al.* 2009; Wurl *et al.* 2009; Wurl *et al.* 2011; Van Osterende *et al.* 2012). The ML is a layer between the ocean surface and a depth usually ranging between 10 and 150 m with more uniform water properties, i.e., density and temperature. Due to the typically higher abundance of phytoplankton in the ML, pre-cursor material, e.g., polysaccharides, accumulate in the ML which undergo aggregation to TEP. Indeed, pronounced turbulent conditions in the ML facilitate the self-assembly of dissolved pre-cursor material to polymers and TEP (Chin *et al.* 1998; C in Figure 13.2). TEP accumulates in the ML and sea-surface microlayer (SML), due to a low density of TEP of 0.70 to 0.84 g cm^{-3} (Engel & Schartau, 1999) and the density change between the ML and deeper water masses. It leaves the ML by sticking larger detritus material together forming aggregates sufficiently heavy to sink down (D and F in Figure 13.2). However, TEP may also change the sinking behavior of aggregates. For example TEP adds low density mass to aggregates slowing down their sinking rates or at high ratios of TEP-to-solid particles even stay in the upper water masses (Azetsu-Scott & Passow, 2004). Clogging pores of aggregates is another effect how TEP decreased sinking rates by slowing the equilibration between interstitial and surrounding waters of different densities, or as opposite effect decrease the drag of the sinking aggregate (Ploug *et al.* 2002). However, during massive phytoplankton blooms it is well established that a considerable amount of detritus sinks to deeper water masses, even only a small fraction of few percent reaches the sea bed. Passow *et al.* (2001) reported sedimentation rates of TEP with 10–100 mg Xeq m^{-2} d^{-1} during a 2-year period in the Santa Barbara Channel. In deeper water masses, bacterial production of TEP becomes more important (Ortega-Retuerta *et al.* 2009; Ortega-Retuerta *et al.* 2010), but overall Schuster and Herndl (1995) concluded that turbulence is more important than bacterial activity for the production of TEP consistent with the TEP accumulation in the ML. Beauvais *et al.* (2006) pointed out that turbulence favors the formation of larger TEP (>40 μm), but have no influence on smaller particles. A more uniform distribution (Kodama *et al.* 2014) can be observed, if the ML is less pronounced or in very oligotrophic regions with low primary productivity.

13.2.3 Enrichment of TEP at the sea surface

The sea surface plays a crucial role in the interaction between the atmosphere and ocean, including exchange of climate-relevant gases, heat and aerosols. Sieburth (1983) hypothesized that the sea-surface microlayer (SML), the boundary layer between the ocean and the atmosphere, is a hydrated gel-like layer formed by a complex structure of carbohydrates, proteins, and lipids. His hypothesis was more recently confirmed by finding that TEP are frequently enriched in the SML (Wurl & Holmes, 2008; Cunliffe *et al.* 2009a; Wurl *et al.* 2009; Wurl *et al.* 2011). The enrichment factor of TEP, defined as the ratio of concentrations in the SML sample to that of the corresponding subsurface bulk water sample, can be as high as 8 for a typical sea surface (Wurl *et al.* 2011), but the highest enrichments of up to 45 were found in slicks (Wurl *et al.* 2016). Slicks are sea surface phenomena formed by wave-damping action of surface-active organic material and defined as smooth and visible spots or streaks at the sea surface. Slicks are stable until wind speeds of 5–6 m/s, and very coherent ecosystems considering that 6.6 m/s is the average wind speed over the ocean (Archer & Jakobson, 2005).

As discussed earlier, TEP acts as a glue-matrix for other solid particles (e.g., detritus) forming larger aggregates (e.g., marine snow); however if unballasted, the low-density of TEP become buoyant and rise to the surface (H in Figure 13.2) (Azetsu-Scott & Passow, 2004; Mari, 2008). The enrichment of TEP decreased with increasing wind speed (Wurl *et al.* 2009; Wurl *et al.* 2011), probably due to injection as aerosols into the atmosphere (J in Figure 13.2) or mixing in the underlying subsurface water. However, the generation of bubbles plumes via wave breaking is probably also responsible to maintain an enrichment of TEP in the SML above unity. As discussed in section 13.2.1, TEP forms in filtered seawater on ascending bubbles (Zhou *et al.* 1998). The SML may become disproportionably enriched in high-molecular organic compounds by selective enrichment on bubble surfaces. These organic compounds are presumably more surface-active and therefore prone for polymerisation (Johnson & Cooke, 1980; Gershey, 1983). Bubble-generated TEP were rich in covalently bound sulfate (Zhou *et al.* 1998), which is consistent with the finding that TEP in the SML contain a higher fraction of half-ester sulfate bonding compared to TEP in the subsurface water (Wurl & Holmes, 2008). Wurl and Holmes (2008) suggest that the sulfate-containing bindings in TEP are important for the formation and stability of the SML, as such bindings have an influence on the intrinsic viscosity of polymers (Nichols *et al.* 2005) and sulfur-containing algal carbohydrates have been shown to be insoluble and nonhydrolyzable (Kok *et al.* 2000).

Laboratory studies in Langmuir–Blodgett systems indicate that compression and dilation of air-water interfaces may have a positive effect on the aggregation of high molecular weight compounds (Zhao *et al.* 2009; Madivala *et al.* 2009; Garofalakis & Murray, 2002). Only Wheeler (1975) attempted to compress microlayers to investigate the transformation of DOM into POM using Langmuir–Blodgett system and in-situ through glass barriers. He demonstrated the conversion of DOM to POM through compression of microlayers from lakes, but productive coastal waters needed to be compressed by 90% of the original area. However, he was unable to observe the potential aggregation of smaller-sized (10–200 μm) and transparent gel-like particles such as TEP. Indeed, Wurl *et al.* (2011) reported significant higher TEP formation rates (mean 31.1 ± 40.7 µmol C L^{-1} d^{-1}) in the SML, by an average factor of 5, than in the underlying bulk water using an aggregation model described by Engel *et al.* (2004) (see section 13.2.1). Constant compression and dilation of the SML may facilitate the aggregation process by enhancing collisions between polysaccharides and TEP (K in Figure 13.2).

It has been suggested that biofilm matrixes offers considerable UV protection for microorganisms (Elasri & Miller, 1999). A biofilm transmits only a small amount of UV radiation (13% of UV C, 31% of UV B, and 33% of UV A) protecting the cells from exposure and suggesting that the excretion of biofilm material (e.g., exopolymeric material such as TEP) may be a natural defense mechanism used to attenuate UV light exposure in nature. Tilstone *et al.* (2010) observed high concentrations of mycosporine-like amino acids in the SML, and those amino acids are effective in absorbing UV radiation (310 to 350 nm). Such a protective mechanism could therefore significantly contribute to the gelatinous nature of the SML. In the absence of microorganisms, Ortega-Retuerta *et al.* (2009) reported losses of TEP exposed to UV-B via photolysis (L in Figure 13.2), but not in samples with microbes. Furthermore, Young *et al.* (1978) observed excretion of considerable amounts of polymeric materials from bacterial communities collected from the SML. Cunliffe *et al.* (2011) reported the aggregation of bacterial communities in the SML, whereas cells in the bulk seawater were more dispersed. It indicates that microbial communities in the SML are important in the maintenance of the enrichment of TEP.

13.2.4 TEP in sea ice

A number of studies have reported high concentrations of exopolymeric substances, including TEP, in sea ice (Krembs *et al.* 2002; Meiners *et al.* 2003; Riedel *et al.* 2006; Collins *et al.* 2008). Median concentrations in ice (up to 4.9×10^6 particles L^{-1}) were higher by one order of magnitude to the abundances found in the underlying water column (0.56×10^6 particles L^{-1}) (M in Figure 13.2). The abundances were highest in the interior of the sea ice and were significantly correlated with chl-a, autotrophs and bacteria (Meiners *et al.* 2003; van der Merwe *et al.* 2009). During a sea ice algal bloom, Riedel *et al.* (2006) found the highest concentrations at the bottom of first-year sea ice, two-order of magnitudes higher than in the underlying seawater. It is likely that sea ice algae actively produce TEP as part of their survival strategy (see below). However, evidence from a number of studies has indicated that the bacterial community is also likely to affect carbon cycling and EPS concentration and distribution within sea ice (Collins *et al.* 2008; Deming, 2010; Krembs *et al.* 2011; Miller *et al.* 2011). Seasonal studies confirmed that TEP abundance increased throughout the winter (Krembs *et al.* 2002). It is consistent with low TEP concentrations found in seawater underneath Arctic sea ice during the autumn (10.5 ± 5.8 µmol C L^{-1}), but concentrations increased up to 74.3 µmol C L^{-1} in spring as sea ice began to melt Wurl *et al.* (2011). Correlations between DOC and TEP indicate a linkage between the carbon pools in sea ice, including reversible transitions (Riedel *et al.* 2008). At the end of a sea ice algal bloom, TEP can account for 70% of the total POC in sea ice, which subsequently is released into the underlying water during brine exchange or ice melt. In April, Wurl *et al.* (2011) observed a rapid release on TEP into the underlying water, within 8 days, and subsequent sinking of TEP down to 200 m in the High Arctic. Riedel *et al.* (2006) observed that TEP remained in sea-ice longer in late spring (May to June) at a coastal site on the Mackenzie shelf, while algal POC was released to the underlying water. It indicates that TEP may get released sporadically, and the release process to the underlying water may vary with season and location.

TEP plays a key role in carbon dynamic and microbial cycles in sea ice. Krembs *et al.* (2002) suggested that TEP act as a buffer and cryoprotectant for diatoms trapped in high saline sea ice. It includes the support to diatoms to stick to brine channels, and, at the same time, change the viscosity of the brine fluid limiting diffusion of high salt and gas concentration towards the cell. TEP acts as a substrate for bacteria

(Meiners *et al.* 2004), nutrients retention (Becquevort *et al.* 2009) and limit grazing pressure to algal communities by heterotrophs (Riedel *et al.* 2007). It further highlights the multiple ecological function of TEP in the marine environment, which is further discussed in the next section.

13.3 ROLE OF TEP IN THE MARINE MICROBIAL LOOP

Microorganisms are the most abundant and diverse components of marine ecosystems, they sustain biogeochemical cycles and underpin life-supporting ecosystem services. Through an ecological process known as the microbial loop, dissolved organic carbon (DOC) is assimilated by heterotrophic bacteria, which are preyed upon by small protists that are subsequently the prey of larger protists. Larger components of the microbial loop, such as ciliates, are the prey of zooplankton, such as copepods, transferring organic carbon up through the food web to higher trophic levels (Azam *et al.* 1983). At each stage of the food web, DOC is released by a variety of mechanisms, such as cell leakage and grazing, and the released DOC is returned to and recycled by the heterotrophic bacteria. The 'classic' microbial loop concept was developed in the early 1980s and focused on the transfer and recycling of DOC (Azam *et al.* 1983). Researchers have now reconsidered the microbial loop to also include particulate organic carbon (Verdugo *et al.* 2004), and that TEP may potentially restructure the flow of carbon through the base of the marine food web (Passow, 2002).Previously in this chapter we have discussed microbial production of TEP. Here we focus on the role of TEP in supporting the ecology and biology of the three major groups that constitute the microbial loop; bacteria, microbial eukaryotes (protists) and zooplankton.

13.3.1 TEP and bacteria

Early research on how bacteria might interact with TEP focused on bacteria that attached to the particles. In their seminal paper, Alldredge *et al.* (1993) assessed bacterial attachment to TEP collected off the Californian coast by dual staining the particles with DAPI and alcian blue. They showed that 22–68% of bacteria in the samples were attached to TEP (Alldredge *et al.* 1993). Importantly, before the discovery of TEP, the TEP-attached bacteria would have been considered 'free living' and their association unacknowledged. Subsequent studies have shown that the proportion of bacteria attached to TEP is probably lower, including TEP collected off the Californian coast and off Bermuda (2–25% attached to TEP) (Passow & Alldredge, 1994), TEP collected in the Adriatic Sea (0–20% attached to TEP) (Schuster & Herndl, 1995) and in the Kattegat (0.5–20% attached to TEP) (Mari & Kiørboe, 1996). Using the double staining approach, Mari and Kiørboe (1996) showed that bacteria are attached to the outer surface of TEP and within TEP, and also that TEP attachment to TEP is related to TEP size.

The potential implications of bacteria attached to TEP extend beyond the microbial loop and into other ecosystem functions. A recent study has shown that bacteria attached to TEP act as a 'protobiofilm' and, because TEP are sticky, readily attach to surfaces in the marine environment (Bar-Zeev *et al.* 2012). The bacteria colonized TEP causes the rapid development of biofilms on the surface, and may be the key controlling factor in biofilm formation (Bar-Zeev *et al.* 2012). Biofilm formation by TEP has wide implications, including impacting the water industry.

The enrichment of TEP at the sea surface, an in particular, within the sea surface microlayer, has a defining impact on the bacterioneuston communities that thrive there (Cunliffe & Murrell, 2009; Cunliffe *et al.* 2011). Because TEP within the water column are readily colonized by bacterioplankton (Alldredge *et al.* 1993; Passow & Alldredge, 1994; Schuster & Herndl, 1995; Mari & Kiørboe, 1996), and TEP ascend from the water column up into the microlayer (Azetsu-Scott & Passow, 2004), it is likely that this process has a major influence on bacterial colonization and subsequent diversity within the microlayer (Cunliffe *et al.* 2011). Comparison of bacterioneuston and bacterioplankton communities using fluorescent *in situ* hybridization has shown that the neuston communities are more aggregated, in keeping with a biofilm-like nature caused by the enrichment of TEP in the microlayer (Cunliffe & Murrell, 2009).

Links between bacterial diversity and TEP have been made at the L4 time-series station in the Western English Channel. The study followed the progression of bacterioplankton and bacterioneuston community diversity using 16S rRNA gene high-throughput sequencing in relation to the production of TEP by the spring diatom bloom. Distinct changes in both bacterioplankton and bacterioneuston community diversity occurred as the diatom bloom progressed and coincided with the rapid decline in TEP concentration after the peak of the bloom. Analysis of bacterial diversity showed a significant relationship between the decline in TEP concentration and the increase in abundance of the bacterial orders Flavobacteriales and Rhodobacterales, suggesting that both groups utilize TEP (Taylor *et al.* 2014). Flavobacteriales and Rhodobacterales have been associated with phytoplankton, and frequently shown to increase in abundance during phytoplankton blooms (Tada *et al.* 2011; Teeling *et al.* 2012). Both groups have been particularly associated with marine diatoms (Amin *et al.* 2012).

Metagenome and metaproteome analysis of bacterioplankton communities in the North Sea following a spring diatom bloom have shown that the metabolic capabilities to degrade and utilize TEP are potentially present within Flavobacteriales (Teeling *et al.* 2012). The study showed that the Flavobacteriales enzymes sulfatases and fucosidases were abundant during the diatom bloom. TEP are sulfated polysaccharides that have been shown to include fucose (Passow, 2002). Sulfatases and fucosidases would therefore be capable of degrading TEP. Extracellular glycoside hydrolases have been shown to degrade TEP (Smith *et al.* 1995), and Flavobacteriales are also known to have high numbers of glycoside hydrolase encoding genes (Fernández-Gómez *et al.* 2013) that are expressed during diatom blooms (Teeling *et al.* 2012).

Rhodobacterales may be able to utilize TEP indirectly via cross-feeding from Flavobacteriales activity (Taylor *et al.* 2014). In response to the diatom bloom in the North Sea, Rhodobacterales (principally Marine *Roseobacters*) expressed low molecular weight substrate transporters (Teeling *et al.* 2012), which would allow them to utilize the monosaccharides produced by the Flavobacteriales-derived extracellular enzymes degrading TEP.

13.3.2 TEP and microbial eukaryotes

A diverse array of microbial eukaryotes (protists) thrive in the marine environment, and have several important roles in the functioning of marine ecosystems. Phototrophic and mixotrophic protists (phytoplankton) assimilate inorganic carbon and produce both dissolved and particulate organic carbon, including TEP and TEP precursors (see section 13.2.1). Bacteria are the prey of heterotrophic and mixotrophic bacterivorous protists that form the next stage in the microbial loop. The importance of TEP for protists can be broadly divided across three processes; protists feeding on bacteria attached to TEP, attachment of protists to TEP and the direct consumption of TEP by protists. Clearly these processes will overlap. Protists feeding on TEP-attached bacteria will generally also be attached themselves.

Phaeocystis globosa is a bloom-forming haptophyte that produces significant amounts of TEP which frequently result in surface foams washing up on the shoreline. Mesocosm studies that used North Sea water to experimentally induce *P. globosa* blooms showed that the succession of protists attached to thehaptophyte-produced TEP is complex (Mari *et al.* 2004). During early growth phases of the bloom, when TEP is relatively 'new' and at lower concentrations (<20 ppm), the oligotrichous ciliate *Strombidium* spp. dominated the protist communities. At later stages of the *P. globosa* blooms when TEP was more concentrated and started to form larger aggregates the hypotrichous ciliate *Euplotes* spp. became dominant. The large TEP-based aggregates created a physical structure for the *Euplotes* spp. to attach to and graze on the TEP-attached bacteria and appeared to inhibit feeding by *Strombidium* spp. Mari *et al.* (2004) show that the TEP quantity and quality has a clear impact on the structure and function of the microbial loop, by controlling which bacterivorous protists dominant and how they feed on their prey.

As discussed above, the air-sea interface is characteristically enriched in TEP (Wurl & Holmes, 2008). Early, pioneering work by John McN. Sieburth observed that protists, including amoeba, flagellates and ciliates, in the SML are physically associated with the surface film (Sieburth, 1983). Large-scale mesocosm studies have been conducted using fjord water to determine the interactions between planktonic and neustonic microbial communities (Cunliffe *et al.* 2009b). Under phytoplankton bloom conditions, TEP concentration was increased compared to control mesocosms, and in all mesocosms TEP was significantly enriched in the SML (Cunliffe *et al.* 2009a). Using the molecular tool denaturing gradient electrophoresis (DGGE), eukaryote 18S rRNA genes were profiled to compare the diversity of protists in the planktonic and neustonic microbial communities. DGGE analysis showed that distinct protist communities formed in the TEP-enriched SML of the mesocosms (Cunliffe & Murrell, 2010). As with Mari *et al.* (2004), neustonic protist communities changed at different stages of the phytoplankton bloom. During early stages, *Cercozoa* spp., which were not detected in the plankton communities, dominated the microlayer. At later stages of the bloom, other protists were enriched in the SML, including ciliates. It is probable that the enrichment of TEP in the SML creates a greater surface area for protists that feed by attachment to particles and therefore promotes the development of alternate niches compared to those in the underlying water column (Cunliffe & Murrell, 2010).

Consumption of TEP by protists is also probable, however, unlike studies that have shown protists attached to TEP and feeding on TEP-attached bacteria (Mari *et al.* 2004), direct evidence has yet to be shown. Earlier studies have shown that heterotrophic flagellates can uptake and utilize colloidal material, including carbohydrates (Tranvik *et al.* 1993). Thraustochytrids were originally classified as fungi, but have since been reclassified and are now considered to be within the kingdom Chromalveolata. Incubation experiments showed positive correlations between Thraustochytrids and diatom-derived TEP in the presence of antibiotics, indicating that Thraustochytrids utilize TEP directly (Damare & Raghukumar, 2012). A novel picoeukaryote lineage was discovered in 2007 called the picobiliphytes and has been shown to be widely distributed in marine ecosystems (Not *et al.* 2007). Their first culture representative *Picomonas judraskeda* has been studied in detail and appears to feeding directly on colloids (Seenivasan *et al.* 2013). Clearly, research effort must be put into determining if protists can feed directly on TEP.

13.3.3 TEP and zooplankton

Laboratory feeding experiments with *Calanus pacificus* have shown that the copepod is able to consume TEP, even in the presence of alternate food items such as diatoms (Ling & Alldredge, 2003). Similar experiments have been conducted with the euphausiid *Euphausia pacifica* which also feed on TEP (Passow & Alldredge, 1999).

TEP alone presents a nutritional problem for zooplankton because TEP has a C:N stoichiometry typically at 26 (Engel & Passow, 2001), which is above the Redfield ratio (Redfield, 1934). Ling and Alldredge (2003) proposed that TEP alone would be a poor resource for zooplankton because of the relatively low nitrogen content. However, TEP that was colonized by bacteria would have a greater C:N stoichiometry and therefore would be more nutritious. Zooplankton feeding on bacterially-colonized TEP could access components of the microbial loop that would otherwise be too small to consume individually, and TEP may therefore act as a 'gel shunt' by transferring carbon more rapidly through the food web and bypassing other trophic levels (Figure 13.3).

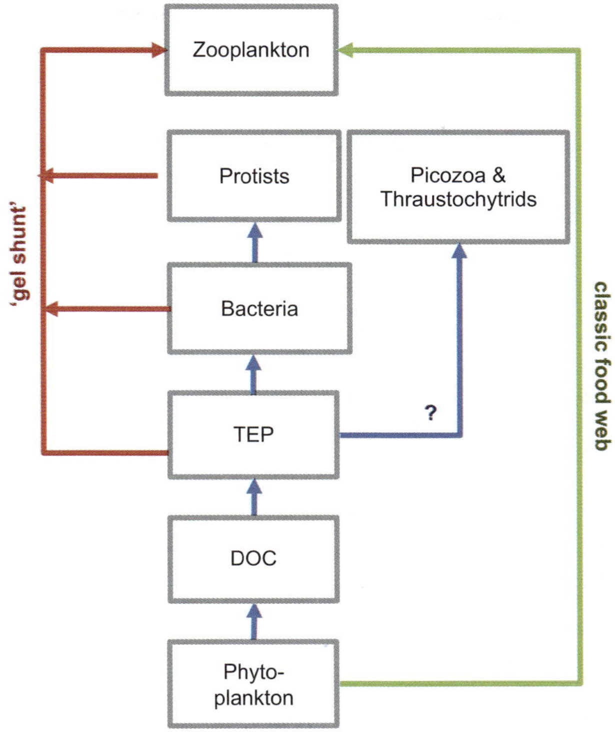

Figure 13.3 Diagram of the possible route of carbon transfer from phytoplankton-derived TEP through the base of the marine food web (modified from Passow, 2002). Specific carbon pathways have been highlight; the classic food web, zooplankton feed directly on phytoplankton, and the microbial loop; phytoplankton-derived carbon is sequentially transferred through different microbial groupings. We suggest the consideration of the 'gel shunt', where carbon could be transferred from TEP and smaller components of the microbial loop directly to zooplankton, such as copepods. Some groups of protist, such as the Thraustochytrids and the recently discovered phylum Picozoa, may be able to consume TEP directly and not in-directly via bacteria that graze on TEP.

13.3.4 TEP and viruses

Viruses are not part of the microbial loop *sensu stricto*, but do have the potential to modify the microbial loop by releasing DOC from virus-lysed cells (Suttle, 2007). Marine viruses have been shown to readily attach to TEP, and in some regions TEP-associated viruses can form a significant proportion of the total virus abundance (Mari *et al.* 2007). Mari *et al.* (2007) have proposed that TEP may be 'hotspots' for viral infection because TEP-attached bacteria and viruses would be in close proximity and so promoting infection rates.

13.4 CONCLUSIONS

TEPs are ubiquitious EPS particles in the ocean and play key roles in marine biogeochemical processes, like carbon sequestration to the deep, food source for plankton, and growth of sea-ice algae. Especially the direct consumption of TEP by zooplankton is an additional pathway in the marine carbon cycle requiring further research The enrichment of TEP in the SML may affect the chemical composition of aerosol, but direct evidence has yet to be proven. If TEP can be transferred to the atmosphere through sea spray, the increased organics on aerosols can directly affect the formation of cloud condensation nuclei, and, therefore, cloud properties. More reliable techniques in determining TEP and other EPS particles in seawater are needed to advance our understand how those particle drives the microbial life in the ocean.

ACKNOWLEDGEMENT

Oliver Wurl would like to thank the European Research Council for financial support (GA336408 PassMe). Michael Cunliffe is supported by a Marine Biological Association (MBA) Research Fellowship.

REFERENCES

Alldredge A. L., Passow U. and Logan B. E. (1993). The abundance and significance of a class of large, transparent organic particles in the ocean. *Deep Sea Res. Part I*, **40**, 1131–1140.

Amin S. A., Parker M. S. and Armbrust E. V. (2012). Interactions between diatoms and bacteria. *Microbiol. Mol. Biol. R.*, **76**, 667–694.

Archer C. L. and Jacobson M. Z. (2005). Evaluation of global wind power. *J. Geophys. Res.*, **110**, D12110.

Azam F., Fenchel T., Field J. G., Gray J. S., Meyer-Reil L. A. and Thingstad F. (1983). The ecological role of water-column microbes in the sea. *Mar. Ecol. Prog. Ser.*, **10**, 257–263.

Azetsu-Scott K. and Passow U. (2004). Ascending marine particles: significance of transparent exopolymer particles (TEP) in the upper ocean. *Limnol. Oceanogr.*, **49**, 741–748.

Bar-Zeev E., Bermann-Frank I., Stambler N., Vázquez Domínguez E., Zohary T., Capuzzo E., Meeder E., Suggett D. J., Iluz D., Dishon G. and Berman T. (2009). Transparent exopolymer particles (TEP) link phytoplankton and bacterial production in the Gulf of Aqaba. *Aquat. Microb. Ecol.*, **56**, 217–225.

Beauvais S., Pedrotti M. L., Egge J., Iversen K. and Marrasé C. (2006). Effects of turbulence on TEP dynamics under contrasting nutrient conditions: implications for aggregation and sedimentation processes. *Mar. Ecol. Prog. Ser.*, **323**, 47–57.

Becquevort S., Dumont I., Tison J.-L., Lannuzel D., Sauvée M.-L., Chou L. and Schoemann V. (2009). Biogeochemistry and microbial community composition in sea ice and underlying seawater off East Antarctica during early spring. *Polar Biol.*, **32**, 879–895.

Berman-Frank I., Rosenberg G., Levitan O., Haramaty L. and Mari X. (2007). Coupling between autocatalytic cell death and transparent exopolymeric particle production in the marine cyanobacterium *Trichodesmium*. *Environ. Microbiol.*, **9**, 1415–1422.

Bhaskar P. V. and Bhosle N. B. (2006). Dynamics of transparent exopolymeric particles (TEP) and particle-associated carbohydrates in the Dona Paula bay, west coast of India. *J. Earth Syst. Sci.*, **115**, 403–413.

Chin W.-C., Orellana M. V. and Verdugo P. (1998). Spontaneous assembly of marine dissolved organic matter into polymer gels. *Nature*, **391**, 568–572.

Claquin P., Probert I., Lefebvre S. and Veron B. (2008). Effects of temperature on phytosynthetic parameters and TEP production in eight species of marine microalgae. *Aquat. Microb. Ecol.*, **51**, 1–11.

Collins R. E., Crapenter S. D. and Deming J. W. (2008). Spatial heterogeneity and temporal dynamics of particles, bacteria, and pEPS in Arctic winter sea ice. *J. Mar. Syst.*, **74**, 902–917.

Corzo A., Morillo J. A. and Rodríguez S. (2000). Production of transparent exopolymer particles (TEP) in cultures of *Chaetoceros calcitrans* under nitrogen limitation. *Aquat. Microb. Ecol.*, **23**(1), 63–72.

Corzo A., Rodríguez-Gálvez S., Lubian L., Sangrá P., Martínez A. and Morillo J. A. (2005). Spatial distribution of transparent exopolymer particles in the Bransfield Strait, Antarctica. *J. Plankton Res.*, **27**, 635–646.

Cunliffe M. and Murrell J. C. (2009). The sea surface microlayer is a gelatinous biofilm. *The ISME Journal*, **3**, 1001–1003.

Cunliffe M. and Murrell J. C. (2010). Eukarya 18S rRNA gene diversity in the sea surface microlayer: insights on the structure of the neustonic microbial loop. *The ISME Journal*, **4**, 455–458.

Cunliffe M., Salter M., Mann P. J., Whiteley A. S., Upstill-Goddard R. C. and Murrell J. C. (2009a). Dissolved organic carbon and bacterial populations in the gelatinous surface microlayer of a Norwegian fjord mesocosm. *FEMS Microbiology Letters*, **299**, 248–254.

Cunliffe M., Whiteley A. S., Schäfer H., Newbold L., Oliver A. and Murrell J. C. (2009b). Comparison of bacterioneuston and bacterioplankton dynamics during a phytoplankton bloom in a fjord mesocosm. *Appl. Environ. Microbiol.*, **75**, 7173–7181.

Cunliffe M., Upstill-Goddard R. C. and Murrell J. C. (2011). Microbiology of aquatic surface microlayers. *FEMS Microbiol.*, **35**, 233–246.

Damare V. S. and Raghukumar S. (2012). Marine aggregates and transparent exopolymeric particles (TEPs) as substrates for the stramenopilan fungi, the thraustochytrids: roller table experimenal approach. *KAVAKA*, **40**, 22–31.

Dupouy C., Neveux J., Subrammaniam A., Mulholland M. R., Montoya J. P., Campbell L., Carpenter E. J. and Capone D. G. (2000). Satellite captures trichodesmium blooms in the southwestern tropical Pacific. *EOS*, **81**, 13–16.

Dutz J., Klein Breteler W. C. M. and Kramer G. (2005). Inhibition of copepod feeding by exudates and transparent exopolymer particles (TEP) derived from a Phaeocystis globosa dominated phytoplankton community. *Harmful Algae*, **4**, 929–940.

Elasri M. O. and Miller R. V. (1999). Study of the response of a biofilm bacterial community to UV radiation. *Appl. Environ. Microbiol.*, **65**, 2025–2031.

Engel A. (2004). Distribution of transparent exopolymer particles (TEP) in the northeast Atlantic Ocean and their potential significance for aggregation processes. *Deep-Sea Res. I*, **51**, 83–92.

Engel A. and Passow U. (2001). Carbon and nitrogen content of transparent exopolymer particles (TEP) in relation to their Alcian Blue adsorption. *Mar. Ecol. Prog. Ser.*, **219**, 1–10.

Engel A. and Schartau M. (1999). Influence of transparent exopolymer particles (TEP) on sinking velocity of *Nitzschia closterium* aggregates. *Mar. Ecol. Prog. Ser.*, **182**, 69–76.

Engel A., Meyerhöfer M. and von Bröckel K. (2001). Chemical and biological composition of suspended particles and aggregates in the Baltic Sea in Summer (1999). *Est. Coastal Shelf Sci.*, **55**, 729–741.

Engel A., Thoms S., Riebesell U., Rochelle-Newall E. and Zondervan I. (2004). Polysaccharride aggregation as a potential sink of marine dissolved organic carbon. *Nature*, **428**, 929–932.

Fernández-Gómez B., Richter M., Schüler M., Pinhassi J., Acinas S. G., González J. M. and Pedrós-Alió C. (2013). Ecology of marine Bacteroidetes: a comparative genomics approach. *The ISME Journal*, **7**, 1026–1037.

Gaerdes A., Iversen M. H., Grossart H.-P., Passow U. and Ullrich M. (2011). Diatom associated bacteria are required for aggregation of Thalassiosira weissflogii. *ISME J.*, **5**(3), 436–445.

Garofalakis G. and Murray B. S. (2002). Surface pressure isotherms, dilational rheology, and brewster angle microscopy of insoluble monolayers of sugar monoesters. *Langmuir*, **18**, 4765–4774.

Gershey R. M. (1983). Characterization of seawater organic matter carried by bubble-generated aerosols. *Limnol. Oceanogr.*, **28**, 309–319.

Grossart H. P., Czub G. and Simon M. (2006). Algae-bacteria interactions and their effects on aggregation and organic matter flux in the sea. *Environ. Microbiol.*, **8**, 1074–1084.

Harlay J., Borges A. V., Van Der Zee C., Delille B., Godoi R. H. M., Schiettecatte L.-S., Roevros N., Aerts K., Lapernat P.-E., Rebreanu L., Groom S., Daro M.-H., Van Grieken R. and Chou L. (2010). Biogeochemical study of a coccolithophore bloom in the northern Bay of Biscay (NE Atlantic Ocean) in June 2004). *Prog. Oceanogr.*, **86**, 317–336.

Hong Y., Smith W. O. and White A-M. (1997). Studies on transparent exopolymer particles (TEP) produced in the Ross Sea (Antarctica) and by Phaeocystis Antarctica (prymnesiophyceae). *J. Phycol.*, **33**, 368–376.

Jackson G. A. (1995). TEP and coagulation during a mesocosm experiment. *Deep-Sea Res. II*, **42**, 215–222.

Jackson G. A. and Burd A. B. (1998). Aggregation in the marine environment. *Environ. Sci. Technol.*, **32**, 2805–2814.

Jiao N., Herndl G. J., Hansell D. A., Benner R., Kattner G., Wilhelm S. W., Kirchman D. L., Weinbauer M. G., Luo T., Chen F. and Azam F. (2010). Microbial production of recalcitrant dissolved organic matter: long-term carbon storage in the global ocean. *Nature Rev.*, **8**, 593–599.

Johnson B. D. and Cooke R. C. (1980). Organic particle and aggregate formation resulting from the dissolution of bubbles in seawater. *Limnol. Oceanogr.*, **25**, 653–661.

Kepkay P. E. (1994). Particle aggregation and the biological reactivity of colloids. *Mar. Ecol. Prog. Ser.*, **109**, 293–304.

Kiørboe T. (1997). Small-scale turbulence, marine snow formation, and planktivorous feeding. *Sci. Mar.*, **61**(Supl. 1), 141–158.

Kok M., Schouten S. and Sinninghe Damsté J. S. (2000). Formation of insoluble, nonhydrolyzable, sulfur-rich macromolecules via incorporation of inorganic sulfur species into algal carbohydrates. *Geochim. Cosmochim. Acta*, **64**, 2689–2699.

Krembs C., Eicken H., Junge K. and Deming J. W. (2002). High concentrations of exopolymeric substances in Arctic winter sea ice: implications for the polar ocean carbon cycle and cryoprotection of diatoms. *Deep Sea Res. I*, **49**, 2163–2181.

Krembs C., Eicken H. and Deming J. W. (2011). Exopolymer alteration of physical properties of sea ice and implications for ice habitability and biogeochemistry in a warmer Arctic. *Proc. Natl. Acad. Sci. USA*, **108**(9), 3653–3658.

Lechtenfeld O., Hertkonr N., Shen Y., Witt M. and Benner R. (2015). Marine sequestration of carbon in bacterial Metabolites. *Nature Commun.*, **6**, 6711.

Leck C. and Bigg E. K. (2005). Biogenic particles in the surface microlayer and overlaying atmosphere in the central Arctic Ocean during summer, Tellus, Ser. *B*, **57**, 305–316.

Ling S. C. and Alldredge A. L. (2003). Does the marine copepod Calanus pacificus consume transparent exopolymer particles (TEP). *J. Plankton Res.*, **25**(5), 507–515.

Logan B. E., Grossart H.-P. and Simon M. (1994). Direct observation of phytoplankton, TEP and aggregates on polycarbonate filters using brightfield microscopy. *J. Plankton Res.*, **16**, 1811–1815.

Ma L., Chen M., Guo L., Lin F. and Tong J. (2012). Distribution and source of transparent exopolymer particles in the northern Bering Sea. *Acta Oceanol. Sin.*, **5**, 010.

Madivala B., Fransaer J. and Vermant J. (2009). Self-assembly and rheology of ellipsoidal particles at interfaces. *Langmuir*, **25**, 2718–2728.

Malpezzi M. A., Sanford L. P. and Crump B. C. (2013). Abundance and distribution of transparent exopolymer particles in the estuarine turbidity maximum of Chesapeake Bay. *Mar. Ecol. Prog. Ser.*, **486**, 23–35.

Mari X. (2008). Does ocean acidification induce an upward flux of marine aggregates? *Biogeosci.*, **5**, 1023–1031.

Mari X. and Kiørboe T. (1996). Abundance, size distribution and bacterial colonization of transparent exopolymer particles (TEP) during spring in the Kattegat. *J. Plankton Res.*, **18**, 969–986.

Mari X., Rassoulzadegan F. and Brussaard C. P. D. (2004). Role of TEP in the microbial food web structure. II. Influence on the ciliate community structure. *Mar. Ecol. Prog. Ser.*, **279**, 23–32.

Mari X., Kerros M.-E. and Weinbauer M. G. (2007). Virus attachment to transparent exopolymeric particles along trophic gradients in the southwestern lagoon of New Caledonia. *Appl. Environ. Microbiol.*, **73**, 5245–5252.

Mazilla F. F. M., Shapiro K. and Silver M. W. (2013). A new pathogen transmission mechanism in the ocean: the case of sea Otter exposure to the land-parasite Toxoplasma gondii. *PLOS ONE*, **8**, e82477.

Meiners K., Gradinger R., Fehling J., Civitarese G. and Spindler M. (2003). Vertical distribution of exopolymer particles in sea ice of the Fram Strait (Arctic) during autumn. *Mar. Ecol. Prog. Ser.*, **248**, 1–13.

Middelboe M., Jørgensen N. O. G. and Kroer N. (1996). Effects of viruses on nutrient turnover and growth efficiency of noninfected marine bacterioplankton. *Appl. Environ. Microbiol.*, **62**, 1991–1997.

Miller L. A., Papkyriakou T. N., Collins R. E., Deming J. W., Ehn J. K., Macdonald R. W., Mucci A., Owens O., Raudsepp M. and Sutherland N. (2011). Carbon dynamics in sea ice: a winter flux time series. *J. Geophys. Res.*, **116**, C02028.

Møller E. F. (2003). Sloppy feeding in marine copepods: prey-size-dependent production of dissolved organic carbon. *J. Plankton Res.*, **27**(1), 27–35.

Myklestad S. M. (1995). Release of extracellular products by phytoplankton with special emphasis on polysaccharides. *Sci. Total Environ.*, **165**, 155–164.

Nichols C. M., Lardière S. G., Bowman J. P., Nichols P. D., Gibson J. A. E. and Guézennec J. (2005). Chemical characterization of exopolysaccharide from Antarctic marine bacteria. *Microb. Ecol.*, **49**, 578–589.

Noble R. T., Middelboe M. and Fuhrman J. A. (1999). The effects of viral enrichment on the mortality and growth of heterotrophic bacterioplankton. *Aquat. Microb. Ecol.*, **18**, 1–13.

Not F., Valentin K., Romari K., Lovejoy C., Massana R., Töbe K., Vaulot D. and Medlin L. K. (2007). Picobiliphytes: a marine picoplanktonic algal group with unknown affinities to other eukaryotes. *Science*, **315**, 253–255.

Ortega-Retuerta E., Passow U., Duarte C. M. and Reche I. (2009b). Effects of ultraviolet B radiation on (not so) transparent exopolymer particles. *Biogeosci.*, **6**, 3071–3080.

Ortega-Retuerta E., Reche I., Pulido-Villena E., Agustí S. and Duarte C. M. (2009a). Uncoupled distributions of transparent exopolymer particles (TEP) and dissolved carbohydrates in the Southern Ocean. *Mar. Chem.*, **115**, 59–65.

Ortega-Retuerta E., Duarte C. M. and Reche I. (2010). Significance of bacterial activity for the distribution and dynamics of transparent exopolymer particles in the Mediterranean Sea. *Microb. Ecol.*, **59**, 808–818.

Passow U. (2000). Formation of Transparent Exopolymer Particles, TEP, from dissolved precursor material. *Mar. Ecol. Prog. Ser.*, **192**, 1–11.

Passow U. (2002). Transparent exopolymer particles (TEP) in aquatic environments. *Prog. Oceanogr.*, **55**, 287–333.

Passow U. and Alldredge A. L. (1994). Distribution, size and bacterial colonization of transparent exopolymer particles (TEP) in the ocean. *Mar. Ecol. Prog. Ser.*, **113**, 185–198.

Passow U. and Alldredge A. L. (1995). A dye-binding assay for the spectrophotometric measurements of transparent exopolymer particles (TEP) in the ocean. *Limnol. Oceanogr.*, **40**, 1326–1335.

Passow U. and Alldredge A. L. (1999). Do transparent exopolymer particles (TEP) inhibit grazing by the euphausiid Euphausia pacifica? *J. Plankton Res.*, **21**(11), 2203–2217.

Passow U. Shipe R. F., Murray A., Pak D. K., Brzezinski M. A. and Alldredge A. L. (2001). The origin of transparent exopolymer particles (TEP) and their role in the sedimentation of particulate matter. *Cont. Shelf Res.*, **21**, 327–346.

Plough H., Hietanen S. and Kuparinen J. (2002). Diffusion and advection within and around sinking, porous diatom aggregates. *Limnol. Oceanogr.*, **47**, 1129–1136.

Prieto L. and Cowen J. P. (2007). Transparent exopolymer particles in a deep-sea hydrothermal system: Guaymas Basin, Gulf of California. *Mar. Biol.*, **150**, 1093–1101.

Prieto L., Sommer F., Stibor H. and Koeve W. (2001). Effects of planktonic copepods on transparent exopolymeric particles (TEP) abundance and size spectra. *J. Plankton Res.*, **23**, 515–525.

Prieto L., Navarro G., Cózar A., Echevarría F. and García C. M. (2006). Distribution of TEP in the euphotic and upper mesopelagic zones of the southern Iberian coasts. *Deep Sea Res. II*, **53**, 1314–1328.

Radić T., Kraus R., Fuks D., Radić J. and Pečar O. (2005). Tranparent exopolymeric particles distribution in the northern Adriatic and their relation to microphytoplankton biomass and composition. *Sci. Total Environ.*, **353**, 151–161.

Ramaiah N. and Furuya K. (2002). Seasonal variations in phytoplankton composition and transparent exopolymer particles in a eutrophicated coastal environment. *Aquat. Microb. Ecol.*, **30**, 69–82.

Ramaiah N., Takeda S., Furuya K., Yoshimura T., Nishioka J., Aono T., Nojiri Y., Imai K., Kudo I., Saito H. and Tsuda A. (2005). Effects of iron enrichment on the dynamics of transparent exopolymer particles in the western subarctic Pacific. *Prog. Oceanogr.*, **64**, 253–261.

Redfield A. C. (1934). On the proportions of organic derivations in sea water and their relation to the composition of plankton. In: James Johnstone Memorial Volume, R. J. Daniel (ed.), University Press of Liverpool, pp. 177–192.

Riedel A., Michel C. and Gosselin M. (2006). Seasonal study of sea-ice exopolymeric substances on the Mackenzie shelf: implications for transport of sea-ice bacteria and algae. *Aquat. Microb. Ecol.*, **45**, 195–206.

Riedel A., Michel C. and Gosselin M. (2007). Grazing of large-sized bacteria by sea-ice heterotrophic protists on the Mackenzie Shelf during the winter–spring transition. *Aquat. Microb. Ecol.*, **50**, 25–38.

Riedel A., Michel C., Gosselin M. and LeBlanc B. (2008). Winter-spring dynamics in sea-ice carbon cycling in the coastal Arctic Ocean. *J. Mar. Sys.*, **74**, 918–932.

Schuster S. and Herndl G. J. (1995). Formation and significance of transparent exopolymeric particles in the northern Adriatic Sea. *Mar. Ecol. Prog. Ser.*, **124**, 227–236.

Seenivasan R., Sausen N., Medlin L. K. and Melkonian M. (2013). Picomonas judraskeda Gen. Et Sp. Nov.: the first identified member of the *Picozoa phylum Nov.*, a widespread group of picoeukaryotes, formerly known as 'Picobiliphytes'. *PLOS One*, **8**, e59565.

Sieburth J. McN. (1983). Microbiological and organic-chemical processes in the surface and mixed layers. In: Air–Sea Exchange of Gases and Particles, P. S. Liss and W. G. N. Slinn (eds), Reidel Publishers Co., Hingham, MA, pp. 121–172.

Smith D. C., Steward G. F., Long R. A. and Azam F. (1995). Bacterial mediation of carbon fluxes during a diatom bloom in a mesocosm. *Deep-Sea Res. Part II*, **42**, 75–97.

Smoluchowski M. (1917). Versuch einer mathematischen Theorie der Koagulationskinetic kolloider Lösungen. *Z. Phys. Chem.*, **92**, 129–168.

Sun C.-C., Wang Y. S., Wu M. L., Li N., Lin L., Song H., Wang Y.-T., Deng C., Peng Y. L., Sun F.-L. and Li C.-L. (2010). Distribution of transparent exopolymer particles in the Pearl River Estuary in summer. *J. Topical Oceanogr.*, **29**, 81–87.

Suttle C. A. (2007). Marine viruses—major players in the global ecosystem. *Nature Rev. Microbiol.*, **5**, 801–812.

Tada Y., Taniguchi A., Nagao I., Miki T., Uematsu M., Tsuda A. and Hamasaki K. (2011). Differing growth responses of major phylogenetic groups of marine bacteria to natural phytoplankton blooms in the western North Pacific Ocean. *Appl. Environ. Microbiol.*, **77**, 4055–4065.

Taylor J. D., Cottingham S. D., Billinge J. and Cunliffe M. (2014). Seasonal microbial community dynamics correlate with phytoplankton-derived polysaccharides in surface coastal waters. *The ISME Journal*, **8**, 245–248.

Teeling H., Fuchs B. M., Becher D., Klockow C., Gardebrecht A., Bennke C. M., Kassabgy M., Huang S., Mann A. J., Waldmann J., Weber M., Klindworth A., Otto A., Lange J., Bernhardt J., Reinsch C., Hecker M., Peplies J., Bockelmann F. D., Callies U., Gerdts G., Wichels A., Wiltshire K. H., Glöckner F. O., Schweder T. and Amann R. (2012). Substrate-controlled succession of marine bacterioplankton populations induced by a phytoplankton bloom. *Science*, **336**, 608–811.

Thornton D. (2002). Diatom aggregation in the sea: mechanisms and ecological implications. *Eur. J. Phycol.*, **37**(2), 149–161.

Tilstone G. H., Airs R. L., Martinez-Vicente V., Widdicombe C. and Llewellyn C. (2010). High concentrations of mycosporine-like amino acids and colored dissolved organic matter in the sea surface microlayer off the Iberian Peninsula. *Limnol. Oceanogr.*, **55**(5), 1835–1850.

Tranvik L. J., Sherr E. B. and Sherr B. F. (1993). Uptake and tuliszation of 'colloidal DOM' by heterotrophic flagellates in seawater. *Mar. Ecol. Prog. Ser.*, **92**, 301–309.

Van der Merwe P., Lannuzel D., Mancuso Nichols C. A., Meiners K., Heil P., Norman L., Thomas D. N. and Bowie A. R. (2009). Biogeochemical observations during the winter-spring transition in East Antarctic sea ice. Evidence of iron and exopolysaccharide controls. *Mar. Chem.*, **115**, 163–175.

Van Oostende N., Harlay J., Vanelslander B., Chou L., Vyverman W. and Sabbe K. (2012). Phytoplankton community dynamics during late spring coccolithophore blooms at the continental margin of the Celtic Sea (North East Atlantic, 2006-2008). *Prog. Oceanogr.*, **104**, 1–16.

Van Oostende N., Moerdijk-Poortvliet T. C. W., Boschker H. T. S., Vyverman W. and Sabbe K. (2013). Release of dissolved carbohydrates by *Emiliania huxleyi* and formation of transparent exopolymer particles depend on algal life cycle and bacterial activity. *Environ. Microbiol.*, **15**, 1514–1531.

Verdugo P., Alldredge A. L., Azam F., Kirchman D. L., Passow U. and Santschi P. H. (2004). The oceanic gel phase: a bridge in the DOM-POM continuum. *Mar. Chem.*, **92**, 67–85.

Villacorte L. O., Ekowati Y., Calix-Ponce H. N., Schippers J. C., Amy G. L. and Kennedy M. D. (2015). Improved method for measuring transparent exopolymer particles (TEP) and their precursors in fresh and saline water. *Water Res.*, **70**, 300–312.

Wheeler J. R. (1975). Formation and collapse of surface films. *Limnol. Oceanogr.*, **20**, 338–342.

Wild C., Huettel M., Klueter A., Kremb S. G., Rasheed M. Y. M. and Jørgensen B. B. (2004). Coral mucus functions as an energy carrier and particle trap in the reef ecosystem. *Nature*, **428**, 66–70.

Wurl O. and Holmes M. (2008). The gelatinous nature of the sea-surface microlayer. *Mar. Chem.*, **110**, 89–97.

Wurl O., Miller L., Röttgers R. and Vagle S. (2009). The distribution and fate of surface-active substances in the sea-surface microlayer and water column. *Mar. Chem.*, **115**, 1–9.

Wurl O., Miller L. and Vagle S. (2011). Production and fate of transparent exopolymer particles in the ocean. *J. Geophys. Res.*, **116**, C00H13.

Wurl O., Stolle C., Thuoc C. V., Thu P. T. and Mari X. (2016). Biofim-like properties of the sea surface and predicted effects on air-sea CO_2 exchange. *Prog. Oceanogr.*, **144**, 15–24.

Young L. Y. (1978). Bacterioneuston examined with critical point drying and transmission electron microscopy. *Microb. Ecol.*, **4**, 267–277.

Young S., Palm M., Grover J. P. and McKee D. (1997). How Daphnia cope with algae selected for inedibility in long-running microcosms. *J. Plankton Res.*, **19**, 391–397.

Zhao L., Goodman M. D., Bowden N. B. and Lin Z. (2009). Self-assembly of an ultra-high-molecular-weight comb block copolymer at the air-water interface. *Soft Matter*, **5**, 4698–4703.

Zhou J., Mopper K. and Passow U. (1998). The role of surface-active carbohydrates in the formation of transparent exopolymer particles by bubble adsorption of seawater. *Limnol. Oceanogr.*, **43**(8), 1860–1871.

Chapter 14

Snapshots of fungal extracellular matrices

Romy Breitenbach[1], Jörg Toepel[1], Polina Dementyeva[1], Nicole Knabe[1] and Anna A. Gorbushina[1,2]

[1]Federal Institute for Materials Research and Testing (Bundesanstalt für Materialforschung und -prüfung, BAM), Department 4 "Materials & Environment", Berlin, Germany

[2]Freie Universität Berlin, Department of Earth Sciences and Department of Biology, Chemistry and Pharmacy, Berlin, Germany; e-mail: anna.gorbushina@bam.de

14.1 SUMMARY

Fungal extracellular materials reinforce a constant interaction between their cell wall and the environment. A dynamic mixture of chitin, glucans, mannans, glycoproteins, glycolipids and pigments supports the success of all fungal life styles – from symbiotic to the free-living and pathogenic. Fungi are perfectly adapted to grow on surfaces and in porous environments, where they form medically and geochemically relevant biofilms. Fungal EPS are critical in adhesion to other fungi, other cells or substratum as well as in the following interaction with the host immune system or material they attack, degrade and deteriorate respectively. Characterisation of extracellular compounds and understanding of its function is necessary to limit damage caused by fungal activity. All necessary methodology from chemical characterisation to complete genetic analyses has been developed for medically important fungi. Now it is time to apply this knowledge to the numerous, largely aerobic and very active organisms that occupy a wide range of atmosphere-exposed habitats in the upper lithosphere. One can expect that analogies between medically- and environmentally-relevant model fungal species will help us to address the dynamics of the fungal cell EPS matrix in much more efficient and widely applicable ways.

Keywords: fungal cell wall, melanin, fungal life styles, environmental and pathogenic fungi, model fungal biofilms, biogenic weathering

14.2 THE DIVERSE MATRIX OF BIOFILMS

Fungi, like all living matter in natural and thus constantly changing environments, are exposed to biotic and abiotic stressors with which they can only cope at considerable metabolic costs. Individual stress response result, but frequently community organisation in biofilms as the oldest, most successful and widespread form of life on Earth (Flemming & Wingender, 2010) provides a protective co-operative setting for many organisms including fungi.

Life styles in matrix-embedded biofilms are characteristic of all microbial groups and various fungi are known to contribute to diverse biofilms (Blankenship & Mitchell, 2006; Gorbushina & Broughton, 2009; Bonhomme & d'Enfert, 2013; Latge & Beauvais, 2014). Biofilm development is supported by fungi, since they can grow in porous environments, enhance ion release from materials and produce protective extracellular polymeric substances (EPS). Fungi cause devastating diseases of animals and plants (Perez-Nadales *et al.* 2014), which are partially supports by the fungal ability to effectively form biofilms (e.g., Kojic & Darouiche, 2004; Blankenship & Mitchell, 2006; Bonhomme & d'Enfert, 2013). On surfaces and in biofilms fungi often cooperate with heterotrophic and photosynthetic microbes (e.g., Turian, 1977; Gadd, 2007; Gorbushina & Broughton, 2009). Such fungal-influenced biofilm consortia are particularly problematic when growing on medical device (Lynch & Robertson, 2008; Espinosa & Rivera, 2016), but in natural environments fungi-dominated biofilms promote mineral weathering and soil formation on bare rock surfaces (Gorbushina & Broughton, 2009). In all these niches, the fungal EPS matrix enhances biofilm substrate-interactions and is therefore important for understanding processes of substrate modification and degradation.

Fungi with a diversity of life styles – from symbiotic to free-living and pathogenic- effectively colonise all manner of surfaces. In doing so, the outer most layer of a fungal cell and the adherent EPS modulates their interactions with the external environment. Extracellular fungal components are critical in adhesion to other living partners, for the colonisation of the substrate, in the subsequent usage/damage of the colonised surface and help define the outcome of the interaction with the host immune system. as well as function as a protective layer around the fungal cell. Extracellular polymers also form a physical contact between the cells and the inanimate solid material – and subsequently influence chemical interactions with the substratum.

As evident from the above, fungi are significant components of environmental microbiota, especially in aerobic terrestrial locations where they can make up the majority of the biogeochemistry-active microbial biomass (Gadd, 2007). Here we have reviewed the characteristics, properties and functions of fungal EPS in microbial biofilms – and would herewith like to compensate for the relative imbalances in knowledge between fungal and bacterial producers and users of EPS.

14.3 FUNGAL CELL WALL AND BEYOND

An outermost barrier of any fungal cell starts just outside a plasma membrane as a structurally unique cell wall that confers both protective and aggressive properties (Latge, 2007). Cell walls maintain the functional shape of any fungal cell from hyphae to yeast. These rigid structures support a forceful penetrating for the invasion of hosts tissues and lifeless organic and inorganic substrates (Bechinger *et al.* 1999). Cell walls protect the fungal cell from the environment and participate in adhesion of the cells to each other and the underlying substrate. Cell wall of yeasts accounts for up to 30% of the dry weight biomass and, thus, represent a substantial metabolic investment of the organism (Klis *et al.* 2014). Disruptions of the cell wall structure have been proven to affect the growth and morphology of the fungal cell, often making it susceptible to lysis and death (Bulawa, 1993; Kapteyn *et al.* 1999; Bowman & Free, 2006; Latge, 2007). Any fungal cell is thus surrounded by a complex cross-linked network of chitin, branched glucans, mannans and glycoproteins (Bowman & Free, 2006; Reilly & Doering, 2009; Mahapatra & Banerjee, 2013). The inner rigid core in the plasma membrane proximity is formed by chitin and β-glucans (Bartnicki-Garcia, 1968; Figueiredo & Barreto-Bergter, 2014). External layers are more varied and harbour many loosely anchored and structurally complex polysaccharides, glycoproteins, enzymes and lipids. Also aggressive hydrolytic and toxic molecules of fungi are often literally "embedded" into their cell wall, as, most of them being just in transit to the extracellular space and required for a fungus to invade its ecological niche (Latge, 2007).

14.3.1 Polysaccharides

Polysaccharides are well-known biopolymers and fungal cell wall and EPS contain a high variety of them, while the location and mobility of individual substances depends on the environment as well as on the chemical composition of these polymers. For instance, some of fungal cell wall polysaccharides are water-soluble and thus can reach the extracellular space (Reilly & Doering, 2009), especially if abundantly produced. The major fungal polysaccharides are defined through their monosaccharide composition and the degree of cross-linking and branching of their chains. Diverse methods for extraction and characterisation of the fungal glucans have been developed and were recently summarised (Donot *et al.* 2012; Kozarski *et al.* 2014; Ruthes *et al.* 2015). It is interesting that the majority of fungal polysaccharides are known by their "proper names" like botryosphaeran, lentinan, pullulan, pustulan, schizophyllan, scleroglucan, and thelebolan (Manzi & Pizzoferrato, 2000; Fukuda *et al.* 2009; Synytsya & Novak, 2014), with the names frequently reflecting the species or genus of the main producer, in which the polysaccharide was described for the first time. A catalogue of specific fungal polysaccharides is given in Table 14.1.

Chitin, a linear homopolymer of β-1,4-N-acetylglucosamine, is a very important and probably best known cell wall component in fungi. Crystalline microfibrils of chitin are formed through inter-chain hydrogen bonding and are essential to the overall integrity and stability of the cell wall (Bowman & Free, 2006). Chitin is generated from cytosolic UDP-N-acetylglucosamine by membrane-associated synthases that translocate the growing chain across the cell membrane (Bowman & Free, 2006; Latge, 2007; Doering, 2009). In contrast to chitin, which is insoluble in water and nearly all common organic solvents, a deacetylated chitin derivative is termed chitosan. This polymer with β-1,4-linked D-glucosamine (GlcN) and N-acetyl-D-glucosamine (GlcNAc) presents a more soluble and flexible variant and is frequently present in the fungal cell wall. The solubility depends on the degree of N-acetylation – and the terminology on the solubility: with less than 50%, chitin becomes soluble in aqueous solutions and is called chitosan. An exact nomenclature with respect to the degree of N-deacetylation has not been defined (Kumirska *et al.* 2010). Understandably, formation of chitin and chitosan may be physically linked, adding a layer to the already complex synthesis of fungal chitin (Doering, 2009).

Glucans are the dominant (approximately 50 to 60% of the dry weight) structural as well as soluble polysaccharides of the fungal cell wall. Beta-1,3-glucan, along with β-1,6-glucan, serves as a skeleton to which other cell wall components are covalently attached. For almost all fungi, the central core of the cell wall are branched β-1,3-glucan and β-1,6-glucan linked to chitin via a β-1,4-linkage (Kapteyn *et al.* 1999; Bowman & Free, 2006; Latge, 2007).

Glucans are known to be formed from nucleotide sugars (e.g., UDP-glucose) by the multi-subunit enzyme glucan synthase (Mazur *et al.* 1995; Latge & Beauvais, 2014). Biosynthetic steps of glucans are probably fairly similar in all fungi. For example, for scleroglucan synthesis glucose is taken up by glucose transporter(s) and phosphorylated to glucose-6-phosphate via a hexokinase reaction followed by isomerisation to glucose-1-phosphate by a phosphoglucose mutase. Uridine diphosphoglucose (UDPG) is formed via an UDPG-1-phosphate uridylyltransferase. A β-1,3 glucansynthase uses UDPG as substrate for polymer formation. Finally, a β-1,3;1,6-glucosyltransferase is assumed to incorporate β-1,6-linked glucosyl side residues into the continuously elongating β-1,3-glucan backbone (Batra *et al.* 1969; Survase *et al.* 2007; Schmid *et al.* 2011). The enzyme complex of glucan synthase is membrane-bound and possesses multiple transmembrane domains therefore extruding the product into the cell wall space (Bowman &

Table 14.1 Overview of important exopolysaccharides in fungi. A ascomycetes, B basidiomycetes, Z zygomycetes.

Type of Exopolysaccharide	EPS Composition	M_W (kDa)	Exopolysaccharides (Proper Name)	Main Fungal Strain	Phylum	References
Homopolysaccharides						
α-Glucan	linear α-1,4; α-1,6 glucan	5–9000	Pullulan	*Aureobasidium pullulans*	A	(Bender *et al.* 1959; Prajapati *et al.* 2013)
	linear α-1,4; α-1,3 glucan	200–500	Elsinan	*Elsinoe leucospila*	A	(Tsumuraya *et al.* 1978)
β-Glucan	linear β-1,3 glucose chain with β-1,6-linked residue on every third unit	450	Schizophyllan	*Schizophyllum commune*	B	(Kikumoto *et al.* 1971; Schmid *et al.* 2011; Zhang *et al.* 2013)
		5000–6000	Scleroglucan	*Sclerotium sp.*	B	(Schmid *et al.* 2011)
	linear β-1,3 glucose chain with two β-1,6 branches on every fifth unit	500–1000	Lentinan	*Lentinus edodes*	B	(Sasaki & Takasuka, 1976; Zhang *et al.* 2011)
	β-1,3 with β-1,6 branching	1000	Grifolan	*Grifola frondosa*	B	(Bae *et al.* 2005)
	β-1,3 with β-1,6 branching	2000	Pestalotan	*Pestalotia sp.*	A	(Misaki *et al.* 1984)
	linear β-1,3 glucose chain with one β-1,6 branch on every fifth unit	2000–5000	Botryosphaeran	*Botryosphaeria rhodina*	A	(Barbosa *et al.* 2003; Giese *et al.* 2008)
	β-1,3 with β-1,6 branching	500	Thelebolan	*Thelebolus sp.*	A	(Mukhopadhyay *et al.* 2014)
	linear β-1,6 glucan	200	Pustulan	*Umbilicaris pustulata*	A	(Gorin & Barreto-Bergter, 1983; Sassaki *et al.* 2002)
	linear β-1,4-linked D-glucosamine and N-acetyl-D-glucosamine	30–1400	Chitosan	*Mucorales*	Z	(Pochanavanich & Suntornsuk, 2002; Singh *et al.* 2011)
				Cryptococcus neoformans	B	(Doering, 2009)

(*Continued*)

Table 14.1 Overview of important exopolysaccharides in fungi. A ascomycetes, B basidiomycetes, Z zygomycetes (*Continued*).

Type of Exopolysaccharide	EPS Composition	M_W (kDa)	Exopolysaccharides (Proper Name)	Main Fungal Strain	Phylum	References
	linear β-1,3 glucose chain with β-1,6 residues at every second/third unit	10.000	Pleuran	*Pleurotus sp.*	B	(Gutierrez *et al.* 1996; Bergendiova *et al.* 2011)
Heteropolysaccharides						
Galactan	α-1,4-linked galactose and α-1,4-linked N-acetylgalactosamine	10–1000	galactosaminogalactan	*Aspergillus fumigatus*	A	(Fontaine *et al.* 2011)
	β-1,4 galactose with β-1,2 and terminal rhamnose	187	rhamnogalactan	*Fusarium solani*	A	(Mahapatra & Banerjee, 2012)
Mannan	linear α-1,3 mannose with β-1,2 glucuronic acid residues and xylose substituted β-1,2 or β-1,4 to mannose units	1210	glucuronoxylomannan	*Cryptococcus neoformans*	B	(Frases *et al.* 2008)
	α-1,2 mannose with β-1,6 galactose branches and α-glucose residues	18.3	glucogalactomannan	*Penicillium commune*	A	(Chen *et al.* 2013)
	α-1,2-linked mannose with non-reducing ends of β-galactose and α-mannose	120	galactomannan	*Exophiala jeanselmei*	A	(Sassaki *et al.* 2011)
	1,2;1,6-linked mannose chain with 1,3-linked side units	35–40	glucomannan	*Saccharomyces cerevisiae*	A	(Beauvais *et al.* 2009; Faria-Oliveira *et al.* 2015)
	α-1,6 mannose with α-1,2-linked side chains complexed to linear β-1,6 glucose chain	300	mannan-glucan complex	*Candida albicans*	A	(Zarnowski *et al.* 2014)
polyhexNAc	1,4-linked β-N-acetylmannosamine with 1,3-linked branches of β-N-acetylgalactosamine	6–30	poly-N-acetylhexosamine	*Cordyceps sinensis*	A	(Chen *et al.* 2013)

Free, 2006; Latge & Beauvais, 2014) or further out to the cell surface. Genetic analysis in yeasts has demonstrated that their glucan synthase complex includes two functionally redundant catalytic subunits (Mazur *et al.* 1995). On the other hand, in *Exophiala dermatitidis* a single gene encodes the catalytic subunit of the β-1,3-glucan synthase, but reduction of the same gene retards growth and causes the loss of the cell wall integrity of this black yeast (Guo & Szaniszlo, 2011). While the number of genes involved in glucan synthesis may vary, the membrane localisation of these enzyme complexes bounds glucan synthase either to the Golgi or to the cytoplasmic membrane (Latge & Beauvais, 2014). The biosynthesis pathway is similar between different glucans but is not yet completely clarified.

14.3.1.1 α-Glucans

Alpha-glucans are polysaccharides composed of D-glucose monomer units linked mainly with α-glycosidic bonds. Pseudonigeran and nigeran from *Aspergillus niger*, pullulan from *Aureobasidium pullulans* and elsinan from *Elsinoe leucospila* belong to this group. However, the water-insoluble polysaccharides pseudonigeran and nigeran are components of the fungal cell wall and only pullulan and elsinan occur as exopolysaccharides in the extracellular matrix of the corresponding fungi and will be therefore described in detail.

14.3.1.1.1 Pullulan

Pullulan is an extracellular α-glucan with a long-existing biotechnological fame (Bender *et al.* 1959). It is obtained from the fermentation broth of the black ascomycetous yeast *Aureobasidium pullulans*. The water-soluble, linear neutral homopolysaccharide is secreted out to the cell surface to form a slimy layer consisting of maltotriose trimers connected by α-1,6 glycosidic linkages. The internal glucose units within maltotriose trimers are connected by α-1,4 glycosidic bonds (Bender *et al.* 1959; Cheng *et al.* 2011). Catley (1971) observed, that 6.6% of maltotetraose units exist randomly within the pullulan maltotriose chain. The alteration of α-1,4 and α-1,6-linkages results in structural flexibility and enhanced solubility (Leathers, 1993). However, other linkage types, besides α-1,4, like α-1,3 and even β-1,3 and β-1,6 were found by several research groups in the pullulan chain produced by different strains (Bouveng *et al.* 1962; Sowa, 1963; Fujii, 1984). Many additional polysaccharides have been described from cultures of *A. pullulans* (Bernier, 1958; Taguchi *et al.* 1973). Elinov *et al.* (1975) reported that the *A. pullulans* produces the homopolysaccharide aubasidan, with a β-1,3-linked backbone and α-1,4-linked side chains attached by β-1,6-linkages (Leathers, 2002). Yurlova and de Hoog (1997) identified strains that produce both pullulan and aubasidan-like polysaccharides. As reported by Leathers (2002) pullulan is also produced by other fungi like the saprophytic fungus *Tremella mesenterica*, the tree parasitic fungi *Cyttaria harioti*, *C. darwinii* and *Cryphonectria parasitica*, the fungal agent of chestnut blight.

14.3.1.1.2 Elsinan

An exopolysaccharide called elsinan was found in 1978 (Tsumuraya *et al.* 1978). The linear homopolysaccharide has been isolated from the culture filtrate of the phytopathogenic ascomycete *Elsinoe leucospila* and is also produced by other *Elsinoe* species. Like pullulan, elsinan is composed of maltotriose and maltotetraose residues and the internal glucose units within trimers and tetramers are connected by α-1,4-glycosidic bonds. However, the maltotriose units are connected by α-1,3 glycosidic linkages, whereas in pullulan chains they are linked by α-1,6 bonds (Tsumuraya *et al.* 1978). Moreover the elsinan molecule has a five-fold helical conformation with M_W 2–6 × 10^5. Elsinan shows a high viscosity and is capable of forming oxygen impermeable film. Furthermore elsinan films are resistant to dilute acids, but are partially digestible with human pancreatic α-amylase. Therefore a possible application for elsinan could be in food industry as an edible packaging film for fresh foods (Misaki *et al.* 2000).

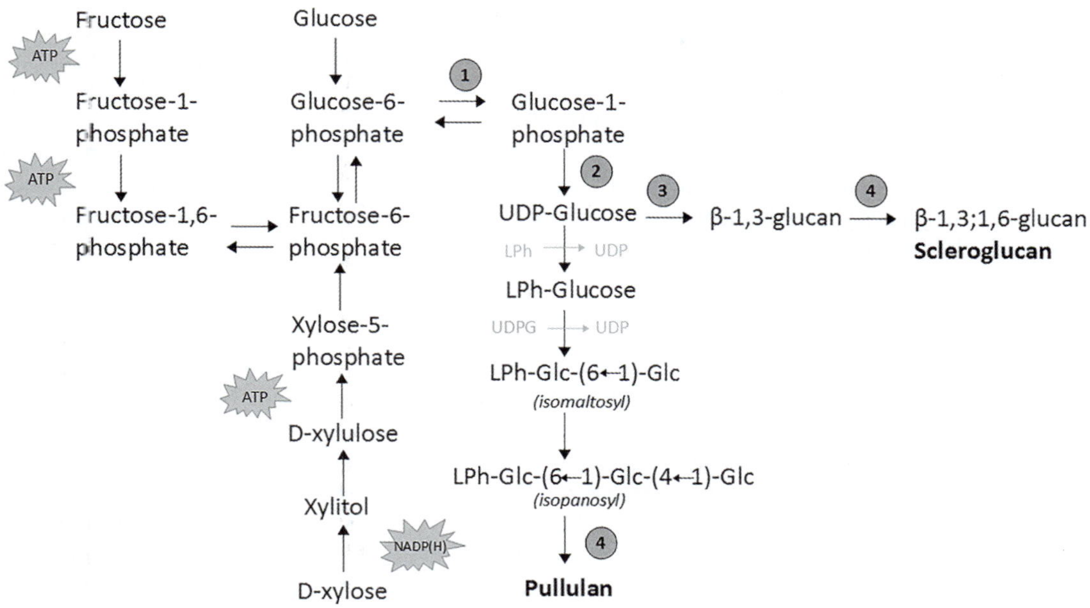

Figure 14.1 Assumed pathway of pullulan/scleroglucan biosynthesis. 1) phosphoglucose mutase, 2) UDPG pyrophosphorylase,3) β-1,3-glucan synthase 4) glycosyltransferase [modified from Duan et al. (2008), Cheng et al. (2011) and Schmid et al. (2011).

14.3.1.2 ß-glucans

All polysaccharides consisting of D-glucose monomers or its derivatives that are mainly linked with β-glycosidic bonds belong to the group of β-glucans, which are important structural components of the fungal cell wall. The most important secreted β-glucans botryosphaeran, lentinan, schizophyllan and scleroglucan will be described in detail.

14.3.1.2.1 Botryosphaeran

Botryosphaeran was first described from the ascomyceteous fungus *Botryosphaeria rhodina*. Fungi belonging to the genus *Botryosphaeria* are well recognised as plant pathogens causing various diseases in plants of commercial importance, including storage rots in preharvested fruits (Biggs, 1995; Ramos, 1997). Botryosphaeran consists of a linear backbone of β-1,3 glucopyranosyl groups to which branched chains of β-1,6-linked glucosyl and diglucosyl residues are attached with a frequency (approx 22%) of 1 branch point to every 5 glucose residues (Barbosa *et al.* 2003). The purified polysaccharide has a molecular weight of 2–5 × 10^6 Da (Selbmann *et al.* 2003; Weng *et al.* 2011) and is present in a triple helix conformation (Giese *et al.* 2008). Investigations of da Silva *et al.* (2005) showed an influence on branch formation when growing on different carbon sources. Growing on sucrose produced less branching (21%) than fructose (31%). The polymer was found to be non-mutagenic by the micronucleus test and exhibited strong antimutagenic activity in animal studies (Miranda *et al.* 2008).

14.3.1.2.2 Lentinan

Lentinan is found in the fruiting bodies of *Lentinus edodes*, or shiitake mushrooms, with the chemical structure of a β-1,3-glucan with two β-1,6 glucopyranoside branches for every five β-1,3 glucopyranoside linear linkages (Saito *et al.* 1979; Zhang *et al.* 2005; Zhang *et al.* 2011). The molecular weight of lentinan is between 500 and 1000 kDa and thus large for a polysaccharide. Lentinan forms triple helix structures in aqueous solutions and random helixes in DMSO (Sasaki & Takasuka, 1976; Bluhm & Sarko, 1977; Zhang *et al.* 2008). The glucose based lentinan is not a real exopolysaccharide once in the fruiting body, however, mycelia cultures excrete lentinan.

14.3.1.2.3 Schizophyllan and scleroglucan

Schizophyllan (or sizofiran, sonifilan) and scleroglucan are water-soluble polysaccharides initially discovered respectively in filamentous basidiomycetes *Schizophyllum commune* (Kikumoto *et al.* 1971) and *Sclerotium rolfsii* (Johnson *et al.* 1963). The repeating tetrasaccharide unit of schizophyllan consists of three β-1,3-linked glucose molecules together with a β-1,6-linked glucose residue. The structure of schizophyllan is that of a triple helix with a persistence length between 150–200 nm and a molecular weight of approximately 450 kDa (Norisuye *et al.* 1980; Yanaki *et al.* 1980; Kashiwagi *et al.* 1981). Scleroglucan forms a linear chain of β-1,3-linked glucopyranosyl residues with single D-glucopyranosyl groups linked β-1,6 to about every third residue of the main chain (Bluhm *et al.* 1982; Lecacheux *et al.* 1986). Dissolved scleroglucan chains assume a rod-like triple helical structure (Yanaki *et al.* 1981; Yanaki & Norisuye, 1983; Palleschi *et al.* 2005) with a molecular weight of $5–6 \times 10^6$ Da (Lecacheux *et al.* 1986; Pretus *et al.* 1991), but can differ depending on the cultivation media and strains used (Survase *et al.* 2007). Farina *et al.* (2001) identified two scleroglucan varieties from *S. rolfsii* cultures, which showed differences in rheological parameters, hydrogel topography, salt tolerance, imbibition as well as emulsifying and suspending properties. It has been considered that scleroglucan is more diverse than schizophyllan in its branch distribution and molecular weight. Both polysaccharides exhibit a discernible difference in physico-chemical properties. Native schizophyllan in a culture medium separates spontaneously from the mycelium and migrates freely into the liquid phase (Kikumoto, 1970), whereas scleroglucan adheres to the mycelium as a gel-like aggregate and cannot be dispersed in the liquid phase unless the culture medium is heated and homogenized (Yanaki & Norisuye, 1983).

14.3.1.3 Mannans: specific in yeasts

The biofilm-forming and pathogenic yeast *Candida albicans* has been intensively studied and produces β-1,3-linked glucans just like a majority of fungi (Nobile & Mitchell, 2006; Bonhomme & d'Enfert, 2013; Klis *et al.* 2014). However, the major fraction of the EPS in this very well studied yeast is built by α-1,2 branched α-1,6-mannans (87%) associated with unbranched β-1,6-glucans (13%) in an apparent mannan-glucan complex (MGCx) (Zarnowski *et al.* 2014). The molecular weight of this soluble exopolysaccharide is around 300 kDa with glucose, mannose, rhamnose and N-acetyl glucosamine as the main subunits. EPS production was seen to increase dramatically when developing biofilms were grown under liquid flow compared to static conditions; however, both types of biofilms were equally resistant to antifungal drugs (Ramage *et al.* 2005). Another interesting finding was that EPS in *Candida* biofilms contained significantly less total carbohydrate (41%) and protein (5%) than EPS from planktonic *Candida*. Differences in monosaccharide composition (presence of galactose and a higher glucose concentration) between EPS in biofilms versus planktonic cells suggest the existence of matrix-bound components that are unique to biofilms (Baillie & Douglas, 2000).

The expressed ability of *C. albicans* to form biofilms contrasts with that of *Saccharomyces cerevisiae*, a model species for all less tractable fungi. But while unable to form a mature biofilm, the brewer's yeast can nevertheless adhere to surfaces and form clumps or "flocs" consisting of many cells (Verstrepen & Klis, 2006). It is for these aggregates of "beer snow" that EPS production in flocculating cells *S. cerevisiae* was reported (Beauvais *et al.* 2009). The extracellular matrix of flocculating *S. cerevisiae* was isolated and characterised as containing mono- and polysaccharides of glucose, galactose and mannose, the latter in the form of highly branched mannan. Mannans are frequently connected to protein glycosylation and are inherent parts of glycoproteins that are found in yeast (Figueiredo & Barreto-Bergter, 2014; Faria-Oliveira *et al.* 2015).

14.4 DIVERSE FUNCTIONS OF FUNGAL EPS

14.4.1 Natural functions and their applications

Components secreted into the extracellular compartment are responsible for the interaction with the environment. Fungal extracellular components determine the efficiency of interaction with hosts as well as protection against environmental stresses. Additionally described soluble α- and β-glucans have been identified as important components of the biofilm matrix of numerous fungi, supporting the interaction with hosts and environment. The biochemical properties of these EPS yielded in a huge interest in technical and biomedical applications. Pullulan is probably the most important glucan followed by lentinan, both used for a wide variety of biotechnological and especially medical applications. Several publications described the biotechnological usability of fungal EPS (Jeanes, 1976; Leathers, 2003; Singh *et al.* 2008; Nwe *et al.* 2011; Mahapatra & Banerjee, 2013; Kagimura *et al.* 2015). The molecular weight of pullulan decreases during extensive fermentation to one-tenth of its maximum molecular weight of $2–3 \times 10^6$ (Steiner, 1993). Due to its enhanced viscosity pullulan can be used as an adhesive. Pullulan's ability to form fibers and transparent, oxygen-impermeable films has many applications, especially in pharmaceutical and food industry. Pullulan derivatives are used as blood plasma substitutes, drug/gene delivery carriers or for food preservation (Forabosco *et al.* 2006; Cheng *et al.* 2011).

Anti-inflammatory (Nikitina *et al.* 2007; McCormack *et al.* 2010; Nishitani *et al.* 2013), anticancer (Higashi *et al.* 2012; Ren *et al.* 2013) functions and usage for other health purposes were described for lentinan (Zhang *et al.* 2011; Wang *et al.* 2012; Ina *et al.* 2013; Sun *et al.* 2015), however systematic analyses are rare. It is proposed that β-glucans do not attack cancer cells directly but stimulate the cell internal immune response (Kozarski *et al.* 2014). The linkage of the side chains seems to be important for the medical effect (da Silva *et al.* 2006; Kozarski *et al.* 2014). Reviews for medical applications are provided by Chen and Seviour (2007) and Zhang *et al.* (2011).

Soluble β-glucans have also been identified as important components of the biofilm matrix of *C. albicans in vivo* and *in vitro*. Elevated β-1,3-glucan levels were also found in the surrounding biofilm milieu and these substances led to an increased fungicide resistance (Nett *et al.* 2007). Nobile and Mitchell (2006) put the role of EPS regarding adhesion and fungicide resistance into perspective. A general hypothesis is that multiple matrix components contribute to antifungal drug resistance, but experimental proof in this research area is needed.

In pathogenic yeasts a similar effect to bacterial EPS was detectable, e.g. *Candida* EPS may adsorb metal cations from their surroundings and contribute to metal resistance (Harrison *et al.* 2006). As metal availability might be decisive for performance of metabolic processes both of host and the pathogen, an ability to scavenge these might determine the outcome of fungi/host interaction. Biotechnological applications EPS produced by different fungi are considered in regard to heavy metal sorption in sludge, bioleaching of metals and remediation of contaminated soils by fungi. EPS of various extremophilic fungi are potential candidates to use for biotechnological purposes, especially for cosmetic applications (Pavlova *et al.* 2011; Vlaev *et al.* 2013).

Wang and McNeil (1996) showed that the β-glucan scleroglucan has remarkable rheological properties and a high tolerance to a broad range of temperature, pH-value and salt concentrations. The polysaccharide has various industrial applications but was initially used in the oil industry for thickening, discharge of drilling mud and enhanced oil recovery (Davison & Mentzer, 1982). Scleroglucan has become increasingly interesting in recent years due to its potential applications in food industry as a thickener and as gelling agent or stabilizing agent. Refined scleroglucan is mainly used for cosmetics as it can be applied in hair control compositions and in various skin care preparations, creams and protective lotions (Schmid *et al.* 2011). The biological activity of scleroglucan as an antitumor, antiviral and antimicrobial compound has also been investigated (Singh *et al.* 1974; Jong & Donovick, 1989; Prets, 1991).

Extensive studies on schizophyllan have revealed the wide potential application of this polysaccharide, such as petroleum recovery, a thickener for cosmetic lotions, oxygen- impermeable films for food preservation, and high-value pharmaceutical applications (Zhang *et al.* 2013). Schizophyllan is a non-specific stimulator of the immune system. It is reported that only triple helical schizophyllan is biologically active (Kojima *et al.* 1986). Schizophyllan with lower molecular weight (< 50 000 Da) has neither the triple helix structure nor any antitumor activity (Tabata *et al.* 1981; Yanaki & Norisuye, 1983). Other applications of schizophyllan include the ability to form hybrid complexes with nucleic acids (Sakurai & Shinkai, 2000; Sakurai *et al.* 2005) and the use in the formulation of matrices for drug delivery (Coviello *et al.* 1999).

The chitin derivative chitosan stands out due to its broad application in biotechnological, medicinal and biochemical industry. Cell-wall-stabilising chitosan adsorbs ionic material by complex formation and protects chitin from enzymatic attack (Singh *et al.* 2011). It is used for removal of metal ions in wastewater treatment, in cosmetics, photography, food industry and for drug delivery (Kumar, 2000).

Research regarding increased yield of the produced EPS demonstrated a highly diverse response to varying growth conditions. The main conditions are summarized in Mahapatra and Banerjee (2013), with carbon source, nitrogen availability, temperature, stirring, pH value and salt concentration as the main factors influencing EPS production.

14.4.2 EPS support interactions with solid substrates including minerals

As mentioned most of the polysaccharides produced by fungi are functionally involved in cell wall stability, especially α- and β-glucans are important for the integrity of the fungal cell wall. However, extracellular matrix on the fungi/environment/substrate interface provides a protective and stabilised environment to support these mutual and rather complex interactions. This self-created habitat becomes especially important under desiccated conditions, where the EPS contribute essentially to the water maintenance. EPS possess a high potential for water retention and favour desiccation tolerance, as demonstrated in cyanobacteria (Hill *et al.* 1994; Potts, 1999). Additionally EPS contribute to the stress resistance of Antarctic fungi (Pavlova *et al.* 2011; Kuncheva *et al.* 2013; Vlaev *et al.* 2013), supporting the capacity to grow under high osmotic pressure, matrix stress, low temperatures and low water availability. Production of various EPS enables some fungal specialists to survive almost every stress, even the outer space (Onofri *et al.* 2012; Scalzi *et al.* 2012; Selbmann *et al.* 2014).

EPS enhance the adhesion to the substrate, reduce water stress and increase the survival rate during freezing and thawing cycles. For these reasons, EPS synthesis seems to be an important survival adaptation in soil- or rock-inhabiting fungi, excreted EPS can mediate the cell-substrate interaction, e.g., to increase the availability of minerals and ions for themselves as well for their associated partner. EPS increase metal adsorption to cells and can facilitate mineral dissolution by complexation of cations (Warscheid & Braams, 2000). Biosorption of metals is one function of EPS used also for biotechnological purposes (Li & Yu, 2014). The research in this field is still mainly focused on bacteria (Tourney & Ngwenya, 2014), with a clear emerging need to investigate the role of fungi in these processes (Das *et al.* 2009). However, the capability of fungi to dwell and form biofilms on mineral surfaces and to enhance mineral accessibility for their symbiotic partners lead to a worldwide distribution of such fungi. Soil and rock-inhabiting fungi can be found in almost all environments (Cantrell *et al.* 2011; Cantrell & Duval-Perez, 2013). It became clear that bacterial and fungal EPS are involved in weathering processes of rocks and mineral building materials and influence soil development and soil structure (Tourney & Ngwenya, 2014). This is known as "biological weathering" and deserves to be investigated in a greater detail, especially regarding the function of the EPS in this process. While an active role of EPS in biological weathering is still unproven, EPS components certainly provide the reaction space for the maintenance of chemical processes. Further active factors of fungi-mineral and fungi-hosts interaction might be siderophores (Renshaw *et al.* 2002). Siderophores are also considered to be important pathogenicity elements, and given the general importance for fungi they can be expected to be produced by numerous pathogenic species of *Aspergillus* (Fedorova *et al.* 2008; Abad *et al.* 2010; Haas, 2014), but probably also by non-pathogenic e.g., rock-inhabiting species. EPS-layers also enable other microbes to join the biofilm by direct adhesion promoting contacts with photobionts and stimulating growth of accompanying bacteria (Figure 14.2, Gorbushina *et al.* (2005)).

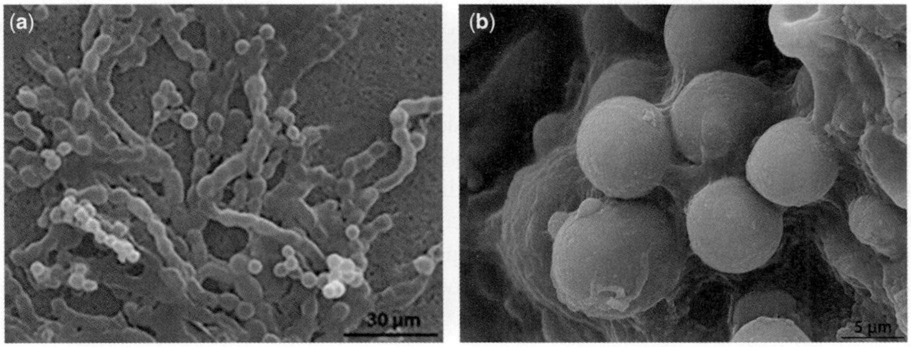

Figure 14.2 Concentrated EPS in desiccation-stressed subaerial biofilms (SABs) as demonstrated in laboratory experiments with a model genetically tractable rock biofilm. Rock-inhabiting fungi that dominate SABs (Gorbushina & Broughton, 2009) produce a substantial amount of EPS matrix around their cells. A - A combined *Knufia petricola/Nostoc punctiforme* biofilm grown under atmospheric exposure on a nitrogen-deficient but carbon-amended medium. Extracellular polymeric matrix of both partners surrounds the confluent biofilms. B - Close-up of a non-fixed (frozen and sublimated) *K. petricola* colony surface with a shrunken and thus very concentrated EPS matrix.

Additionally, a connection of EPS with excreted substances with increased UV light protection was demonstrated. In pathogenic fungi, EPS production lead to a higher resistance against antifungal agents (Ajesh & Sreejith, 2012). EPS of phytopathogenic fungi are also important for attachment to the host plant, protection of spores and ensuring the survival of the fungus (Forabosco *et al.* 2006). Furthermore pullulan from *Aureobasidium pullulans* is suggested to be important for cell protection and adhesion to leaf surfaces and wood (Leathers, 2002). Biofilm formation and EPS production enhance the cohesion of the film and influence the grazing of fungal dominated biofilms.

Cell adhesion and biofilm formation is not a trivial process. Cell adhesion to the surface is supported both by cell surface proteins (e.g., adhesin, flocculin) and EPS. High EPS production begins shortly after cell adhesion to the surface. Atomic Force Microscopy was used for imaging EPS produced by the fungus *Paxillus involutus* on phyllosilicate surfaces. Hyphal tips initially deposit EPS which assumes the shape of a halo surrounding hyphae. The fusion of adjacent EPS halos is likely responsible for the creation of EPS monolayers covering mineral surfaces (Gazze *et al.* 2013). The final stage is the formation of a mature biofilm with defined EPS structure. However, fungal EPS production, structure and composition are highly variable. EPS composition can change in response to water and nutrient availability (Reeslev *et al.* 1996), pH value and temperature (Ajesh & Sreejith, 2012; Smirnou *et al.* 2014) or salt concentration (Breierova *et al.* 2005). Salt stress induced EPS production and led to a different chemical composition. A higher degree of branching with single xylose side chains of the heteromannan components was detectable in *Cryptoccus laurentii*, leading to the assumption that EPS effectively help to survive high osmotic pressure. Additionally in *C. laurentii* increase of the pH value in the media from pH 3 to pH 6 lead to an increase in glucuronic acid (GluAc) content, while galactose, xylose, and glucose content decreased in EPS (Smirnou *et al.* 2014). Also important for the EPS production and composition is the carbon to nitrogen ratio (Imshenetskii *et al.* 1981; Reeslev *et al.* 1996; Elisashvili *et al.* 2009): nitrogen surplus changes EPS composition as well as cross-linking (Fraga *et al.* 2014). Additionally, EPS production, composition and structural properties vary in response in dependency to nitrogen and carbon sources and also depend on combination of both. In general biofilm formation strongly depends on the environmental conditions and can exhibit different phenotypes as a result of gene expression of different regulatory pathways and transcription factors. Additionally, EPS production, composition and structural properties vary in response to nitrogen and carbon sources and also depend on combination of both.

Data from a single contextual framework e.g., a single species biofilm cannot easily be extrapolated to all biofilms. A better reflection of reality demands the use of more complex but precisely defined *in vitro* biofilms. By expanding model fungal biofilm systems to at least two relevant and well-characterised organisms will significantly improve our understanding of natural systems. Use of such a system might help to clarify the degree of mutual dependence/succession of phototrophs and fungal heterotrophs in the microbial colonisation of rocks and have been developed (Gorbushina & Broughton, 2009). This particular *in vitro* biofilm comprises the rock-inhabiting and rock-weathering ascomycete *Knufia petricola* and the cyanobacterium *Nostoc punctiforme* that fixes both carbon and nitrogen (Nai *et al.* 2013) (Figure 14.3). A recently improved transformation protocol for *K. petricola* has permitted generation of gene-specific mutants (Noack-Schonmann *et al.* 2014). Detailed studies of such model fungi-dominated biofilms will shed light not only on evolution of fungal/cyanobacterial symbioses, but also as to how these rock-inhabiting communities enhance rock weathering, influence the degradation of building materials as well as cope with environmental and nutrient stresses.

For soil crusts and rock biofilms the production of EPS not only increases the stability, it enhances the harvest of organic particles from air, as a result of the inherent stickiness of any EPS-layer. The influence of anthropogenic carbon input, e.g., by aerosols, is not known, but a few investigations demonstrated the stimulating effect of aerosols on fungi ecosystems studied in urban environments (Isola *et al.* 2013). Almost every organic carbon source can be assimilated by biofilm forming organisms, for which fungi and bacteria actively excrete enzymes that are stabilised as a part of EPS matrix. The capacity of fungi to degrade and to assimilate organic carbon sources depends on the activity of such excreted enzymes. EPS stabilize the functionality of such enzymes and enhance the transport of products into the cells. Laccases, cellulases, proteases, lipases and peroxidases can be found in the secretome of fungi (Burns *et al.* 2013; Alfaro *et al.* 2014). In fungi it is proposed that extracellular polysaccharides (i) form a supporting network on which some of the excreted enzymes adsorb as well as (ii) play a role in the degradation of lignin as an indirect source of hydrogen peroxide (Krcmar *et al.* 1999). Additionally, it is assumed that the pH value for the optimal functionality of the enzymes is adjusted by EPS. EPS can also be utilised by other organisms and thus support the rock-inhabiting community as a whole. In biological soil crusts and in rock biofilms EPS that condition the substrate may serve as the main organic carbon source for the fungi as well as for other heterotrophic microbes.

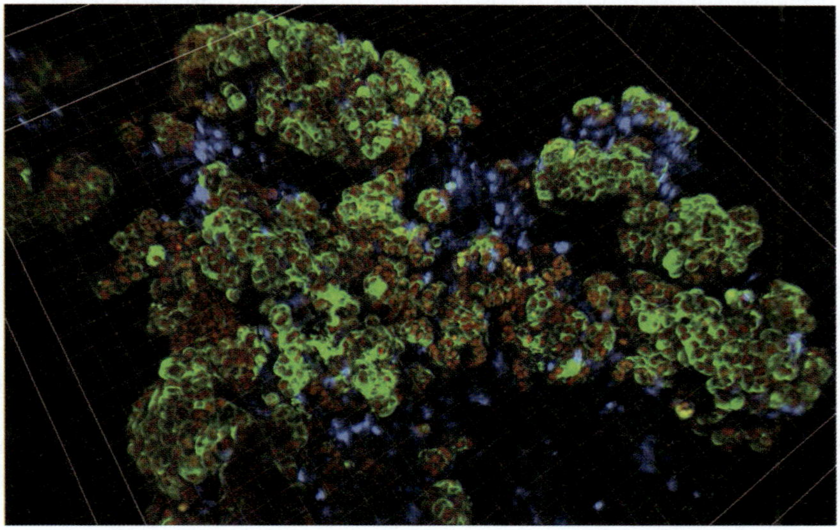

Figure 14.3 Three-dimensional volume projection of a model rock biofilm growing on Posta sandstone. A combined *Knufia petricola/Nostoc punctiforme* biofilm was cultivated under fluctuating conditions on the surface of a periodically desiccated mineral slab. The 90 µm thick biofilm was stained for polysaccharides and glycoconjugates by using Calcofluor white and AAL-Alexa568 lectin respectively. Colour allocation: autofluorescence = orange-red; lectin-specific glycoconjugates = green; β-D-glucans with 1,3- or 1,4-linkage = blue. Grid size = 20 µm. Sample was supplied by courtesy of Franz Seiffert. Laser microscopy was done by Thomas R. Neu at the River Ecology Department, Helmholtz Centre for Environmental Research – UFZ, Magdeburg.

14.4.3 Function of EPS in yeasts and melanised yeast-like fungi

EPS production was reported for several yeasts and yeast-like growing fungi, from the brewers's yeast *S. cerevisiae* to a very special group of melanised (black) yeast-like fungi. Isolated EPS of flocculating S. cerevisiae contains mono- and polysaccharides of glucose, galactose and mannose, along with a highly branched mannan (Beauvais *et al.* 2009 and glycosylated proteins (Faria-Oliveira *et al.* 2015). Black yeasts are called so in reference to their melanisation and growth form (Gueidan *et al.* 2008) and are best known for the animal and human opportunistic pathogens like *Wangiella/Exophiala dermatitidis* (Chaetothyriales, Winka *et al.* 1998). But though the pathogenic representatives of black yeasts might be best known, the proportion of environmental black fungi with extraordinary survival capabilities but without pathogenic traits is overwhelmingly high in comparison to a relatively low number of pathogens (de Hoog *et al.* 2013). In addition to cell wall and EPS the black yeasts additionally possess protective and rather prominent dark pigmentation, which is also located in the outer layer of the cell wall. Spores or other "differentiated-for-survival" structures of fungi (e.g., vegetative and dedicated to overwintering or survival of other stresses sclerotia, chlamydospores or microcolonies) are resistant against extreme environmental constraints. Frequently the stress tolerance of pathogenic and environmental black yeasts is supported by protective pigments in their cell wall and extracellular matrices. Dark pigmented fungi frequently prevail in the environment but our knowledge of the diversity of black yeast-like fungi is only very recent (de Hoog *et al.* 2013). This is largely due to the fact that their compact tissue-like colonies are difficult to isolate into culture and thus their existence has been long overlooked. These morphologically simplified ascomycetes possess round cells embedded in extracellular polymeric substances (Figure 14.2), and effectively encrust their cell wall with UV-absorbing melanin that convey multiple stress resistance to desiccation, temperature, and irradiation changes (Gorbushina, 2003). The ancestral ecological niche of these melanised fungi are rock surfaces (Gueidan *et al.* 2008; Ruibal *et al.* 2009).

Exophiala sp., *Aureobasidium pullulans*, *Hortaea* sp. and *Knufia* sp. are a few representatives of black yeast, this phylogenetically very heterogeneous group (Ruibal *et al.* 2008; Gueidan *et al.* 2011) that extensively employ the strategy of protective EPS synthesis. *Exophiala* and other pathogenic and non-pathogenic black yeast strains need a high measure of protection form the environmental stress: they were found in dishwashers and sauna facilities and the EPS probably help them to survive such hot and moist environments (Matos *et al.* 2002; Gostincar *et al.* 2011; Zalar *et al.* 2011). The general assumption is that black yeasts, as a result of melanin and EPS production, are very resistant against stress factors was demonstrated for *Exophiala* species (Zalar *et al.* 2011). *Exophiala* species excrete EPS or form a capsule (Yurlova & de Hoog, 2002), while the outer, galactomannan-rich wall layer is sometimes being shed off (Geis & Jacobs, 1985). It is suggested that the polysaccharides are necessary for this pathogenic fungus to adhere and infect cells. However, a detailed biochemical characterization of the produced EPS is only partially available. For *E. dermatiditis* a β-1,6-glucan was proposed as an EPS constituent (Montijn *et al.* 1997). In *E. jeanselmei* two major exopolysaccharides were found, with mannose, galactose and glucose as major monosaccharides, forming glycogen-like structure or characterized by galactofuranose units (Sassaki *et al.* 2011). It is known that glucopyranose, rhamnopyranose residues and galactofuranosyl units can be found in cell wall of yeasts and hyphae and those residues are widespread in glycolipids, proteoglycans, cell-wall polysaccharides and exopolysaccharides of fungi and yeasts.

Thick melanised walls and also the possibility to switch between yeast-like and filamentous forms are other important factors in stress resistance (Gostincar *et al.* 2011). EPS production is considered to be an important adaptation strategy in yeast and yeast-like fungi from extreme habitats, for instance, from Antarctic soils, salterns, deserts or from high altitudes. According to Selbmann *et al.* (2002) and Selbmann *et al.* (2014) EPS play a fundamental role in increasing resistance to cold conditions by protecting cells from freeze–thaw damage. Moreover, EPS create buffered physiochemical conditions around the cells and their hygroscopic nature increases success under dry conditions (de los Rios *et al.* 2003). Also osmotic pressure can be reduced and xerophilic fungi produce EPS under stress conditions, e.g., EPS are covering the outer cell-wall layer of *Wallemia ichthyophaga* at low glucose and low honey concentrations (Kralj Kuncic *et al.* 2013). In these extreme habitats, where low temperatures, low humidity and nutrient deficiency are predominant, EPS-forming microbes persist, that produce high molecular weight and cryoprotecting polysaccharides. Indeed several Antarctic yeasts were found to produce EPS (Pavlova *et al.* 2009; Cantrell *et al.* 2011; Gunde-Cimerman *et al.* 2012). Also a filamentous and spore-forming fungus *Phoma herbarum*, isolated from continental Antarctica soil, was shown to produce a homopolysaccharide composed of glucose units linked by β-1,3 and β-1,6 bonds, which has a cryoprotective effect (Selbmann *et al.* 2002). In a systematic survey Poli *et al.* (2010) found 24 antarctic yeast strains producing EPS. *Sporobolomyces salmonicolor* AL1, which had the highest EPS production rate, was found to produce a mannan-like exopolysaccharide with a β-backbone structure and a maximum yield of 5.64 g L^{-1} achieved at 22°C. Because of its ability to inhibit cytotoxic effects and a high biocompatibility, they suggested that this EPS could be useful for further applications. The plant-associated yeast *Pseudozyma* also produced EPS (Sajna *et al.* 2013) as many other soil-growing yeasts do.

14.4.3.1 Capsule-forming fungi: *Cryptococcus* species

Some fungi sport a real halo around their cells – a cell-wall-bound polysaccharide capsule. Intertwined fibers of capsular material in *Cryptococcus* species are characteristics used for identifications of this fungal infection in a clinical setting (Doering, 2009). The name *Cryptococcus* is used when referring to the yeast state of these well-known pathogenic filamentous fungi that likely diverged from *Saccharomyces* some billion years ago (Hedges *et al.* 2004). *C. neoformans* is the major causative agent of fungal meningoencephalitis in both immunocompromised and immunocompetent individuals (Lin, 2009). During such infections, the polysaccharide capsule seems to be the most important virulence factor (Zaragoza *et al.* 2009). Other factors are the melanin production, temperature and pH values. The capsule is composed primarily of two polysaccharides, glucuronoxylomannan (GXM) (90%) and galactoxylomannan (GalXM), in addition to a smaller proportion of mannoproteins (MPs) (Zaragoza *et al.* 2009). GXM is a linear α-1,3-mannan with a β-1,2-glucuronic acid residue attached at every third mannose. GalXM is a α-1,6-galactan that contains branches of β-1,3-galactose–α-1,4-mannose–α-1,3-mannose (Doering, 2009; Kumar *et al.* 2011). The capsule is produced intracellular and transferred to the outside via vesicle transport (Rizzo *et al.* 2014). Under normal conditions, the capsule is 1–2 µm thick; stress factors (virulent conditions) induce enhanced capsule production up to a size thickness of 30 µm. The enlargement of the capsule depends on the environmental conditions, is coordinated by the cell cycle (Doering, 2014; Garcia-Rodas *et al.* 2014; Li *et al.* 2014) and not reversible. Factors inducing the capsule production are CO_2, iron limitation, pH-value, temperature and nutrient availability (McFadden *et al.* 2006; Ajesh & Sreejith, 2012; Frazzitta *et al.* 2013; Smirnou *et al.* 2014). *C. neoformans* biofilms (with capsule) were more resistant than planktonic cells to oxidative stress and fungicides. These results suggest that the biofilm phenotype shows increased tolerance against biocides and the host immune mechanisms (Martinez & Casadevall, 2006). In general, fungal α-1,3-glucans play an essential role in the virulence of human fungal pathogens. Mutants of *Cryptococcus neoformans*, *Histoplasma capsulatum*, *Paracoccidioides brasiliensis* and *Blastomyces dermatitidis* lacking α-1,3-glucan are avirulent. The main functions of glucans regarding virulence are to anchor extracellular substances (e.g., capsule) and to mask antigens. Especially 1,3-linked glucans seem to be necessary to enhance linkage between cell and capsule (Doering, 2009).

A second pathogenic *Cryptococcus* species, *C. laurentii* has a high capacity to produce exopolysaccharides, with similar chemical structure compared to *C. neoformans*. The major constituent of *C. laurentii* EPS is acidic heteropolysaccharide glucuronoxylomannan (GXM) composed of α-1,3-linked mannan backbone, with *O*-6 and *O*-2 side chains of D-xylose and D-glucuronic acid residues (Pavlova *et al.* 2011; Ajesh & Sreejith, 2012; Kuncheva *et al.* 2013). It is important to note, that several non-pathogenic *Cryptococcus* strains isolated from different natural habitats, also produce EPS and capsule (Pavlova *et al.* 2009; Buzzini *et al.* 2012). EPS production was found frequently in extreme habitats, e.g., in antarctic yeasts. *C. flavus* strains produced EPS with production rates up to 5.75 g L^{-1} at 24°C. The composition of these exopolysaccharides (molecular mass = 1.01 MDa) was determined: 55.1% mannose, 26.1% glucose, 9.6% xylose and 1.9% galactose (Pavlova *et al.* 2009).

14.5 NON-POLYSACCHARIDE CONPOUNDS IN THE EXTRACELLULAR MATRIX

14.5.1 Proteins

For nutrition the fungal heterotrophic life style requires the uptake of organic metabolites from the environment. Fungi are decomposers that release enzymes into the extracellular space to break down biomass and then absorb metabolites (van Tilburg & Thomas, 1993). Because of various habitat conditions quite diverse carbon and nitrogen sources could be available. Hence fungi have to adjust their metabolism to prevalent conditions and must produce several extracellular enzymes (Girard *et al.* 2013). Filamentous fungi of both basidiomycetes and ascomycetes play an important role in the recycling of lignocellulosic biomass. Especially ascomycetes of the genus *Aspergillus* and *Trichoderma* and the basidiomycete *Phanerochaete chrysosporium* secrete large amounts of lignocellulose–degrading enzymes (Alfaro *et al.* 2014; Adav *et al.* 2015). *A. fumigatus* was shown to secrete mainly cellulases, hemicellulases and lignin degrading enzymes but also proteases, lipases and amylases (Adav *et al.* 2015). *T. reesei* excrete also mainly cellobiohydrolases and endoglucanases (He *et al.* 2013; Alfaro *et al.* 2014). In the class of basidiomycetes extracellular enzymes of white and brown rot basidiomycetes show distinct differences, e.g., white rot basidiomycetes mostly secrete glycoside hydrolases followed by peptidases, lignin-degrading enzymes and oxidoreductases and thus use both extracellular hydrolytic and oxidative reactions for cellulose degradation, and quinone reductase was only identified in brown rot basidiomycetes (Hori *et al.* 2013). Besides enzymes that play a crucial role in lignocellulose degradation in pathogenic ascomycete species, e.g., *A. fumigatus*, *A. terreus* and *Penicillium expansum* extracellular proteins with allergenic potential under infectious conditions were identified (Qin *et al.* 2007; Han *et al.* 2010; Beauvais & Latge, 2015). Many of the fungal enzymes are associated with extracellular vesicles, which are normally present in fungal cell walls (Brown *et al.* 2015) and are important interactive components of cell walls and EPS. Numerous fungal enzymes are also of deep interest for bioremediation because fungi possess the capacity to degrade environmental chemicals and thus decrease the risk of several pollutants (Harms *et al.* 2011).

A protein family that is unique to the fungal kingdom are hydrophobins. They play a key role in the growth and development of the majority of filamentous fungi (Kershaw & Talbot, 1998; Wosten & de Vocht, 2000; Wosten, 2001). These small secreted proteins (70–350 amino acids in length) self-assemble at the surface of the cell wall at hydrophilic-hydrophobic interfaces allowing hyphae to breach the medium-air interface but also assist the attachment of hyphae to hydrophobic surfaces (Wosten & de Vocht, 2000; Wosten, 2001; Wosten & Scholtmeijer, 2015). These normally abundant proteins are found not only in aerial hyphae, but also on the surface of fungal spores and infection structures, or on the surface of fruiting bodies The spores that develop on the end of the aerial structures are coated by an amphipathic hydrophobin layer that renders their surface hydrophobic and resists wetting, thus facilitating effective dispersal in air (Wang *et al.* 2005) but they also protect conidia from being recognized by the immune system in mammals (Aimanianda *et al.* 2009).

Additionally numerous lectin-like surface proteins termed flocculins, by a family of FLO genes (Teunissen & Steensma, 1995) and highly glycosyltated proteins were described at first for the brewer's yeast *S. cerevisiae* (Faria-Oliveira *et al.* 2015). This group of substances is especially deeply studied in pathogenic fungi, where mannoproteins are know to be involved in the adhesion of *Cryptococcus neoformans* to host cells (Teixeira *et al.* 2014).

14.5.2 Glycolipids

Microorganisms, including bacteria and fungi, are known to produce so called biosurfactants – amphipathic substances that increase the solubility of hydrophilic substances and facilitate the solution of hydrophobic compounds to lighten their absorption and consumption by microorganisms (Kulakovskaya *et al.* 2009; Santos *et al.* 2016). Substances belonging to the group of biosurfactants are lipopetides, glycolipids, fatty acids, neutral lipids and phospholipids. The most important extracelllular biosurfactants in fungi, especially in yeasts, are glycolipids, consisting of carbohydrates and fatty acids or hydroxyl acids (Kulakovskaya *et al.* 2009). Three most-known and well characterized glycolipids in yeasts of the genus *Candida*, *Pseudozyma* and *Ustilago* should be mentioned: cellobiose lipids, mannosylerythritol lipids and sophorolipids. It was shown that cellobiose lipids of *Pseudozyma fusiformata* and *Cryptococcus humicola*, that are composed of disaccharide cellobiose linked to a fatty acid residue (Kulakovskaya, 2014), have a broad antifungal activity (Puchkov *et al.* 2002; Kulakovskaya *et al.* 2005). Besides cellobiose lipids, mannosylerythritol lipids (MELs) were identified as a considerable component in the extracellular matrix of yeasts, especially in *Pseudozyma* sp., but also in *Ustilago* sp., *Schizonella melanogramma* and different *Candida* species (Arutchelvi & Doble, 2011). MELs consist of a mannose residue as the hydrophilic group and one or two fatty acid residues as the hydrophobic portion (Kulakovskaya, 2014). The sugar moiety of extracellular MELs has an antibacterial activity (Kitamoto *et al.* 1993). Furthermore MEL-containing liposomes exhibit the features for gene and drug delivery in DNA encapsulation and membrane fusion (Ueno *et al.* 2007) and have growth inhibition and differentiation-inducing activities against human leukemia cells, and therefore are of interest regarding medical applications (Arutchelvi & Doble, 2011). The third important extracellular yeast glycolipid – sophorolipid – was first described in *Torulopsis magnoliae* by (Gorin *et al.* 1961), and can also be found in other *Torulopsis* species, several *Candida* species and some other fungi (Van Bogaert & Soetaert, 2011). Sophorolipids are similar to cellobiose lipids, but contain the disaccharide sophorose as hydrophilic and a hydroxylated fatty acid as the hydrophobic part (Van Bogaert & Soetaert, 2011). They can be used as cleaning agents and detergents (Van Bogaert & Soetaert, 2011), in bioremediation of hydrocarbon contaminated sites having limited solubility in water and for increasing the bioavailability of microbial consortia for biodegradation (Kang *et al.* 2010). Because of their multiple biological activities, their non-toxicity and biodegradability, related to their special composition with a hydrophilic carbohydrate and a hydrophobic lipid part, fungal extracellular glycolipids have a huge benefit as biosurfactants in a broad application field.

14.5.3 Fungal *quorum sensing* molecules

Various microorganisms that normally exist as single cells are nevertheless capable of efficient coordination of their group activities. The use of an intercellular signalling system in a cell-density-dependent manner, called *quorum sensing* (QS) (Keller & Surette, 2006; Li & Nair, 2012) is broadly spread among different biofilm-forming microbes. Fungi are no exception from these rule and Hornby *et al.* (2001) first described QS in eukaryotes. The quorum sensing factor farnesol was discovered then as a component in high-cell density cultures of *Candida albicans*. Hornby *et al.* (2001) reported that farnesol prevented the development of mycelia, resulting in a developmental stage of budding yeasts but without influencing cellular growth rates. However, farnesol does not block the growth of already existing mycelia (Mosel *et al.* 2005). As a consequence thereof farnesol has the ability to inhibit biofilm formation. Ramage *et al.* (2002) showed that the rate of inhibition depends on the cells initial adherence time before farnesol was added. Already adherent cells that started to germinate were not blocked by the QS molecule. Further Westwater *et al.* (2005) reported an involvement of farnesol in the oxidative stress response against hydrogen peroxide and superoxide anion-generating agents. Farnesol is reported to control expression of genes important for hyphal development, drug resistance and cell wall integrity (Cao *et al.* 2005). Farnesol also has effects on other bacteria and fungi like *S. cerivisae*, *Aspergillus niger* or *Pseudomonas aeruginosa* (Albuquerque & Casadevall, 2012). A second QS molecule identified in *C. albicans* is tyrosol, a derivative of tyrosine, which is released continuously into the medium (Chen *et al.* 2004). It decreases the length of the lag phase but had no effect on the exponential growth of *C. albicans*. Moreover tyrosol accelerates the conversion from yeast cells to the filamentous form and stimulates biofilm formation (Chen *et al.* 2004). In biofilms there is 50% more tyrosol secreted than in planktonic cells. Therefore it seems, that morphogenesis in *C. albicans* is under complex positive and negative control by actions of the QS molecules tyrosol and farnesol (Alem *et al.* 2006). In early stage biofilms treated with low concentrations of the biofilm inhibiting QS molecule farnesol and higher concentrations of tyrosol, the latter can abolish the effects of farnesol. However, when biofilms are treated with higher amounts of farnesol the inhibiton of biofilm formation was unaffected by tyrosol and biofilms mostly composed of yeast cells. Thus it is supposed that there is a QS-mediated control in biofilm formation including farnesol and tyrosol as QS molecules, but farnesol has a dominant effect on cell morphology (Alem *et al.* 2006). In addition to *C. albicans*, QS systems are also described or hypothesised in other fungi. However, the molecules which are responsible for these activities are still unknown (Albuquerque & Casadevall, 2012).

14.6 DYNAMICS OF THE FUNGAL CELL WALL DEFINES THE EXTRACELLULAR MATRIX IN A CONSTANT INTERPLAY BETWEEN ORGANISMS AND THE ENVIRONMENT

Cell wall composition is strongly dependant on the environmental conditions and fungi rapidly respond to changing environments by adapting the composition of the newly formed walls. Especially prominent are modifications such as changes in thickness of the skeletal layer of the new walls (Ene *et al.* 2012), and by variation of the incorporation of chitin (Orlean, 2012). There is no clear evidence that the composition of existing walls is drastically changed in such cases, except probably for increased chitin incorporation upon cell surface stress (Orlean, 2012; Klis *et al.* 2014).

These major and multifunctional components are arranged into an interacting and highly dynamic network termed a fungal cell wall (Bowman & Free, 2006; Latge, 2007; Figueiredo & Barreto-Bergter, 2014). A paradigm change in mycology of the last decade is connected to a broadly accepted opinion that fungal cell wall in its chemical as well as structural properties is in a constant process of re-organisation. Water-soluble polysaccharides in fungal cell walls can easily reach the extracellular space i.e., migrate into the cell proximity and form extracellular matrix. The abundance as well as location of polysaccharides in the fungal cell wall is highly dependent on their production rates. Almost every component in the fungal cell wall is being reshuffled in response to environmental changes (Latge & Beauvais, 2014), and this is exactly the fact that contributes to the dynamic border between the structural and the extracellular components of the fungal outer layers. The major fungi-derived polysaccharides might contribute to the matrix and the outer border of the fungal cell, and thus sometimes it might be difficult to distinguish between the extracellular and structure-related polysaccharides.

Fungal cell walls provide rigidity and tensile strength for withstanding environmental changes while retaining plasticity to allow for the cell growth and division. Fungal cell walls are rigid structures – but by no mean static. Recently extracellular vesicles were discovered, in which molecules like pigments, polysaccharides, lipids, and enzymes are efficiently transferred through the cell wall in vesicles that efficiently pass this only seemingly rigid barrier and accumulate in the extracellular space (Albuquerque *et al.* 2008; Brown *et al.* 2015). It was demonstrated that production of such vesicles is directly and positively connected to pathogenicity (Rodrigues *et al.* 2014) – but it remains to look at these vesicles in environmentally and geochemically relevant fungi.

Fungal cell walls are capable of regulating their permeability, are flexible and allow for dramatic morphological changes. All fungal (i.e., yeast and hyphal) walls are elastic and support the processes of keeping cellular internal osmotic pressure equal to the pressure outside. In the process of turgor maintenance cell-wall-covered fungal cells can shrink or expand considerably following the interplay of osmotic conditions inside and outside the cell (Klis *et al.* 2014).

Fibrillar layer on the outside of the cell membrane and soluble β-glucans are critical components of the biofilm matrix that trap antifungal compounds (Nobile & Mitchell, 2006). Fungi secrete glucans into the cell-surrounding space – where those can either get interlinked and bound into the cell wall network or further migrate into the extracellular space. The secreted polysaccharides can have the same structure as the cell wall polysaccharides or become modified and further cross-linked with other interacting components. The wall proteome is another active participant, as it also reflects morphotype and contains yeast- and hypha-specific wall proteins (Heilmann *et al.* 2012). The dynamic combination of the cell wall and extracellular matrix shows an amazing adaptability that helps the fungal cell to adequately respond to changes in, for example, pH value, temperature, hypoxic conditions, carbon source, and iron availability and to plasma membrane and wall stress (Munro & Richard, 2012).

14.7 CONCLUSIONS: HEIGHTENED AWARENESS OF FUNGAL COMPLEXITY AND THEIR ENVIRONMENTAL RELEVANCE

Like all living organisms, fungi produce many different extracellular metabolites which function as mediators of interactions with inanimate and animate partners including animals and plants but also hard, inert surfaces. Extracellular matrices are in constant exchange with the environment and this dynamic has been studied in different model species. Nevertheless more questions remain open than have already been answered. Fungal diversity precludes establishment of a universal principle of how the extracellular matrix functions – but in the search for general rules we will discover interesting and variegated features.

By teaming-up with photosynthetic organisms, free-living and lichenised fungi effectively overgrow underlying substrates fostering colonisation of every exposed surface. Fungi are especially well adapted to growth in porous environments, form special biofilms and significantly contribute to almost all the others including geochemical roles. All these functions are supported by the characteristics of the matrix they create and secrete. Thus the influence of fungal biofilms is impossible to understand without much deeper understanding of extracellular polymeric substances.

Characterisation of extracellular compounds supports measures to prevent the damage caused by fungal activity. Mannoproteins of *Cryptococcus neoformans* are immunoreactive antigens that are discussed as potential cryptococcosis vaccine candidates (Teixeira *et al.* 2014). Once we know the composition of extracellular matrices that are actively weathering minerals, we may address the problem of monument and material weathering in much more efficient ways.

Despite that fact that terrestrially relevant fungi are omnipresent, most studies of fungal EPS have been conducted on medically important or plant pathogenic fungi. Largely aerobic fungi occupy the upper portion of the lithosphere and are therefore important settlers on a wide range of atmosphere-exposed rocks and other solid substrates. Extracellular-matrix-producing ascomycetes involved in mineral weathering and material deterioration are gradually developing into laboratory tools including all necessary methodological instruments that permit complete genetic analyses (Nai *et al.* 2013; Noack-Schönmann *et al.* 2014).

Important questions related to the dynamics of the fungal cell/the surrounding matrix interactions are:

- How do fungi retain substances in vicinity of the cell? Is the EPS matrix a molecular sieve for enzymes and other survival-enhancing compounds?
- How heterogeneous and hydrophobic are fungal extracellular matrices?
- Do the porosity, structure and diversity of extracellular matrices change in natural environments?

Answers to these questions will be found by comparing medically relevant fungi with environmentally-relevant model fungal species.

ACKNOWLEDEMENTS

Laser microscopy work on rock-inhabiting fungal biofilms was performed by Thomas R. Neu (UFZ, Magdeburg). We thank Renate Kort (Carl von Ossietzky University of Oldenburg) and Ines Feldmann and Steffi Noack-Schönmann (BAM Berlin) for their help with the electron microscopy. We are grateful to the editors of "The perfect slime", especially Hans-Curt Flemming for the stimulating discussions and friendly encouragement, as well as Franz Seiffert and W.J. Broughton for valuable suggestions on the manuscript.

REFERENCES

Abad A., Fernandez-Molina J. V., Bikandi J., Ramirez A., Margareto J., Sendino J., Hernando F. L., Ponton J., Garaizar J. and Rementeria, A. (2010). What makes *Aspergillus fumigatus* a successful pathogen? Genes and molecules involved in invasive aspergillosis. *Rev Iberoam Micol*, **27**(4), 155–182.

Adav S. S., Ravindran A. and Sze S. K. (2015). Quantitative proteomic study of *Aspergillus fumigatus* secretome revealed deamidation of secretory enzymes. *J Proteomics*, **119**, 154–168.

Aimanianda V., Bayry J., Bozza S., Kniemeyer O., Perruccio K., Elluru S. R., Clavaud C., Paris S., Brakhage A. A., Kaveri S. V., Romani L. and Latge, J. P. (2009). Surface hydrophobin prevents immune recognition of airborne fungal spores. *Nature*, **460**(7259), 1117–1121.

Ajesh K. and Sreejith K. (2012). *Cryptococcus laurentii* biofilms: structure, development and antifungal drug resistance. *Mycopathologia*, **174**(5–6), 409–419.

Albuquerque P. and Casadevall A. (2012). Quorum sensing in fungi - a review. *Medical Mycology*, **50**(4), 337–345.

Albuquerque P. C., Nakayasu E. S., Rodrigues M. L., Frases S., Casadevall A., Zancope-Oliveira R. M., Almeida I. C. and Nosanchuk, J. D. (2008). Vesicular transport in Histoplasma capsulatum: an effective mechanism for trans-cell wall transfer of proteins and lipids in ascomycetes. *Cellular Microbiology*, **10**(8), 1695–1710.

Alem M. A. S., Oteef M. D. Y., Flowers T. H. and Douglas L. J. (2006). Production of tyrosol by *Candida albicans* biofilms and its role in quorum sensing and biofilm development. *Eukaryotic Cell*, **5**(10), 1770–1779.

Alfaro M., Oguiza J. A., Ramirez L. and Pisabarro A. G. (2014). Comparative analysis of secretomes in basidiomycete fungi. *Journal of Proteomics*, **102**, 28–43.

Arutchelvi J. and Doble M. (2011). Mannosylerythritol lipids: microbial production and their applications. *Biosurfactants: From Genes to Applications*, **20**, 145–177.

Bae J. T., Sim G. S., Lee D. H., Lee B. C., Pyo H. B., Choe T. B. and Yun J. W. (2005). Production of exopolysaccharide from mycelial culture of *Grifola frondosa* and its inhibitory effect on matrix metalloproteinase-1 expression in UV-irradiated human dermal fibroblasts. *FEMS Microbiology Letters*, **251**(2), 347–354.

Baillie G. S. and Douglas L. J. (2000). Matrix polymers of *Candida* biofilms and their possible role in biofilm resistance to antifungal agents. *Journal of Antimicrobial Chemotherapy*, **46**(3), 397–403.

Barbosa A. M., Steluti R. M., Dekker R. F., Cardoso M. S. and Corradi da Silva M. L. (2003). Structural characterization of Botryosphaeran: a (1 -> 3;1 -> 6)-beta-D-glucan produced by the ascomyceteous fungus, *Botryosphaeria* sp. *Carbohydrate Research*, **338**(16), 1691–1698.

Bartnicki-Garcia S. (1968). Cell wall chemistry, morphogenesis, and taxonomy of fungi. *Annual Review of Microbiology*, **22**, 87–108.

Batra K. K., Nordin J. H. and Kirkwood S. (1969). Biosynthesis of beta-D-glucan of *Sclerotium rolfsii* Sacc. direction of chain propagation and insertion of branch residues. *Carbohydrate Research*, **9**(2), 221–229.

Beauvais A. and Latge J. P. (2015). *Aspergillus* biofilm in vitro and in vivo. *Microbiol Spectr*, **3**(4), 149–161.

Beauvais A., Loussert C., Prevost M. C., Verstrepen K. and Latge J. P. (2009). Characterization of a biofilm-like extracellular matrix in FLO1-expressing *Saccharomyces cerevisiae* cells. *FEMS Yeast Research*, **9**(3), 411–419.

Bechinger C., Giebel K. F., Schnell M., Leiderer P., Deising H. B. and Bastmeyer M. (1999). Optical measurements of invasive forces exerted by appressoria of a plant pathogenic fungus. *Science*, **285**(5435), 1896–1899.

Bender H., Lehmann J. and Wallenfels K. (1959). Pullulan, a extracellular glucan from *Pullularia pullulans. Biochimica Et Biophysica Acta*, **36**(2), 309–316.

Bergendiova K., Tibenska E. and Majtan J. (2011). Pleuran (beta-glucan from *Pleurotus ostreatus*) supplementation, cellular immune response and respiratory tract infections in athletes. *European Journal of Applied Physiology*, **111**(9), 2033–2040.

Bernier B. (1958). The production of polysaccharide by fungi active in the decomposition of wood and forest litter. *Canadian Journal of Microbiology*, **4**(3), 195–204.

Biggs A. R. (1995). Detection of latent infections in apple fruit with paraquat. *Plant Disease*, **79**(10), 1062–1067.

Blankenship J. R. and Mitchell A. P. (2006). How to build a biofilm: a fungal perspective. *Curr Opin Microbiol*, **9**(6), 588–594.

Bluhm T. L. and Sarko A. (1977). Triple helical structure of Lentinan, a Linear Beta-(1-]3)-D-Glucan. *Canadian Journal of Chemistry-Revue Canadienne De Chimie*, **55**(2), 293–299.

Bluhm T. L., Deslandes Y., Marchessault R. H., Perez S. and Rinaudo M. (1982). Solid-state and solution conformation of scleroglucan. *Carbohydrate Research*, **100**(Mar), 117–130.

Bonhomme J. and d'Enfert C. (2013). *Candida albicans* biofilms: building a heterogeneous, drug-tolerant environment. *Curr Opin Microbiol*, **16**(4), 398–403.

Bouveng H. O., Mckay J., Lindberg B. and Kiessling H. (1962). Polysaccharides elaborated by *Pullularia pullulans* .1. Neutral glucan synthesised from sucrose solutions. *Acta Chemica Scandinavica*, **16**(3), 615–622.

Bowman S. M. and Free S. J. (2006). The structure and synthesis of the fungal cell wall. *Bioessays*, **28**(8), 799–808.

Breierova E., Hromadkova Z., Stratilova E., Sasinkova V. and Ebringerova A. (2005). Effect of salt stress on the production and properties of extracellular polysaccharides produced by *Cryptococcus laurentii. Zeitschrift Fur Naturforschung C-a Journal of Biosciences*, **60**(5–6), 444–450.

Brown L., Wolf J. M., Prados-Rosales R. and Casadevall A. (2015). Through the wall: extracellular vesicles in Gram-positive bacteria, mycobacteria and fungi. *Nature Reviews Microbiology*, **13**(10), 620–630.

Bulawa C. E. (1993). Genetics and molecular biology of chitin synthesis in fungi. *Annual Review of Microbiology*, **47**, 505–534.

Burns R. G., DeForest J. L., Marxsen J., Sinsabaugh R. L., Stromberger M. E., Wallenstein M. D., Weintraub M. N. and Zoppini, A. (2013). Soil enzymes in a changing environment: current knowledge and future directions. *Soil Biology & Biochemistry*, **58**, 216–234.

Buzzini P., Branda E., Goretti M. and Turchetti B. (2012). Psychrophilic yeasts from worldwide glacial habitats: diversity, adaptation strategies and biotechnological potential. *FEMS Microbiology Ecology*, **82**(2), 217–241.

Cantrell S. A. and Duval-Perez L. (2013). Microbial mats: an ecological niche for fungi. *Frontiers in Microbiology*, **4**.

Cantrell S. A., Dianese J. C., Fell J., Gunde-Cimerman N. and Zalar P. (2011). Unusual fungal niches. *Mycologia*, **103**(6), 1161–1174.

Cao Y. Y., Cao Y. B., Xu Z., Ying K., Li Y., Xie Y., Zhu Z. Y., Chen W. S. and Jiang, Y. Y. (2005). cDNA microarray analysis of differential gene expression in *Candida albicans* biofilm exposed to farnesol. *Antimicrob Agents Chemother*, **49**(2), 584–589.

Catley B. J. (1971). Role of Ph and nitrogen limitation in elaboration of extracellular polysaccharide pullulan by *Pullularia pullulans. Applied Microbiology*, **22**(4), 650–654.

Chen J. Z. and Seviour R. (2007). Medicinal importance of fungal beta-(1 -> 3), (1 -> 6)-glucans. *Mycological Research*, **111**, 635–652.

Chen H., Fujita M., Feng Q. H., Clardy J. and Fink G. R. (2004). Tyrosol is a quorum-sensing molecule in *Candida albicans. Proceedings of the National Academy of Sciences of the United States of America*, **101**(14), 5048–5052.

Chen S. G., Siu K. C., Wang W. Q., Liu X. X. and Wu J. Y. (2013a). Structure and antioxidant activity of a novel poly-N-acetylhexosamine produced by a medicinal fungus. *Carbohydrate Polymers*, **94**(1), 332–338.

Chen Y., Mao W., Wang J., Zhu W., Zhao C., Li N., Wang C., Yan M., Guo T. and Liu, X. (2013b). Preparation and structural elucidation of a glucomannogalactan from marine fungus *Penicillium commune. Carbohydrate Polymers*, **97**(2), 293–299.

Cheng K. C., Demirci A. and Catchmark J. M. (2011). Pullulan: biosynthesis, production, and applications. *Applied Microbiology and Biotechnology*, **92**(1), 29–44.

Coviello T., Grassi M., Rambone G., Santucci E., Carafa M., Murtas E., Riccieri F. M. and Alhaique, F. (1999). Novel hydrogel system from scleroglucan: synthesis and characterization. *Journal of Controlled Release*, **60**(2–3), 367–378.

da Silva M. D. C., Izeli N. L., Martinez P. F., Silva I. R., Constantino C. J. L., Cardoso M. S., Barbosa A. M., Dekker R. F. H. and da Silva G. V. J. (2005). Purification and structural characterisation of (1 -> 3; 1 -> 6)-beta-D-glucans (botryosphaerans) from *Botryosphaeria rhodina* grown on sucrose and fructose as carbon sources: a comparative study. *Carbohydrate Polymers*, **61**(1), 10–17.

da Silva M. D. C., Martinez P. F., Izeli N. L., Silva I. R., Vasconcelos A. F. D., Cardoso M. D., Stelutti R. M., Giese E. C. and Barbosa A. D. (2006). Chemical characterization and biotechnology applications of fungal glucans. *Quimica Nova*, **29**(1), 85–92.

Das B. K., Roy A., Koschorreck M., Mandal S. M., Wendt-Potthoff K. and Bhattacharya J. (2009). Occurrence and role of algae and fungi in acid mine drainage environment with special reference to metals and sulfate immobilization. *Water Research*, **43**(4), 883–894.

Davison P. and Mentzer E. (1982). Polymer flooding in North-Sea reservoirs. *Society of Petroleum Engineers Journal*, **22**(3), 353–362.

de Hoog G. S., Vicente V. A. and Gorbushina A. A. (2013). The bright future of darkness–the rising power of black fungi: black yeasts, microcolonial fungi, and their relatives. *Mycopathologia*, **175**(5–6), 365–368.

de los Rios A., Wierzchos J., Sancho L. G. and Ascaso C. (2003). Acid microenvironments in microbial biofilms of Antarctic endolithic microecosystems. *Environmental Microbiology*, **5**(4), 231–237.

Doering T. L. (2009). How sweet it is! cell wall biogenesis and polysaccharide capsule formation in *Cryptococcus neoformans*. *Annual Review of Microbiology*, **63**, 223–247.

Doering T. L. (2014). Regulation of cryptococcal capsule synthesis. *Mycoses*, **57**, 26–27.

Donot F., Fontana A., Baccou J. C. and Schorr-Galindo S. (2012). Microbial exopolysaccharides: main examples of synthesis, excretion, genetics and extraction. *Carbohydrate Polymers*, **87**(2), 951–962.

Duan X. H., Chi Z. M., Wang L. and Wang X. H. (2008). Influence of different sugars on pullulan production and activities of alpha-phosphoglucose mutase, UDPG-pyrophosphorylase and glucosyltransferase involved in pullulan synthesis in *Aureobasidium pullulans* Y68. *Carbohydrate Polymers*, **73**(4), 587–593.

Elinov N. P., Marikhin V. A., Dranishnikov A. N., Miasnikov L. P. and Mariukhta Iu B. (1975). Characteristics of glucan, produced by a culture of *Aureobasidium (Pullularia) pullulans*. *Dokl Akad Nauk SSSR*, **221**(1), 213–216.

Elisashvili V. I., Kachlishvili E. T. and Wasser S. P. (2009). Carbon and nitrogen source effects on basidiomycetes exopolysaccharide production. *Applied Biochemistry and Microbiology*, **45**(5), 531–535.

Ene I. V., Adya A. K., Wehmeier S., Brand A. C., MacCallum D. M., Gow N. A. and Brown A. J. (2012). Host carbon sources modulate cell wall architecture, drug resistance and virulence in a fungal pathogen. *Cellular Microbiology*, **14**(9), 1319–1335.

Espinosa V. and Rivera A. (2016). First line of defense: innate cell-mediated control of pulmonary aspergillosis. *Frontiers in Microbiology*, **7**, 272.

Faria-Oliveira F., Carvalho J., Belmiro C. L. R., Ramalho G., Pavao M., Lucas C. and Ferreira C. (2015). Elemental biochemical analysis of the polysaccharides in the extracellular matrix of the yeast *Saccharomyces cerevisiae*. *Journal of Basic Microbiology*, **55**(6), 685–694.

Farina J. I., Sineriz F., Molina O. E. and Perotti N. I. (2001). Isolation and physicochemical characterization of soluble scleroglucan from *Sclerotium rolfsii*. Rheological properties, molecular weight and conformational characteristics. *Carbohydrate Polymers*, **44**(1), 41–50.

Fedorova N. D., Khaldi N., Joardar V. S., Maiti R., Amedeo P., Anderson M. J., Crabtree J., Silva J. C., Badger J. H., Albarraq A., Angiuoli S., Bussey H., Bowyer P., Cotty P. J., Dyer P. S., Egan A., Galens K., Fraser-Liggett C. M., Haas B. J., Inman J. M., Kent R., Lemieux S., Malavazi I., Orvis J., Roemer T., Ronning C. M., Sundaram J. P., Sutton G., Turner G., Venter J. C., White O. R., Whitty B. R., Youngman P., Wolfe K. H., Goldman G. H., Wortman J. R., Jiang B., Denning D. W. and Nierman W. C. (2008). Genomic islands in the pathogenic filamentous fungus *Aspergillus fumigatus*. *PLoS Genet*, **4**(4), e1000046.

Figueiredo R. T. and Barreto-Bergter E. B. (2014). Fungal glycans and the innate immune recognition. *Frontiers in Cellular and Infection Microbiology*, **4**, Article Number 145.

Flemming H. C. and Wingender J. (2010). The biofilm matrix. *Nature Reviews Microbiology*, **8**(9), 623–633.

Fontaine T., Delangle A., Simenel C., Coddeville B., van Vliet S. J., van Kooyk Y., Bozza S., Moretti S., Schwarz F., Trichot C., Aebi M., Delepierre M., Elbim C., Romani L. and Latge J. P. (2011). Galactosaminogalactan, a new immunosuppressive polysaccharide of *Aspergillus fumigatus*. *Plos Pathogens*, **7**(11), Article Number: e1002372.

Forabosco A., Bruno G., Sparapano L., Liut G., Marino D. and Delben F. (2006). Pullulans produced by strains of *Cryphonectria parasitica* - I. Production and characterisation of the exopolysaccharides. *Carbohydrate Polymers*, **63**(4), 535–544.

Fraga I., Coutinho J., Bezerra R. M., Dias A. A., Marques G. and Nunes F. M. (2014). Influence of culture medium growth variables on *Ganoderma lucidum* exopolysaccharides structural features. *Carbohydrate Polymers*, **111**, 936–946.

Frases S., Nimrichter L., Viana N. B., Nakouzi A. and Casadevall A. (2008). *Cryptococcus neoformans* capsular polysaccharide and exopolysaccharide fractions manifest physical, chemical, and antigenic differences. *Eukaryotic Cell*, **7**(2), 319–327.

Frazzitta A. E., Vora H., Price M. S., Tenor J. L., Betancourt-Quiroz M., Toffaletti D. L., Cheng N. and Perfect J. R. (2013). Nitrogen source-dependent capsule induction in human-pathogenic *Cryptococcus* Species. *Eukaryotic Cell*, **12**(11), 1439–1450.

Fujii N., Shinohara, S., Ueno, H., Imada, K. (1984). Polysaccharides produced by *Aureobasidium sp.* (black yeast). *Kenkyu Hokoku—Miyazaki Daigaku Nogakubu*, **31**, 253–262.

Fukuda E. K., Vasconcelos A. F. D., Matias A. C., Barbosa A. D., Dekker R. F. H. and da Silva M. D. C. (2009). Fungal cell wall polysaccharides: purification and characterization. *Semina-Ciencias Agrarias*, **30**(1), 117–133.

Gadd G. M. (2007). Geomycology: biogeochemical transformations of rocks, minerals, metals and radionuclides by fungi, bioweathering and bioremediation. *Mycological Research*, **111**(Pt 1), 3–49.

Garcia-Rodas R., Cordero R. J., Janbon G., Moyrand F., Casadevall A. and Zaragoza O. (2014). Capsule growth in *Cryptococcus neoformans* is coordinated with cell cycle progression. *Mycoses*, **57**, 94–95.

Gazze S. A., Saccone L., Smits M. M., Duran A. L., Leake J. R., Banwart S. A., Ragnarsdottir K. V. and McMaster T. J. (2013). Nanoscale observations of extracellular polymeric substances deposition on phyllosilicates by an ectomycorrhizal fungus. *Geomicrobiology Journal*, **30**(8), 721–730.

Geis P. A. and Jacobs C. W. (1985). Polymorphism of *Wangiella dermatitidis*. In: Fungal Dimorphism, P. J. Szaniszlo (ed.), Plenum, New York, pp. 205–233.

Giese E. C., Dekker R. F. H., Barbosa A. M. and da Silva R. (2008). Triple helix conformation of botryosphaeran, a (1 -> 3;1 -> 6)-beta-D-glucan produced by *Botryosphaeria rhodina* MAMB-05. *Carbohydrate Polymers*, **74**(4), 953–956.

Girard V., Dieryckx C., Job C. and Job D. (2013). Secretomes: the fungal strike force. *Proteomics*, **13**(3–4), 597–608.

Gorbushina A. (2003). Microcolonial fungi: survival potential of terrestrial vegetative structures. *Astrobiology*, **3**(3), 543–554.

Gorbushina A. A. and Broughton W. J. (2009). Microbiology of the atmosphere-rock interface: how biological interactions and physical stresses modulate a sophisticated microbial ecosystem. *Annual Review of Microbiology*, **63**, 431–450.

Gorbushina A. A., Beck A. and Schulte A. (2005). Microcolonial rock inhabiting fungi and lichen photobionts: evidence for mutualistic interactions. *Mycological Research*, **109**(Pt 11), 1288–1296.

Gorin P. A. J. and Barreto-Bergter E. (1983). 6 - The chemistry of polysaccharides of fungi and lichens. In: *The Polysaccharides*, G. Aspinall (ed.), Vol. **2**, Academic Press, pp. 365–409.

Gorin P. A., Spencer J. F. T. and Tulloch A. P. (1961). Hydroxy fatty acid glycosides of sophorose from torulopsis magnoliae. *Canadian Journal of Chemistry-Revue Canadienne De Chimie*, **39**(4), 846–855.

Gostincar C., Grube M. and Gunde-Cimerman N. (2011). Evolution of fungal pathogens in domestic environments? *Fungal Biology*, **115**(10), 1008–1018.

Gueidan C., Villasenor C. R., de Hoog G. S., Gorbushina A. A., Untereiner W. A. and Lutzoni F. (2008). A rock-inhabiting ancestor for mutualistic and pathogen-rich fungal lineages. *Stud Mycol*, **61**, 111–119.

Gueidan C., Ruibal C., De Hoog G. S. and Schneider H. (2011). Rock-inhabiting fungi originated during periods of dry climate in the late Devonian and middle Triassic. *Fungal Biology*, **115**(10), 987–996.

Gunde-Cimerman N., Wagner D. and Haggblom M. (2012). Polar and alpine microbiology. *Fems Microbiology Ecology*, **82**(2), 215–216.

Guo P. and Szaniszlo P. J. (2011). RNA interference of WdFKS1 mRNA expression causes slowed growth, incomplete septation and loss of cell wall integrity in yeast cells of the polymorphic, pathogenic fungus *Wangiella (Exophiala) dermatitidis*. *Medical Mycology*, **49**(8), 806–818.

Gutierrez A., Prieto A. and Martinez A. T. (1996). Structural characterization of extracellular polysaccharides produced by fungi from the genus *Pleurotus*. *Carbohydr Res*, **281**(1), 143–154.

Haas H. (2014). Fungal siderophore metabolism with a focus on *Aspergillus fumigatus*. *Nat Prod Rep*, **31**(10), 1266–1276.

Han M. J., Kim N. J., Lee S. Y. and Chang H. N. (2010). Extracellular proteome of *Aspergillus terreus* grown on different carbon sources. *Current Genetics*, **56**(4), 369–382.

Harms H., Schlosser D. and Wick L. Y. (2011). Untapped potential: exploiting fungi in bioremediation of hazardous chemicals. *Nat Rev Microbiol*, **9**(3), 177–192.

Harrison J. J., Rabiei M., Turner R. J., Badry E. A., Sproule K. M. and Ceri H. (2006). Metal resistance in *Candida* biofilms. *Fems Microbiology Ecology*, **55**(3), 479–491.

He J., Han G. Y. and Chen D. W. (2013). Insights into enzyme secretion by filamentous fungi: comparative proteome analysis of *Trichoderma reesei* grown on different carbon sources. *J Proteomics*, **89**, 191–201.

Hedges S. B., Blair J. E., Venturi M. L. and Shoe J. L. (2004). A molecular timescale of eukaryote evolution and the rise of complex multicellular life. *BMC Evol Biol*, **4**, Article Number 2.

Heilmann C. J., Sorgo A. G. and Klis F. M. (2012). News from the fungal front: wall proteome dynamics and host–pathogen interplay. *Plos Pathogens*, **8**(12), e1003050.

Higashi D., Seki K., Ishibashi Y., Egawa Y., Koga M., Sasaki T., Hirano K., Mikami K., Futami K., Maekawa T. and Sudo M. (2012). The effect of Lentinan combination therapy for unresectable advanced gastric cancer. *Anticancer Research*, **32**(6), 2365–2368.

Hill D. R., Peat A. and Potts M. (1994). Biochemistry and structure of the glycan secreted by desiccation-tolerant *Nostoc commune* (Cyanobacteria). *Protoplasma*, **182**(3–4), 126–148.

Hori C., Gaskell J., Igarashi K., Samejima M., Hibbett D., Henrissat B. and Cullen D. (2013). Genomewide analysis of polysaccharides degrading enzymes in 11 white- and brown-rot Polyporales provides insight into mechanisms of wood decay. *Mycologia*, **105**(6), 1412–1427.

Hornby J. M., Jensen E. C., Lisec A. D., Tasto J. J., Jahnke B., Shoemaker R., Dussault P. and Nickerson K. W. (2001). Quorum sensing in the dimorphic fungus *Candida albicans* is mediated by farnesol. *Applied and Environmental Microbiology*, **67**(7), 2982–2992.

Imshenetskii A. A., Kondrateva T. F. and Smutko A. N. (1981). Influence of carbon and nitrogen-sources on pullulan biosynthesis by polyploid strains of *Pullularia pullulans*. *Microbiology*, **50**(1), 73–76.

Ina K., Kataoka T. and Ando T. (2013). The use of lentinan for treating gastric cancer. *Anti-Cancer Agents in Medicinal Chemistry*, **13**(5), 681–688.

Isola D., Selbmann L., de Hoog G. S., Fenice M., Onofri S., Prenafeta-Boldu F. X. and Zucconi L. (2013). Isolation and screening of black fungi as degraders of volatile aromatic hydrocarbons. *Mycopathologia*, **175**(5–6), 369–379.

Jeanes A. (1976). Dextrans and pullulan - industrially significant alpha-D-glucans. *Abstracts of Papers of the American Chemical Society*, **172**(Sep3), 43–43.

Johnson J., Kirkwood S., Misaki A., Nelson T. E., Scaletti J. V. and Smith F. (1963). Structure of a new glucan. *Chemistry & Industry*, (20), 820–822.

Jong S. C. and Donovick R. (1989). Antitumor and antiviral substances from fungi. *Advanced Applied Microbiology*, **34**, 183–262.

Kagimura F. Y., da Cunha M. A., Barbosa A. M., Dekker R. F. and Malfatti C. R. (2015). Biological activities of derivatized D-glucans: a review. *International Journal of Biological Macromolecules*, **72**, 588–598.

Kang S. W., Kim Y. B., Shin J. D. and Kim E. K. (2010). Enhanced biodegradation of hydrocarbons in soil by microbial biosurfactant, sophorolipid. *Appl Biochem Biotechnol*, **160**(3), 780–790.

Kapteyn J. C., Van Egmond P., Fau - Sievi E., Van Den Ende H., Makarow M. and Klis F. M. (1999). - The contribution of the O-glycosylated protein Pir2p/Hsp150 to the construction of the yeast cell wall in wild-type cells and beta 1,6-glucan-deficient mutants. *Molecular Microbiology*, **31**(6), 1835–1844.

Kashiwagi Y., Norisuye T. and Fujita H. (1981). Triple helix of *Schizophyllum commune* polysaccharide in dilute-solution .4. Light-scattering and viscosity in dilute aqueous sodium-hydroxide. *Macromolecules*, **14**(5), 1220–1225.

Keller L. and Surette M. G. (2006). Communication in bacteria: an ecological and evolutionary perspective. *Nature Reviews Microbiology*, **4**(4), 249–258.

Kershaw M. J. and Talbot N. J. (1998). Hydrophobins and repellents: proteins with fundamental roles in fungal morphogenesis. *Fungal Genetics and Biology*, **23**.

Kikumoto S., Miyajima, T., Yoshizumi S., Fujimoto S. and Kimura, K. (1970). Polysaccharide produced by *Schizophyllum commune*. 1. Formation and some properties of an extracellular polysaccharide. *Journal of the Agricultural Chemical Society of Japan*, **44**, 337–342.

Kikumoto S., Miyajima T., Kimura K., Okubo S. and Komatsu N. (1971). Polysaccharide produced by *Schizophyllum commune* .2. Chemical structure of an extracellular polysaccharide. *Journal of the Agricultural Chemical Society of Japan*, **45**(4), 162–168.

Kitamoto D., Yanagishita H., Shinbo T., Nakane T., Kamisawa C. and Nakahara T. (1993). Surface-active properties and antimicrobial activities of mannosylerythritol lipids as biosurfactants produced by *Candida antarctica*. *Journal of Biotechnology*, **29**(1–2), 91–96.

Klis F. M., de Koster C. G. and Brul S. (2014). Cell wall-related bionumbers and bioestimates of *Saccharomyces cerevisiae* and *Candida albicans*. *Eukaryot Cell*, **13**(1), 2–9.

Kojima T., Tabata K., Itoh W. and Yanaki T. (1986). Molecular-weight dependence of the antitumor-activity of schizophyllan. *Agricultural and Biological Chemistry*, **50**(1), 231–232.

Kozarski M. S., Klaus A. S., Niksic M. P., Van Griensven L. J. L. D., Vrvic M. M. and Jakovljevic D. M. (2014). Polysaccharides of higher fungi: biological role, structure and antioxidative activity. *Hemijska Industrija*, **68**(3), 305–320.

Kralj Kuncic M., Zajc J., Drobne D., Pipan Tkalec Z. and Gunde-Cimerman N. (2013). Morphological responses to high sugar concentrations differ from adaptation to high salt concentrations in the xerophilic fungi *Wallemia* spp. *Fungal Biology*, **117**(7–8), 466–478.

Krcmar P., Novotny C., Marais M. F. and Joseleau J. P. (1999). Structure of extracellular polysaccharide produced by lignin-degrading fungus *Phlebia radiata* in liquid culture. *International Journal of Biological Macromolecules*, **24**(1), 61–64.

Kulakovskaya E. K. and Kulakovskaya T. (2014). *Extracellular Glycolipids of Yeasts; Biodiversity, Biochemistry, and Prospects*. Academic Press.

Kulakovskaya T. V., Shashkov A. S., Kulakovskaya E. V. and Golubev W. I. (2005). Ustilagic acid secretion by *Pseudozyma fusiformata* strains. *Fems Yeast Research*, **5**(10), 919–923.

Kulakovskaya T., Shashkov A., Kulakovska E., Golubev W., Zinin A., Tsvetkov Y., Grachev A. and Nifantiev N. (2009). Extracellular cellobiose lipid from yeast and their analogues: structures and fungicidal activities. *Journal of Oleo Science*, **58**(3), 133–140.

Kumar M. N. V. R. (2000). A review of chitin and chitosan applications. *Reactive & Functional Polymers*, **46**(1), 1–27.

Kumar P., Yang M., Haynes B. C., Skowyra M. L. and Doering T. L. (2011). Emerging themes in cryptococcal capsule synthesis. *Current Opinion in Structural Biology*, **21**(5), 597–602.

Kumirska J., Czerwicka M., Kaczynski Z., Bychowska A., Brzozowski K., Thoming J. and Stepnowski P. (2010). Application of spectroscopic methods for structural analysis of chitin and chitosan. *Mar Drugs*, **8**(5), 1567–1636.

Kuncheva M., Panchev I., Pavlova K., Russinova-Videva S., Georgieva K. and Dimitrova S. (2013). Functional characteristics of an exopolysaccharide from antarctic yeast strain *Cryptococcus laurentii* Al(62). *Biotechnology & Biotechnological Equipment*, **27**(5), 4098–4102.

Latge J. P. (2007). The cell wall: a carbohydrate armour for the fungal cell. *Molecular Microbiology*, **66**(2), 279–290.

Latge J. P. and Beauvais A. (2014). - Functional duality of the cell wall. *Curr Opin Microbiol*, **20**, 111–117.

Leathers T. D. (1993). Substrate regulation and specificity of amylases from *Aureobasidium* strain Nrrl Y-12,974. *Fems Microbiology Letters*, **110**(2), 217–222.

Leathers T. D. (2002). Pullulan. In: *Biopolymers, Polysaccharides II: Polysaccharides from Eukariotes*, E. J. Vandamme, S. De Baets, A. Steinbüchel (eds), Vol. **6**, Wiley-VHC, Weinheim, Germany, pp. 1–35.

Leathers T. D. (2003). Biotechnological production and applications of pullulan. *Applied Microbiology and Biotechnology*, **62**(5–6), 468–473.

Lecacheux D., Mustiere Y., Panaras R. and Brigand G. (1986). Molecular-weight of scleroglucan and other extracellular microbial polysaccharides by size-exclusion chromatography and low-angle laser-light scattering. *Carbohydrate Polymers*, **6**(6), 477–492.

Li Z. and Nair S. K. (2012). Quorum sensing: how bacteria can coordinate activity and synchronize their response to external signals? *Protein Science*, **21**(10), 1403–1417.

Li W. W. and Yu H. Q. (2014). Insight into the roles of microbial extracellular polymer substances in metal biosorption. *Bioresource Technology*, **160**, 15–23.

Li L. X., Wang Z. and Doering T. L. (2014). Nucleotide sugar transporters: key players in capsule glycan biosynthesis. *Mycoses*, **57**, 78–78.

Lin X. R. (2009). *Cryptococcus neoformans*: morphogenesis, infection, and evolution. *Infection Genetics and Evolution*, **9**(4), 401–416.

Lynch A. S. and Robertson G. T. (2008). Bacterial and fungal biofilm infections. *Annual Reviews of Medicine* **59**, 415–428.

Mahapatra S. and Banerjee D. (2012). Structural elucidation and bioactivity of a novel exopolysaccharide from endophytic *Fusarium solani* SD5. *Carbohydrate Polymers*, **90**(1), 683–689.

Mahapatra S. and Banerjee D. (2013). Fungal exopolysaccharide: production, composition and applications. *Microbiology Insights*, **6**, 1–16.

Manzi P. and Pizzoferrato L. (2000). Beta-glucans in edible mushrooms. *Food Chemistry*, **68**(3), 315–318.

Martinez L. R. and Casadevall A. (2006). *Cryptococcus neoformans* cells in biofilms are less susceptible than planktonic cells to antimicrobial molecules produced by the innate immune system. *Infection and Immunity*, **74**(11), 6118–6123.

Matos T., de Hoog G. S., de Boer A. G., de Crom I. and Haase G. (2002). High prevalence of the neurotrope *Exophiala dermatitidis* and related oligotrophic black yeasts in sauna facilities. *Mycoses*, **45**(9–10), 373–377.

Mazur P., Morin N., Baginsky W., el-Sherbeini M., Clemas J. A., Nielsen J. B. and Foor F. (1995). Differential expression and function of two homologous subunits of yeast 1,3-beta-D-glucan synthase. *Mol Cell Biol*, **15**(10), 5671–5681.

McCormack E., Skavland J., Mujic M., Bruserud O. and Gjertsen B. T. (2010). Lentinan: hematopoietic, immunological, and efficacy studies in a syngeneic model of acute myeloid leukemia. *Nutrition and Cancer-an International Journal*, **62**(5), 574–583.

McFadden D., Zaragoza O. and Casadevall A. (2006). The capsular dynamics of *Cryptococcus neoformans. Trends in Microbiology*, **14**(11), 497–505.

Miranda C. C. B. O., Dekker R. F. H., Serpeloni J. A., Fonseca E. A. L., Couls L. M. S. and Barbosa A. M. (2008). Anticlastogenic activity exhibited by botryosphaeran, a new exopolysaccharide produced by *Botryosphaeria rhodina* MAMB-05. *International Journal of Biological Macromolecules*, **42**(2), 172–177.

Misaki A., Kawaguchi K., Miyaji H., Nagae H., Hokkoku S., Kakuta M. and Sasaki T. (1984). Structure of Pestalotan, a highly branched (1-]3)-Beta-D-glucan elaborated by *Pestalotia* Sp-815, and the enhancement of Its antitumor-activity by polyol modification of the side-chains. *Carbohydrate Research*, **129**(Jul), 209–227.

Misaki A., Tsumuraya Y. and Kakuta M. (2000). Elsinan, a potential food hydrocolloid produced by elsinoe species; properties and enzymatic degradation. *Hydrocolloids*, **Pt 1**, 205–214.

Montijn R. C., Van Wolven P., De Hoog S. and Klis F. M. (1997). beta-Glucosylated proteins in the cell wall of the black yeast *Exophiala (Wangiella) dermatitidis. Microbiology*, **143**(Pt 5), 1673–1680.

Mosel D. D., Dumitru R., Hornby J. M., Atkin A. L. and Nickerson K. W. (2005). Farnesol concentrations required to block germ tube formation in *Candida albicans* in the presence and absence of serum. *Applied and Environmental Microbiology*, **71**(8), 4938–4940.

Mukhopadhyay S. K., Chatterjee S., Gauri S. S., Das S. S., Mishra A., Patra M., Ghosh A. K., Das A. K., Singh S. M. and Dey S. (2014). Isolation and characterization of extracellular polysaccharide Thelebolan produced by a newly isolated psychrophilic Antarctic fungus *Thelebolus*. *Carbohydrate Polymers*, **104**, 204–212.

Munro C. and Richard M. L. (2012). The cell wall: glycoproteins, remodeling, and regulation. In: *Candida and Candidiasis*, Second Edition (ed.), Vol. American Society of Microbiology, pp. 197–224.

Nai C., Wong H. Y., Pannenbecker A., Broughton W. J., Benoit I. de Vries R. P., Gueidan C. and Gorbushina A. A. (2013). Nutritional physiology of a rock-inhabiting, model microcolonial fungus from an ancestral lineage of the Chaetothyriales (Ascomycetes). *Fungal Genet Biol*, **56**, 54–66.

Nett J., Lincoln L., Marchillo K., *et al.* (2007). Putative role of beta-1,3 glucans in *Candida albicans* biofilm resistance. *Antimicrob Agents Chemother*, **51**(2), 510–520.

Nikitina V. E., Tsivileva O. M., Pankratov A. N. and Bychkov N. A. (2007). *Lentinula edodes* biotechnology - From lentinan to lectins. *Food Technology and Biotechnology*, **45**(3), 230–237.

Nishitani Y., Zhang L., Yoshida M., Azuma T., Kanazawa K., Hashimoto T. and Mizuno M. (2013). Intestinal Anti-Inflammatory Activity of Lentinan: Influence on IL-8 and TNFR1 Expression in Intestinal Epithelial Cells. *Plos One*, **8**(4), Article Number: e62441.

Noack-Schonmann S., Bus T., Banasiak R., Knabe N., Broughton W. J., Den Dulk-Ras H., Hooykaas P. J. and Gorbushina, A. A. (2014). Genetic transformation of *Knufia petricola* A95 - a model organism for biofilm-material interactions. *AMB Express*, **4**, 80.

Nobile C. J. and Mitchell A. P. (2006). Genetics and genomics of *Candida albicans* biofilm formation. *Cellular Microbiology*, **8**(9), 1382–1391.

Norisuye T., Yanaki T. and Fujita H. (1980). Triple Helix of a *Schizophyllum commune* Polysaccharide in Aqueous-Solution. *Journal of Polymer Science Part B-Polymer Physics*, **18**(3), 547–558.

Nwe N., Furuike T. and Tamura H. (2011). Production, Properties and Applications of Fungal Cell Wall Polysaccharides: Chitosan and Glucan. *Advanced Polymer Science*, **244**, 187–207.

Onofri S., de la Torre R., de Vera J. P., Ott S., Zucconi L., Selbmann L., Scalzi G., Venkateswaran K. J., Rabbow E., Sanchez Inigo F. J. and Horneck G. (2012). Survival of rock-colonizing organisms after 1.5 years in outer space. *Astrobiology*, **12**(5), 508–516.

Orlean P. (2012). Architecture and biosynthesis of the *Saccharomyces cerevisiae* cell wall. *Genetics*, **192**(3), 775–818.

Palleschi A., Bocchinfuso G., Coviello T. and Alhaique F. (2005). Molecular dynamics investigations of the polysaccharide scleroglucan: first study on the triple helix structure. *Carbohydrate Research*, **340**(13), 2154–2162.

Pavlova K., Panchev I., Krachanova M. and Gocheva M. (2009). Production of an exopolysaccharide by Antarctic yeast. *Folia Microbiol (Praha)*, **54**(4), 343–348.

Pavlova K., Rusinova-Videva S., Kuncheva M., Kratchanova M., Gocheva M. and Dimitrova S. (2011). Synthesis and characterization of an exopolysaccharide by antarctic yeast strain *Cryptococcus laurentii* AL(1)(0)(0). *Applied Biochemistry and Biotechnology*, **163**(8), 1038–1052.

Perez-Nadales E., Almeida Nogueira M. F., Baldin C., Castanheira S., El Ghalid M., Grund E., Lengeler K., Marchegiani E., Mehrotra P. V., Moretti M., Naik V., Oses-Ruiz M., Oskarsson T., Schäfer K., Wasserstrom L., Brakhage A. A., Gow N. A. R., Kahmann R., Lebrun M.-H., Perez-Martin J., Di Pietro A., Talbot N. J., Toquin V., Walther A. and Wendland J. (2014). Fungal model systems and the elucidation of pathogenicity determinats. *Fungal Genetics and Biology*, **70**, 42–67.

Pochanavanich P. and Suntornsuk W. (2002). Fungal chitosan production and its characterization. *Letters in Applied Microbiology*, **35**(1), 17–21.

Poli A., Anzelmo G., Tommonaro G., Pavlova K., Casaburi A. and Nicolaus B. (2010). Production and chemical characterization of an exopolysaccharide synthesized by psychrophilic yeast strain *Sporobolomyces salmonicolor* AL(1) isolated from Livingston Island, Antarctica. *Folia Microbiol (Praha)*, **55**(6), 576–581.

Potts M. (1999). Mechanisms of desiccation tolerance in cyanobacteria. *European Journal of Phycology*, **34**(4), 319–328.

Prajapati V. D., Jani G. K. and Khanda S. M. (2013). Pullulan: an exopolysaccharide and its various applications. *Carbohydr Polym*, **95**(1), 540–549.

Prets H. E. H., McNamee, R., Jones, E., Browder, I. And Williams, D. (1991). Isolation, physicochemical characterisation and pre-clinical efficacy evaluation of a soluble scleroglucan. . *J Pharmacol Exp Ther*, **257**.

Pretus H. A., Ensley H. E., McNamee R. B., Jones E. L., Browder I. W. and Williams D. L. (1991). Isolation, physicochemical characterization and preclinical efficacy evaluation of soluble scleroglucan. *Journal of Pharmacology and Experimental Therapeutics*, **257**(1), 500–510.

Puchkov E. O., Zahringer U., Lindner B., Kulakovskaya T. V., Seydel U. and Wiese A. (2002). The mycocidal, membrane-active complex of *Cryptococcus humicola* is a new type of cellobiose lipid with detergent features. *Biochimica Et Biophysica Acta-Biomembranes*, **1558**(2), 161–170.

Qin G., Tian S., Chan Z. and Li B. (2007). Crucial role of antioxidant proteins and hydrolytic enzymes in pathogenicity of *Penicillium expansum*: analysis based on proteomics approach. *Molecular & Cellular Proteomics*, **6**(3), 425–438.

Ramage G., Saville S. P., Wickes B. L. and Lopez-Ribot J. L. (2002). Inhibition of *Candida albicans* biofilm formation by farnesol, a quorum-sensing molecule. *Applied and Environmental Microbiology*, **68**(11), 5459–5463.

Ramage G., Saville S. P., Thomas D. P. and Lopez-Ribot J. L. (2005). *Candida* biofilms: an update. *Eukaryotic Cell*, **4**(4), 633–638.

Ramos L. J. Davenport T. L., McMillan R. T. and Lara S. (1997). The resistance of mango (*Mangifera indica*) cultivars to tip dieback disease in Florida. *Pl. Dis*, **81**, 509–514.

Reeslev M., Jorgensen B. B. and Jorgensen O. B. (1996). Exopolysaccharide production and morphology of *Aureobasidium pullulans* grown in continuous cultivation with varying ammonium-glucose ratio in the growth medium. *Journal of Biotechnology*, **51**(2), 131–135.

Reilly M. C. and Doering T. L. (2009). Biosynthesis of fungal and yeast glycans. *Microbial Glycobiology: Structures, Relevance and Applications*, 393–412.

Ren L., Reynisson J., Perera C. and Hemar Y. (2013). The physicochemical properties of a new class of anticancer fungal polysaccharides: A comparative study. *Carbohydrate Polymers*, **97**(1), 177–187.

Renshaw J. C., Robson G. D., Trinci A. P. J., Wiebe M. G., Livens F. R., Collison D. and Taylor R. J. (2002). Fungal siderophores: structures, functions and applications. *Mycological Research*, **106**, 1123–1142.

Rizzo J., Oliveira D. L., Joffe L. S., Hu G. G., Gazos-Lopes F., Fonseca F. L., Almeida I. C., Frases S., Kronsta J. W. and Rodrigues M. L. (2014). Role of the Apt1 Protein in Polysaccharide Secretion by *Cryptococcus neoformans*. *Eukaryotic Cell*, **13**(6), 715–726.

Rodrigues M. L., Nakayasu E. S., Almeida I. C. and Nimrichter L. (2014). The impact of proteomics on the understanding of functions and biogenesis of fungal extracellular vesicles. *Journal of Proteomics*, **97**, 177–186.

Ruibal C., Platas G. and Bills G. F. (2008). High diversity and morphological convergence among melanised fungi from rock formations in the Central Mountain System of Spain. *Persoonia*, **21**, 93–110.

Ruibal C., Gueidan C., Selbmann L., Gorbushina A. A., Crous P. W., Groenewald J. Z., Muggia L., Grube M., Isola D., Schoch C., L. Staley J. T., Lutzoni F. and de Hoog, G. S. (2009). Phylogeny of rock-inhabiting fungi related to Dothideomycetes. *Stud Mycol*, **64**, 123–133S127.

Ruthes A. C., Smiderle F. R. and Iacomini M. (2015). D-Glucans from edible mushrooms: a review on the extraction, purification and chemical characterization approaches. *Carbohydrate Polymers*, **117**, 753–761.

Saito H., Ohki T. and Sasaki T. (1979). C-13-nuclear magnetic-resonance study of polysaccharide gels - molecular architecture in the gels consisting of fungal, branched (1 -] 3)-beta-D-glucans (Lentinan and Schizophyllan) as manifested by conformational-changes induced by sodium-hydroxide. *Carbohydrate Research*, **74**(Sep), 227–240.

Sajna K. V., Sukumaran R. K., Gottumukkala L. D., Jayamurthy H., Dhar K. S. and Pandey A. (2013). Studies on structural and physical characteristics of a novel exopolysaccharide from *Pseudozyma* sp NII 08165. *International Journal of Biological Macromolecules*, **59**, 84–89.

Sakurai K. and Shinkai S. (2000). Molecular recognition of adenine, cytosine, and uracil in a single-stranded RNA by a natural polysaccharide: Schizophyllan. *Journal of the American Chemical Society*, **122**(18), 4520–4521.

Sakurai K., Uezu K., Numata M., Hasegawa T., Li C., Kaneko K. and Shinkai S. (2005). beta-1,3-glucan polysaccharides as novel one-dimensional hosts for DNA/RNA, conjugated polymers and nanoparticles. *Chemical Communications*, (35), 4383–4398.

Santos D. K., Rufino R. D., Luna J. M., Santos V. A. and Sarubbo L. A. (2016). Biosurfactants: multifunctional biomolecules of the 21st century. *Int J Mol Sci*, **17**(3).

Sasaki T. and Takasuka N. (1976). Further study of the structure of lentinan, an anti-tumor polysaccharide from *Lentinula edodes*. *Carbohydrate Research*, **47**(1), 99–104.

Sassaki G. L., Ferreira J. C., Glienke-Blanco C., Torri G., De Toni F., Gorin P. A. J. and Iacomini M. (2002). Pustulan and branched beta-galactofuranan from the phytopathogenic fungus *Guignardia citricarpa*, excreted from media containing glucose and sucrose. *Carbohydrate Polymers*, **48**(4), 385–389.

Sassaki G. L., Czelusniak P. A., Vicente V. A., Zanata S. M., Souza L. M., Gorin P. A. J. and Iacomini M. (2011). Some biomolecules and a partially O-acetylated exo-galactomannan containing beta-Galf units from pathogenic *Exophiala jeanselmei*, having a pronounced immunogenic response. *International Journal of Biological Macromolecules*, **48**(1), 177–182.

Scalzi G., Selbmann L., Zucconi L., Rabbow E., Horneck G., Albertano P. and Onofri S. (2012). LIFE experiment: isolation of cryptoendolithic organisms from Antarctic colonized sandstone exposed to space and simulated Mars conditions on the international space station. *Origins of Life and Evolution of the Biosphere: the Journal of the International Society for the Study of the Origin of Life*, **42**(2–3), 253–262.

Schmid J., Meyer V. and Sieber V. (2011). Scleroglucan: biosynthesis, production and application of a versatile hydrocolloid. *Applied Microbiology and Biotechnology*, **91**(4), 937–947.

Selbmann L., Onofri S., Fenice M., Federici F. and Petruccioli M. (2002). Production and structural characterization of the exopolysaccharide of the Antarctic fungus *Phoma herbarum* CCFEE 5080. *Research in Microbiology*, **153**(9), 585–592.

Selbmann L., Stingele F. and Petruccioli M. (2003). Exopolysaccharide production by filamentous fungi: the example of *Botryosphaeria rhodina*. *Antonie Van Leeuwenhoek*, **84**(2), 135–145.

Selbmann L., Zucconi L., Isola D. and Onofri S. (2014). Rock black fungi: excellence in the extremes, from the Antarctic to space. *Current Genetics*.

Singh P. P., Whistler R. L., Tokuzen R. and Nakahara W. (1974). Scleroglucan, an antitumor polysaccharide from sclerotium-glucanicum. *Carbohydrate Research*, **37**(1), 245–247.

Singh R. S., Saini G. K. and Kennedy J. F. (2008). Pullulan: microbial sources, production and applications. *Carbohydrate Polymers*, **73**(4), 515–531.

Singh R., Shivaprakash M. R. and Chakrabarti A. (2011). Biofilm formation by zygomycetes: quantification, structure and matrix composition. *Microbiology-Sgm*, **157**, 2611–2618.

Smirnou D., Hrubosova D., Kulhanek J., Svik K., Bobkova L., Moravcova V., Krcmar M., Franke L. and Velebny V. (2014). *Cryptococcus laurentii* extracellular biopolymer production for application in wound management. *Applied Biochemistry and Biotechnology*, **174**(4), 1344–1353.

Sowa W., Blackwood, A. C., Adams, G. A. (1963). Neutral extracellular glucan of *Pullularia pullulans* (de Bary) Berkhout. *Canadian Journal of Chemistry*, **41**, 2314–2319.

Steiner W., Haltrich, D. and Lafferty, R. M. (1993). Production, properties, and applications of fungal polysaccharides. In: *Biosurfactants: Production: Properties: Applications*, N. Kosaric (ed.), Vol. Marcel Dekker, Inc., New York, USA, pp. 175–204.

Sun M., Zhao W., Xie Q., Zhan Y. and Wu B. (2015). Lentinan reduces tumor progression by enhancing gemcitabine chemotherapy in urothelial bladder cancer. *Surgical Oncology-Oxford.*, **24**(1), 28–34.

Survase S. A., Saudagar P. S. and Singhal R. S. (2007a). Enhanced production of scleroglucan by *Sclerotium rolfsii* MTCC 2156 by use of metabolic precursors. *Bioresource Technology*, **98**(2), 410–415.

Survase S. A., Saudagar P. S., Bajaj I. B. and Singhal R. S. (2007b). Scleroglucan: Fermentative production, downstream processing and applications. *Food Technology and Biotechnology*, **45**(2), 107–118.

Synytsya A. and Novak M. (2014). Structural analysis of glucans. *Annals of Translational Medicine*, **2**(2), 17.

Tabata K., Ito W., Kojima T., Kawabata S. and Misaki A. (1981). Ultrasonic degradation of Schizophyllan, an anti-tumor polysaccharide produced by *Schizophyllum commune* fries. *Carbohydrate Research*, **89**(1), 121–135.

Taguchi R., Kikuchi Y., Sakano Y. and Kobayash T. (1973). Polysaccharide production by *Pullularia pullulans* .1. Structural uniformity of pullulan produced by several strains of *Pullularia pullulans*. *Agricultural and Biological Chemistry*, **37**(7), 1583–1588.

Teixeira E. A., Penha L. L., Mendonca-Previato L. and Previato J. O. (2014). Mannoprotein MP84 mediates the adhesion of *Cryptococcus neoformans* to epithelial lung cells. *Front Cell Infect Microbiol*, **4**, 106.

Teunissen A. W. and Steensma H. Y. (1995). Review: the dominant flocculation genes of *Saccharomyces cerevisiae* constitute a new subtelomeric gene family. *Yeast*, **11**(11), 1001–1013.

Tourney J. and Ngwenya B. T. (2014). The role of bacterial extracellular polymeric substances in geomicrobiology. *Chemical Geology*, **386**, 115–132.

Tsumuraya Y., Misaki A., Takaya S. and Torii M. (1978). New fungal alpha-D-Glucan, Elsinan, elaborated by *Elsinoe leucospila*. *Carbohydrate Research*, **66**(Oct), 53–65.

Turian, G. (1977). *Coniosporium aeroalgicolum* sp. nov. - a dematiaceous fungus living in balanced parasitism with aerial algae. *Bulletin de la Societe Botanique Suisse*, **87**(1–2), 19–24.

Ueno Y., Hirashima N., Inoh Y., Furuno T. and Nakanishi M. (2007). Characterization of biosurfactant-containing liposomes and their efficiency for gene transfection. *Biological & Pharmaceutical Bulletin*, **30**(1), 169–172.

Van Bogaert I. N. A. and Soetaert W. (2011). Sophorolipids. *Biosurfactants: From Genes to Applications*, **20**, 179–210.

van Tilburg A. U. and Thomas M. D. (1993). Production of extracellular proteins by the biocontrol fungus *Gliocladium virens*. *Applied Microbiology and Biotechnology*, **59**(1), 236–242.

Verstrepen K. J. and Klis F. M. (2006). Flocculation, adhesion and biofilm formation in yeasts. *Molecular Microbiology*, **60**(1), 5–15.

Vlaev S., Rusinova-Videva S., Pavlova K., Kuncheva M., Panchev I. and Dobreva S. (2013). Submerged culture process for biomass and exopolysaccharide production by Antarctic yeast: some engineering considerations. *Applied Microbiology and Biotechnology*, **97**(12), 5303–5313.

Wang Y. C. and McNeil B. (1996). Scleroglucan. *Critical Reviews in Biotechnology*, **16**(3), 185–215.

Wang X., Shi F. X., Wosten H. A. B., Hektor H., Poolman B. and Robillard G. T. (2005). The SC3 hydrophobin self-assembles into a membrane with distinct mass transfer properties. *Biophysical Journal*, **88**(5), 3434–3443.

Wang J. L., Bi Z., Zou J. W. and Gu X. M. (2012). Combination therapy with lentinan improves outcomes in patients with esophageal carcinoma. *Molecular Medicine Reports*, **5**(3), 745–748.

Warscheid T. and Braams J. (2000). Biodeterioration of stone: a review. *International Biodeterioration & Biodegradation*, **46**(4), 343–368.

Weng B. B., Lin Y. C., Hu C. W., Kao M. Y., Wang S. H., Lo D. Y., Lai T. Y., Kan L. S. and Chiou R. Y. (2011). Toxicological and immunomodulatory assessments of botryosphaeran (beta-glucan) produced by *Botryosphaeria rhodina* RCYU 30101. *Food and Chemical Toxicology*, **49**(4), 910–916.

Westwater C., Balish E. and Schofield D. A. (2005). *Candida albicans*-conditioned medium protects yeast cells from oxidative stress: a possible link between quorum sensing and oxidative stress resistance. *Eukaryotic Cell*, **4**(10), 1654–1661.

Winka K., Eriksson O. E., Bång Å. (1998). Molecular evidence for recognizing the Chaetothyriales. *Mycologia*, **90**, 822–830.

Wosten H. A. (2001). Hydrophobins: multipurpose proteins. *Annual Review of Microbiology*, **55**, 625–646.

Wosten H. A. and Scholtmeijer K. (2015). Applications of hydrophobins: current state and perspectives. *Applied Microbiology and Biotechnology*, **99**(4), 1587–1597.

Wosten H. A. and de Vocht M. L. (2000). Hydrophobins, the fungal coat unravelled. *Biochimica Et Biophysica Acta*, **1469**(2), 79–86.

Yanaki T. and Norisuye T. (1983). Triple helix and random coil of scleroglucan in dilute-solution. *Polymer Journal*, **15**(5), 389–396.

Yanaki T., Norisuye T. and Fujita H. (1980). Triple helix of *Schizophyllum commune* polysaccharide in dilute-solution .3. Hydrodynamic properties in water. *Macromolecules*, **13**(6), 1462–1466.

Yanaki T., Kojima T. and Norisuye T. (1981). Triple helix of scleroglucan in dilute aqueous sodium-hydroxide. *Polymer Journal*, **13**(12), 1135–1143.

Yurlova N. A. and de Hoog G. S. (1997). A new variety of *Aureobasidium pullulans* characterized by exopolysaccharide structure, nutritional physiology and molecular features. *Antonie Van Leeuwenhoek*, **72**(2), 141–147.

Yurlova N. A. and de Hoog G. S. (2002). Exopolysaccharides and capsules in human pathogenic *Exophiala* species. *Mycoses*, **45**(11–12), 443–448.

Zalar P., Novak M., de Hoog G. S. and Gunde-Cimerman N. (2011). Dishwashers–a man-made ecological niche accommodating human opportunistic fungal pathogens. *Fungal Biology*, **115**(10), 997–1007.

Zaragoza O., Rodrigues M. L., De Jesus M., Frases S., Dadachova E. and Casadevall A. (2009). The capsule of the fungal pathogen *Cryptococcus neoformans*. *Advances in Applied Microbiology*, **68**, 133–216.

Zarnowski R., Westler W. M., Lacmbouh G. A., Marita J. M., Bothe J. R., Bernhardt J., Lounes-Hadj Sahraoui A., Fontaine J., Sanchez H., Hatfield R. D., Ntambi J. M., Nett J. E., Mitchell A. P. and Andes D. R. (2014). Novel entries in a fungal biofilm matrix encyclopedia. *MBio*, **5**(4), e01333–01314.

Zhang L., Li X., Xu X. and Zeng F. (2005). Correlation between antitumor activity, molecular weight, and conformation of lentinan. *Carbohydrate Research*, **340**(8), 1515–1521.

Zhang Y. Y., Xu X. J. and Zhang L. (2008). Gel formation and low-temperature intramolecular conformation transition of a triple-helical polysaccharide lentinan in water. *Biopolymers*, **89**(10), 852–861.

Zhang Y. Y., Li S., Wang X. H., Zhang L. N. and Cheung P. C. K. (2011). Advances in lentinan: isolation, structure, chain conformation and bioactivities. *Food Hydrocolloids*, **25**(2), 196–206.

Zhang Y., Kong H., Fang Y., Nishinari K. and Phillips G. O. (2013). Schizophyllan: a review on its structure, properties, bioactivities and recent developments. *Bioactive Carbohydrates and Dietary Fibre*, **1**(1), 53–71.

Chapter 15

Biofilms X-treme: composition of extracellular polymeric substances in Archaea

Jens C. Benninghoff[1], *Jost Wingender*[2,3], *Hans-Curt Flemming*[2,3,4] *and Bettina Siebers*[1]

[1]*Molecular Enzyme Technology and Biochemistry (MEB), Biofilm Centre, Centre for Water and Environmental Research (CWE), University of Duisburg-Essen, Essen, Germany*

[2]*Aquatic Microbiology, Biofilm Centre, Centre for Water and Environmental Research (CWE), University of Duisburg-Essen, Essen, Germany*

[3]*IWW Water Centre, Mülheim, Moritzstrasse 26, 45456 Mülheim an der Ruhr, Germany*

[4]*Singapore Center for Life Science Engineering (SCELSE), Nanyang Technological University, 60 Nanyang Drive, SBS-01N-27, Singapore 637551*

ABSTRACT

Archaeal biofilms and their extracellular polymeric substances (EPS) have been investigated only marginally over the last decades compared to their bacterial counterparts. Recently, the focus on Archaea and archaeal biofilms is increasing, because they offer great potential for industrial applications. Although Archaea are globally ubiquitous, especially extremophilic members belonging to Crenarchaea (e.g., *Sulfolobales*) or Euryarchaea (e.g., Haloarchaea) are of increasing interest due to their adaptation to extreme habitats. More and more strains are genetically tractable, enabling basic molecular approaches to address biofilm and EPS formation. So far, archaeal biofilms and EPS are mostly investigated using methods, which are already established for decades in the bacterial field. However, in contrast to Bacteria, studies on archaeal biofilms are predominantly performed with static incubation systems and not with flow-through systems, and as incubation systems, often microtiter plates and Petri dishes are used. In general, modified plastic surfaces and glass are used as substrata for attachment and biofilm formation. The characterization of archaeal biofilms is typically performed by microscopic methods like light and epifluorescence microscopy, confocal laser scanning microscopy and scanning electron microscopy. So far EPS isolation, quantification and analysis has been limited to *Sulfolobus* species. Research on archaeal biofilms indicated that the architecture and composition of biofilms formed by Archaea and Bacteria are comparable. As was found for Bacteria, archaeal biofilms have a complex architecture and EPS matrix, which consists of diverse carbohydrates and proteins as main components.

Keywords: Archaea, biofilm, extracellular polymeric substances, isolation, composition, polysaccharide, lectin, *Sulfolobus*

15.1 ARCHAEA

Archaea have been established as the third domain of life besides Bacteria and Eukarya by Carl Woese and co-workers based on their phylogenetic comparison of small subunit ribosomal RNA (rRNA) sequences culminating in the three domain tree of life (Woese *et al.* 1990). Most culturable species of the Archaea are either adapted to extreme environments or possess unique metabolic properties (i.e., methanogenesis). Some archaeal species occur in environments with low pH and high temperature (*Thermoacidophiles*), high salt concentration (*Halophiles*) or under anaerobic conditions (methanogens) (DeLong, 1998). Methanogens occur in anaerobic marine and freshwater environments as well as in the gastrointestinal tracts of animals (Cavicchioli, 2011). In hypersaline environments such as salterns or lakes containing high salt concentrations, halophiles (haloarchaea) can be found, e.g., *Haloferax volcanii*, which was first isolated from sediments of the Dead Sea (Mullakhanbhai & Larsen, 1975). *Thermoacidophiles* thrive in volcanic springs or hydrothermal vents in the deep-sea (Cavicchioli, 2011). One of the best studied *Thermoacidophiles* are members of the *Sulfolobales*, isolated from solfatara fields (Grogan, 1989). Besides these extreme habitats, which represent only a small bandwidth of archaeal habitats, many Archaea thrive in "normal" environments and are everywhere. However, most of this universally distributed Archaea are difficult to be cultured. Archaea are divided into different phyla such as *Crenarchaeota, Euryarchaeota, Thaumarchaeota, Korarchaeota, Nanoarchaeota*, *Aigarchaeota* and *Geoarchaeota* (Auchtung *et al.* 2006; Cavicchioli, 2011; Hedlund *et al.* 2015; Kozubal *et al.* 2013; Pester *et al.* 2011). The major archaeal phyla studied under laboratory conditions are the *Euryarchaeota* and *Crenarchaeota. Euryarchaeota* comprise halophilic, thermoacidophilic, and hyperthermophilic, psychrophilic and methanogenic, species, whereas only (hyper)thermophilic and thermoacidophilic species belong to the *Crenarchaeota*. Examples for well characterized organisms belong to the genus *Sulfolobus*. Although Archaea were named initially as Archaeabacteria, they are quite distinct from Bacteria as well as Eukarya and possess archaea-specific features (Woese *et al.* 1990). Archaea are single-celled and look in size and shape similar to Bacteria. Also their DNA structure is similar and they possess one circular chromosomes (double-stranded DNA), plasmids and operon structure are commonly found. Interestingly, the information processing (i.e., transformation from DNA to protein) is eukaryal-like but processes like replication, transcription and translation are often, less complex. In regard to metabolic complexity Archaea resemble Bacteria and lower Eukarya. However, biochemical studies revealed that they possess novel pathways and even classical pathways like glycolysis are modified by harbouring novel enzymes that show no similarity to their bacterial and eukaryal counterparts (Bräsen *et al.* 2014). Absolutely unique for Archaea is the absence of murein, one of the key feature of bacterial cell walls, and their membrane. The archaeal membrane lipids are composed of isoprenoid chains ether-linked to *sn*-glycerol 1-phosphate head groups (i.e., dibiphytanyltetraethersor biphytanylethers) rather than fatty acids ester-linked to *sn*-glycerol 3-phosphate, as found in Bacteria and Eukarya (Albers & Meyer, 2011)

The archaeal cell envelope, important for biofilm formation, is in many Archaea formed by a two-dimensional protein crystal layer, called "surface layer" (S-layer) (Kates, 1996) (Albers & Meyer, 2011). The S-layer forms a barrier to the often harsh environment and can respond to environmental stress by changes in the post-translational modification of the glycoproteins (Kandiba & Eichler, 2014).

Archaea harbour different appendages like the archaellum (the archaeal flagellum, type IV pilus), cannulaea, hami, different pili and bindosome (Jarrell *et al.* 2013; Ng *et al.* 2008). Cell appendages play a major roll for colonization by attachment on surfaces, knowledge about their importance for biofilm formation is important. Genetic tools are available for some model organisms, detailed information about these cell appendages exists (Jarrell *et al.* 2013). Since cell appendages play a major roll for colonization by attachment on surfaces, knowledge about their importance for biofilm formation is important. Thus, knock-out experiments for *Sulfolobus acidocaldarius* and *Haloferax volcanii* confirmed the importance of cell appendages for archaeal biofilm formation (Henche *et al.* 2012; Tripepi *et al.* 2010). The involvement of type IV pili in biofilm formation was summarized by Pohlschröder and Esquivel (Pohlschröder & Esquivel, 2015).

15.2 ARCHAEAL BIOFILMS

Archaeal biofilms are ubiquitous, like bacterial ones, and in the majority of observed biofilms members of both domains of life occur together, but so far mainly monospecies biofilms were studied under laboratory conditions in the Archaea. As the review title "Archaeal Biofilms: The Great Unexplored" by Orell *et al.* (2013a) indicates, not much is known about archaeal biofilms in contrast to bacterial biofilms (Fröls, 2013). The few studied biofilm-forming species predominantly belong to the *Crenarchaeota* (e.g., *Sulfolobales*) or *Euryarchaeota* (e.g., *Haloarchaea*). Biofilms of both phyla were found in natural environments and have been studied under laboratory conditions to get deeper insights into the biofilm mode of life.

Mixed-population biofilms containing Archaea and Bacteria are found in different habitats. On the basis of FISH (fluorescence in situ hybridization), t-RFLP (terminal restriction fragment length polymorphism), clone libraries, PhyloChip or denaturing gradient gel electrophoresis (DGGE) analysis the bacterial and archaeal co-occurrence in environmental biofilms was demonstrated (Briggs *et al.* 2011; Ionescu *et al.* 2012; Perras *et al.* 2014; Probst *et al.* 2013; Vetriani *et al.* 1999; Ziegler *et al.* 2013). Communities of Bacteria und Archaea were found all over the world, for example attached to mineral surfaces at thermal vents located in the deep sea, on rocks in streams in the Austrian Alps, on the outflow channels of hot springs, on sediments and rocks around the springs of the Dead Sea and biofilms grown on glass slides in Antarctic marine water (Battin *et al.* 2016; Battin *et al.* 2001; Ionescu *et al.* 2012; Schrenk *et al.* 2003; Schubotz *et al.* 2013; Webster & Negri, 2006). Most of the published literature on environmental biofilms of Archaea and Bacteria focusses on phylogenetic analysis of the biofilm and does not analyse the amount, structure or composition of their EPS in detail. As the biofilms were often found in extreme habitats, sampling is quite difficult, however, the phylogenetic composition of mixed biofilm forms an important basis for future analysis.

Bacterial cells living as biofilm are embedded in a matrix of various biopolymers such as proteins, carbohydrates, extracellular DNA or lipids, defined as "extracellular polymeric substances" (EPS) and well described for Bacteria (Flemming & Wingender, 2010). Knowledge of archaeal EPS, however, is limited since research of archaeal biofilms only started some years ago with a limited number of model organisms like *Sulfolobus* spp. and some Haloarchaea like *Haloferax volcanii* belonging to the *Crenarchaeota* and *Euryarchaeota*, respectively (Fröls *et al.* 2012; Koerdt *et al.* 2010). It was found that most of the investigated Archaea of different phyla and habitats were able to form biofilms (Bang *et al.* 2014; Fröls *et al.* 2012; Koerdt *et al.* 2010). In general, many of the studied Archaea form three dimensional multilayer biofilms, often with a complex architecture, and the cells are embedded in EPS, which is composed of different carbohydrates, extracellular DNA, proteins and lipids (Orell *et al.* 2013a; Zhang *et al.* 2015).

Cultivation of Archaea, however, is often more difficult than of Bacteria with respect to growth conditions like incubation temperature or media. Therefore, the cultivation systems established for biofilm formation for Bacteria frequently do not match the cultivation requirements, e.g., for (hyper)thermophilic or thermoacdophilic species and can not be easily adopted. For mesophilic Archaea (e.g., halophiles, methanogens), often already established systems for Bacteria can be applied. Surprisingly, research on archaeal biofilms is not as intensively performed compared to bacterial biofilms, maybe because no pathogenic Archaeon has been reported so far (Cavicchioli *et al.* 2003). However, Archaea have been identified on the skin, the intestinal tract and in the oral cavity of humans, and a possible opportunistic and/or pathogenic function is under discussion (Horz & Conrads, 2010).

So far only little is known about the regulation of archaeal biofilm formation. By knock-out of six transcriptional regulator genes for proteins belonging to the Lrs14-like protein family, the effect of these proteins on biofilm formation of *Sulfolobus acidocaldarius* was analysed (Orell *et al.* 2013b). It was found that the gene Saci_0446, later named as *abfR1* (Archaeal Biofilm Regulator 1), modulates the biofilm formation of *S. acidocaldarius* as well as EPS production. By deletion of Saci_0446 gene, carbohydrate and protein amounts of isolated EPS were much higher and thus, biofilm formation was increased.

The *Sulfolobus solfataricus* mutant PBL2025, which lacks 50 genes compared to *S. solfataricus* P2, was compared to the P2 wild type strain in respect to surface attachment, growth and EPS formation (Koerdt *et al.* 2012; Zolghadr *et al.* 2010). Biofilms of PBL2025 contained significantly more carbohydrates and proteins compared to the wild-type strain. The cell amount and distribution within the biofilm was not found to be affected. Complementation studies indicated that the α-mannosidase (SSO3006) might be involved in the modulation of EPS composition (Koerdt *et al.* 2012).

The first described biofilm formation with two different archaeal species was performed by incubating *Pyrococcus furiosus* and *Methanopyrus kandleri* as co-culture (Schopf *et al.* 2008). It was speculated that EPS might contribute to interactions between both species.

An overview of laboratory investigations on archaeal biofilms and EPS analysis is given in Table 15.1.

15.3 EPS ANALYSIS

For EPS quantification and analysis, various destructive and non-destructive methods are available. Using non-destructive or qualitative methods like microscopic analysis, cell and EPS distribution can be visualized under environmental conditions. Advanced microscopic methods and lectin binding analysis allow the visualization of spatial distribution and arrangement of binding sugar residues (Neu, 2016). Procedures for EPS isolation and extraction are destructive, because the biofilm gets disrupted by chemical and/or physical methods to separate cells and EPS components for later quantification (Wingender *et al.* 2001). However, such methods provide only information about quantities of different EPS components but not about their distribution. For a comprehensive analysis of biofilms, both approaches offer advantages and complement each other.

15.3.1 Static incubation systems

Archaeal biofilms have been analysed predominantly by lectin binding or other staining-dependent microscopic methods. For the purpose of analysing growing biofilms, mostly static systems and only very few flow-through systems were employed (Baker-Austin *et al.* 2010; Di Meglio *et al.* 2014). Therefore, most information about archaeal biofilms are based on static incubation systems like 96-well microtiter plates, Petri dishes or dishes suitable for confocal laser scanning microscopy. Moreover, there are mainly experiments performed with single species cultures. In general, mostly glass (cover slips) and different kind of polymers (e.g., polystyrene) were used as a substratum for biofilm formation. Especially for thermo(acido)philic archaea, the cultivation system is crucial with respect to thermostability of the equipment. Compounds released from the material of the cultivation system can influence or even impair cell growth. Static incubation of planktonic cells in presence of different substrata can provide information about attachment and biofilm formation in response to different support materials, and the influence of

Table 15.1 Analysis of archaeal biofilms and EPS under laboratory conditions.

Organism **Crenarchaeota (C)** **Euryarchaeota (E)**	**Growth Conditions and Substratum**	**Methods for Biofilm and EPS Analysis**	**Major Results**	**Reference**
Archaeoglobus fulgidus (E)	Anaerobic growth inside glass vessels with biofilm formation on interior glass wall surface. Incubation under stress conditions (high temperature and pH, UV light, metals, antibiotics)	Photometric Congo red-binding assay 1D-SDS-PAGE Inductively coupled plasma analysis SEM	Non-physiological extreme conditions (elevated temperature > 90 °C, high pH, antimicrobials, metal ions) induce biofilm formation Heat-shocked biofilms produce enhanced levels of polysaccharides Metal ion accumulation in biofilms Cells inside biofilm show enhanced tolerance to stress	(Lapaglia & Hartzell, 1997)
Ferroplasma acidarmanus Fer1 (E)	Growth on pyrite in flow cell Growth on glass surface in continuous cultures	CLSM 2DE MALDI-TOF mass spectrometry	Multi-layered biofilm on glass as well on pyrite Ten up-regulated proteins of biofilms compared to planktonic cells	(Baker-Austin *et al.* 2010)
Methanosarcina mazei strain Gö1 (DSM3647) (E) *Methanosphaera stadtmanae (DSM, 3091)* (E) *Methanobrevibacter smithii (DSM861)* (E)	Static incubation with biofilm formation on mica plates (hydrophilic, smooth), and uncoated µ-dishes (hydrophobi, rough)	SEM CLSM Phase-contrast microscopy	Biofilm formation of different structure and thickness (bilayered, multilayered) on both mica and µ-dish surfaces; first report of biofilm formation of methanogenic Archaea SEM analysis indicated high EPS production of *M. stadtmanae* and *M. smithii* Staining with lectins ConA and IB4 indicated production of glycosylated proteins and/or exopolysaccharides containing glucose, mannose and galactose	(Bang *et al.* 2014)

(*Continued*)

Table 15.1 Analysis of archaeal biofilms and EPS under laboratory conditions (*Continued*).

Organism ***Crenarchaeota* (C)** ***Euryarchaeota* (E)**	**Growth Conditions and Substratum**	**Methods for Biofilm and EPS Analysis**	**Major Results**	**Reference**
Ferroplasma acidiphilum DSM 28986 (E) *Sulfolobus metallicus DSM 6482* (C) *Acidianus sp. DSM 289099* (C)	Static incubation and biofilm formation on pyrite grains or sulphur coupons	Lectin staining and CLSM Lectin library and DNA staining with CLSM Atomic force microscopy combined with epifluorescence microscopy	22 lectins bound to the archaeal biofilms on pyrite, and 21 lectins bound to *Acidianus* sp. biofilms on elemental sulphur Tightly bound and loosely bound EPS detected in biofilms on pyrite and sulphur All strains produced various EPS glycoconjugates on pyrite surfaces	(Zhang *et al.* 2014)
Sulfolobus metallicus DSM 6482 (C)	Static incubation, biofilm formation on elemental sulphur (S^0)	Lectin staining and CLSM Combined atomic force microscopy and epifluorescence microscopy EPS extraction and analysis Fourier transformed infrared spectroscopic assay	Heterogeneous distribution of cells, single cells as well as clusters up to hundred μm in diameter The EPS from biofilms contained proteins (major components), carbohydrates (binding of 21 from 72 lectins to biofilms) and eDNA	(Zhang *et al.* 2015)
Thermococcus litoralis (E)	Biofilms on polycarbonate filters and glass slides	SEM Biofilm visualization by Congo red staining for EPS and carbol fuchsin staining for cells EPS analysis by high-performance liquid chromotography (HPLC)	Biofilm formation on hydrophilic surfaces like polycarbonate filters and glass slides Light microscopy revealed polymeric material covering biofilm cells Thin EPS retain Congo red stain Mannose detected as only monomer of isolated EPS	(Rinker & Kelly, 1996)
20 different *Haloarchaea* including species of *Halobacterium*, *Haloferax* and *Halorubrum* (E)	Biofilms on plastic and glass surfaces	Differential interference contrast microscopy Lectin and DNA staining Fluorescence microscopy CLSM TEM	EPS of strongly adhesive haloarchaeal strains contained eDNA and glycoconjugates Cell appendages visible (TEM)	(Fröls *et al.* 2012)

Halobacterium sp. (E) *Halorubrum sp.* (E)	Cultivation of biofilms on glass surfaces in a flow cell consisting of polymethyl-methacrylate	SEM Phase contrast microscopy	Three dimensional structure of biofilm was microorganism dependent *Halobacterium* sp.: Clusters, covered about 30% of supporting surface Thickness 86 ± 4 μm *Halorubrum* sp.: Surface coverage less than 20% Thickness 41 ± 1 μm	(Di Meglio *et al.* 2014)
Haloferax volcanii (E)	Growth on polystyrene surface (microtiter plate and Petri dish), chamber slides, borosilicate glass coupons or coverglass	DNA and EPS staining, visualization by CLSM Time-lapse macroscopic photography	Biofilm development of *H. volcanii* from microcolonies (24 h), large clusters (48 h) to flake-like towers often greater than 100 μm (7 days) EPS and cells were co-localized and EPS contained polysaccharides, eDNA and amyloid protein	(Chimileski *et al.* 2014)
Pyrococcus furiosus (E) and *Methanopyrus kandleri* (E)	Biofilm formation on carbon-coated grids, glass cover slips and glassy carbon	CLSM TEM SEM	Formation of bi-species biofilm on all tested surfaces First attachment of *M. kandleri* to abiotic surfaces followed by attachment of *P. furiosus* on biotic surface	(Schopf *et al.* 2008)
Sulfolobus acidocaldarius MW001 (C)	Biofilm formation on glass slides in shaking cultures and static in μ-dishes Growth on carbon-coated nickel grids	Light microscopy of biofilm grown on glass slides and cell counting Lectin and DNA staining combined with CLSM of biofilm grown in μ-dish Transmission electron microscopy of cells grown on carbon-coated nickel grids	Deletion of different surface appendages effect colonization of surfaces EPS detected by lectins increases continusly during incubation Clouds of EPS on top of biofilm, change in lectin binding during incubation	(Henche *et al.* 2012)

(*Continued*)

Table 15.1 Analysis of archaeal biofilms and EPS under laboratory conditions (*Continued*).

Organism ***Crenarchaeota* (C)** ***Euryarchaeota* (E)**	**Growth Conditions and Substratum**	**Methods for Biofilm and EPS Analysis**	**Major Results**	**Reference**
Sulfolobus acidocaldarius MW001 (C)	Biofilm formation in 96-well microtiter plates, μ-dishes and Petri dishes	Quantification of biofilm in 96-well microtiter plates by crystal violet staining Lectin and DNA staining of biofilms in μ-dishes, CLSM and differential interference contrast microscopy EPS extraction and quantification of biofilms grown in Petri dishes	Deletion of *Saci_0446* leads to highly increased EPS production and robust biofilm structure Background strain MW001 produces mostly carbohydrates of low molecular weight (≤3.5 kDa), Δ*Saci0446* results in fivefold higher carbohydrate concentration	(Orell *et al.* 2013b)
Sulfolobus solfataricus (C)	Biofilm formation on glass slides, pyrite, mica and carbon-coated gold grids in shaking cultures	Lectin staining of biofilm grown on glass slides and fluorescence microscopy and differential interference contrast TEM of cells grown on carbon-coated gold grids	Attachment of *S. solfataricus* on glass, mica, pyrite and carbon-coated gold grids Flagella and pili were essential for initial attachment EPS contained mannose, galactose and N-acetylglucosamine residues	(Zolghadr *et al.* 2010)
Sulfolobus solfataricus (C)	Biofilm formation on glass slides in shaking cultures and static in μ-dishes Growth on solid medium	Lectin staining of biofilm grown on glass slides, fluorescence microscopy and differential interference contrast Lectin and DNA staining of biofilm grown in μ-dishes and CLSM EPS extraction of biofilm grown on solid medium	LacS expression in PBL2025 plays a role in modulation of EPS production by reduction of carbohydrate content PBL2025 showed enhanced secretion of mannose and glucose residues compared to P2 *S. solfataricus* P2 revealed low amount of carbohydrates and proteins Highest concentration of carbohydrates and proteins in natural mutant PBL2025	(Koerdt *et al.* 2012)

Sulfolobus acidocaldarius (C) *S. solfataricus* (C) *S. islandicus* (C)	Biofilm formation in 96-well microtiter plates, μ-dishes (IBIDI) and polylysin treated glass coverslips	Quantification of biofilm formation in 96-well microtiter plates by crystal violet Lectin and DNA staining of biofilms combined with CLSM SEM of biofilms grown on glass	The tested strains formed morphologically different biofilms with carpet-like to high density tower-like structures All strains produced extracellular polysaccharides containing glucose, galactose, mannose and N-acetylglucosamine	(Koerdt *et al.* 2010)
S. acidocaldarius DSM 639 (C)	Growth on solid medium	Comparison of different EPS isolation methods, 2DE, zymograpic analysis of EPS proteins, identifcation of EPS proteins by LC-MS/MS	Cation exchange resin was best suited for EPS isolation, majority of EPS proteins of cytoplasmic origin (2DE, MS/MS), functional analysis of EPS proteins demonstrated activity of diverse enzyme classes, including proteases, lipases, esterases, phosphatases and glucosidases.	(Jachlewski *et al.* 2015)
Natrialba aegyptiaca (E)	Growth on solidified maintenance medium	EPS extraction and analysis	Extracellular polymer, composed predominantly of glutamic acid (85% [w/w], representing poly(glutamic acid), carbohydrates (12.5% [w/w] and unidentified compounds (2.5% [w/w])	(Hezayen *et al.* 2001)

Notes: SEM, scanning electron microscopy; CLSM, confocal laser scanning microscopy; TEM, transmission electron microscopy.

different media and growth conditions can be easily tested (Fröls *et al.* 2012; Koerdt *et al.* 2010). For *Crenarchaeota* like *Sulfolobus*, biofilm growth can be quantified by using crystal violet staining of biofilms adapted from bacterial biofilm research (Orell *et al.* 2013b). Crystal violet method includes staining of cells and EPS, thus allows determination of biofilm biomass. Quantification of biofilms of haloarchaeal species was performed by acridine orange staining of biofilm cells (Fröls *et al.* 2012). However, crystal violet binds to EPS components, therefore, cell numbers can be overestimated. Nevertheless, this method is suitable to visualize cells and their distribution as well changes in EPS production in parallel (Esquivel *et al.* 2013).

15.3.2 Flow-through systems

In contrast to Bacteria, archaeal biofilm formation under flow-through conditions is less established. Substrata used for biofilm formation were either pyrite or glass embedded in flow-cells consisting of polymethylmethacrylate (PMMA) or a combination of polycarbonate and glass (Baker-Austin *et al.* 2010; Di Meglio *et al.* 2014). The flow-through systems were applied so far for extreme acidophilic or halophilic archaeal species. For mesophilic Archaea adapted bacterial systems can be used for biofilm analysis under constant medium flow. For thermoacidophilic Archaea, like the crenarchaeal *Sulfolobales*, so far no flow-through systems have been published, probably because of the high incubation temperature, which comes along, for example, with medium evaporation, air bubble formation and instability of materials.

The iron-oxidizing acidophilic mesophile *Ferroplasma acidarmanus* (*Thermoplasmatales*, *Euryarchaeota*) was one of the first Archaea incubated in a flow-through system (Baker-Austin *et al.* 2010). The system was adapted from an earlier study on iron-oxidizing Bacteria (Crundwell, 1996). In this system, a polished pyrite chip was placed in a flow cell consisting of a polycarbonate base and a glass cover slip. Cells were visualized by addition of fluorescein into the growth medium. Pyrite surface and biofilm formation was imaged by using a confocal laser scanning microscope.

The extreme halophilic Archaea *Halorubrum* sp. and *Halobacterium* sp. were incubated in a flow cell of a transparent PMMA plate sealed on top and bottom by silicone grease fixed glass coverslips (Di Meglio *et al.* 2014). The glass coverslips served as a substratum for the cells and phase contrast microscopy was used for *in vivo* observation of biofilm formation. The cells grown on the glass slides were fixed by glutaraldehyde and analysed by scanning electron microscopy.

15.3.3 Microscopic techniques

Mostly analysis of EPS composition and distribution of archaeal biofilms was performed by microscopic methods based on detection of fluorescent dyes to visualize cells, carbohydrates and eDNA by either confocal laser scanning microscopy (CLSM) or epifluorescence microscopy. Besides this, atomic force microscopy (AFM), transmission electron microscopy (TEM) and (scanning) electron microscopy have been employed (Anton *et al.* 1988; Koerdt *et al.* 2010; Lapaglia & Hartzell, 1997; Probst *et al.* 2014; Schopf *et al.* 2008; Zhang *et al.* 2014). As the magnification and image quality of electron microscopy is higher compared to CLSM, electron microscopy was predominantly used to visualize cell surface appendages and lesser to visualize the distribution of EPS. As electron microscopy only results in an image in grey scale, differentiation and identification of structures or molecules by dyes is not possible and based on speculation. Furthermore, as desiccation is a prerequisite for electron microscopy and the original three-dimensional structure of a biofilm can only be viewed in its collapsed form.

For CLSM, thin glass or synthetic polymer surfaces as well as natural substrates were mostly used for visualization of archaeal biofilms. Images generated by electron microscopy mostly visualized cells grown on carbon grids. AFM of samples containing high amount of EPS leads to problems, because the EPS may stick to the cantilever and disturbs microscopy. Therefore, scanning electron microscopy is more suitable for images of biofilms containing a higher amount of EPS. Images of *S. acidocaldarius* biofilms on glass slides visualized by atomic force and scanning electron microscopy are shown in Figure 15.1.

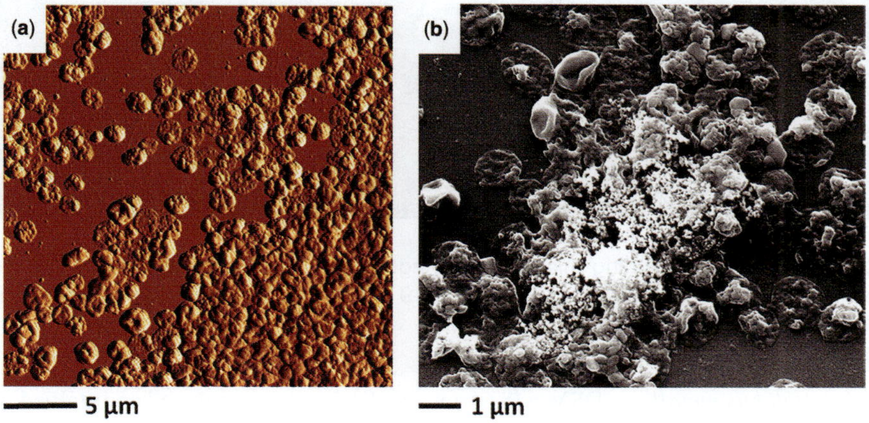

Figure 15.1 Biofilm formation of *S. acidocaldarius* on cover slides (glass) after incubation at 78°C for four days. (a) AFM of cells after fixation with glutardialdehyde. (b) Scanning electron microscopic image after fixation with glutardialdehyde and dehydration.

15.3.3.1 Fluorescence microscopy

For 2D- and 3D-images of archaeal biofilms stained with different fluorescent dyes, epifluorescence or confocal laser scanning microscopy were used. For visualization of cells and eDNA, a broad range of fluorescent dyes binding to chromosomal and extracellular DNA are used. Visualization of EPS components is performed by staining with lectins selective for different carbohydrate residues. Visualization of EPS components other than eDNA and carbohydrates was not performed with archaeal biofilms so far, except for *Sulfolobus metallicus* DSM 6482 by application of lipid (FM1–43 and Nile red) and protein (SyproRed and SyproOrange) binding dyes (Zhang *et al.* 2015).

Since lectins are sensitive proteins which denature at high ionic strength, high temperatures or low pH values, staining and visualization has to be performed adapted to medium and growth conditions necessary for cultivation of thermoacidophilic archaea. Furthermore, the stability and fluorescence spectrum of the applied fluorophores needs to be tested under the respective application conditions. For crenarchaeal and haloarchaeal species, detection of carbohydrates by CLSM, predominantly Concanavalin A (ConA), wheat germ agglutinin (WGA) and GS-II/IB4 (from *Griffonia simplicifolia*) were used (Fröls *et al.* 2012; Koerdt *et al.* 2010; Orell *et al.* 2013b). Biofilms of *S. acidocaldarius*, stained by lectins IB4 and ConA and SYTO 9 for visualization of cells, is shown in Figure 15.2.

Screening with 75 commercially available lectins was performed for acidophilic archaeal biofilms of *Ferroplasma acidiphilum*, *S. metallicus* and *Acidianus* sp. (Zhang *et al.* 2014). Biofilms of these three organisms grown on pyrite surfaces produce various EPS glycoconjugates bound by lectins. The major binding pattern of lectin-binding was detected in biofilms formed on pyrite as well as elemental sulfurand various EPS glycoconjugates were detected on pyrite surface produced by all three strains. For *Acidianus* sp. it was found, that biofilm structures of cells and EPS glycoconjugates are dependent of the substratum.

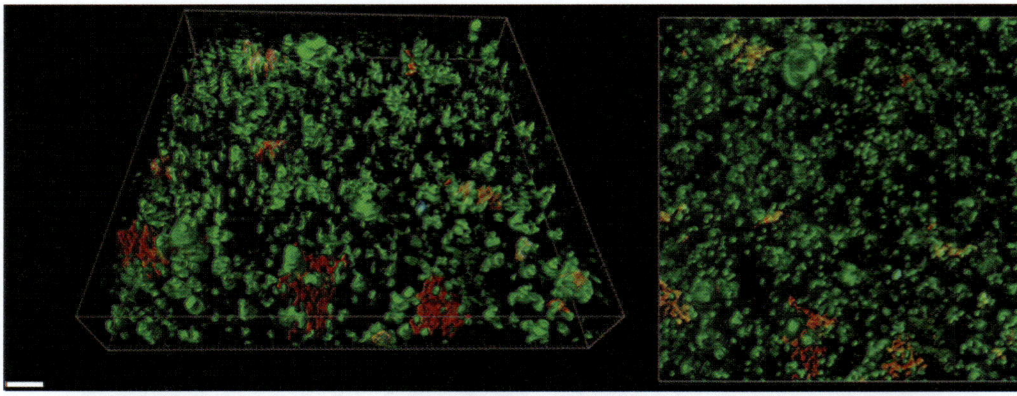

Figure 15.2 Confocal laser scanning microscope image of *S. acidocaldarius* biofilms grown in μ-dishes (IBIDI) for four days at 78°C. Staining of cells by SYTO 9 (green signal) and carbohydrates by lectins ConA (red signal) and IB4 (blue signal). Scale bar: 10 μm.

Biofilms of *Sulfolobus metallicus* DSM 6482 grown on S^0 were also analyzed and stained with different lectins and dyes binding to proteins and eDNA (Zhang *et al.* 2015). Lectin binding was carried out by fluorescence analysis (FLBA). For 21 out of 72 commercially available lectins binding to the EPS matrix was demonstrated (Zhang *et al.* 2015).

15.4 EPS ISOLATION AND COMPOSITION

Common methods for EPS isolation of bacterial biofilms are based on the application of chemical or physical methods or a combination of both. Chemical methods for EPS extraction are based on biofilm treatment with NaOH, EDTA, crown ether or formaldehyde (Aguilera *et al.* 2008; Brown & Lester, 1980; Tapia *et al.* 2009; Wuertz *et al.* 2001). Shaking, stirring, ultrasound, filtration and centrifugation are commonly used methods for physical extraction of EPS (Brown & Lester, 1980; Liu & Fang, 2002). In addition cation exchange resins are commonly used for EPS extraction (Jahn & Nielsen, 1995). EPS extraction methods for bacterial biofilms are established, but it was not known if these extraction methods are also applicable for archaeal biofilms. EPS extraction of archaeal biofilms often has some difficulties compared to EPS extraction of bacterial biofilms. Haloarchaea grow under high salt concentrations, ranging from 1 M up to saturation at 5.3 M (Pfeifer, 2015). Especially for chemical isolation methods, high salt concentrations may reduce the extraction efficiency, e.g., for EDTA and cation exchange resins. Pretreatment by dialysis is necessary to reduce the salt concentrations for later isolation. Archaea living in acidic environments such as *Sulfolobales* or *Thermoplasmatales* grow in environments at a pH value around 3. Since most of the chemical extraction methods require a neutral pH value, it has to be adjusted. As the EPS isolation using NaOH leads to a strong pH shift for the cells, cell lysis may occur and intracellular components will be released and contaminate the EPS fraction.

15.4.1 EPS isolation of archaeal biofilms

Until now, EPS analysis based on destructive isolation techniques has been performed for biofilms of only a few archaeal species. EPS extraction and analysis were conducted with *Sulfolobus* spp., belonging to the Crenarchaeota (Jachlewski *et al.* 2015; Koerdt *et al.* 2012; Orell *et al.* 2013b).

15.4.1.1 Crenarchaeota

For the thermoacidophilic *S. acidocaldarius* the EPS isolation from unsaturated biofilms, grown on solid nutrient media, was performed and the efficiency of five different EPS extraction methods commonly used for Bacteria was compared and adapted (Jachlewski *et al.* 2015). This is the first publication with detailed comparison of different EPS extraction methods for Archaea with respect to isolation efficiency, compatibility with quantification assays and cell viability. Experiments revealed that the use of the cation exchange resin method after Frølund *et al.* (1996) leads to the highest EPS yield and low cell lysis after EPS isolation of *S. acidocaldarius* biofilms grown on semisolid growth medium for 4 d was performed by using different extraction methods. For EPS isolated by cation exchange resin, 4.7 fg/cell of carbohydrates, 4.2 fg/cell of proteins and 0.91 fg/cell of eDNA were quantified (Jachlewski *et al.* 2015). Based on analysis of the extracellular proteome, the majority of EPS proteins were suggested to be cytoplasmatic. Functional analysis showed activity of various enzyme classes within the EPS (Jachlewski *et al.* 2015).

Sulfolobus solfataricus, a close relative of *S. acidocaldarius*, was incubated in polystyrene Petri dishes with submerged biofilm formation on the plastic surface (Koerdt *et al.* 2012). The *S. solfataricus* P2 mutant strain PBL2025 (lacking 50 genes) and several complemented mutants were compared in terms of their EPS composition. EPS were extracted using the cation exchange resin method, quantified and both cell number and number of colony forming units were determined. The carbohydrate and protein amounts were quantified by colorimetric assays (Dubois, 1956; Peterson, 1977). EPS of *S. solfataricus* P2 showed 11 fg/cell proteins and 4.82 fg/cell carbohydrates, whereas the amount of EPS was significantly enhanced in the mutant strain PBL2025. The complementation of PBL2025 with α-mannosidase and *β*-galactosidase genes altered the amount of carbohydrates and proteins of the EPS, indicating that both enzymes are involved in EPS formation. Proteins and carbohydrates were predominantly composed of low molecular weight (<14 kDa) molecules.

With the same isolation method, the EPS of *S. acidocaldarius* MW001, a deletion strain of *S. acidocaldarius* lacking the pyrEF-cassette for selection, were analyzed and the results were compared to six mutants lacking lrs14-like transcriptional regulators (Orell *et al.* 2013b). Deletion of gene *Saci_0446* led to significant changes in the EPS amount. EPS isolation and quantification showed higher abundance of carbohydrates and proteins in the Δsaci0446 mutant, mostly of low molecular weight. This finding in conjunction with CLSM/lectin staining confirmed the effect of the transcriptional regulator Saci_0446 on EPS biosynthesis.

15.4.1.2 Euryarchaeota

For EPS isolation from biofilms of the hyperthermophilic *Archaeoglobus fulgidus*, cells grown on the walls of culture vessels were washed multiple times with ethanol and water while gentle vortexing and isolated by brief centrifugation. (Lapaglia & Hartzell, 1997). Chemical extraction by ethanol and phenol was used to separate the EPS from the cells (Lapaglia & Hartzell, 1997). The protein concentration of the ethanol-phenol-extracted fraction was determined by the method of Bradford (Bradford, 1976). Polysaccharides were determined by incubation of samples with Congo red and spectrophotometrically quantification. A problem with this approach might be that possible important EPS components have been removed by the harsh washing procedure.

Natrialba, a genus of the *Halobacteriaceae* and extremely halophilic Archaeon, was grown on solidified maintenance medium for growth and EPS formation (Hezayen *et al.* 2001). After harvesting the cells by ultracentrifugation, the EPS in the supernatant were precipitated by addition of 1 volume ethanol. The precipitate was resuspended in water, dialysed against water and lyophilized. *Natrialba aegyptiaca* produces, when growing on solidified media, an extracellular polymer consisting predominantly of glutamic acid (85% w/w), representing poly(glutamic acid), carbohydrates (12.5% w/w) and unidentified compounds (2.5%) (Hezayen *et al.* 2001). However, proteinaceous components of the EPS will be lost upon alcohol precipitation and, thus, missed.

15.5 CONCLUSIONS

Archaeal biofilms were less investigated than those of Bacteria but gain now increasing interest as they were shown to be ubiquitous and play important roles in the geochemical cycle. As Archaea are essentially not pathogenic in contrary to bacterial species with pathogenic properties, they received much less attention and funding. However, with the finding of Archaea (mainly methanogens) in human microbiomes and an increasing interest in novel and sustainable production processes and organisms, Archaea receive increasing interest. Extremophiles as well as their enzymes, 'Extremozymes', are stable and active under harsh process conditions commonly used in industries (e.g. biofuel production) a therefore possess significant application potential. Increasingly more attention is also put on their biofilm mode of life. Interestingly, archaeal biofilm research already had an active decade between the mid-eighties (e.g., (Robinson *et al.* 1985; Sowers & Gunsalus, 1988)) and the mid-nineties of the last century (e.g., (Lapaglia & Hartzell, 1997; Rinker & Kelly, 1996)), but then it was neglected for about 15 years and has only recently gained momentum. It has become evident that, as has been known for the bacterial biofilm mode of life for decades, archaea are also regular inhabitants of microbial multi-species biofilms in diverse environments, are able to produce monospecies biofilms on different substrata under laboratory cultivation conditions and produce EPS consisting of biopolymers similar to those of bacterial EPS. Thus, archaeal biofilms also seem to be characterized by the EPS-mediated formation of a biofilm matrix, which is considered the key property for structure and functions of bacterial biofilms (Flemming & Wingender, 2010). Still, there is a long way to go for deeper understanding of archaeal biofilms and the functions of EPS in the formation and properties of these biofilms, especially under extreme environmental conditions, which are not relevant or have not yet been considered in bacterial biofilm research.

For cultivation and analysis of archaeal biofilms and EPS mostly systems and techniques established for Bacteria are applied, but have to be further optimized for Archaea. Static incubation systems like 96-well microtiter plates, Petri dishes for incubation and polystyrene and glass as substrata are used for studies of bacterial and archaeal biofilms, but many of these systems are not suited for thermophilic, acidophilic or halophilic Archaea. Suitable systems have to be developed and validated practically from the scratch. In addition microscopic, physical and molecular biological methods applied in medical and environmental bacterial biofilm research are adopted and applied.

For the time being, most studies on archaeal biofilms have been done with euryarchaeal and crenarchaeal model species, which are genetically tractable and thus enable studies on biofilm as well as EPS formation. Further progress in genetic manipulations and biofilm studies may give further basic information about biofilm formation, EPS synthesis and the regulation behind these processes. As a matter of fact, in almost all archaeal phyla, species have been discovered, which form biofilms, embedded in an EPS matrix. Therefore, it is appropriate to predict that this matrix is equally central for archaeal biofilms as it is to bacterial biofilms.

With this review, the authors want to stimulate attention to deeper research on archaeal biofilms and the potential this research provides.

ACKNOWLEDGEMENTS

The authors thank the Mercator foundation for support with a Mercur startup grant (Pr-2011–0058).

REFERENCES

Aguilera A., Souza-Egipsy V., San Martin-Uriz P. and Amils R. (2008). Extraction of extracellular polymeric substances from extreme acidic microbial biofilms. *Appl Microbiol Biotechnol*, **78**(6), 1079–1088.

Albers S. V. and Meyer B. H. (2011). The archaeal cell envelope. *Nat Rev Microbiol*, **9**(6), 414–426.

Anton J., Meseguer I. and Rodriguez-Valera F. (1988). Production of an extracellular polysaccharide by *Haloferax mediterranei*. *Appl Environ Microbiol*, **54**(10), 2381–2386.

Auchtung T. A., Takacs-Vesbach C. D. and Cavanaugh C. M. (2006). 16S rRNA phylogenetic investigation of the candidate division "Korarchaeota". *Appl Environ Microbiol*, **72**(7), 5077–5082.

Baker–Austin C., Potrykus J., Wexler M., Bond P. L. and Dopson M. (2010). Biofilm development in the extremely acidophilic archaeon 'Ferroplasma acidarmanus' Fer1. *Extremophiles*, **14**(6), 485–491.

Bang C., Ehlers C., Orell A., Prasse D., Spinner M., Gorb S. N., Albers S. V. and Schmitz R. A. (2014). Biofilm formation of mucosa-associated methanoarchaeal strains. *Front Microbiol*, **5**, 353.

Battin T. J., Wille A., Sattler B. and Psenner R. (2001). Phylogenetic and functional heterogeneity of sediment biofilms along environmental gradients in a glacial stream. *Appl Environ Microbiol*, **67**(2), 799–807.

Battin T. J., Besemer K., Bengtsson M. M., Romani A. M. and Packmann A. I. (2016). The ecology and biogeochemistry of stream biofilms. *Nat Rev Microbiol*, **14**(4), 251–263.

Bradford M. M. (1976). A rapid and sensitive method for the quantitation of microgram quantities of protein utilizing the principle of protein-dye binding. *Anal Biochem*, **72**, 248–254.

Bräsen C., Esser D., Rauch B. and Siebers B. (2014). Carbohydrate metabolism in Archaea: current insights into unusual enzymes and pathways and their regulation. *Microbiol Mol Biol Rev*, **78**(1), 89–175.

Briggs B. R., Pohlman J. W., Torres M., Riedel M., Brodie E. L. and Colwell F. S. (2011). Macroscopic biofilms in fracture-dominated sediment that anaerobically oxidize methane. *Appl Environ Microbiol*, **77**(19), 6780–6787.

Brown M. J. and Lester J. N. (1980). Comparison of bacterial extracellular polymer extraction methods. *Appl Environ Microbiol*, **40**(2), 179–185.

Cavicchioli R. (2011). Archaea–timeline of the third domain. *Nat Rev Microbiol*, **9**(1), 51–61.

Cavicchioli R., Curmi P. M., Saunders N. and Thomas T. (2003). Pathogenic archaea: do they exist?. *Bioessays*, **25**(11), 1119–1128.

Chimileski S., Franklin M. J. and Papke R. T. (2014). Biofilms formed by the archaeon *Haloferax volcanii* exhibit cellular differentiation and social motility and facilitate horizontal gene transfer. *BMC Biol*, **12**, 65.

Crundwell F. (1996). The formation of biofilms of iron-oxidising bacteria on pyrite. *Minerals Engineering*, **9**(10), 1081–1089.

DeLong E. F. (1998). Everything in moderation: archaea as 'non-extremophiles'. *Curr Opin Genet Dev*, **8**(6), 649–654.

Di Meglio L., Busalmen J. P., Pastore J. I., Ballarin V. L. and Nercessian D. (2014). Hyperhalophilic archaeal biofilms: growth kinetics, structure, and antagonistic interaction in continuous culture. *Biofouling*, **30**(2), 237–245.

Dubois M., Gilles K. A., Hamilton J. K., Rebers P. A. and Smith F. (1956). Colorimetric method for determination of sugars and related substances. *Analytical Chemistry*, **28**, 350–356.

Esquivel R. N., Xu R. and Pohlschroder M. (2013). Novel archaeal adhesion pilins with a conserved N terminus. *J Bacteriol*, **195**(17), 3808–3818.

Flemming H. C. and Wingender J. (2010). The biofilm matrix. *Nat Rev Microbiol*, **8**(9), 623–633.

Fröls S. (2013). Archaeal biofilms: widespread and complex. *Biochem Soc Trans*, **41**(1), 393–398.

Fröls S., Dyall-Smith M. and Pfeifer F. (2012). Biofilm formation by haloarchaea. *Environ Microbiol*, **14**(12), 3159–3174.

Frølund B., Palmgren R., Keiding K. and Nielsen P. H. (1996). Extraction of extracellular polymers from activated sludge using a cation exchange resin. *Water Res*, **30**(8), 1749–1758.

Grogan D. W. (1989). Phenotypic characterization of the archaebacterial genus *Sulfolobus*: comparison of five wild-type strains. *J Bacteriol*, **171**(12), 6710–6719.

Hedlund B. P., Murugapiran S. K., Alba T. W., Levy A., Dodsworth J. A., Goertz G. B., Ivanova N. and Woyke T. (2015). Uncultivated thermophiles: current status and spotlight on 'Aigarchaeota'. *Curr Opin Microbiol*, **25**, 136–145.

Henche A. L., Koerdt A., Ghosh A. and Albers S. V. (2012). Influence of cell surface structures on crenarchaeal biofilm formation using a thermostable green fluorescent protein. *Environ Microbiol*, **14**(3), 779–793.

Hezayen F. F., Rehm B. H., Tindall B. J. and Steinbuchel A. (2001). Transfer of *Natrialba asiatica* B1T to *Natrialba taiwanensis* sp. nov. and description of *Natrialba aegyptiaca* sp. nov., a novel extremely halophilic, aerobic, non-pigmented member of the Archaea from Egypt that produces extracellular poly(glutamic acid). *Int J Syst Evol Microbiol*, **51**(Pt 3), 1133–1142.

Horz H. P. and Conrads G. (2010). The discussion goes on: what is the role of Euryarchaeota in humans?. *Archaea*, **2010**, 967271.

Ionescu D., Siebert C., Polerecky L., Munwes Y. Y., Lott C., Hausler S., Bizic-Ionescu M., Quast C., Peplies J., Glockner F. O., Ramette A., Rodiger T., Dittmar T., Oren A., Geyer S., Stark H. J., Sauter M., Licha T., Laronne J. B. and de Beer D. (2012). Microbial and chemical characterization of underwater fresh water springs in the Dead Sea. *PLoS One*, **7**(6), e38319.

Jachlewski S., Jachlewski W. D., Linne U., Brasen C., Wingender J. and Siebers B. (2015). Isolation of extracellular polymeric substances from biofilms of the thermoacidophilic Archaeon *Sulfolobus acidocaldarius*. *Front Bioeng Biotechnol*, **3**, 123.

Jahn A. and Nielsen P. H. (1995). Extraction of extracellular polymeric substances (EPS) from biofilms using a cation exchange resin. *Water Science and Technology*, **32**(8), 157–164.

Jarrell K. F., Ding Y., Nair D. B. and Siu S. (2013). Surface appendages of archaea: structure, function, genetics and assembly. *Life (Basel)*, **3**(1), 86–117.

Kandiba L. and Eichler J. (2014). Archaeal S-layer glycoproteins: post-translational modification in the face of extremes. *Front Microbiol*, **5**, 661.

Kates M. (1996). Structural analysis of phospholipids and glycolipids in extremely halophilic archaebacteria. *J Microbiol Methods*, **25**(2), 113–128.

Koerdt A., Godeke J., Berger J., Thormann K. M. and Albers S. V. (2010). Crenarchaeal biofilm formation under extreme conditions. *PLoS One*, **5**(11), e14104.

Koerdt A., Jachlewski S., Ghosh A., Wingender J., Siebers B. and Albers S. V. (2012). Complementation of *Sulfolobus solfataricus* PBL2025 with an alpha-mannosidase: effects on surface attachment and biofilm formation. *Extremophiles*, **16**(1), 115–125.

Kozubal M. A., Romine M., Jennings R., Jay Z. J., Tringe S. G., Rusch D. B., Beam J. P., McCue L. A. and Inskeep W. P. (2013). Geoarchaeota: a new candidate phylum in the Archaea from high-temperature acidic iron mats in Yellowstone National Park. *ISME J*, **7**(3), 622–634.

Lapaglia C. and Hartzell P. L. (1997). Stress-induced production of biofilm in the hyperthermophile *Archaeoglobus fulgidus*. *Appl Environ Microbiol*, **63**(8), 3158–3163.

Liu H. and Fang H. H. (2002). Extraction of extracellular polymeric substances (EPS) of sludges. *J Biotechnol*, **95**(3), 249–256.

Mullakhanbhai M. F. and Larsen H. (1975). *Halobacterium volcanii* spec. nov., a Dead Sea halobacterium with a moderate salt requirement. *Arch Microbiol*, **104**(3), 207–214.

Neu T. R. and Lawrence J. (2016). The extracellular matrix – an inextractable part of biofilm systems. In: The Perfect Slime, H.-C. Flemming, T. Neu, J. Wingender (eds), IWA Publ., London.

Ng S. Y., Zolghadr B., Driessen A. J., Albers S. V. and Jarrell K. F. (2008). Cell surface structures of archaea. *J Bacteriol*, **190**(18), 6039–6047.

Orell A., Fröls S. and Albers S. V. (2013a). Archaeal biofilms: the great unexplored. *Annu Rev Microbiol*, **67**, 337–354.

Orell A., Peeters E., Vassen V., Jachlewski S., Schalles S., Siebers B. and Albers S. V. (2013b). Lrs14 transcriptional regulators influence biofilm formation and cell motility of Crenarchaea. *ISME J*, **7**(10), 1886–1898.

Perras A. K., Wanner G., Klingl A., Mora M., Auerbach A. K., Heinz V., Probst A. J., Huber H., Rachel R., Meck S. and Moissl-Eichinger C. (2014). Grappling archaea: ultrastructural analyses of an uncultivated, cold-loving archaeon, and its biofilm. *Front Microbiol*, **5**, 397.

Pester M., Schleper C. and Wagner M. (2011). The Thaumarchaeota: an emerging view of their phylogeny and ecophysiology. *Curr Opin Microbiol*, **14**(3), 300–306.

Peterson G. L. (1977). A simplification of the protein assay method of Lowry et al. which is more generally applicable. *Anal Biochem*, **83**(2), 346–356.

Pfeifer F. (2015). Haloarchaea and the formation of gas vesicles. *Life (Basel)*, **5**(1), 385–402.

Pohlschröder M. and Esquivel R. N. (2015). Archaeal type IV pili and their involvement in biofilm formation. *Front Microbiol*, **6**, 190.

Probst A. J., Holman H. Y., DeSantis T. Z., Andersen G. L., Birarda G., Bechtel H. A., Piceno Y. M., Sonnleitner M., Venkateswaran K. and Moissl-Eichinger C. (2013). Tackling the minority: sulfate-reducing bacteria in an archaea-dominated subsurface biofilm. *ISME J*, **7**(3), 635–651.

Probst A. J., Birarda G., Holman H. Y., DeSantis T. Z., Wanner G., Andersen G. L., Perras A. K., Meck S., Volkel J., Bechtel H. A., Wirth R. and Moissl-Eichinger C. (2014). Coupling genetic and chemical microbiome profiling reveals heterogeneity of archaeome and bacteriome in subsurface biofilms that are dominated by the same archaeal species. *PLoS One*, **9**(6), e99801.

Rinker K. D. and Kelly R. M. (1996). Growth physiology of the hyperthermophilic archaeon *Thermococcus litoralis*: development of a sulfur-free defined medium, characterization of an exopolysaccharide, and evidence of biofilm formation. *Appl Environ Microbiol*, **62**(12), 4478–4485.

Robinson R. W., Aldrich H. C., Hurst S. F. and Bleiweis A. S. (1985). Role of the cell surface of *Methanosarcina mazei* in cell aggregation. *Appl Environ Microbiol*, **49**(2), 321–327.

Schopf S., Wanner G., Rachel R. and Wirth R. (2008). An archaeal bi-species biofilm formed by *Pyrococcus furiosus* and *Methanopyrus kandleri*. *Arch Microbiol*, **190**(3), 371–377.

Schrenk M. O., Kelley D. S., Delaney J. R. and Baross J. A. (2003). Incidence and diversity of microorganisms within the walls of an active deep-sea sulfide chimney. *Appl Environ Microbiol*, **69**(6), 3580–3592.

Schubotz F., Meyer-Dombard D. R., Bradley A. S., Fredricks H. F., Hinrichs K. U., Shock E. L. and Summons R. E. (2013). Spatial and temporal variability of biomarkers and microbial diversity reveal metabolic and community flexibility in streamer biofilm communities in the Lower Geyser Basin, Yellowstone National Park. *Geobiology*, **11**(6), 549–569.

Sowers K. R. and Gunsalus R. P. (1988). Adaptation for growth at various saline concentrations by the Archaebacterium *Methanosarcina thermophila*. *J Bacteriol*, Vol. **170**(2), 998–1002.

Tapia J. M., Munoz J. A., Gonzalez F., Blazquez M. L., Malki M. and Ballester A. (2009). Extraction of extracellular polymeric substances from the acidophilic bacterium *Acidiphilium* 3.2Sup(5). *Water Sci Technol*, **59**(10), 1959–1967.

Tripepi M., Imam S. and Pohlschroder M. (2010). *Haloferax volcanii* flagella are required for motility but are not involved in PibD-dependent surface adhesion. *J Bacteriol*, **192**(12), 3093–3102.

Vetriani C., Jannasch H. W., MacGregor B. J., Stahl D. A. and Reysenbach A. L. (1999). Population structure and phylogenetic characterization of marine benthic Archaea in deep-sea sediments. *Appl Environ Microbiol*, **65**(10), 4375–4384.

Webster N. S. and Negri A. P. (2006). Site-specific variation in Antarctic marine biofilms established on artificial surfaces. *Environ Microbiol*, **8**(7), 1177–1190.

Wingender J., Strathmann M., Rode A., Leis A. and Flemming H. C. (2001). Isolation and biochemical characterization of extracellular polymeric substances from *Pseudomonas aeruginosa*. *Methods Enzymol*, **336**, 302–314.

Woese C. R., Kandler O. and Wheelis M. L. (1990). Towards a natural system of organisms: proposal for the domains Archaea, Bacteria, and Eucarya. *Proc Natl Acad Sci USA*, **87**(12), 4576–4579.

Wuertz S., Spaeth R., Hinderberger A., Griebe T., Flemming H. C. and Wilderer P. A. (2001). A new method for extraction of extracellular polymeric substances from biofilms and activated sludge suitable for direct quantification of sorbed metals. *Water Sci Technol*, **43**(6), 25–31.

Zhang R. Y., Neu T. R., Bellenberg S., Kuhlicke U., Sand W. and Vera M. (2014). Use of lectins to in situ visualize glycoconjugates of extracellular polymeric substances in acidophilic archaeal biofilms. *Microb Biotechnol*, **8**(3), 448–461.

Zhang R., Neu T. R., Zhang Y., Bellenberg S., Kuhlicke U., Li Q., Sand W. and Vera M. (2015). Visualization and analysis of EPS glycoconjugates of the thermoacidophilic archaeon *Sulfolobus metallicus*. *Appl Microbiol Biotechnol*, **99**(17), 7343–7356.

Ziegler S., Dolch K., Geiger K., Krause S., Asskamp M., Eusterhues K., Kriews M., Wilhelms-Dick D., Goettlicher J., Majzlan J. and Gescher J. (2013). Oxygen-dependent niche formation of a pyrite-dependent acidophilic consortium built by archaea and bacteria. *ISME J*, **7**(9), 1725–1737.

Zolghadr B., Klingl A., Koerdt A., Driessen A. J., Rachel R. and Albers S. V. (2010). Appendage-mediated surface adherence of *Sulfolobus solfataricus*. *J Bacteriol*, **192**(1), 104–110.

Index